Remote Sensing of the African Seas

Vittorio Barale • Martin Gade
Editors

Remote Sensing of the African Seas

Springer

Editors
Vittorio Barale
European Commission
Joint Research Centre
Ispra
Italy

Martin Gade
Institut für Meereskunde
Universität Hamburg
Hamburg
Germany

ISBN 978-94-017-7855-8 ISBN 978-94-017-8008-7 (eBook)
DOI 10.1007/978-94-017-8008-7
Springer New York Heidelberg Dordrecht London

Softcover reprint of the hardcover 1st edition 2014

Cover illlustrations: Pictures were taken from
* Chapter 8, Kostianoy et al.: Fig. 3
* Chapter 3, Hoepffner et al.: Fig. 4B
* Chapter 11, Brandt et al.: Fig. 5B
* Chapter 2, Gower and King: Fig. 15
* Chapter 14, Magelhaes et al.: Fig. 9

Printed on acid-free paper

Springer is part of Springer Science+Business Media (www.springer.com)

To Hatshepsut
Fifth Pharaoh of the Eighteenth Dynasty
of Ancient Egypt
Patron of the Land of Punt "Oceanographic
Expedition"

Preface

Queen Hatshepsut—or sometimes "King" Hatshepsut, in formal circumstances—was the daughter of Pharaoh Thutmosis I, the wife (and half-sister) of Pharaoh Thutmosis II, the stepmother and co-regent of Pharaoh Thutmosis III. She ruled Ancient Egypt in her own right as the fifth Pharaoh of the Eighteenth Dynasty of the New Kingdom, approximately from 1473 (perhaps 1479) to 1458 BC, with the throne name of Maatkare. She is regarded as one of the most successful Pharaohs, preserving power longer than any other woman of an indigenous Egyptian dynasty. Hatshepsut is known to have been engaged in military campaigns early in her reign, but generally is credited to be a monarch who initiated a long era of peace. She re-established international trading relationships, disrupted during the Hyksos occupation of the country, and brought great wealth to Egypt.

During her reign, Hatshepsut organized and funded several missions to foreign lands, the most famous being that to the Land of Punt. This region, also called Pwenet or Pwene, was located to the southeast of Egypt, on the littoral of either, or both, the Horn of Africa and Southern Arabia (most likely Eritrea and neighbouring Ethiopia). The expedition was very ambitious—Hatshepsut sent a fleet of five ships, each measuring about 25 m length, 7 m beam and 2 m draft, bearing several sails and 30 rowers, which left Kosseir, on the Red Sea, with a total crew of about 250–300 men—and turned into a remarkable commercial, as well as diplomatic and scientific success. Many invaluable goods were bought back—Hatshepsut herself informed us in lengthy inscriptions that "*the ships were laden with the costly products of the Land of Punt and with its many valuable woods, with very much sweet-smelling resin and frankincense, with quantities of ebony and ivory*"—including precious myrrh and 31 live myrrh trees, the roots of which we know were kept in baskets for the duration of the voyage.

While the trees were planted in the courts of her mortuary temple complex at Dayr el-Bahari, near Luxor in the Valley of the Kings, Hatshepsut had all stages of the expedition commemorated on the walls of her tomb's Djeser-Djeseru, the Holy of Holies, the Sublime of Sublimes—a colonnaded structure of perfect harmony designed and implemented by Senemut, royal steward and architect (probably royal lover as well) to serve for her posthumous worship and to honor the glory of Amun. A full account of the expedition returned treasures and findings was recorded in the wall reliefs, showing scenes of village life, exotic plants, birds and animals, and men carrying off heaps of goods. Elaborately rigged sailboats are shown, getting ready to bring the tribute back to Egypt. . . while many types of fish swim in the water below!

While other voyages to the Land of Punt had preceded, and others followed later, over the centuries, only the remaining accounts of the expedition set forth by Hatshepsut describe in detail the flora and fauna typical of the Red Sea and Indian Ocean (i.e. saltwater species such as turtles, parrot-fish, scorpion-fish, soldier-fish, triggerfish, wrasse, squid and spiny lobster) collected on that occasion. The rich details of these pictorial descriptions are so accurate that—as reported by oceanographer Sayed El-Sayed—the late Carl Hubbs, famed ichthyologist of the Scripps Institution of Oceanography, claimed he was "able to identify the fish to the species level" from the carvings.

Given the mixed commercial and non-commercial scope of the enterprise, and the resulting (quasi) scientific reports above, the one organized by Hatshepsut can be considered as the first real oceanographic cruise ever to be sent into the African Seas. The ensuing accurate knowledge of geography, hydrography and meteorology, derived from precise observations of the environment, enabled ancient Egyptians to undertake long seafaring ventures. After Hatshepsut, other Egyptian monarchs reached southern Africa and beyond. Herodotus informs us that Necho II, King of Egypt (circa 600 BC) sent Phoenician sailors down to the Red Sea and along the coast of Lybia, modern Africa[1]. After 3 years, they returned through the Pillars of Hercules, the Strait of Gibraltar, and reached Egypt via the Mediterranean Sea. According to William A. Herdmann, author of *The Founders of Oceanography and Their Work* (1923), it is doubtful that the circumnavigation of Africa was repeated again, until Vasco da Gama rounded the Cape of Good Hope (from the west) 2000 years later, in the fifteenth century.

This Volume, focusing on the marginal, semi-enclosed and enclosed water bodies of Africa—herein referred to as the "African Seas"—is dedicated to the extraordinary achievements of Hatshepsut's original, albeit embryonic, oceanographic mission. However, it is on modern expeditions, which use Earth orbiting satellites to map the waters surrounding the African continental landmass, that the present review concentrates. The following collection of topical papers span the vast potential of remote sensing techniques to tackle the issues, peculiarities and special challenges posed by the African coastal and enclosed water bodies. The assessment of surface parameters by means of both passive and active techniques is addressed, focusing on the use of reflected visible and near-infrared sunlight, of surface emissions at thermal infrared or microwave frequencies, or again of the return of transmitted impulses of microwave radiation.

[1] Libya clearly is bounded by the sea, except where it borders on Asia. Nekhau king of Egypt first discovered this and made it known. When he had abandoned the digging of the canal which leads from the Nile to the Arabian Gulf, he sent Phoènicians in ships, with orders to sail on their return voyage past the Pillars of Heracles until they came into the northern sea and so to Egypt. *Herodotus, Histories 4.42*

Studying and understanding the natural history of the African Seas requires integrated observation systems, which ought to combine up-to-date remote sensing techniques with thorough *in situ* observations. As the latter are frequently lacking, due to the continent's sheer size and continuing socio-economic plight, quite often satellite remote sensing from Earth's orbit is the only choice available to environmental scientists. The peer-reviewed papers collected in this volume provide a summary of the most recent results achieved in this field, and of the new scientific ground broken by some of the observations, offering an unprecedented insight into some of the dynamical and bio-geo-chemical complexity of the African Seas.

This book—which targets researchers working in the Earth and Marine Sciences, but also teachers, as well as students, in the same fields, particularly in Africa—includes a brief overview of the African marginal seas, followed by general papers on Visible, Infrared and Microwave (passive and active) remote sensing applications, focussing on African waters. In depth assessments of the main African marginal basins, in both the Atlantic Ocean and Indian Ocean, their prevailing currents and resulting key environmental traits, are grouped in two dedicated sections. Several examples are provided of multi-sensor techniques, which exploit the synergies of complementary sensors and enhance the value of their combined views. In the section on marginal and enclosed water bodies, the Red Sea is discussed, while for the Mediterranean Sea (already covered, at the basin scale, by several papers appearing in a previous volume[2] of this series) only near-coastal and lagoon waters have been considered. A final review paper is dedicated to the African Great Lakes, in order to complete the spectrum of presentations on the most relevant water bodies of the continent. The breath of the environmental themes covered, and of the diverse techniques dealt with, required the collaboration of a truly international (global in fact) group of scientists. Their fundamental contributions, their patience and endurance in resisting the editors' constant harassment is acknowledged by the inclusion of all authors' and reviewers' names and affiliations in the list of contributors. Sincere thanks are due to all of them.

July 2013 Vittorio Barale and Martin Gade

[2] Barale V, Gade M (2008) Remote sensing of the European seas. Springer Science+Business Media BV, p 514.

Acknowledgment

This volume has been prepared and published with the support of the Joint Research Centre, the European Commission's in-house science service.

As the Commission's in-house science service, the Joint Research Centre's mission is to provide EU policies with independent, evidence-based scientific and technical support throughout the whole policy cycle. Working in close cooperation with policy Directorates-General, the JRC addresses key societal challenges while stimulating innovation through developing new methods, tools and standards, and sharing its know-how with the Member States, the scientific community and international partners.

Contents

Authors and Reviewers

James G. Acker NASA GES DISC, Greenbelt, MD, USA

Kouadio Affian Centre Universitaire de Recherche et d'application en Télédétection (CURAT), Université de Cocody, Abidjan, Côte d'Ivoire
Also at: Laboratoire de Géologie Marine et de Sédimentologie, U.F.R. des Sciences de la Terre et des Ressources Minières, Université de Cocody, Abidjan, Côte d'Ivoire

Mahmoud Ahmed Marine Sciences Department, National Authority Remote Sensing and Space Sciences, Cairo, Egypt

Werner Alpers Institut für Meereskunde, Universität Hamburg, Hamburg, Germany

Serge Andréfouët Institut de Recherche pour le Développement, Nouméa, New Caledonia, France

Roy Armstrong Department of Marine Sciences, University of Puerto Rico, Mayaguez, Puerto Rico

Robert Arnone US Naval Research Lab, Bio-Optical/Physical Processes and Remote Sensing Section, Stennis Space Center, MS, USA

Björn Backeberg Nansen-Tutu Centre, University of Cape Town, Cape Town, South Africa

Thomas J. Ballatore Lake Basin Action Network, Moriyama, Japan

Christelle Balt Graduate School of Oceanography, University of Rhode Island, Narragansett, RI, USA

Viva Banzon National Climate Data Center, NOAA NESDIS, Asheville, NC, USA

Vittorio Barale European Commission, Joint Research Centre, Institute for Environment and Sustainability, Ispra, Italy

Helen Beggs Centre for Australian Weather and Climate Research, Bureau of Meteorology, Melbourne, Australia

Simon Belanger Laboratoire d'optique Aquatique et de Télédétection (AquaTel), Université de Quebec à Rimousky (UQAR), Rimousky (Québec), Canada

Mireille Bella Centre Universitaire de Recherche et d'application en Télédétection (CURAT), Université de Cocody, Abidjan, Côte d'Ivoire

Stewart Bernard Council for Scientific and Industrial Research, Natural Resources and Ecosystems, Cape Town, South Africa
Also at: Department of Oceanography, University of Cape Town, Cape Town, South Africa

Shane R. Bradt Lake Basin Action Network, Moriyama, Japan
Also at: University of New Hampshire, Durham, NH, USA

Peter Brandt GEOMAR Helmholtz-Zentrum für Ozeanforschung Kiel, Kiel, Germany

Carsten Brockmann Brockmann Consult, Geesthacht, Germany

Miguel Bruck Institut für Methodik der Fernerkundung, SAR-Signalverarbeitung, Deutsches Zentrum für Luft- und Raumfahrt e.V. (DLR), Bremen Flughafen, Bremen, Germany

Stephan Brusch Institut für Methodik der Fernerkundung, SAR-Signalverarbeitung, Deutsches Zentrum für Luft- und Raumfahrt e.V. (DLR), Bremen Flughafen, Bremen, Germany

Hans Burchard Leibniz Institut für Ostseeforschung Warnemünde, Rostock, Germany

Valborg Byfield Ocean Observing & Climate Group, National Oceanography Centre,Southampton, UK

Bertrand Chapron Ifremer, Plouzané, France

Paolo Cipollini Marine Physics and Ocean Climate, National Oceanography Centre, Southampton, UK

Josep Coca SeaSNET Canarias ground station. Departamento de Biología, Universidad de Las Palmas de Gran Canaria (ULPGC), Las Palmas, Spain

Fabrice Collard Ifremer, Plouzané, France

Andrés Cózar Universidad de Cádiz, Puerto Real, Spain

José C. B. da Silva Department of Geosciences, Environment, and Spatial Planning, Interdisciplinary Centre of Marine and Environmental Research, CIMAR, Porto, Portugal
Also at: Faculty of Sciences, University of Porto, Porto, Portugal

Dominique Dagorne US Imago, Institut de Recherche pour le Développement, Plouzané, France

Pascal Derycke European Commission, Joint Research Centre, Institute for Environment and Sustainability, Ispra, Italy

Eric V. Djagoua Centre Universitaire de Recherche et d'application en Télédétection (CURAT), Université de Cocody, Abidjan, Côte d'Ivoire

Samuel Djavidnia European Maritime Safety Agency, Lisbon, Portugal

Craig Donlon Met Office Hadley Centre, Exeter, UK

Cassandre Ziné Dro Centre Universitaire de Recherche et d'application en Télédétection (CURAT), Université de Cocody, Abidjan, Côte d'Ivoire

Francois Dufois Oceanography Department, Mare Institute, University of Cape Town, Cape Town, South Africa
Also at: CSIRO Marine and Atmospheric Research, Floreat, Australia

Cecile Dupouy LOPB, Centre IRD de Nouméa, Nouméa, Nouvelle Calédonie

Stanislav A. Ermakov Institute of Applied Physics, Russian Academy of Sciences, Nizhny Novgorod, Russia

Robert Evans RSMAS/MPO, University of Miami, Miami, FL, USA

Hayley Evers-King Department of Oceanography, University of Cape Town, Cape Town, South Africa

Ralph Foster Applied Physics Laboratory, Seattle, WA, USA

Martin Gade Institut für Meereskunde, Universität Hamburg, Hamburg, Germany

Elisa Garcia-Gorriz European Commission, Joint Research Centre, Institute for Environment and Sustainability, Ispra, Italy

Luis García-Weil Departamento de Física, Universidad de Las Palmas de Gran Canaria (ULPGC), Las Palmas, Spain

Monika Gerth Leibniz Institut für Ostseeforschung Warnemünde, Rostock, Germany

Jim Gower Institute of Ocean Sciences, Fisheries and Oceans Canada, Sidney, BC, Canada

Harm Greidanus European Commission, Joint Research Centre, Institute for the Protection and Security of the Citizen, Ispra, Italy

Nicolas Hoepffner European Commission, Joint Research Centre, Institute for Environment and Sustainability, Ispra, Italy

Benjamin M. Holt NASA Jet Propulsion Laboratory, Pasadena, CA, USA

Jochen Horstmann Institut für Küstenforschung, Helmholtz-Zentrum Geesthacht, Geesthacht, Germany

Andrey Ivanov PP Shirshov Institute of Oceanology, Russian Academy of Sciences, Moscow, Russia

Johnny A. Johannessen Nansen Environmental and Remote Sensing Center, Bergen, Norway

Mati Kahru Scripps Institution of Oceanography, Unversity of California at San Diego, San Diego, CA, USA

Svetlana S. Karimova Space Research Institute of the Russian Academy of Sciences, Moscow, Russia
Now at: Institut für Küstenforschung, Helmholtz-Zentrum Geesthacht, Geesthacht, Germany

Jean B. Kassi Centre Universitaire de Recherche et d'application en Télédétection (CURAT), Université de Cocody, Abidjan, Côte d'Ivoire

Oleg V. Kopelevich PP Shirshov Institute of Oceanology, Russian Academy of Sciences, Moscow, Russia

Stephanie King Institute of Ocean Sciences, Fisheries and Oceans Canada, Sidney, BC, Canada

Andrey G. Kostianoy P. P. ShirshovInstitute of Oceanology, Russian Academy of Sciences, Moscow, Russia
Also at: I. Kant Baltic Federal University, Kaliningrad, Russia

Adam Krężel Institute of Oceanography, University of Gdansk, Gdynia, Poland

Jeanne M. Kouadio Centre Universitaire de Recherche et d'application en Télédétection (CURAT), Université de Cocody, Abidjan, Côte d'Ivoire

Vladimir Kudryavtsev Nansen International Environmental and Remote Sensing Centre, St. Petersburg, Russia

Marjolaine Krug Ecosystem Earth Observations, Council for Scientific and Industrial Research, Cape Town, South Africa
Also at: Nansen-Tutu Center for Marine Environmental Research, Oceanography Department, University of Cape Town, Cape Town, South Africa

Pierre Larouche Maurice Lamontagne Institute, Fisheries and Oceans Canada, Mont Joli (Québec), Canada

Olga Yu. Lavrova Space Research Institute, Russian Academy of Sciences, Moscow, Russia

Susanne Lehner Institut für Methodik der Fernerkundung, SAR-Signalverarbeitung, Deutsches Zentrum für Luft- und Raumfahrt e.V. (DLR), Bremen Flughafen, Bremen, Germany

Steven A. Loiselle Università degli Studi di Siena, Siena, Italy

Jorge M. Magalhaes Department of Geosciences, Environment, and Spatial Planning, Interdisciplinary Centre of Marine and Environmental Research, CIMAR, Porto, Portugal
Also at: Faculty of Sciences, University of Porto, Porto, Portugal

Ernst Maier-Reimer† (passed away) Max-Planck-Institut für Meteorologie, Hamburg, Germany

Salvatore Marullo Centro Ricerche Casaccia, ENEA, Casaccia, Italy

Mark Matthews Department of Oceanography, University of Cape Town, Cape Town, South Africa

Elena Mauri Istituto Nazionale di Oceanografia e Geofisica Sperimentale (OGS), Trieste, Italy

John Melack University of California, Santa Barbara, CA, USA

Christian Melsheimer Institut für Umweltphysik, Universität Bremen, Bremen, Germany

Peter Minnet RSMAS, University of Miami, Miami, FL, USA

Leonid M. Mitnik Pacific Oceanology Institute, Far-Eastern Branch, Vladivostok, Russia

Brice A. Mobio Centre Universitaire de Recherche et d'application en Télédétection (CURAT), Université de Cocody, Abidjan, Côte d'Ivoire

Frank M. Monaldo Applied Physics Laboratory, Johns Hopkins University, Laurel, MD, USA

Mira Morovic Institute of Oceanography and Fisheries, Split, Croatia

Adrian L. New Marine Systems Modelling Group, National Oceanography Centre, Southampton, European Way, Southampton, UK

Leo Nykjaer European Commission, Joint Research Centre, Institute for Environment and Sustainability, Ispra, Italy

Thomas Ohde Leibniz Institut für Ostseeforschung Warnemünde, Rostock, Germany

Lydia Olaka Lake Basin Action Network, Moriyama, Japan
Also at: University of Nairobi, Nairobi, Kenya

Grant Pitcher Fisheries Research and Development, Department of Agriculture, Forestry and Fisheries, Cape Town, South Africa

Andrey Pleskachevsky Institut für Methodik der Fernerkundung, SAR-Signalverarbeitung, Deutsches Zentrum für Luft- und Raumfahrt e.V. (DLR), Bremen Flughafen, Bremen, Germany

Dmitry V. Pozdnyakov Nansen International Environmental and Remote Sensing Centre, St. Petersburg, Russia

Detlef Quadfasel Institut für Meereskunde, Universität Hamburg, Hamburg, Germany

Andy Rabagliati Department of Oceanography, University of Cape Town, Cape Town, South Africa

Antonio G. Ramos SeaSNET Canarias ground station. Departamento de Biología, Universidad de Las Palmas de Gran Canaria (ULPGC), Las Palmas, Spain

Alexandre Redondo SeaSNET Canarias ground station. Departamento de Biología, Universidad de Las Palmas de Gran Canaria (ULPGC), Las Palmas, Spain

Lisl Robertson Department of Oceanography, University of Cape Town, Cape Town, South Africa

Roland Romeiser RSMAS, University of Miami, Miami, FL, USA

Herbert Siegel Leibniz Institut für Ostseeforschung Warnemünde, Rostock, Germany

Tim Smyth Plymouth Marine Laboratory, Plymouth, UK

Matteo Soccorsi Institut für Methodik der Fernerkundung, SAR-Signalverarbeitung, Deutsches Zentrum für Luft- und Raumfahrt e.V. (DLR), Bremen Flughafen, Bremen, Germany

Dmytro M. Solovyov Marine Hydrophysical Institute, National Academy of Sciences, Sevastopol, Ukraine

Sergey Stanichny Marine Hydrophysical Institute, National Academy of Sciences of Ukraine, Sevastopol, Ukraine

Adolf Stips European Commission, Joint Research Centre, Institute for Environment and Sustainability, Ispra, Italy

Konstantinos Topouzelis Department of Marine Sciences, University of the Aegean, Mitilene, Greece

Chuck Trees NATO Undersea Research Center, La Specia, Italy

Domenico Velotto Institut für Methodik der Fernerkundung, SAR-Signalverarbeitung, Deutsches Zentrum für Luft- und Raumfahrt e.V. (DLR), Bremen Flughafen, Bremen, Germany

Acronyms

AATSR	Advanced Along-Track Scanning Radiometer
ADEOS	Advanced Earth Observing Satellite
AGW	Atmospheric Gravity Waves
AIS	Automatic Identification System
ALOS	Advanced Land Observing Satellite
AMESD	African Monitoring of the Environment for Sustainable Development
AMI	Active Microwave Instrument
AMSR-E	Advanced Scanning Microwave Radiometer
AMSSA	African Maritime Safety and Security Agency
AMSU-A	Advanced Microwave Sounding Unit
AOD	Aerosol Optical Depth
AOML	Atlantic Oceanographic and Meteorological Laboratory
AOP	Apparent Optical Properties
ASAR	Advanced Synthetic Aperture Radar
ASCAT	Advanced Scatterometer
ASTER	Advanced Spaceborne Thermal Emission Reflection Radiometer
ATCOR	Atmospheric & Topographic Correction (model)
ATCOR2	Atmospheric Correction for Flat Terrain (model)
ATI	Along-Track Interferometry
AVHRR	Advanced Very High Resolution Radiometer
AVISO	Archiving, Validation and Interpretation of Satellite Oceanographic (data center, of France)
BC	Before Christ
BMBF	Bundesministerium für Bildung und Forschung (Federal Ministry of Education and Research, Germany)
BPAC	Bright Pixel Atmospheric Correction (algorithm)
BT	Brightness Temperature
CASI	Compact Airborne Spectrographic Imager
CDOM	Coloured Dissolved Organic Material
CDOP	(empirical function relating the) C-band Doppler shift (to the wind field)

CERSAT	Centre ERS d'Archivage et de Traitement
CFAR	Constant False Alarm Rate
CHAMP	Challenging Minisatellite Payload
chl	chlorophyll-like pigment concentration
Chl-a	Chlorophyll-a
CHL	chlorophyll-like pigments of phytoplankton
CHRIS	Compact High Resolution Imaging Spectrometer
CMOD	C-band geophysical MODel function
CNA	Clean Nigeria Associates (consortium)
CNES	Centre National d'Etudes Spatiales
CREPAD	Centro de Recepción, Proceso, Archivo y Distribución
CTD	Conductivity-Temperature-Depth
CZCS	Coastal Zone Color Scanner
DET	(suspended) detritus
DISC	Data and Information Services Center
DLR	Deutsches Zentrum für Luft- und Raumfahrt (German Aerospace Center)
DMC	Disaster Monitoring Constellation
DN	Digital Number
DPR	Department of Petroleum Resources (of Nigeria)
DSP	Diarrhetic Shellfish Poisoning
DRA	Dual Receive Antenna
DUACS	Data Unification and Altimeter Combination System
DWD	Deutscher Wetterdienst (German Weather Service)
EAMNet	Europe-Africa Marine Earth Observation Network
EAP	Equivalent Algal Population (algorithm)
EC	European Commission
ECMWF	European Centre for Medium-Range Weather Forecast
EKE	Eddy Kinetic Energy
ELISA	Enzyme-Linked Immunosorbent Assay
EO	Earth Observation
EOF	Empirical Orthogonal Function
EOL	Earth Observation Laboratory
EOS	Earth Observation Satellite (AM and PM)
EOS	Earth Observing System (of NASA)
ERS	European Remote Sensing (Satellite, 1 and 2)
ESA	European Space Agency
ETM+	Enhanced Thematic Mapper (on Landsat satellites)
EU	European Union
EUMETSAT	European Organisation for the Exploitation of Meteorological Satellites
EVI	Enhanced Vegetation Index

FLH	Fluorescence Line Height
FP7	Framework Programme (for Research and Development) 7
FPSO	Floating Production, Storage And Offloading (vessel)
FR	Full Resolution
GAFRD	General Authority for Fishery Resources Development (Egypt)
GCLME	Guinea Current Large Marine Ecosystem (project)
GCS	great cool season (of the Côte d'Ivoire)
GDP	Gross Domestic Product
GEF	Global Environment Facility
GES-DISC	Goddard Earth Sciences Data and Information Services Center
GEOSS	Global Earth Observation System of Systems
GFO	Geosat Follow-On
GHRSST	Group on High Resolution Sea Surface Temperature
GIOVANNI	Goddard Interactive Online Visualization ANd aNalysis Infrastructure
GIS	Geographic Information System
GK	Generalized-K (distribution)
GM	Global Mode
GMF	Geophysical Model Function
GMIS	Global Marine Information System
GOCE	Gravity field and steady-state Ocean Circulation Explorer
GODAE	Global Ocean Data Assimilation Experiment
GOOS	Global Ocean Observing System
GPCC	Global Precipitation Climatology Centre
G-POD	Grid Processing on Demand (facility of ESA)
GRACE	Gravity Recovery and Climate Experiment
GWS	great warm season (of the Côte d'Ivoire)
HAB	Harmful Algal Bloom
HH	horizontal co-polarisation
HPLC	High Performance Liquid Chromatography
HRG	High Resolution Geometry (sensor, on SPOT-5)
HRV	High-Resolution Visible (sensor, on SPOT 2/3)
HRV-IR	High-Resolution Visible-Infrared (sensor, on SPOT 4)
HS	High-Resolution Spotlight
H-TSRB	Hyperspectral Tethered Surface Radiometer Buoy
HV	horizontal/vertical cross-polarisation
HY	HY (HaiYang = Ocean) (ocean observing satellite series, China)
HYCOM	Hybrid Coordinate Ocean Model
IBU	Intense Benguela Upwelling (index)
ICOL	Improve Contrast between Ocean & Land (technique)
IES	Institute for Environment and Sustainability (of JRC EC)
IM	Image Mode
INTA	Instituto Nacional de Tecnica Aeroespacial
IOC	Indian Ocean Commission

IOCCG	International Ocean-Colour Coordinating Group
IOD	Indian Ocean Dipole
IOP	Inherent Optical Properties
IPCC	Intergovernmental Panel on Climate Change
IR	Infra-Red
ISAR	Inverse Synthetic Aperture Radar
ISWs	Internal Solitary Wave
ITs	Internal Tide
JAXA	Japan Aerospace Exploration Agency
JERS	Japanese Earth Resources Satellite
JRC	Joint Research centre (of the European Commission)
LAADS	Level 1 and Atmospheres Archive and Distribution System
LEGOS	Laboratoire d'Etude en Géophysique et Océanographie Spatiale
LEO	Low Earth Orbit
LRIT	Long Range Identification and Tracking
MADT-DT	Maps of Absolute Dynamic Topography – Delayed Time
MARISS	Maritime Security Information Services
MBR	Maximum Band Ratio
MC	Mozambique Channel
MCI	Maximum Chlorophyll Index
MDT	Mean Dynamic Topography
MERIS	Medium Resolution Imaging Spectrometer
METOP	METeorological OPerational Satellite
MODIS	Moderate Resolution Imaging Spectometer
MOI	Mauritius Oceanography Institute
MPH	Maximum Peak Height (algorithm)
MPH	Moving Polar High (pressure)
MSG	Meteosat Second Generation
MSS	Multiple Solution Scheme
MSS	Multiple Spectral Scanner (on Landsat satellites)
NARSSS	National Authority Remote Sensing and Space Sciences (Egypt)
NASA	National Aeronautics and Space Administration
NASRDA	National Space Research and Development Agency (of Nigeria)
NCAR	National Center for Atmospheric Research
NCDC	National Climatic Data Center (of NOAA)
NCEP	National Center for Environmental Prediction
NDBC	National Data Buoy Center
NDDC	Niger Delta Development Commission
NDVI	Normalized Difference Vegetation Index
NDWI	Normalized Difference Water Index
NESDIS	National Environmental Satellite, Data, and Information Service
NIR	Near Infra-Red
NFLH	Normalized Fluorescence Line Height

NMISA	Nigerian Maritime and Safety Agency
NOAA	National Oceanic and Atmospheric Administration
NOSDRA	National Oil Spill Detection and Response Agency (of Nigeria)
NRCS	Normalized Radar Cross Section
NRT	Near Real Time
NSCAT	NASA Scatterometer
NSF	National Science Foundation
OCM	Ocean Colour Monitor (1 and 2)
OIW	Oceanic Internal Waves
OLCI	Ocean Land Colour Instrument
OSCAT	Oceansat-2 Scatterometer
OSI-SAF	Ocean and Sea Ice Satellite Application Facility
OSTM	Ocean Surface Topography Mission
PALSAR	Phased Array type L-band Synthetic Aperture Radar
PAR	Photosynthetically Available Radiation
PCA	Principle Component Analysis
PFZ	Potential Fishing Zone
PolSAR	Polarimetric SAR
PRF	Pulse Repetition Frequency
PROBA	Project for On-Board Autonomy (satellite)
PROCAS	Protection and Conservation of the European Seas (JRC Action)
PSP	Paralytic Shellfish Poisoning
psu	practical salinity units
PUMA	Preparation for Use of MSG in Africa
QuikSCAT	Quick Scatterometer (EO satellite carrying the SeaWinds scatterometer)
RA	Radar Altimeter
RE	Radiometrically Enhanced (product variant)
RGB	Red-Green-Blue (colour composite)
RS	Remote Sensing
RSS	Remote Sensing Systems
RR	Reduced Resolution
SAINT	SAR AIS INtegrated Toolbox
SAR	Synthetic Aperture Radar
SARAL	Satellite with ARgos and AltiKa
SASS	Seasat-A Satellite Scatterometer
SatAIS	Satellite-based system (for monitoring) AIS (equipped ships)
SCAT	Scatteromenter
SCS	small cool season (of the Côte d'Ivoire)
SDD	Secchi Disk Depth
SDT	Secchi Disk Transparency
SeaDAS	SeaWiFS Data Analysis System

SeaWiFS	Sea-viewing Wide Field-of-view Sensor
SEC	South Equatorial Current
SEM	Scanning Electronic Microscopy
SEVIRI	Spinning Enhanced Visible and Infra-Red Imager
SLA	Sea Level Anomaly
SMAC	Simplified Method for Atmospheric Correction (processor)
SOPRAN	Surface Ocean Processes in the Anthropocene (project)
SPM	Suspended Particulate Matter
SPOT	Système Pour l'Observation de la Terre
SRTM	Shuttle Radar Topography Mission
SSALTO	Segment Sol multi-missions dALTimetrie, d'orbitographie et de localisation précise
SSH	Sea Surface Height
SSHA	Sea Surface Height Anomaly
SSM/I	Special Sensor Microwave / Imager
SSM/T2	Special Sensor Microwave / Temperature & Humidity Profile
SSS	Sea Surface Salinity
SST	Sea Surface Temperature
SSTL	Surrey Satellite Technology Ltd
S&R	Search & Rescue
SRC	Slant Range Complex
SWBD	Shuttle radar topography mission Waterbody Dataset
SWOT	Surface Water and Ocean Topography (instrument)
SWS	small warm season (of the Côte d'Ivoire)
TCT	Tasseled-Cap-Transformation
TCW	Tasseled-Cap Wetness
TD-X	TanDEM-X, *i.e.* TerraSAR-X add-on for Digital Elevation Measurement
TG	Taylor-Goldstein (model)
TIR	Thermal Infrared
TM	Thematic Mapper (on Landsat satellites)
TMI	TRMM Microwave Imager
TRMM	Tropical Rainfall Measuring Mission
TSM	Total Suspended Matter
TS-X	TerraSAR-X
UNESCO	United Nations Educational, Scientific and Cultural Organization
USDA	US Department of Agriculture
US	United States
UTC	Coordinated Universal Time
UV	Ultraviolet
VH	vertical/horizontal cross-polarisation
VIIRS	Visible/Infrared Imager Radiometer Suite
VMS	Vessel Monitoring System

VTS	Vessel Traffic System
VV	vertical co-polarisation
WI	Water Index
ws	wind speed
WS	Wide Swath
WSM	Wide Swath Mode
XBT	eXpendable Bathy-Thermograph
XMOD	X- Band Geophysical Model Function (GMF) algorithm

Part I
Introduction to Remote Sensing of the African Seas

Chapter 1
The African Marginal and Enclosed Seas: An Overview

Vittorio Barale

Abstract The African Seas include marginal basins of two major oceans, the Atlantic and the Indian, a miniature ocean, the Mediterranean Sea, and an infant ocean, the Red Sea. The main geographical traits of all near-coastal regions surrounding the African continent are reviewed, focusing in particular on the surface current systems of the Atlantic and Indian Oceans, which influence physical as well as ecological processes of local areas. These are the Canary, Guinea, Angola and Benguela Currents, in the Atlantic African waters; the Monsoon Currents, the Somali, South Equatorial and East African Currents, the Mozambique Current and/or Channel Eddies, the East Madagascar Current and the Agulhas Current system, including the Agulhas Return Current and the Agulhas Leakine, in the Indian Ocean African waters. Further, the Mediterranean Sea and Red Sea (and the main African Great Lakes) are also reviewed, in order to introduce their environmental peculiarities and special challenges. While some of these basins have been sailed since ancient times, others remain surprisingly uncharted, in particular from the scientific point of view. Systematic and synoptic application of remote sensing techniques can provide clues to help compose the unique mosaic of dynamical and bio-geo-chemical features of the African Seas.

1.1 Introduction

The heterogeneity of the marine regions surrounding Africa matches that of the geographic and cultural scenes typical of this vast continent. The African Seas include marginal basins of two major oceans, the Atlantic and the Indian, a miniature ocean, the Mediterranean Sea, and an infant ocean, the Red Sea (Fig. 1.1). To the west, the Guinea Basin, bounded around the equator by the coast of the Gulf of Guinea, separates the Canary Basin and the Sierra Leone Basin in the North Atlantic from the Angola Basin and the Cape Basin in the South Atlantic. South of the continent, the Agulhas Basin, borders on both the Atlantic and Indian Oceans. To the east, the Mozambique Channel, between the continent and Madagascar, connects the Natal

V. Barale (✉)
European Commision, Joint Research Centre,
Institute for Environment and Sustainability, Ispra, Italy
e-mail: vittorio.barale@jrc.ec.europa.eu

V. Barale, M. Gade (eds.), *Remote Sensing of the African Seas,*
DOI 10.1007/978-94-017-8008-7_1,

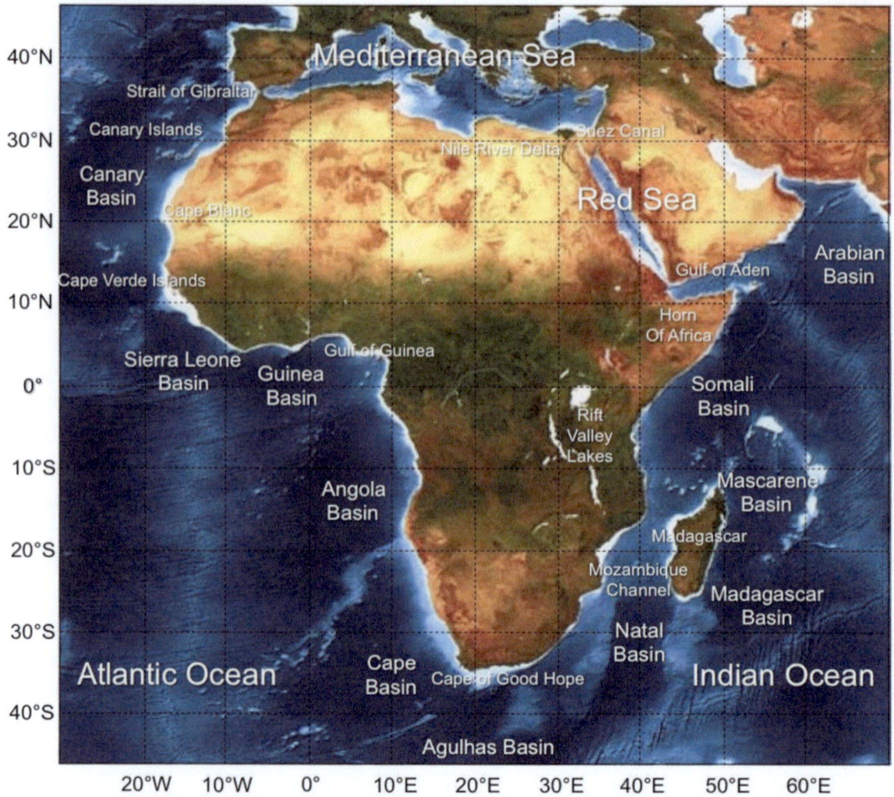

Fig. 1.1 Marginal and enclosed seas surrounding the African continent. The colour coding shows bathymetry (*blue tones*, darker means deeper waters), and vegetated to desert areas (*green* to *yellow* and *brown tones*) over land (courtesy of JRC EC)

Basin in the south and the Somali Basin in the north. Further east, the Mascarene Basin and the Madagascar Basin, edged by a score of island archipelagos; and to the north, the Arabian Basin. This connects, though the Gulf of Aden, with the semi-enclosed Red Sea. North of the continent, finally, is the Mediterranean Sea, connected with the Red Sea via the Suez canal, in the (south) east; with the Marmara Sea and then the Black Sea via the Dandanelles and the Bosphorus, in the (north) east); and with the Atlantic Ocean via the Strait of Gibraltar, in the west.

Most of the marine basins that surround the African continent can be considered as true marginal seas, partially enclosed by peninsulas, island chains or submarine ridges, but contiguous to the ocean proper. Major wind-driven oceanic currents play a key role in shaping the environmental characteristics of their near-coastal waters. Others, *i.e.* the Mediterranean Sea and the Red Sea, are (almost totally) enclosed seas, separated from the outer ocean, with which they have limited exchanges, and where thermohaline processes play an even bigger role in determining environmental peculiarities. Evaporation far exceeds precipitation and continental runoff in this kind of “concentration” basins. Hence, salinity is higher than in the outer ocean, while water exchange in the connecting channels consists of an upper layer inflow of fresher oceanic water and a lower layer outflow of saltier water. In addition, a number of inland basins, the African Great Lakes, or Rift Valley Lakes, should also be included among the major water bodies of the African continent. By the same token, some of the areas impacted by the Nile River, connecting the Great Lakes in the highlands of sub-equatorial east Africa and the Mediterranean Sea, across the eastern Sahara desert, and forming a wide delta dotted by brackish lagoons, should also be numbered among the “coastal” features of interest here.

The African Seas comprise a wide spectrum of different ecosystems, spanning most of the Earth climatic zones, and ranging from unspoiled conditions to pollution hotspots heavily impacted by human activities. Exploring the wide spectrum of environmental features and processes of such a varied collection of marine and coastal (and inland) regions is vital for the assessment of the abundant natural resources that the African Seas have to offer – of those in dire need of being more sustainably exploited, for the well being of continental economies and societies, and of the others that should be protected and maintained in their still pristine conditions.

Due to their vast area, great depth and efficient water circulation, the open oceans are relatively unaffected by human activities, compared with marginal and enclosed basins. Hence, degradation is limited to near-coastal zones, while most of the Atlantic and Indian ecosystems remain undisturbed, even though with notable exceptions. The deterioration of fish stocks, the reasons for which are not yet fully understood but could include overfishing and climatic variability, is an example of the severe problems hampering the development potential offered by existing oceanic resources. The state of the Red Sea and the (southern) Mediterranean Sea, which have limited exchanges with the adjacent oceans, is extremely varied and strongly influenced by local processes.

In the following, the main geographic features of the oceanic areas surrounding the African continent will be recalled, focusing in particular on some key oceanographic features of the marginal basins dominated by major surface current systems. The latter, in fact, influence physical as well as ecological processes of entire near-coastal regions, providing at the same time a critical contribution to regulation of the world’s climate. In addition, the environmental traits of the Mediterranean Sea and of the Red Sea (and of the main Great Lakes as well) shall be reviewed in some detail, in order to introduce the peculiarities and special challenges of these basins that require more in-depth presentations.

1.2 Atlantic Ocean African Waters

The Atlantic Ocean, connecting all of the marginal basins west of the African continent, is the second largest of the Earth’s oceanic divisions. It began to form about 150 million years ago, in the Jurassic period, when a rift opened up in the supercontinent of Gondwana, resulting in the separation of Africa and South America. The separation continues today along the Mid-Atlantic Ridge at a rate of about 2.5 cm/year.

The Atlantic Ocean occupies about 16.2 % of the planetary surface, representing an area of approximately 82,400,000 km^2, with an average depth of 3,926 m, and a volume of 323,600,000 km^3 (figures that rise to 20.8 %, 106,400,000 km^2, 3,338 m and 354,700,000 km^3 when including the major adjacent seas). Its deepest point, still uncertain but possibly just over 8,600 m below the surface, is located in the Puerto Rico Trench. The basin extends in latitude from the Arctic to the Antarctic (where the boundary with the Southern Ocean has been set at 60°S), and varies in longitudinal width from 2,848 km, between South America and Africa, to about 4,830 km, between North America and Africa. Surface salinity ranges between 33 (just north of the equator, because of heavy tropical rainfall, at high latitudes and along coasts impacted by large rivers) and 37 psu (at about 25°N and S, in subtropical regions with low rainfall and high evaporation). Average values between 35 and 36 psu make it the saltiest of the world's oceans. Surface water temperatures vary from −2 °C, in polar regions, to about 29 °C, near the equator.

In a schematic representation of the Atlantic Ocean water column (the hydrographic structure of which actually varies considerably with latitude), based on combined temperature and salinity characteristics, Upper Water at the surface is followed by Central Water, formed by subduction at the tropics (surface to 500 m). Below this, Antarctic Intermediate Water from the south, formed by deep convection east and west of the southern tip of the American continent and spreading into all oceans with the Circumpolar Current; Arctic Intermediate Water from the north, formed at high latitudes by either convection or subduction; Atlantic Subarctic Water and Mediterranean Sea Water sinking from the Strait of Gibraltar, in the northeast (500–1500 m). Still deeper in the water colum, North Atlantic Deep Water, the dominant water mass in the deep Atlantic Ocean, formed by complex processes involving deep convection in the Greenland Sea, Labrador Sea and Arctic Ocean; and the underlying Antarctic Bottom Water, formed by deep convection in the Weddell Sea and Ross Sea, occupying the deepest basins (below 3000–4000 m and to the bottom).

The general thermohaline circulation of the Atlantic Ocean, part of the global Conveyor Belt system, is composed by a prevalent northward flow in surface layers, and by a prevalent southward flow in deeper layers (primarily in the western basins). At the surface, the wind-driven circulation pattern consists of two main basin-wide gyres, circulating in an anticyclonic direction, *i.e.* clockwise in the North Atlantic and counter-clockwise in the South Atlantic. In between, as the trade winds drive waters of the inter-tropical region from east to west, lies what is known as the equatorial current system (see Fig. 1.2 for a sketch of these systems' African components). In the northern gyre, the Canary Current, flowing southward off northwest Africa, feeds the westward North Equatorial Current. South of it, the North Equatorial Countercurrent moves eastward towards the Gulf of Guinea, to form (with the contribution of the Canary Current) the Guinea Current, which is followed, along the coast, by the southward flowing Angola Current. In the southern gyre, the Benguela Current, flowing northward off southwest Africa, feeds the westward South Equatorial Current, bounded to the south by the South Equatorial Countercurrent.

The **Canary Current**, named after the Canary Islands archipelago – which partially block its flow – is the eastern boundary current within the North Atlantic Gyre.

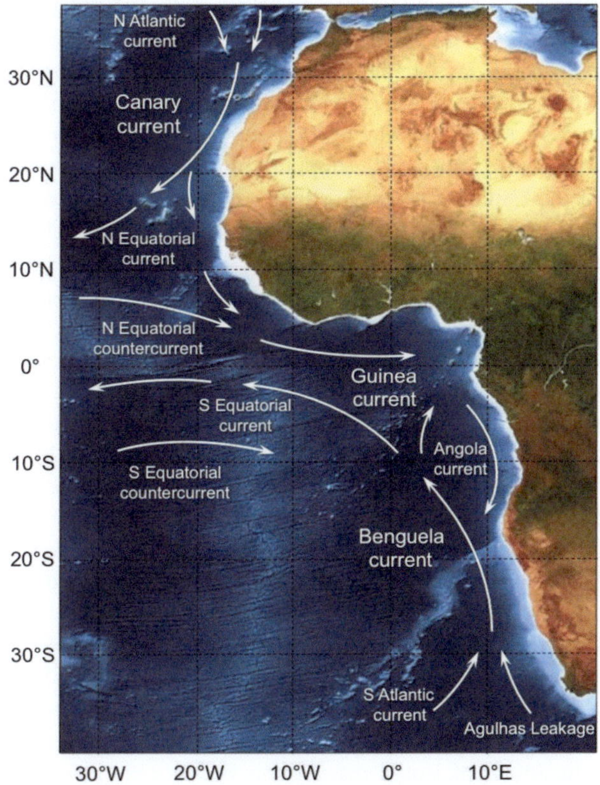

Fig. 1.2 Atlantic Ocean surface currents in the African marginal basins. The colour coding shows bathymetry (*blue tones*, darker means deeper waters), and vegetated to desert areas (*green* to *yellow* and *brown tones*) over land (courtesy of JRC EC)

It branches south from the North Atlantic Current, flows southwest along the African coast between 30°N and 10°N, past the Cape Verde archipelago, and then turns west to join the North Equatorial Current. The Canary Current system is characterized by filaments, eddies and intense coastal upwelling, driven by persistent northerly winds along the coast of northwest Africa. Upwelled waters are drawn offshore by Ekman drift in the Canary Current and also by current jets protruding transversally from the near-coastal area. Upwelling can occur year-round at Cap Blanc, at about 21°N, and northward, in particular between 23°N and 25°N. South of Cap Blanc, upwelling expands in winter and spring and shrinks in summer and fall, due to the northward migration of the Azores high (which is responsible for driving equatorward winds). The upwelling zone begins its southern advance in October and migrates meridionally as the season progresses. It reaches its maximum southward extent in January-March, then retreats northward, until late summer, early fall. Alternating upwelling events of cold, nutrient-rich, deep waters, and relatively calm periods have a key role in stimulating local primary productivity. If upwelling is sustained for a long period of time, phytoplankton are eventually removed from the euphotic zone. Hence, calm periods, during which stratification can develop, allow phytoplankton to bloom in a stable mixed layer. Upwelling events and increased primary production follow the onset of strong winds within a few days. Secondary producers, *i.e.* zooplankton, take longer to respond to the increased availability of food, because they have life cycles of weeks rather than days. In the Canary Current, secondary production reaches its peak in fall, when upwelling intensity decreases, allowing zooplankton to remain over the shelf and exploit the available food supply. Due to the rapid response of phytoplankton to upwelled nutrients, zooplankton are seldom food-limited. The high productivity of the Canary Current has led to the development of sizeable coastal fisheries (*e.g.* in western Morocco). Four types of fish comprise about 75 % of total catches in the Cap Blanc area. Clupeids are the most abundant, with *Sardina pilchardus*, the common sardine, dominating in cooler northern waters and *Sardinella aurita* in warmer southern waters.

The **Guinea Current** flows (south) eastward at approximately 3°N along the coast of equatorial West Africa and into the Gulf of Guinea. This warm, highly saline water body is always located north of the equator, but changes position and intensity with the seasons. Its (westernmost) northern reach may vary between a boreal winter minimum and summer maximum. The seasonal instability of its major sources, the North Equatorial Countercurrent and the Canary Current, can affect this variability. Like other boundary currents of the eastern Atlantic Ocean, the Guinea Current is characterized by areas of upwelling and increased productivity. Cooler waters at the surface are associated with the (boreal) summer intensification of the current. The current is geostrophically balanced, with isotherms that slope upwards towards the coast to the north. As the current intensifies, the slope steepens, bringing the thermocline closer to the surface near the coast. Thus, in the Guinea Current (in which the seasonal shallowing of the thermocline may be linked to geostrophic adjustment of isotherms, Kelvin waves and cyclonic turbulent eddies, rather than local wind stress) summer intensification and coastal upwelling are related, but the apparent lack of correlation between surface temperature and wind patterns on a seasonal scale renders the Gulf of Guinea quite unusual among upwelling regions. Further south, the **Angola Current** forms the eastern section of a large, anticyclonic gyre in the Gulf of Guinea. It is fed by the southward-turning waters from the north branch of the Benguela Current and by various components of the intricate equatorial current system, rather than the Guinea Current itself.

The **Benguela Current** is a broad, northward flowing ocean current that forms the eastern portion of the South Atlantic Ocean gyre. The current extends roughly from Cape of Good Hope in the south, at almost 35°S, to the position of the Angola-Benguela Front in the north, at around 16°S. Where the cold Benguela Current and the warm, south-flowing Agulhas Current mix, off the the sothwestern African shelf, a turbulent, stormy, but highly productive marine ecosystem, is formed. Its western, offshore edge is ill-defined, and modulated by many eddies and meanders, but there is a distinct thermal front between the waters associated with the current and those of the near-coastal area. Inshore of the Benguela Current proper, the prevailing south-easterly winds drive coastal upwelling, forming the Benguela Upwelling System. The cold, nutrient rich waters fuel high rates of phytoplankton growth, and sustain the productive Benguela ecosystem. Variations in wind strength cause pulses of up-welling, similar to Kelvin waves, except on a reduced scale (of 10–100 km instead of

1000 km), which propagate to the south along the coast and around the Cape of Good Hope, depending on the prevailing wind systems. These upwelling pulses, providing a period of upwelling followed by a period of stratification and relatively calm waters, can induce very high productivity. In the Benguela system, phytoplankton blooms usually lag upwelling events by 1–4 days, and continue for 4–10 days. In order for zooplankton to have a continuous food supply, the blooms must not occur too far apart. Pulses of upwelling in the Benguela system regularly have duration of 10 days, an optimal period for biological production. It is estimated that the Benguela system is 30–60 times more productive per unit area than the global ocean average. While upwelling generates abundant primary and secondary production in the upper water column and near the coast, hypoxic areas called oxygen minimum zones can occur in deeper waters with limited oxygen exchange, at the coastal shelf and upper coastal slope. The Benguela oxygen minimum zone starts around a depth of 100 m and is a few hundred meters thick. Bacteria that use sulphur rather than oxygen reside in the oxygen minimum zone. The most abundant fishes in the Benguela system are sardines (*Sardinops ocelata*) and anchovies (*Engraulis capensis*). The sardine was intensely fished beginning in the 1950s, with peak catches in 1968, but has declined since then to the point that the anchovy fishery has taken over. Similar to the Pacific Ocean El Niño, about once per decade warm, nutrient poor surface waters enter the northern part of the Benguela upwelling system off the Namibia coast. During the so-called Benguela Niño, the Angola Current moves southward, from 15°S to 25°S, extending 150 km offshore and to 50 m depth. Temporal proximity to the Pacific El Niño, heavy rains and changes in fish abundance have been observed. However, the causes and effects of the Benguela Niño are not well understood. The phenomenon has been ascribed to the effect of winds in the western equatorial Atlantic Ocean that propagates as subsurface temperature anomalies to the African coast; but the importance of local winds in its development off the coast of Namibia and Angola has also been documented. This local occurrence and the remote signal from the equatorial regions form the basis of a generation mechanism in which both processes can reinforce each other.

1.3 Indian Ocean African Waters

The Indian Ocean, connecting the marginal basins east of the African continent, is the third largest of the Earth's oceanic divisions. It is the youngest of the major oceans, but its origin and evolution are the most complicated. Its formation is a consequence of the breakup, about 150 million years ago, of the southern supercontinent Gondwana, followed by the movement to the northeast of the Indian subcontinent (started about 125 million years ago), which began colliding with Eurasia about 50 million years ago; by the western movement of Africa and the separation of Australia from Antarctica some 53 million years ago. By 36 million years ago, the Indian Ocean had taken on its present configuration.

The Indian Ocean occupies about 14.4 % of the planetary surface, representing an area of approximately 73,556,000 km^2, with an average depth of 3,890 m, and a volume of 292,131,000 km^3 (including all its adjacent seas). Its deepest point, more than 8,000 m below the surface, is in the Diamantina Trench. The basin extends in latitude from the Tropic of Cancer to the Antarctic (where the boundary with the Southern Ocean has been set at 60°S), and reaches a longitudinal width of about 10,000 km, between the southern parts of Africa and Australia. Surface salinity ranges between 32 (in areas of monsoon rains, at high latitudes and along coasts impacted by large rivers) to 37 psu (in the Arabian Sea and in a belt between southern Africa and south-western Australia). Surface water temperatures vary between 22 °C and 28 °C, northward of the Tropic of Capricorn, but drop rather quickly southward.

In a schematic representation of the Indian Ocean water column, based on combined temperature and salinity characteristics, Upper Water at the surface is followed by Central Water, formed by subduction at the tropics, as well as Indonesian, Bengal Bay and Arabian Sea Water (surface to 500 m). Below this, Antarctic Intermediate Water from the south, formed by deep convection east and west of the southern tip of the American continent and spread into all oceans with the Circumpolar Current, and by Indonesian, Red Sea and Persian Gulf Intermediate Water in the north (500–1500 m). Still deeper in the water column, Circumpolar Deep Water, a mixture of North Atlantic Deep Water and Antarctic Bottom Water, formed by deep convection in the Weddell Sea and Ross Sea, occupy the deepest basins (1500 m to bottom).

The general thermohaline circulation of the Indian Ocean, as part of the global Conveyor Belt system, is composed by a prevalent northeastward flow in lower layers (primarily in the western basins), and by a prevalent southwestward flow in the upper layers. At the surface, the wind-driven circulation pattern consists of variable currents, dominated by the monsoon wind reversals, north of the equator, and a main basin-wide gyre, circulating in an anticyclonic direction, *i.e.* counter-clockwise, south of the equator (see Fig. 1.3 for a sketch of these systems' African components). During the summer monsoon, the northern component of the South Equatorial Current (the East African Coastal Current) feeds a northward Somali Current (recirculating also in an anticyclonic gyre off east Africa). This continues northward and eastward into the Southwest Monsoon Current. During the winter monsoon, the North Equatorial Current and the Northeast Monsoon Current, flowing westward and southwestward, feed a southward Somali Current off east Africa. This joins with the northern component of the South Equatorial Current (the East African Coastal Current) into the Equatorial Countercurrent flowing eastward, away from the African continent. In the southern gyre, the southward component of the South Equatorial Current splits west of Madagascar, towards the Mozambique Channel, and east of Madagascar, as the East Madagascar Current, only to re-join southward to form the complex Agulhas Current system.

The **Monsoon Currents** (or Drifts) indicate the seasonally varying current regime found in the tropical regions of the northern Indian Ocean. During summer, the flow of the upper ocean is directed eastward, from the African area, around the Indian sub-continent and towards the Indonesian region. During winter, the direction reverses, with westward flow from the Indonesian region towards the African area. The shift occurs over a time period of several weeks and is due to changes in the wind stress associated with the Indian monsoon: winds blowing from the southwest result

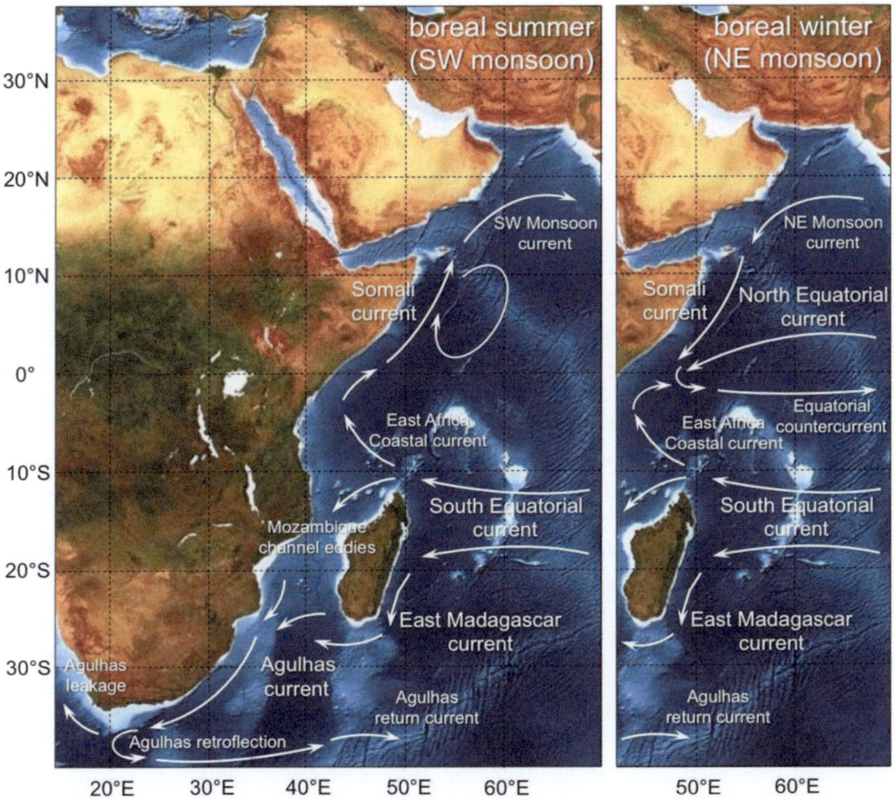

Fig. 1.3 Indian Ocean surface currents in the African marginal basins. The colour coding shows bathymetry (*blue tones*, darker means deeper waters), and vegetated to desert areas (*green* to *yellow* and *brown tones*) over land (courtesy of JRC EC)

in eastward currents, while winds blowing from the northeast result in westward currents. The Southwest (or Summer) Monsoon Current, located between 10°N and 15°N latitude in the Arabian Sea, bends around the Indian sub-continent, and enters the northeastern Indian Ocean. The Current is eastward from April through November. As the current flows toward the northeast, Ekman transport (to the right of the flow in the Northern Hemisphere) is offshore, transporting warmer waters deeper into the Arabian Sea, and permitting upwelling of cooler waters along the coast. This surface temperature pattern (cooler waters west of warmer waters) also reinforces the northward current through geostrophic flow. The Northeast (or Winter) Monsoon Current extends from the northeastern Indian Ocean, around the Indian sub-continent, and across the Arabian Sea at a latitude around 8°N. The current is westward only during the months of January through March, and is strongest in February.

The **Somali Current**, running along the Horn of Africa between the equator and about 9°N, is a boundary current somewhat analogous to the Gulf Stream in the

Atlantic Ocean. However, this current is heavily influenced by seasonal changes in atmospheric forcing, and shifts direction with the monsoon winds. In summer, the Somali Current is fed by the East African Coastal Current and flows to the northeast along the Somali coast, eventually separating from the coastline and turning east as it enters the Arabian Sea (and partly re-circulating in the Great Whirl, a large gyre centered around 10°N and 55°E, present only during summer). As the summer monsoon moves coastal waters northeastward, it induces coastal upwelling, *i.e.* the only major upwelling system that occurs on a western boundary of any ocean. The upwelled water within the current is carried offshore by Ekman transport and merges with water that was brought to the surface by open-ocean upwelling, creating one of the most productive ecosystems in the world's oceans. In winter, the Somali Current flows to the southwest along the Somali coast towards the equator, ultimately contributing with the East African Coastal Current to feed the Equatorial Countercurrent flowing eastward. As the winter monsoon causes a reversal of the Somali current, it damps coastal upwelling, while surface waters move southwestward. However, this causes surface cooling and creates deep mixing, bringing abundant nutrients to the surface.

The **South Equatorial Current**, flowing westward with the Trades north of 22°S, divides into a north-flowing stream that feeds the northward East Africa Coastal Current, and a south-flowing stream that feeds the East Madagascar Current. The **East African Coastal Current** runs northward throughout the year between 11°S and 3°S. The comparison of transport figures with those in the boundary current north of Madagascar suggest that most of the water of the South Equatorial Current northern branch goes into the East African Coastal Current. At depth, there is an excess westward transport north of Madagascar, which probably goes into the Mozambique Channel[1]. In boreal summer, the East Africa Coastal Current merges at its northern end into the northward Somali Current, while in boreal winter it converges with the south-going Somali Current to form the Equatorial Countercurrent. The south-flowing stream of the South Equatorial Current has long been assumed to feed southward currents on both sides of Madagascar. That assumption can be challenged by the alternate view of a recurrent train of large anti-cyclonic eddies, rather than a western boundary current, propagating in the channel, largely dispelling the notion of a steady Mozambique Current within the channel itself. Nevertheless, it is impossible to rule out completely the likelihood that a true **Mozambique Current** may appear intermittently, for short durations[2]. This western boundary anomaly is possibly due to the presence of Madagascar itself[3]. The **Mozambique Channel Eddies** are large cyclonic and anti-cyclonic eddies that propagate southwards in the

[1] The Mozambique Channel is a major feature of the African east coast, about 1,600 km long and 400–1,000 km wide, reaching depths of up to 3,000 m.

[2] Numerical model simulations of the Mozambique Channel show the appearance of a current along the Mozambican Coast, during periods between eddies.

[3] The surface flow in the southeastern Indian Ocean, if there were no Madagascar, would be a normal, continuous western boundary current, constituting a continuum with the Agulhas Current.

Mozambique Channel. The exact causal mechanism and location of the eddy formation are still debated. The frequency of eddy occurrence reduces from around seven per year in the north of the Channel, to around four per year in the south. Once the eddies exit the Mozambique Channel, they contribute to the variability of the Agulhas Current, including the generation of Natal Pulses (meanders of the current itself) and of Agulhas Rings (eddies shed into the South Atlantic by the current, past its retroflection site). On the eastern side of Madagascar, the **East Madagascar Current** flows southward from 20°S to the island southern limit, and subsequently feeds the Agulhas Current, also by virtue of large cyclonic and anti-cyclonic eddies propagating westwards, which induce a strong variability in the current transport.

The **Agulhas Current** is the real western boundary current of the (southern) Indian Ocean – one the largest in the world's oceans, as comparable western boundary currents, *i.e.* the Gulf Stream in the North Atlantic and the Kuroshio in the North Pacific, or the Brasil Current in the South Atlantic, are believed to transport much less water, from approximately one half (Gulf Stream and Kuroshio) to one fourth (Brazil Current), of the Agulhas – flowing southwestward as a narrow jet over a steep continental slope, off the east coast of Africa and further south, between 27°S and 40°S. Large solitary meanders, called Natal pulses, can occur within this region. These extend about 170 km offshore, with downstream propagation rates decreasing as the continental shelf broadens near 34°S. At around 36°S the Agulhas separates from the coast, leaves the continental shelf and develops oscillations of increasing amplitude. In the southeastern Atlantic Ocean, at about 16–20°E, due to shear interactions with the Antarctic Circumpolar Current, eventually it turns back on itself in the Agulhas Retroflection, forming the **Agulhas Return Current** and rejoining the main Indian Ocean Gyre. The retroflection loop usually encloses a pool of Indian Ocean surface water south of Africa, the temperature of which is more than 5 °C warmer than that of South Atlantic surface water at similar latitudes. While most of the warm and salty water transported from the tropical zone to the southern tip of Africa loops around to remain in the Indian Ocean, some leaks into the cooler, fresher Atlantic Ocean via giant eddies (Agulhas Rings) and filaments. Once in the Atlantic, the warm, salty **Agulhas Leakage** eventually flows into the South Atlantic Gyre and ultimately reaches the northern hemisphere, strengthening the Atlantic overturning circulation by enhancing deep-water formation. The Agulhas Current region constitutes a moderately productive ecosystem, with localized, episodic upwelling of colder water in the Mozambique Channel and at the southeastern tip of Madagascar, and in general on the shoreward edge of the current itself (in particular where the current passes from a narrow to a wider shelf). As a consequence of these (and other, poorly sampled and not well understood) factors, the area presents an enhanced primary productivity in comparison to surrounding waters. This is especially notable in the Agulhas Retroflection, where primary productivity tends to be significantly higher than that in the surrounding waters of the South Indian Ocean and South Atlantic Ocean. Conversely, when the current turns back on itself, and the loop of the retroflection pinches off (in general after the passage of a Natal pulse), the eddy released into the South Atlantic Gyre is a warm core ring, having lower primary productivity than the surrounding colder waters. These Agulhas Rings have

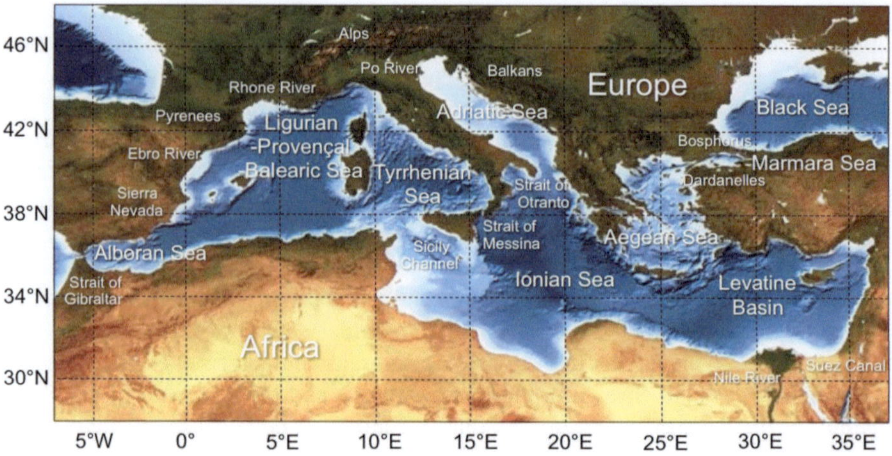

Fig. 1.4 Mediterranean Sea. The colour coding shows bathymetry (*blue tones*, darker means deeper waters), and vegetated to desert areas (*green* to *yellow* and *brown tones*) over land (courtesy of JRC EC)

been observed to carry waters with low phytoplankton concentration into the flow of the Benguela Current (or northwestward across the South Atlantic into the South Equatorial Current) and to remove larval and juvenile fish from the continental shelf (which results in a reduced catch of anchovies (*Engraulis capensis*) in the Benguela system).

1.4 Mediterranean Sea

The Mediterranean Sea is a semi-enclosed inland sea, bounded in the north by southern Europe and the Anatolia peninsula, and in the south by northern Africa. It extends longitudinally between the Atlantic Ocean, to the west, and the Asian landmass, to the east. This elongated and narrow-shaped sea is divided into a western basin, which includes the Alboran Sea, the Ligurian-Provençal-Balearic Sea and the Tyrrhenian Sea, and an eastern basin, which includes the further enclosed Adriatic Sea, the Aegean Sea, the Ionian Sea and the Levatine basin (Fig. 1.4).

The Mediterranean Sea was thought to be a remnant of the ancient Tethys Ocean. Today, it is known to be a structurally younger ocean basin, called Neotethys, which formed during the Late Triassic and Early Jurassic rifting of the African and Eurasian plates. The geologic history of the Mediterranean basin is complex, involving first the break-up and then the collision of the African and Eurasian plates, and the Messinian Salinity Crisis in the late Miocene, when the whole sea dried up almost completely.

The Mediterranean Sea extends from 30°N to 45°N and from 6°W to 36°E, with an area of 2,500,000 km^2 – two orders of magnitude below that of the major oceans, corresponding to about 0.7 % of the global ocean surface, or 0.5 % of the Earth's surface – and a volume of 3,700,000 km^3. It has dimensions of about 3,860 km in the east-west direction and a maximum of 1,600 km (average around 800 km) in the north-south direction. The greater Mediterranean basin consists of a series of deep depressions, connected to each other, with an average depth of 1,500 m. Actual depths vary between 2,500 and 3,500 m in western basin, and between 3,500 and 4,000 m in the eastern basin, where a maximum of 5,267 m is reached in the Calypso Deep of the Ionian Sea. The two main basins are separated by the Sicily Channel (with a 430 m sill depth) and the Strait of Messina (80 m), while the Strait of Otranto (800 m) separates the Adriatic Sea from the eastern basin. The Mediterranean Sea is connected to the Atlantic Ocean by the Strait of Gibraltar (14.3 km wide, at the narrowest point, with a 290 m sill depth), in the west; to the Marmara Sea by the Dardanelles (a strait 61 km long, from 1.2 to 6.0 km wide, averaging 55 m of depth), and subsequently to the Black Sea by the Bosphorus (31 km long, from 0.7 to 3.4 km wide, averaging 65 m of depth), in the northeast; to the Red Sea via the Suez Canal (193.3 km long, 205 m wide and 24 m deep), in the southeast.

The Mediterranean Sea accounts for 46,000 km of coastlines, one-third of which are due to the mainland and islands of the Aegean Sea. Several mountain ranges (*e.g.* the Sierra Nevada, the Pyrenees, the Alps and Apennines, the Balkans) are distributed along the northern side of the basin. Since these mountains mostly slope steeply into the sea, the northern drainage basin is relatively small. And since the southern side is mainly covered by desert, these combined factors tend to limit the input of continental freshwater. Only a few large rivers flow directly into the Mediterranean Sea, *i.e.* the Ebro (length 910 km, average discharge 426 m^3s^{-1}), the Po (682 km, 1540 m^3s^{-1}) and the Rhone (813 km, 1710 m^3s^{-1}), in the north, and the Nile (6650 km, 2830 m^3s^{-1}), in the south. The latter, considered together with its main tributary, the White Nile, originating from the Great Lakes region of central Africa, is generally regarded as the longest river on the planet. Its flow rate has been much altered by the closure of the Aswan High Dam, starting in 1964. The river forms a wide delta on the southeastern Mediterranean coast, dotted by several coastal lagoons of great ecological relevance. Additional contributions of continental freshwater, originating from the Azov Sea – where the main tributaries are the Don (length 1950 km, average discharge 935 m^3s^{-1}) and the Kuban (870 km, 425 m^3s^{-1}) – and then from the Black Sea – main tributaries are the Danube (2860 km, 6500 m^3s^{-1}), the Dnieper (2290 km, 1670 m^3s^{-1}) and the Dniester (1362 km, 310 m^3s^{-1}) – enter the Mediterranean basin from the Marmara Sea.

The typical Mediterranean climate, with low cloudiness and abundant sunshine, is subject to both sub-tropical and mid-latitude weather systems, and is also influenced by the northern mountain ranges. The region's seasonal cycle is characterized by mild winters, when the region receives most of its rainfall, and by relatively hot and dry summers. Strong local winds, such as the cold, dry, northerly Bora and Mistral, and the hot, dry southerly Scirocco, typify the region, particularly in winter. In general, air temperature differences between winter and summer are limited to about 15 °C, although local geographic and meteorological factors can result in extreme conditions such as air temperatures up to 50 °C on the African coast. The mean temperature of surface waters vary between minima of about 14–16 °C (west to east), in winter, and maxima of about 20–26 °C (again, west to east), in summer, but the total excursion can reach up to 20 °C, as in the shallow parts of the Adriatic Sea, where the range is between the 8–10 °C of winter and the 26–28 °C of summer.

Evaporation greatly exceeds precipitation and river runoff (with an estimated freshwater deficit of 2,500 km^3/year), so that the Mediterranean Sea is characterized by very high salinity (mean around 38 psu), a fact that is central to water circulation within the basin. Evaporation is especially high in the eastern Mediterranean, causing the water level to decrease and salinity to increase (up to 39 psu) eastward. The resulting pressure gradient pushes cooler, lower-salinity (about 36 psu) water from the Atlantic Ocean across the entire basin. The less dense Atlantic water flows into the Mediterranean Sea through the Strait of Gibraltar, in the surface layer. The incoming water[4] warms and becomes saltier as it travels east, and is eventually turned into denser Mediterranean waters through evaporation. It sinks in the Levantine basin (as well as in other areas where deep waters are formed), due to winter cooling, moves back westward and ultimately spills over the sill of the Strait of Gibraltar, in the bottom layer, and out into the Atlantic Ocean. This distinct Mediterranean Intermediate Water can be traced in the Atlantic Ocean to the northernmost latitudes, where it contributes to North Atlantic Deep Water formation. The complete cycle determines a residence time in the basin of about 80 to 100 years.

Surface circulation is essentially eastward in the southern part of the basin, and westward along the northern shorelines. In the western basin, the Atlantic Jet moves from the Strait of Gibraltar to the Sicily Channel, while the current transporting waters back flows along the Italian, French, and Spanish coasts. Intermittent and long-lived sub-basin-scale eddies and gyres abound in the Algerian near-coastal region and the Tyrrhenian Sea. In the eastern basin, surface waters are transported eastward by several local currents, but recirculate in numerous eddies and gyres before reaching the northward coastal current off Israel, Lebanon, and Syria and veering westward off Turkey, in the Levantine Basin.

Deep waters can be formed both in the Adriatic Sea and in the Aegean Sea, although with different characteristics, and spill out into the eastern basin over the sills in the Otranto Strait and around Crete, respectively. Deep waters can also be formed in the Ligurian-Provençal Sea, due to the effect of the Mistral wind, which increases the density of surface waters through intense evaporation and cooling, particularly in the Gulf of Lion region. This can lead to a complete overturning of the water column, with deep convection processes taking place over 2000 m of water. The process ventilates the deepest parts of the western Mediterranean and triggers the onset of large algal blooms, sometimes covering the entire northwestern basin. The deep waters of the eastern and western basins do not communicate, due to the shallow sill of the Sicily Channel. While a substantial amount of vertical mixing is provided by strong regional (winter) wind regimes, tidal amplitudes are very small, in the Mediterranean Sea, and the narrow continental shelves prevent tidal amplification along the coast.

---

[4] About 1 Sv of transport. The Sverdrup (Sv) is a unit of measure for large oceanic flows: 1 Sv = 10^6 m^3 s^{-1}.

On average, the Mediterranean Sea is poor in nutrients, with consequent low phytoplankton biomass and primary production. Oligotrophy increases from west to east. Primary production in the open sea is considered to be phosphorus-limited, rather than nitrogen-limited as in most of the world's oceans. By contrast, a rich biodiversity characterizes the Mediterranean ecosystem: the fauna and flora are among the richest in the world, with over 10,000 marine species recorded, highly diverse and with a large proportion (28 %) of endemic species. No species disappearances have been reported, but changes in species composition and richness have occurred in some areas (with some species showing alarming trends, as in the case of the bluefin tuna, *Thunnus thynnus* in particular, due to overfishing of a stock already threatened). The introduction of exotic species – such as that of tropical species from the Red Sea, which occurred after the Suez Canal opening[5] – is a growing concern. Large populations of marine mammals (*e.g.* the fin whale, *Balaenoptera physalus*) have been documented, while other species are in danger (*e.g.* the monk seal, *Monachus monachus*, is critically endangered; the marine turtles *Caretta caretta* and *Chelonya midas* are listed respectively as threatened and endangered).

The state of the Mediterranean's open waters is generally good, but coastal areas are subject to various environmental problems, including eutrophication and heavy metal, organic and microbial pollution. Land-based activities (urbanization, industry and agriculture, particularly in the north-west) are the main sources of pollution. About 30 % of the ever-increasing population of Mediterranean countries, approximately 150 million people, lives on the coast. Adding to the resident population, each year the Mediterranean Sea is also host to 150 million tourists. Population density, almost 100 people *per* km^2, twice that of the region as a whole, and related economic activities place a high pressure on coastal zones. Since the Mediterranean Sea is essentially oligotrophic, eutrophication is limited to near-coastal sites that receive enhanced nutrient loads (of anthropogenic nature) from rivers, or direct discharges of untreated domestic and industrial wastewaters, and to adjacent open waters. However, this is expected to worsen, together with the occurrence of harmful algal blooms. Oil shipping is also a growing concern, since of the 2000 cargos that sail in the Mediterranean at any given time, about 250–300 are oil tankers, transporting 400 million tons/year of petroleum products (*i.e.* 30 % of world traffic). This large traffic gives rise to about 1 million tons/year of illegal oil discharges at sea (*i.e.* 20 % of the total oil pollution in the oceans).

1.5 Red Sea

The Red Sea is a semi-enclosed tropical basin, bounded by northeastern Africa, to the west, and the Arabian Peninsula, to the east (Fig. 1.5). The elongated and narrow-shaped basin extends between the Mediterranean Sea, to the north-west, and

[5] A phenomenon known as the Lessepsian Migration, after Ferdinand de Lesseps, the engineer who oversaw the Canal's construction.

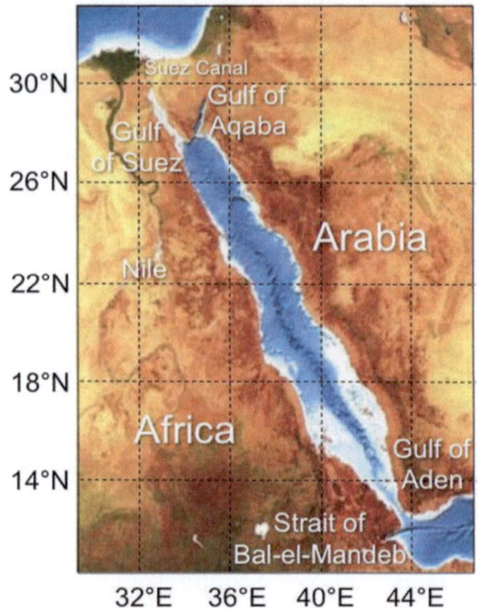

Fig. 1.5 Red Sea. The colour coding shows bathymetry (*blue tones*, darker means deeper waters), and vegetated to desert areas (*green* to *yellow* and *brown tones*) over land (courtesy of JRC EC)

the Indian Ocean, to the south-east. At the northern end, it separates into the Gulf of Aqaba and the Gulf of Suez, which is connected to the Mediterranean Sea *via* the Suez Canal. At the southern end, it is connected to the Gulf of Aden, and the outer Indian Ocean, *via* the Strait of Bal-el-Mandeb.

The formation of the Red Sea begun in the Eocene, when the African and Asian continental plates started to move apart, and accelerated during the Oligocene. The basin acquired its present shape over the past 4–5 million years, by slow seafloor spreading, a fact that makes it a geologically recent opening and one of the youngest oceanic zones on Earth. Today, the basin continues to widen at a rate of 1–2 cm *per* year.

The Red Sea covers a wide span of latitudes, from 12°N to 30°N, and longitudes, from 32°E to 44°E. The basin extends from north to south over a distance of approximately 1,900 km. Its average width from east to west is 280 km, with a maximum of 306 km, in the south, and a minimum of 26 km, in the Bab-el-Mandeb Strait. The two northern gulfs, Aqaba and Suez, are 180 km long and 25 km wide, and 300 km long and 50 km wide, respectively. The basin has a total area of about 440,000 km^2, a volume of 230,560 km^3, and more than 4,000 km of coastlines.

Due to the presence of wide, shallow continental shelves – about 25 % of the basin is less than 50 m deep, 40 % less than 100 m, and only 15 % is more than 1000 m deep – the average depth of the basin is only 524 m. The shelf breaks are marked by extensive coral reefs, while the continental slope has an irregular profile, breaking into a series of steps down to about 500 m of depth. A central deep trench, sandwiched between the eastern and the western shelf, stretches from north to south for almost the entire axis of the basin. The deepest region lies between 14°N and 28°N and has a maximum depth of 2,920 m (3,040 m according to some sources). The deep areas are still geologically active and present numerous hydrothermal vents, emitting hot, salty and metal-rich brines. More than 20 Deeps have been discovered (*e.g.* the Atlantis II Deep, Discovery Deep, Oceanographer Deep), containing abundant metal-bearing sediments and salt deposits[6]. The seafloor rises in the southern region toward the Bab-el-Mandeb Strait, just 137 m deep at the Hanish Sill. The two gulfs in the northern region are similar in shape, but their bottom topography is not. The Gulf of Suez is shallower and has a relatively flat bottom, with depths ranging from 55 to 73 m. The Gulf of Aqaba, instead, comprises a deep basin – separated by a submarine sill into two pits, both deeper than 1,000 m – bordered by a narrow shelf.

Desert or semi-desert areas, with no major freshwater inflow, surround the Red Sea. Given the absence of rivers and permanent streams, occasional runoff is due only to seasonal torrents (wadis). In this arid region, extremely high temperatures characterize the weather, particularly in summer, increasing from the northern to the southern area (which is considered to be among the hottest in the world). Air temperatures range from 6 to 39 °C at the Suez Canal and from 13 to 42 °C along the coast of the Arabian Peninsula. Rainfall is extremely sparse and occurs usually in the form of short, local showers, often associated with thunderstorms and occasionally with dust storms. The annual rainfall over the Red Sea region is extremely low, averaging about 110 mm *per* year (but only half of that, about 60 mm *per* year, for the Red Sea itself and its coastal zones).

The Red Sea climate is characterized by two distinct seasons, under the influence of the northeast monsoon, in winter, and the southwest monsoon, in summer, respectively. Monsoon winds occur because of the differential heating between landmasses and ocean surface. The local geographic and climatic conditions, leading to a low freshwater input and a high evaporation rate, render the Red Sea one of the hottest and saltiest bodies of seawater in the world. The average surface temperature in summer ranges from 26 °C in the north to 30 °C in the south, with variations of only 2 °C in winter. In general, surface temperature declines toward Bab-el-Mandeb, due to exchanges with the Gulf of Aden, and gradually decreases toward the northern region. Deeper waters are stable throughout the region, so that below 300 m the temperature is constantly between 21 and 22 °C. Salinity is also high, due to the combination of high evaporation, low precipitation and lack of river outflow. It ranges between 36 and 39 psu, and is usually lower in the southern region, again due to the exchanges with the Gulf of Aden. Water renewal in the Red Sea is relatively slow, as the exchange with the outer ocean takes approximately 6 years for the 200 m layer above the thermocline, but approximately 200 years for the entire sea.

The Red Sea circulation follows an anti-estuarine pattern, with relatively fresher water entering the basin above an outflowing saltier, denser layer. The fresher Gulf of Aden Surface Water enters the Red Sea via the Bab-el-Mandeb. Within the basin, this water is subject to the aforementioned excess of evaporation over precipitation and runoff, estimated to be 2.06 ± 0.22 m year^{-1}. The net evaporation drives the formation of Red Sea Water in the northern part of the basin. This northern water

[6] Sometime during the Paleocene (Tertiary) period, the Bab-el-Mandeb is thought to have closed – either because of volcanic eruptions at Perim Island or due to sea level lowering during the Ice Ages – causing the early Red Sea to evaporate, and creating the salt beds.

flows southwards as a dense, cool, salty layer, under which resides a relatively stagnant layer of still denser Red Sea Deep Water, formed during the winter months in the Gulf of Suez and Gulf of Aqaba. The Red Sea thermohaline circulation is modified in the upper two layers (Gulf of Aden Surface Water and Red Sea Water) by the action of the wind field and rotation, which helps generate a system of gyres, eddies and boundary currents.

In the Red Sea, detailed current data are lacking, partially because currents are weak and variable, both spatially and temporally. Such space and time variations are essentially wind-driven. In summer, the prevailing winds drive surface waters south, whereas in winter the mean flow is reversed. Over the full annual cycle, the net value of the latter predominates, resulting in an overall drift to the northern end of the Red Sea, where evaporation generates the Red Sea Water, which then moves southward, and ultimately leaves the basin at intermediate depth.

Unlike that of many other marginal seas, the Red Sea overflow is highly seasonal[7], due to the monsoon winds and to variations in buoyancy fluxes. In the colder months (November to May), a two-layer system dominates the exchange over the Hanish Sill: Gulf of Aden Surface Water enters the basin above the Red Sea Water layer (while the underlying Red Sea Deep Water leaves the basin at a much slower rate, via a combination of mixing into the Red Sea Water above it and of Bernoulli aspiration[8], without significantly altering the two-layer exchange). In the warmer months (June to October), the Gulf of Aden Intermediate Water, present under the Gulf of Aden Surface Water, is upwelled in the Gulf of Aden by the southwest monsoon winds. As the Gulf of Aden Intermediate Water upper interface rises, it penetrates into the southern Red Sea as an intermediate layer. The intrusion can last for a period of 3 months and is mixed into the upper layer, inducing an occasional Gulf of Aden Surface Water flow reversal.

Tides are small in the Red Sea, ranging from 0.6 m in the north, near the Gulf of Suez, to 0.9 m in the south, near the Gulf of Aden, but fluctuating between 0.2 and 0.3 m away from the nodal point. The Central Basin is therefore almost tideless, and thus annual water level changes appear to be more significant than tides. In spite of the small tidal range, a thin sheet of water can inundate extensive near-coastal areas and lagoons during high tide, especially along the coast of the Arabian Peninsula.

The Red Sea has low productivity, due to a strong nutrient limitation, given the absence of major continental inflows and the presence of a permanent thermocline, which inhibits benthic nutrients from mixing into surface waters, where most primary production occurs. The main nutrient input is from the Indian Ocean, through the Gulf of Aden and the southern part of the basin. In spite of these conditions, the Red Sea constitutes a rich and diverse ecosystem. More than 1100 species of fish have been recorded, with about 10 % of these being endemic. This rich diversity is in part

[7] Transport varies from a winter maximum of 0.6 Sv to a summer minimum of 0.05 Sv.

[8] The high velocities of the Red Sea outflow in the bab el Mandeb can provide a Bernoulli suction that enables the Red Sea Deep Water to flow up and over the sill, out into the Gulf of Aden. If the speed along the upper streamline is great enough, in the stratified outflow, the lower streamline will be able to rise above the depth of the sill.

due to 2000 km of coral reefs and other particular habitats, such as mangrove forests, seagrass beds and salt marshes, distributed throughout the region.

These unique habitats support a wide range of marine life (including sea turtles, dugongs and other marine mammals). Coral reefs mainly extend along the northern and central coasts, and decrease in abundance towards the southern region, as coastal waters become more turbid. In the central region, corals are mainly found 3–10 km offshore, along a narrow bank, forming a large barrier reef structure that runs parallel to the coastline. The most extensive areas of coral reefs are found along the Arabian Peninsula coast, with over 194 recorded species. Mangrove forests are scattered along much of the coastline, but their highest concentration is in the southern region, due to the soft bottom substrate. Various species of seagrass are widespread in the basin, and are most common in the lagoons and embayments of the southern region.

Although seven countries border the Red Sea, namely Egypt, Sudan, Eritrea, Yemen, Saudi Arabia, Jordan and Israel, the region is sparsely populated, with no more than 5 million people estimated to live along the coast. The area of Jeddah, in Saudia Arabia, with over 2 million residents, has the highest population. The major industries in the region are linked to oil exploration, extraction, processing and transport. In fact, in spite of the fame due to its natural beauty, the Red Sea constitutes a major shipping route for oil tankers and other vessels travelling through the Suez Canal. Fisheries and, more recently, tourism also started to play an important role in the local economy. Though the Red Sea accounts for 0.123 % of the total world ocean area, its contribution to the world fish catch is only 0.07 %. Nevertheless, it has fish resources – including classical fisheries, but also fish collection for aquarium trade, as well as pearl oyster and shrimp farming – that are significant to the countries in the region, providing a good source of protein and livelihood for coastal communities. Due to the high diversity of marine life and favorable climate, tourism has also become a major factor in many Red Sea countries, with over 1 million tourists per year expected in the future.

Although the Red Sea has remained relatively free of pollution, its environment is under increasing threat from a wide range of human activities, both land-based and, primarily, marine-based. Destruction of habitat, due to urban expansion, industrial development and particularly tourism, is on the rise. So is the over-exploitation of some living marine resources, such as lobsters, sharks and sea turtles. The Red Sea has several marine protected areas, the best known being the Ras Mohammed National Park, at the southern tip of the Sinai Peninsula; Hurghada and the adjacent islands along the Egyptian coast; Sanganeb atoll in Sudan; and the Farasan Islands in Saudi Arabia.

1.6 Rift Valley Lakes

The Rift Valley lakes, accounted here among the main water basins of Africa, are a group of lakes in the East African Rift running in the north-south direction, through the eastern part of the continent (Fig. 1.6). The East African Rift came into being

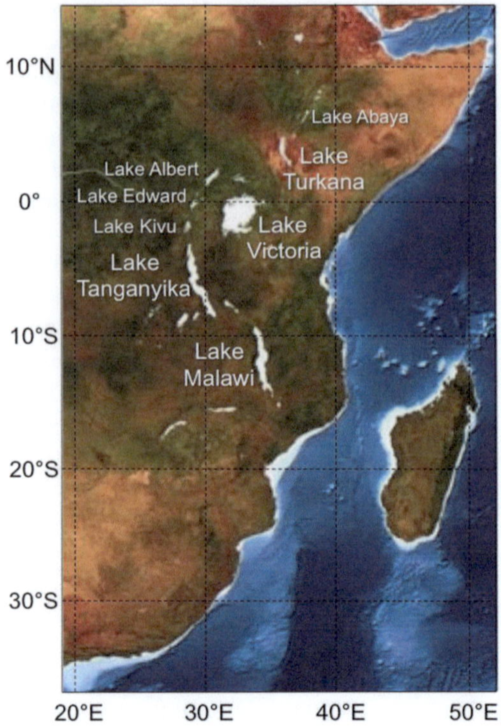

Fig. 1.6 Rift Valley Lakes. The colour coding shows bathymetry (*blue tones*, darker means deeper waters), and vegetated to desert areas (*green* to *yellow* and *brown tones*) over land (courtesy of JRC EC)

approximately 40 million years ago, as the African tectonic plate began to split. The lakes have formed in the valleys of the rift zone, and include some of the largest, oldest and deepest freshwater bodies in the world. They belong to regions of great ecological variability, and great biodiversity, including some alkaline "soda lakes" that support highly specialized organisms.

The Ethiopian Rift Valley lakes are the northernmost water bodies in the rift zone. In central Ethiopia, the Great Rift Valley splits the Ethiopian highlands into northern and southern sections, and the lakes occupy the floor of the depression in between. Several of these lakes do not have an outlet, and most are alkaline. Lake Abaya (1,162 km^2, elevation 1,285 m) is the largest water body of the Ethiopian Rift Valley.

South of the Ethiopian highlands, the rift valley splits into two major troughs. The Eastern Rift is home to the Kenyan Rift Valley lakes, the Western Rift to the Central African Rift Valley lakes. Of the eight lakes located in the Kenyan section of the rift valley, five are alkaline. The shallowest soda lakes have crystallized salt turning the shores white, and are famous for the large flocks of flamingo that feed on crustaceans. Lake Turkana (6,405 km^2, elevation 360 m), on the border between Kenya and Ethiopia, is the largest of the Kenyan lakes.

The lakes of the Westen or Albertine Rift, together with Lake Victoria, include the largest, deepest and oldest of the Rift Valley Lakes. They are also referred to as the Central African lakes. Lake Albert and Lake Edward (as well as Lake Victoria) are part of the (White) Nile River basin. Lake Kivu empties into the Ruzizi River, which flows southwards into Lake Tanganyika. Lake Victoria (68,800 km^2, elevation 1,134 m) is the largest lake in Africa. However, it is not in the rift valley: rather, it occupies a depression between the eastern and western rifts, formed by the uplift of the rifts to either side. Together with Lake Tanganyika and Lake Malawi, all located in the area, these lakes are collectively known as the African Great Lakes. The Western Rift Valley Lakes are all freshwater lakes and home to an extraordinary number of species (including about 1,500 cichlid fish – *Cichlidae* – species). Lake Tanganyika (32,000 km^2, elevation 773 m) is the largest and deepest (more than 1,400 m) of all Rift Valley lakes, and is the second deepest fresh water lake on the planet (after Lake Baikal). Below 200 m, its water is anoxic and devoid of large aquatic life. It is part of the Congo River basin, feeding into the River Congo via the Lukuga River.

The Southern Rift Valley lakes are similar to the Western Rift Valley lakes in that, with one exception, they are all freshwater lakes. Lake Malawi (30,000 km^2, elevation 500 m), also known as Lake Nyasa, is the second largest and second deepest (over 700 m) of the Rift Valley lakes. It is drained by the Shire River, a tributary of the Zambezi River.

1.7 Outlook

The marginal and enclosed basins surrounding the African continent present a great variety of environmental conditions. While some of the African Seas have been sailed since ancient times, others remain surprisingly uncharted, in particular from the scientific point of view. As seen earlier, the dynamics of the marginal seas around Africa are still the subject of different, sometimes contrasting, interpretation (see *e.g.* the issue of western boundary current *vs* train of anticyclonic eddies in the Mozambique Channel). The biological response to such physical forcing also remains to be fully understood (as *e.g.* in the case of the so-called Benguela Niño). The exploration of these marginal seas – aiming to exploit their natural resources and protect their ecological balance, but also to understand their role in shaping the global climate – requires the adoption of appropriate observation systems. Data collection with traditional surface measurements remains an key component of much environmental research dealing with these regions. However, the need to assess the large scales typical of most marine processes, and the prohibitive cost of lengthy, repeated oceanographic expeditions – not to mention the recurrence of piracy and political unrest in several coastal countries – has determined, in recent years, a substantial undersampling of field data.

Remote sensing from Earth's orbit can complement more conventional *in situ* data gathering techniques in mapping the interaction of physical and bio-geo-chemical processes, at the large space scales and over the long time periods required in the African Seas. The environmental indicators measured from satellite remote sensors allow exploring the relationships of natural settings, water dynamics, basic ecological mechanisms and their driving forces, as well as anthropogenic impacts on coastal and marine regions. In the case of the still poorly known African marginal and enclosed basins, the systematic and synoptic application of remote sensing techniques promises to help closing the remaining gaps. An in-depth analysis of the specific merits and drawbacks of each technique and spectral region can provide clues to help compose the unique mosaic of environmental features characterizing the African Seas.

Acknowledgements The present review of the African Seas was carried out at the Joint Research Centre (JRC) of the European Commission (EC), as part of research undertaken in the Framework Programme 7 (FP7) JRC Action PROCAS (later re-named Action SEACOAST). The author wishes to express his gratitude to all JRC Colleagues who contributed to this activity. Thanks are also due to M. Gade, Institute of Oceanography, University of Hamburg, for his editorial review of the final paper; to J.F.R. Gower, Institute of Ocean Sciences, Canada Fisheries and Oceans, for providing valuable suggestions concerning both science and language; and to two anonymous reviewers, whose input allowed the author to improve substantially several sections of the original paper. Images and data shown here were acquired, in part, from the Global Marine Information System (GMIS) of the Institute for Environment and Sustainability (IES), JRC EC.

Bibliography

Introduction, General

Apel JR (1987) Principles of ocean physics. Academic Press, New York
Barber RT, Smith RL (1981) Coastal upwelling ecosystems. In: Longhurst AR (ed) Analysis of marine ecosystems. Academic Press, New York, pp 31–68
Corso W, Joyce PS (1995) Oceanography. Applied science review. Springhouse Corporation, Springhouse
Emery WJ (2001) Water types and water masses. In: Steele JH, Thorpe SA, Turekian KK (eds) Encyclopedia of ocean sciences, vol 4. Academic Press, San Diego, pp 3179–3187
Gill AE (1982) Atmosphere-ocean dynamics. Academic Press, New York
Grant Gross M (1995) Principles of oceanography, 7th edn. Prentice-Hall, Englewood Cliffs
Hance WA (1975) The geography of modern Africa. Columbia University Press, New York
International Hydrographic Organization (1953) Limits of oceans and seas. Special publication 23, 3rd edn. Imp. Monégasque, Monte Carlo, p 40
Longhurst AR (1998) Ecological geography of the sea. Academic Press, San Diego
Mann KH, Lazier JRN (2006) Dynamics of marine ecosystems: biological-physical interactions in the oceans. Blackwell, Oxford
Neumann G, Pierson WJ (1966) Principles of physical oceanography. Prentice-Hall, New Jersey
Pedlosky J (1996) Ocean circulation theory. Springer, Berlin
Pinet PR (1992) Oceanography: an Introduction to the planet Oceanus. West Publishing Company, St. Paul
Rahmstorf S (2003) Thermohaline circulation: the current climate. Nature 421(6924):699–699
Roemmich D (2007) Physical oceanography: super spin in the southern seas. Nature 449(7158): 34–35
Schwartz ML (ed) (2005) Encyclopedia of coastal science. Springer, Dordrecht
Siedler G, Church J, Gould J (2001) Ocean circulation and climate: observing and modeling the global ocean. International Geophysics Series, vol 77. Academic Press, London

Sverdrup HU, Johnson MW, Fleming RH (1942) The oceans: their physics, chemistry and general biology. Prentice-Hall, Englewood Cliffs
Sverdrup KA, Duxbury AC, Duxbury AB (2005) An Introduction to the world's oceans, 8th edn. McGraw-Hill, Boston
Williams J, Higginson JJ, Rohrbough JD (1968) Air and sea: the naval environment. Naval Institution Press, Annapolis
Wunsch C (1996) The ocean circulation inverse problem. University Press, Cambridge
Wunsch C (2002) What is the thermohaline circulation? Science 298(5596):1179–1180

Atlantic Ocean, Canary Current

Barton ED, Basterretxea G, Flament P, Mitchelson-Jacob EG, Jones B, Aristegui J, Herrera F (2000) Lee region of Gran Canaria. J Geophys Res 105:17173–17193
Batten ML, Martinez JR, Bryan DW, Buch EJ (2000) A modeling study of the coastal eastern boundary current system off Iberia and Morocco. J Geophys Res 105:14173–14195
Binet D (1997) Climate and pelagic fisheries in the Canary and Guinea currents 1964–1993: the role of trade winds and the southern oscillation. Oceanologia Acta 20:177–190
Huntsman SA, Barber RT (1977) Primary production off Northwest Africa: the relationship to wind and nutrient conditions. Deep-Sea Res 24:25–33
Mittelstaedt E (1991) The ocean boundary along the northwest African coast: circulation and oceanographic properties at the sea surface. Prog Oceanogr 26:307–355
Stramma L, Siedler G (1988) Seasonal changes in the North Atlantic sub-tropical gyre. J Geophys Res 93:8111–8118
Wooster WS, Bakum A, McLain DR (1976) The seasonal upwelling cycle along the eastern boundary of the North Atlantic. J Mar Res 34:131–140
Zhou M, Paduan JD, Niiler PP (2000) Surface currents in the Canary basin from drifter observations. J Geophys Res 105:21893–21911

Atlantic Ocean, Guinea Current, Angola Current

Bakun A (1978) Guinea current upwelling. Nature 271:147–150
Colin C (1988) Coastal upwelling events in front of Ivory Coast during the FOCAL program. Oceanologia Acta 11:125–138
Hisard P, Henin C, Houghton R, Piton B, Rual P (1986) Oceanic conditions in the tropical Atlantic during 1983 and 1984. Nature 322:243–245
Ingham MC (1970) Coastal upwelling in the northwestern Gulf of Guinea. Bull Mar Sci 20:1–34
Longhurst AR (1962) A review of the oceanography of the gulf of Guinea. Buleltin de l'Institut Fondamental de l'Afrique Noire. Series A. Sciences Naturelles 24(3):633–663
Philander SGH (1979) Upwelling in the Gulf of Guinea. J Mar Res 37:23–33
Picaut J (1983) Propagation of the seasonal upwelling in the eastern equatorial Atlantic. J Phys Oceanogr 13:18–37
Richardson PL, Reverdin G (1987) Seasonal cycle of velocity in the Atlantic north equatorial countercurrent as measured by surface drifters, current meters, and ship drifts. J Geophys Res 92:3691–3708
Verstraete JM (1992) The seasonal upwellings in the Gulf of Guinea. Prog Oceanogr 29:1–60

Atlantic Ocean, Benguela Current

Boyer D, Cole J, Bartholome C (2000) Southwestern Africa: Northern Benguela current region. Mar Pollut Bull 41:123–140

Fu LL (1981) The general circulation and meridional heat transport of the subtropical South Atlantic determined by inverse methods. J Phys Oceanogr 11:1171–1193

Garzoli SL, Gordon AL (1996) Origins and variability of the Benguela current. J Geophys Res 101:897–906

Garzoli SL, Gordon AL, Kamenkovich V, Pillsbury D, Duncombe-Rae C (1996) Variability and sources of the south eastern Atlantic circulation. J Mar Res 54:1039–1071

Garzoli SL, Goni GJ, Mariano AJ, Olson DB (1997) Monitoring the upper southeastern Atlantic transports using altimeter data. J Mar Res 55:453–481

Moroshkin KV, Bunov VA, Bulatov RP (1970) Water circulation in the eastern South Atlantic ocean. Oceanology 10:27–34

Nelson G (1992) Equatorial wind and atmospheric pressure spectra as metrics for primary productivity in the Benguela system. S Afr J Mar Sci 12:19–28

Peterson RG, Stramma L (1991) Upper-level circulation in the South Atlantic Ocean. Prog Oceanogr 26:1–73

Peterson RG, Stramma L, Kortum G (1996) Early concepts and charts of ocean circulation. Prog Oceanogr 37:1–115

Richter I, Behera SK, Matsumoto Y, Taguchi B, Komori N, Yamagata T (2010) On the triggering of Benguela Niños: Remote equatorial versus local influences. Geophys Res Lett 37:L20604

Shannon LV (1985) The Benguela ecosystem, I., Evolution of the Benguela, physical features and processes. Oceanogr Mar Biol 23:105–182

Shannon LV, Boyd AJ, Brundrit GB, Taunton-Clark J (1986) On the existence of an El Nino-type phenomenon in the Benguela system. J Mar Res 44:495–520

Skogen MD (1999) A biophysical model applied to the Benguela upwelling system. S Afr J Mar Sci 21:235–249

Indian Ocean, Monsoon Currents, Somali Current

Hastenrath S, Greischar L (1991) The monsoonal current regimes of the tropical Indian ocean: observed surface flow fields and their geostrophic and wind-driven components. J Geophys Res 96(C7):12619–12633

Lee CM, Jones BH, Brink KH, Fischer AS (2000) The upper-ocean response to monsoonal forcing in the Arabian Sea: seasonal and spatial variability. Deep-Sea Res Part II 47(7/8):1177–1226

Lighthill MJ (1969) Dynamic response of the Indian Ocean to onset of the Southwest Monsoon. Philosophical transactions of the Royal Society of London, Series A. Mathematical Physical Sciences 265(1159):45–92

Quadfasel DR, Schott F (1982) Water-mass distributions at intermediate layers off the Somali coast during the onset of the Southwest Monsoon, 1979. J Phys Oceanogr 12:1358–1372

Quadfasel DR, Schott F (1983) Southward subsurface flow below the Somali Current. J Geophys Res 88 (C12):5973–5979

Shankar D, Vinayachandran PN, Unnikrishnan AS (2002) The monsoon currents in the north Indian Ocean. Prog Oceanogr 52(1):63–120

Indian Ocean, South Equatorial Current, East African Coastal Current

McClanahan TR (1988) Seasonality in East Africa's coastal waters. Mar Ecol Prog Ser 44:191–199
Morrow R, Birol F (1997) Variability in the south-east Indian Ocean from altimetry: forcing mechanisms for the Leeuwin current. J Geophys Res 103(C9):18529–18544
Quadfasel DR, Frische A, Cresswell G (1996) The circulation in the source area of the south equatorial current in the eastern Indian Ocean. J Geophys Res 101(C5):12483–12488
Swallow JC, Schott F, Fieux M (1991) Structure and transport of the East African coastal current. J Geophy Res 96(C12):22245–22257
Woodberry KE, Luther M, O'Brien JJ (1989) The wind-driven seasonal circulation in the southern tropical Indian Ocean. J Geophys Res 94(C12):17985–18002

Indian Ocean, Mozambique Channel

De Ruijter WPM, Ridderinkhof H, Lutjeharms JRE, Schouten M, Veth C (2002) Observations of flow in the Mozambique channel. Geophys Res Lett 29:1401–1403
Penven P, Lutjeharms JRE, Florenchie P (2006) Madagascar: a pacemaker for the Agulhas current system? Geophys Res Lett 33:L17609
Quartly GD, Srokosz MA (2004) Eddies in the southern Mozambique channel. Deep Sea Res Part II 51(1–3):69–83
Schouten M, de Ruijter W, van Leeuwen P, Dijkstra H (2002) An oceanic teleconnection between the equatorial and southern Indian Ocean. Geophys Res Lett 29(16):1812, 59:1–4
Schouten M, Ruijter W de, Leeuwen P van, and R. Ridderinkhof (2003) Eddies and variability in the Mozambique channel. Deep-Sea Res II 50:1987–2003

Indian Ocean, Agulhas Current, Agulhas Return Current

Bakun A, Claude R, Lluch-Cota S (1998) Coastal upwelling and other processes regulating ecosystem productivity and fish production in the Western Indian Ocean. In: Sherman K, Okemwa EN, Ntiba MJ (eds) Large marine ecosystems of the Indian Ocean: assessment, sustainability and management. Blackwell Science, Malden, pp 103–141
Beckley LE (1998) The Agulhas current ecosystem with particular reference to dispersal of fish larvae. In: Sherman K, Okemwa EN, Ntiba MJ (eds) Large marine ecosystems of the Indian Ocean: assessment, sustainability and management. Blackwell Science, Malden, pp 255–276
Boebel O, Rossby T, Lutjeharms JRE, Zenk W, Barron C (2003) Path and variability of the Agulhas return current. Deep-Sea Research PART II 50:35–56
Bryden HL, Beal LM (2001) Role of the Agulhas current in Indian Ocean circulation and associated heat and freshwater fluxes. Deep-Sea Res Part I 48(8):1821–1845
Bryden HL, Beal LM, Duncan LM (2003) Structure and transport of the Agulhas current and its temporal variability. J Oceanogr 61:479–492
Gordon AL (1985) Indian-Atlantic transfer of thermocline water at the Agulhas retroflection. Science 227:1030–1033
Lass HU, Schmidt M, Mohrholz V, Nausch G (2000) Hydrographic and current measurements in the area of the Angola-Benguela front. J Phys Oceanogr 30:2589–2609
Leeuwen PJ, de Ruijter WPM, Lutjeharms JRE (2000) Natal pulses and the formation of Agulhas rings. J Geophys Res 105:6425–6436
Lutjeharms JRE (2006a) The Agulhas current. Springer, Heidelberg

Lutjeharms JRE (2006b) The coastal oceans of southeastern Africa. In: Robinson AR, Brink KH (eds) The sea, vol 14B. The global coastal ocean – interdisciplinary regional studies and syntheses. Harvard University Press, Cambridge, pp 781–832
Lutjeharms JRE, Ansorge IJ (2001) The Agulhas return current. J Mar Syst 30:115–138
Lutjeharms JRE, Boebel O, Rossby HT (2003) Agulhas cyclones. Deep-Sea Res Part II 50:13–34
Lutjeharms JRE, Cooper J (1996) Interbasin leakage through Agulhas current filaments. Deep-Sea Res Part I 43(2):213–238
Lutjeharms JRE, Roberts HR (1988) The Natal Pulse: an extreme transient on the Agulhas current. J Geophys Res 93:631–635
Lutjeharms JRE, van Ballegooyen RC (1988) The Retroflection of the Agulhas current. J Phys Oceanogr 18(11):1570–1583
Matano RP, Simionato CG, Ruijter WP, van Leeuween PJ, Strub PT, Chelton DB, Schlax MG (1998) Seasonal variability in the Agulhas retroflection region. Geophys Res Lett 25(23):4361–4364
Peterson RG, Stramma L (1991) Upper-level circulation in the South Atlantic. Prog Oceanogr 26:1–73
Stramma L, Lutjeharms JRE (1997) The flow field of the subtropical gyre of the South Indian ocean. J Geophys Res 102(C3):5513–5530
Van Ballegooyen RC, Grundlingh ML, Lutjeharms JRE (1994) Eddy fluxes of heat and salt from the southwest Indian Ocean into the southeast Atlantic Ocean: a case study. J Geophys Res 99:14053–14070
Van Leeuwen PJ, de Ruijter WPM, Lutjeharms JRE (2000) Natal pulses and the formation of Agulhas rings. J Geophys Res 105:6425–6436

Mediterranean Sea

Boxer B (1982) Mediterranean pollution: problem and response. Ocean Dev Int Law 10(3–4): 315–356
Boxer B (1983) Environment and regional identity in the Mediterranean. In: Pinkele CF, Pollis A (eds) The contemporary Mediterranean world. Praeger, New York, pp 59–73
Busuttil S (1993) The future of the Mediterranean. Ocean Yearbook 10:240–247
European Environment Agency (2006) Priority issues in the Mediterranean environment. European Environment Agency Report No. 4/2006, Copenhagen, EEA Copenhagen, p 86
Garcés E, Masó M, Vila M, Camp J (2000) HAB events in the Mediterranean Sea: are they increasing? A case study of the last decade in the NW Mediterranean and the genus Alexandrium. Harmful Algal News 20:1–11
Gascard JC (1978) Mediterranean deep water formation, baroclinic instability and oceanic eddies. Oceanol Acta 1(3):315–330
MEDOC Group (1970) Observations of deep water in the Mediterranean Sea. Nature 227: 1037–1040
Millot C (2005) Circulation in the Mediterranean Sea: evidences, debates and unanswered questions. Scientia Marina 69:5–21
Millot C, Taupier-Letage I (2005) Circulation in the Mediterranean Sea. In: Saliot A (ed) The Mediterranean Sea, handbook of environmental chemistry/water pollution, vol 5, Part K. Springer, Berlin, pp 29–66
Robinson AR, Leslie WG, Theocharis A, Lascaratos A (2001) Mediterranean Sea circulation. In: Steele JH et al (eds) Encyclopedia of ocean sciences. Elsevier, New York, pp 1689–1705
Robinson AR, Golnaraghi M (1995) The physical and dynamical oceanography of the Mediterranean Sea. In: Malanotte RP, Robinson AR (eds) Ocean processes in climate dynamics: global and Mediterranean examples. Springer, New York, pp 255–306
White GF (1988) The environmental effects of the high dam at Aswan. Environment 30(7):5–11, 34:40
Williams N (1998) Meeting briefs: the Mediterranean beckons to Europe's oceanographers. Science 279(5350):483–484

Red Sea

Bower AS, Hunt HD, Price JF (2000) Character and dynamics of the Red Sea and Persian Gulf outflows. J Geophys Res 105:6387–6414
Edwards AJ (1987) Climate and oceanography. In: Edwards AJ, Head SM (eds) Red Sea: key environments. Pergamon Press, Oxford, pp 45–69
Eschel G, Cane MA, Blumenthal MB (1994) Modes of subsurface, intermediate, and deep water renewal in the Red Sea. J Geophys Res 99:15941–15952
Gladstone W, Tawfiq N, Nasr D, Andersen I, Cheung C, Drammeh H, Krupp F, Lintner S (1999) Sustainable use of renewable resources and conservation in the Red Sea and Gulf of Aden: issues, needs and strategic actions. Ocean Coast Manag 42:671–697
Grassen G, Kroon D (1991) Evidence for Red Sea surface circulation from oxygen isotopes of modern surface waters and planktonic foraminiferal test. Paleoceanography 6:73–82
Head SM (1987) Red Sea fisheries. In: Edwards AJ, Head SM (eds) Red Sea: key environments. Pergamon Press, Oxford, pp 363–382
Kinder TH, Bryden HL (1990) Aspiration of deep waters through straits. In: Pratt LJ (ed) The physical oceanography of sea straits. Kluwer Academic, Dordrecht, pp 295–319
Morcos SA (1970) Physical and chemical oceanography of the Red Sea. Oceanography Marine Biology Annual Review 8:73–202
Murray SP, Jones W (1997) Direct observations of seasonal exchange through the Bab al Mandab Strait. Geophys Res Lett 24:2557–2560
Quadfasel D, Baudner H (1993) Gyre-scale circulation cells in the Red Sea. Oceanol Acta 16: 221–229
Sheppard C, Prince A, Roberts C (1992) Marine ecology of the Arabian region. Academic Press, London
Sheppard CRC (2000) The Red Sea. In: Sheppard CRC (ed) Seas at the millennium: and environmental evaluation. Pergamon Press, Oxford, pp 35–45
Siedler G (1969) General circulation of the water masses in the Red Sea. In: Degens ET, Ross DA (eds) Hot brines and recent heavy metals deposits in the Red Sea. Springer, Berlin, pp 131–137
Smeed DA (1997) Seasonal variations of the flow in the strait of Bab el Mandab. Oceanol Acta 20:773–781
Smeed DA (2000) Hydraulic control of three-layer exchange flows: application to the Bab el Mandab. J Phys Oceanogr 30:2574–2588
Sofianos S, Johns WE, Murray SP (2002) Heat and freshwater budgets in the Red Sea from direct observations at Bab el Mandab. Deep-Sea Res Part II 49:1323–1340

Outlook

Barale V, Gade M (eds) (2008) Remote sensing of the European seas. Springer, Heidelberg
Bhatt JJ (1978) Oceanography. Exploring the planet ocean. D Van Nostrand Company, New York
Kitchen K (1993) The land of punt. In: Shaw T, Sinclair P, Andah B, Okpoko A (eds) The archaeology of Africa: foods, metals, towns. Routledge, London, pp 587–608
Peterson RG, Stramma L, Kortum G (1996) Early concepts in charts and circulation. Prog Oceanogr 37:1–115
Tyldesley J (1996). Hatchepsut: the female pharaoh. Viking, London
Tomczak M, Godfrey JS (2003) Regional oceanography: an introduction, 2nd Improved edn. Daya Publishing House, Delhi

Chapter 2
Satellite Water Colour Observations in African Seas

Jim Gower and Stephanie King

Abstract Satellite water colour image data can now be used to show the spatial patterns of coastal and ocean water properties, especially surface chlorophyll concentration, on scales covering from kilometers to the entire globe. The repeated coverage shows weekly, seasonal and longer term variability, and provides a unique resource for regions where in-situ data sources are scarce, as is the case for most waters round the continent of Africa. Easy access to extensive data sets of satellite-measured chlorophyll, characterized through the average absorption and scattering properties of phytoplankton, and through natural fluorescence, are now available from Goddard Interactive Online Visualization ANd aNalysis Infrastructure (GIOVANNI), developed by the US National Aeronautics and Space Administration (NASA), and other data systems. In addition, data such as the MCI index from MEdium Resolution Imaging Spectrometer (MERIS), available by the European Space Agency (ESA), derived using spectral bands near 700 nm in the near-infrared part of the spectrum, provide further information on blooms and aquatic vegetation. We present here results of satellite observations of various African seas, showing some of the variety of observations and time series results possible.

2.1 Satellite Ocean Colour Data

"Satellite water colour data" refers to the relatively low (typically 1 km) spatial resolution optical image data of oceans and coastal waters provided by specialized satellite imagers. Data are collected in carefully chosen spectral bands, typically 10–20 nm wide. Land imaging usually requires higher spatial resolution (to 30 m or less) and often uses broad spectral bands (100 nm) or even the full available optical window (panchromatic images). Important requirements for water imaging are frequent repetition, with the satellites providing near-daily coverage, though frequently interrupted by cloud, and precise short and long-term calibration.

The US ocean colour satellite sensors Coastal Zone Color Scanner (CZCS), Sea-viewing Wide Field of view Scanner (SeaWiFS), MODerate Resolution Imaging

J. Gower (✉) · S. King
Institute of Ocean Sciences, Fisheries and Oceans Canada, Sidney, BC V8L 4B2, Canada
e-mail: Jim.Gower@dfo-mpo.gc.ca

V. Barale, M. Gade (eds.), *Remote Sensing of the African Seas*,
DOI 10.1007/978-94-017-8008-7_2,

Spectroradiometer (MODIS) and Visible/Infrared Imager Radiometer Suite (VIIRS), and the European sensor MEdium Resolution Imaging Spectrometer (MERIS) provide global data on ocean and coastal near-surface phytoplankton concentration at spatial resolutions down to 1 km (300 m for MERIS) over image swaths 2,000 km across (3,000 km for VIIRS, 1,100 km for MERIS). These have been joined for short periods by sensors from several other countries. The Indian Ocean Colour Monitor (OCM) has provided some data since launch in 1999 on the Indian Oceansat-1 satellite, and a follow-on instrument (OCM 2) was launched in 2009 on Oceansat-2.

Optical data from these satellites are used primarily to deduce chlorophyll concentrations in near-surface waters. Chlorophyll is a standard indicator of phytoplankton concentration, which in turn represents the primary-producing biomass at the base of the marine food chain. Images can be used to examine spatial patterns of high-chlorophyll in upwelling zones and blooms, or conversely "ocean deserts" in centres of mid-ocean gyres, where chlorophyll concentrations are low. The images show seasonal patterns, for example those due to coastal upwelling driven by seasonal winds, or the spring bloom in extra-tropical waters. On longer time scales, they may also show how bloom frequency or extent are affected by global warming, changing nutrient input, or the spread of new species in ship's ballast water. In some cases blooms are triggered by mixing after intense storms, or by nutrient input from rivers, volcanic eruptions or artificial fertilization. Timing and intensity of patterns of measured chlorophyll can then be related to fish catch and properties of marine ecosystems.

Chlorophyll concentrations are deduced from the satellite optical data by measuring the combined effects of increased scattering near 550 nm (green light) and increased absorption near 450 nm (blue light). Increased phytoplankton concentration causes more scattering from cell walls and other structural parts of the organisms, which increases the amount of green light back-scattered from the ocean. At the same time, more absorption due to higher concentrations of chlorophyll-a and auxiliary pigments responsible for energy gathering for photosynthesis reduces the amount of blue light. This "blue to green" ratio is used to deduce chlorophyll concentrations from observed spectral radiances. The relation is well-determined in "Case 1" waters, defined as those whose varying optical properties are determined by phytoplankton and not by the presence of other constituents such as suspended sediments or dissolved organic material, or again by bottom reflection in shallow water.

In addition to deducing chlorophyll content, optical sensors can measure water reflectance, for example in green light near 550 nm, and use this to deduce the total amount of scattering material in the water. If this is found to be higher than expected for the measured amount of phytoplankton, then the water can be flagged as "Case 2" (essentially anything other than Case 1) and the higher scattering can be related to sediment from rivers, or resuspension from the bottom in shallow water. In some cases, high back-scatter can be due to blooms of species such as *Coccolithophores*, which are considered Case 2 because of the extreme brightness of their blooms. Concentration of Coloured Dissolved Organic Material (CDOM) can also be deduced from measured radiances at the shortest wavelengths, typically 412 nm. Waters with significant concentrations of CDOM are also classed as Case 2.

In order to measure the actual reflectance of water, the water-leaving radiances have to be computed from those measured at the satellite. This involves computing the radiance added by scattering in the atmosphere and the radiance lost due to absorption, that is, atmospheric correction has to be applied. This is conventionally referred to as deriving "Level 2" data (atmospherically corrected radiances, reflectances and derived products), from "Level 1" (radiances measured by the satellite). Level 1 to Level 2 conversion also tries to correct for sun-glint, sky reflection and light transmission at the water surface. "Level 0" refers to the raw data packets transmitted from the satellite, whose redundancies and encodings make them less useful for applications. "Level 3" refers to image composites covering particular time periods (often 8-day, or monthly), or larger, often global, areas. "Level 4" is sometimes used to refer to products derived from satellite data, but with addition of information from other sources. A subdivision of Level 1 to Levels 1A and 1B is used to denote pixel location (latitude and longitude) information added in 1B.

The major component of atmospheric correction is due to Rayleigh scattering by atmospheric gases, which can be computed knowing the viewing geometry of the scene, the sun and the sensor. Precise calculation requires exact calibration of the sensor, including detailed knowledge of its polarization sensitivity, and also accurate atmospheric pressure at the ocean surface. The much more variable component due to aerosol scattering requires added measurements, which over water can be provided by the same sensor using wavelengths longer than 700 nm, where signal from the water is usually negligible. These measured radiances therefore show the strength and spectral properties (Angstrom coefficient, if a simple power-law) of the aerosol component, which can then be extrapolated to shorter wavelengths at which water-leaving radiances are significant. Measurements at the shortest wavelengths are the most problematic, causing problems for estimates of CDOM.

2.2 Chlorophyll Fluorescence Measurements from Satellites

We also present here results from measurements of chlorophyll fluorescence which can be made by more recent sensors such as MODIS and MERIS (but not VIIRS) using bands near 683 nm (Gower and Borstad 1990; Abbott and Letelier 1999; Gower and King 2007a). Measurements are made in the wavelength range where fluorescence is emitted (675–695 nm), and also at nearby wavelengths where fluorescence is absent or much reduced, to give background reference radiances. Oxygen absorption at wavelengths longer than 687 nm needs to be avoided, so MODIS measures fluorescence at 673 nm, and MERIS at 681 nm, shorter wavelengths than otherwise preferred. In both cases, fluorescence is measured as Fluorescence Line Height (FLH), computed as radiance at a central wavelength, above a reference radiance linearly interpolated from two adjacent bands, one at a shorter wavelength and one at a longer. For MODIS, measurements are made at 665, 673 and 748 nm and for MERIS at 665, 681 and 709 nm. For large-area studies, FLH is computed from Level 2 data, but the atmospheric correction for FLH is relatively small, and FLH values are often computed from Level 1 radiances. This has the advantage of avoiding problems with atmospheric correction, which are usually more severe in coastal areas, but requires a method for rejecting data affected by cloud or strong sun glint. For MERIS, a typical relation (Gower and King 2007a) is:

$$\text{if} \quad L_{865} < 15\ \text{mW m}^{-2}\ \text{sr}^{-1}\ \text{nm}^{-1},$$
$$\text{then} \quad \text{FLH} = L_{681} - L_{665} - 0.364\,(L_{709} - L_{665}) \qquad (2.1)$$

where L_{865} represents Level 1 radiances (as measured at the satellite) at a wavelength of 865 nm, and similarly for other wavelengths, and the factor 0.364 represents the wavelength ratio (681–665)/(709–665). For MODIS, bands are different, as noted above, and the factor becomes 0.096.

It is often assumed that the measured fluorescence signal will increase with sun elevation, and this is compensated by normalizing FLH to what would be observed had the sun been at zenith (NFLH), that is NFLH = FLH/cos(Z), where Z is the zenith angle of the sun at the time of observation. In fact, we show below that normalization is not appropriate.

We have suggested that fluorescence is proportional to chlorophyll concentration at low concentrations, but saturates due to absorption of both stimulated and emitted radiation at higher concentrations (Gower et al. 2004; Gower and King 2007a). For FLH computed from Level 1 data, we propose

$$\text{FLH} = (0.18C/(1 + 0.2C)) - 0.24 \qquad (2.2)$$

as a good average relation between chlorophyll (in mg.m^{-3}) and fluorescence in radiance units ($\text{W m}^{-2}\ \text{nm}^{-1}\ \text{ster}^{-1}$). The offset, here given as – 0.24, is variable, depending on viewing and atmospheric conditions. The factor 0.2 is the ratio of absorption of fluorescence by 1 mg m^{-3} of chlorophyll, to the absorption of fluorescence by water. The term appears in equation (2.2) by analogy with the relation $R = k \times b/a$ for water reflectance (Morel and Prieur 1977). This factor expresses the fact that FLH increases less rapidly with C, as C increases. At a value of C = 20 mg m^{-3} the rate is reduced by a factor 5 from the value at low C. At higher values, the combination of water and chlorophyll absorption results in a radiance peak near 700 nm for C = 30 mg m^{-3}, and near 710 nm for C = 300 mg m^{-3}, which prevents observation of fluorescence.

2.3 Detection of Blooms and Vegetation Using MCI

MERIS provides the Maximum Chlorophyll Index (MCI), useful for detecting concentrated surface blooms and floating vegetation. MCI measures a radiance peak at 700–710 nm. The index is unique to MERIS, in that MODIS and VIIRS do not have a band near 709 nm. The peak indicates the presence of a high surface concentration of chlorophyll-a against a scattering background (Gitelson et al. 1992; Yacobi et al. 1995; Gower et al. 1999, 2005, 2008a, b). MCI is high in "red tide" conditions (intense, visible, surface plankton blooms), and also when aquatic vegetation is present leading to a "red edge" step increase in radiance. For some blooms, the peak dominates the spectrum, giving a radiance change due to the bloom that is over ten times greater at 709 nm than at any other wavelength. We have demonstrated use of the MCI to detect plankton blooms (Gower et al. 2005), floating *Sargassum*

(Gower et al. 2006; Gower and King 2011) and "superblooms" of Antarctic diatoms associated with platelet ice (Gower and King 2007b).

The MCI is computed as radiance at 709 nm above a linear baseline defined by radiances at 681 and 753 nm (Gower et al. 2005, 2008b). As for FLH, this is calculated only for pixels for which radiance at 865 nm is less than 15 mW m^{-2} sr^{-1} nm^{-1} to eliminate land pixels and areas of strong sun glint, haze or cloud:

$$\text{if}\quad L_{865} < 15\,\text{mW}\,m^{-2}\,sr^{-1}\,nm^{-1},$$
$$\text{then}\quad MCI = L_{709} - L_{681} - 0.389\,(L_{753} - L_{681}) \qquad (2.3)$$

where L_{865} represents Level 1 radiances (as measured at the satellite) at a wavelength of 865 nm, and similarly for other wavelengths, and the factor 0.389 represents the wavelength ratio (709–681)/(753–681).

The MCI therefore indicates an excess radiance at 709 nm above this baseline, which often indicates a water-leaving radiance spectrum with a peak at 709 nm. Models indicate that this type of spectrum is characteristic of intense surface plankton blooms in which high concentrations of phytoplankton are distributed in near-surface waters. In this case, the absorption by chlorophyll reduces radiance at wavelengths shorter than 700 nm, while absorption by water reduces radiance at wavelengths longer than 720 nm, leading to a radiance peak at the wavelength of minimum absorption, near 709 nm. Models also show that vegetation under a shallow layer of water, including coral reefs, can give rise to a spectral peak near 709 nm, also giving a positive MCI signal.

MCI can also be high due to presence of a "red-edge" in the spectrum. At this "edge," water-leaving radiance shows a step increase near 700 nm with increasing wavelength. This type of spectrum is characteristic of land vegetation, for which the red-edge (maximum rate of increase of observed radiance with wavelength) usually occurs at a longer wavelength of about 720 nm. Over water with MERIS, we observe an apparent step increase between 681 and 709 nm. The MCI as defined in Eq. 2.1 will give a high value in this case as well as in the case of a peak. We interpret red-edge type spectra as showing presence of buoyant slicks of either phytoplankton or macroalgae such as pelagic *Sargassum*.

We make use of daily, global composites of MCI data produced by the Grid Processing on Demand (G-POD) facility of ESA (Gower et al. 2008b). These have a spatial resolution of 5 km. Each composite pixel shows the maximum MCI value computed for any RR MERIS (1.2 km) pixel assigned to that composite pixel on that day. The reduced spatial resolution results in smaller blooms being spatially distorted, but the use of the maximum values in the composite preserves the record of their occurrence.

Daily composites are combined into monthly products, also at 5 km resolution, recording the maximum MCI value measured in the month. This gives complete spatial coverage in most areas and preserves information on detected blooms and vegetation. The monthly products also show presence or absence of water, and so can be used to measure water area.

The monthly composites can also be analyzed for "total MERIS count" by computing the sum of the number of MCI values above a threshold, multiplied by the amount by which MCI exceeds its background value (in mW m^{-2} nm^{-1} sr^{-1}), i.e.

$$\text{MERIS count} = \sum_{m=b+t}^{m=\infty} (m - b)n(m) \tag{2.4}$$

where n(m) is the number of pixels in the area having an MCI value of m, b is the background value of MCI corresponding to open water and t is the threshold value. The MERIS count can then be used to assess annual and interannual variations in bloom intensity. Timing can also be computed in terms of a peak month:

$$\text{Peak month} = \sum_{m=1}^{m=12} m.M(m) / \sum_{m=1}^{m=12} M(m) \tag{2.5}$$

where months are represented by m = 1–12, the total MERIS count in each month is M(m) and the peak month is expressed as a number including fractions of a month. The range of months included in the sums is usually restricted to near the time of peak values in M.

2.4 Data Availability for Regional Studies

While satellite data at Level 2 has all the required information for applications by users, the data are still segmented by the satellite's orbital patterns, large in volume and generally inconvenient to use. Global composites (Level 3 data) are more convenient, but still involve handling large volumes of data. A very convenient tool is the Goddard Interactive Online Visualization ANd aNalysis Infrastructure (GIOVANNI), developed by NASA[1] (Acker and Leptoukh, 2007). Other similar tools are available from NOAA[2] and the Colorado Center for Astrodynamics Research at the University of Colorado[3]. The equivalent tool for MERIS data is ESA's G-POD system as mentioned above, but this is not publicly available on the web.

We present here satellite water colour images and data for seas round the African continent. In Sect. 5, results are derived from GIOVANNI, showing chlorophyll data (derived through standard, green to blue ratio algorithms) and fluorescence (derived as shown in Eq. 2.1) from the MODIS version onboard the Aqua orbital platform (MODIS Aqua). These data show the seasonal patterns of surface chlorophyll, indicating coastal productivity variations in response to upwelling and other mechanisms. Comparison of chlorophyll and fluorescence data show a general correlation between these two types of data, with major differences due to an error in the present way fluorescence is derived and presented, as we discuss below. When this error is removed, the two data types show good agreement in many cases. In others, the differences suggest a problem in one or other data type. These differences need to be understood, and may lead to useful new knowledge.

---

[1] Available at http://disc.sci.gsfc.nasa.gov/giovanni/overview/index.html

[2] Available at http://las.pfeg.noaa.gov/oceanWatch/oceanwatch.php

[3] Available at http://eddy.colorado.edu/ccar/modis/color_global_viewer

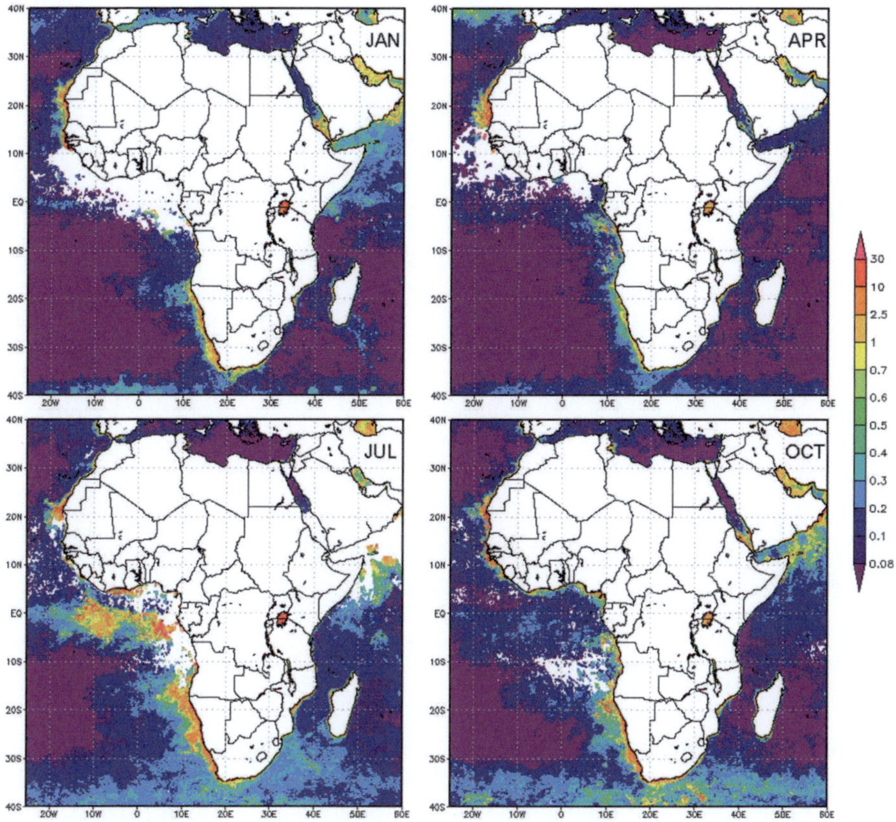

Fig. 2.1 Average surface chlorophyll concentrations round the continent of Africa in January (*top left*), April (*top right*), July (*bottom left*) and October (*bottom right*) of 2011. (MODIS Aqua data)

It has been suggested that wind-blown dust from the Sahara and other African and Asian desert areas may be a problem for remote sensing of some African waters. However, although dust contributes to marine productivity (see e.g. Singh et al. 2008), and hence affects the observed water colour, the signal from dust itself can be compensated in Level 1 to Level 2 processing (see e.g. Moulin et al. 2001). Also, dust events are always intermittent and tend to cause a smaller reduction of satellite coverage than do clouds.

2.5 African Ocean and Coastal Chlorophyll

Figure 2.1 shows the average surface chlorophyll round the continent of Africa as measured by MODIS Aqua (9-km data, downloaded from the NASA GIOVANNI data system; note the colour scale for chlorophyll-a concentration in units of mg m^{-3}

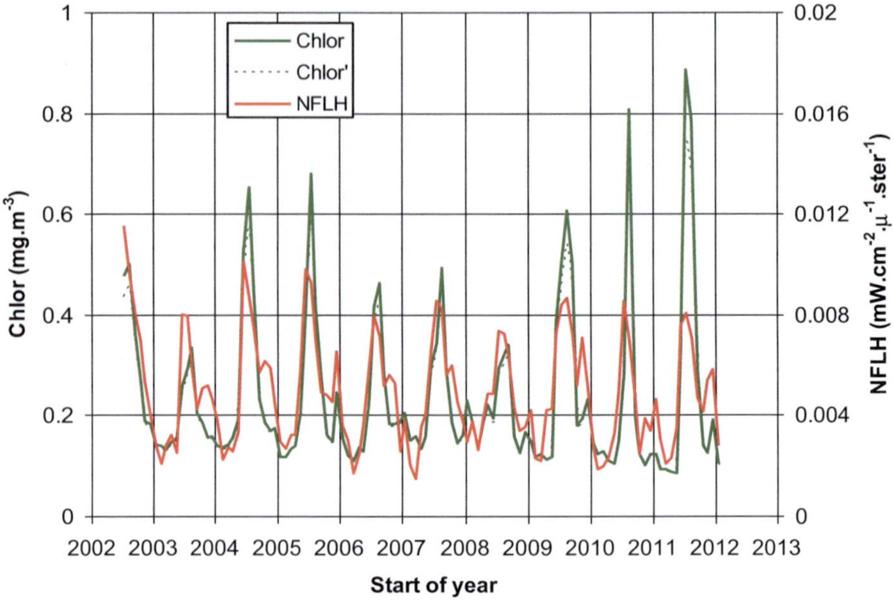

Fig. 2.2 Average surface chlorophyll concentrations (*green*) and normalized fluorescence (*red*) from MODIS Aqua data in the Gulf of Guinea, from 3°N to 3°S and 20°W to 0°

shown at right) in the months of January, April, July and October of 2011. Land is masked as white. Ocean areas which are cloudy for the entire month, such as in the Gulf of Guinea in January and in the Arabian Sea in July, are also masked to white. The bright purple areas represent the lowest surface chlorophyll, less than 0.05 mg m^{-3} (see colour scale at right). A belt of high chlorophyll covers all latitudes at 40°S at all seasons. Areas of high chlorophyll along the equator especially in July in the Gulf of Guinea in the Atlantic and to a lesser extent in the Indian Ocean, indicate the effect of equatorial upwelling. Narrow bands of high chlorophyll near shore indicate coastal upwelling in many areas, such as along the coasts of Namibia and western South Africa near 30°S, 15°E, the Horn of Africa near 10°N, 50°E and west Africa near 20°N, 15°W.

Figure 2.2 shows time series of MODIS Aqua data averaged by GIOVANNI over the area 3°N to 3°S and 20 W to 0 W along the equator in the Gulf of Guinea. Of the nine years for which MODIS Aqua so far provides data, maximum chlorophyll (green line in Fig. 2.2) is in August in 5 years, July in three and September in one, confirming the annual cycle apparent in Fig. 2.1. Fluorescence data (red line) peak slightly earlier, in July of all years except 2003 to 2005 (June) and 2009 (August). At the relative scaling of the two plots (50:1 when units are mg m^{-3} and W m^{-2} nm^{-1} $ster^{-1}$), the green peaks tend to be higher than the red. When comparing the green and red plots, allowance needs to be made for the expected self absorption of fluorescence as expressed in Eq. 2.2. A modified chlorophyll (i.e. Chlor': "fluorescing

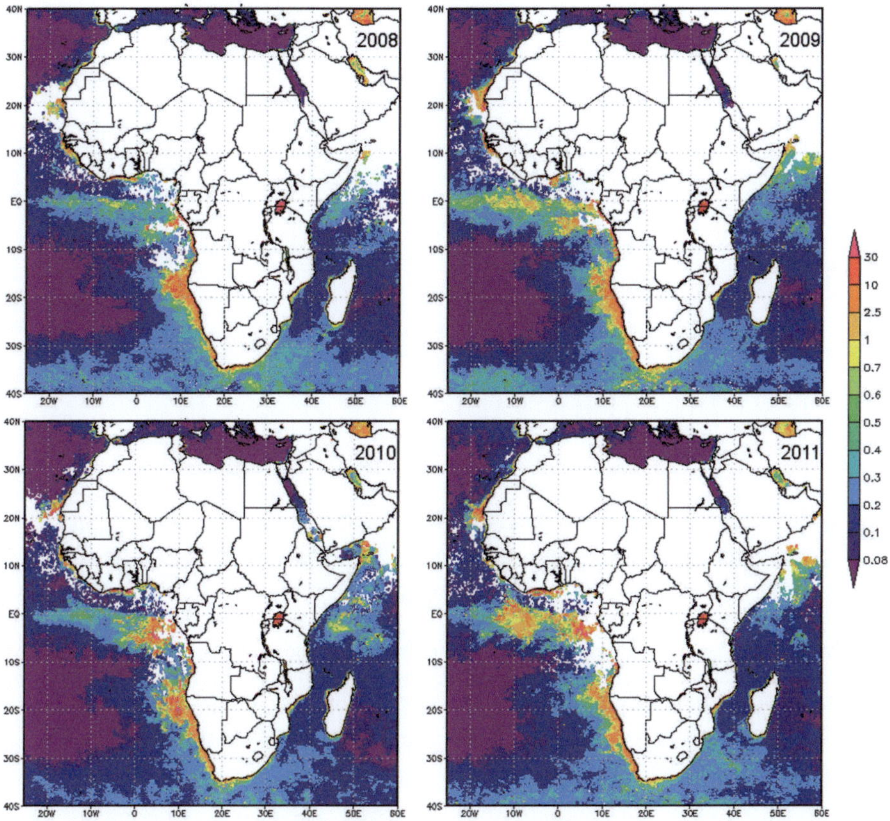

Fig. 2.3 Average surface chlorophyll concentrations round the continent of Africa in July of 2008 (*top left*), 2009 (*top right*), 2010 (*bottom left*) and 2011 (*bottom right*). (MODIS Aqua data)

chlorophyll") is therefore also plotted (green dotted line) for which the chlorophyll value is reduced by the factor C/(1 + 0.2 C), where C is the chlorophyll concentration in mg m^{-3}. In Fig. 2.2, the effect is relatively small.

A second smaller peak appears later in most years in both chlorophyll and fluorescence, but is higher in the fluorescence than in chlorophyll in all years. The relation between these two variables may indicate an average change in fluorescence yield with season, for example, the later peak may be at a time of higher nutrient stress on the phytoplankton, leading to relatively more fluorescence.

Figure 2.3 shows the July distribution in the years 2008, 2009, 2010 and 2011. The major high-chlorophyll areas are present in all years, but with significant changes in extent and concentration. Figure 2.4 shows the series of averages for the Horn of Africa, where high values are seen in Figs. 2.1 and 2.3. Figure 2.4 shows the interannual variability of chlorophyll and hence of coastal primary productivity. The high chlorophyll peaks occur each year from June to September, i.e. the cloudy season

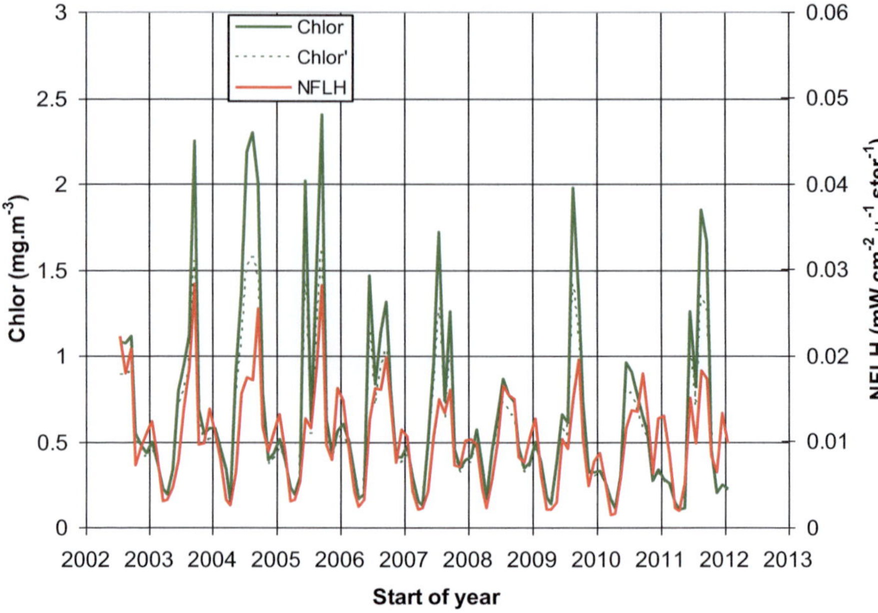

Fig. 2.4 Average surface chlorophyll concentrations (*green*) and normalized fluorescence (*red*) from MODIS Aqua data over the rectangular area off the Horn of Africa (Somalia), defined by 5–15°N and 50–55°E. Relative scaling as in Fig. 2.2

(Fig. 2.1). Variability in the peak heights, in both chlorophyll and fluorescence, will be partly due to varying amounts of cloud cover. The two time series are again very similar, apart from the higher and variable green peaks. These peaks are now significantly reduced when fluorescence absorption is included (dotted lines). Again, later peaks in the year tend to show higher values in fluorescence (red). Figure 2.5 shows results from the area off the west coasts of Angola and South West Africa, for which Figs. 2.1 and 2.3 show high chlorophyll in July and October. Clouds are much less frequent in this area, so the satellite results should give a good representation of true interannual variability. The year 2010 shows a significantly higher peak value than any other year. Peaks are in September of five years, in August of 2003, 2008 and 2010, and June and October of 2005.

The dotted green line shows a significant drop for fluorescing chlorophyll in this case, as per Eq. (2.2), but fluorescence values are still lower by about a factor 5. This may indicate conditions of ample nutrients due to coastal upwelling leading to reduced fluorescence, or it may indicate a problem with one of the two estimates of surface chlorophyll. The area includes coastal waters for which the high "blue-to-green ratio" chlorophyll estimate is not confirmed by the fluorescence data. Removing normalization (dotted and solid red lines) has only a small effect at these equatorial latitudes, as it would do in Figs. 2.2 and 2.4 (not shown).

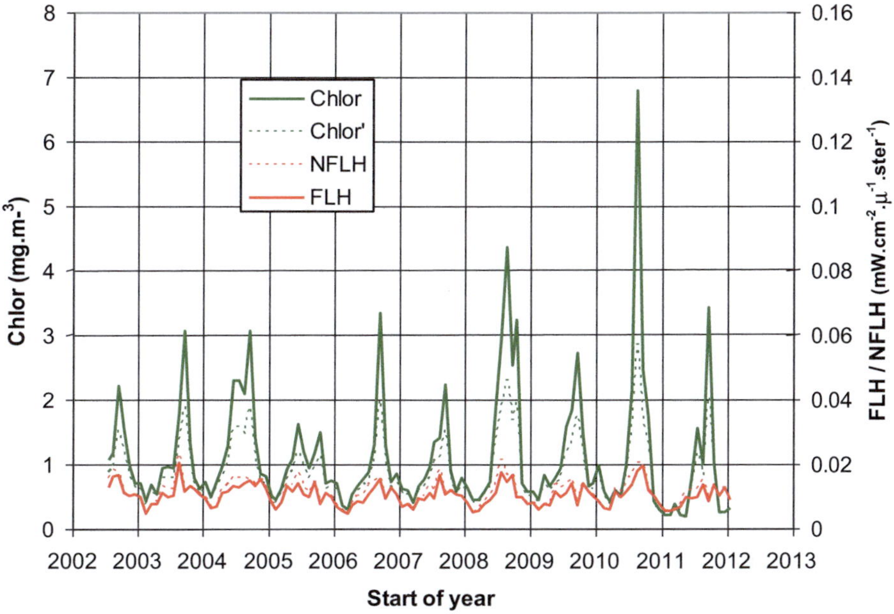

Fig. 2.5 Average surface chlorophyll concentrations (*green*) and normalized fluorescence (*red*) from MODIS Aqua data off the coasts of Angola and South West Africa, from 10 to 20°S and 5 to 15°E. Relative scaling as in Fig. 2.2

Figure 2.6 compares chlorophyll and fluorescence data averaged over 39–41°S, 10–30°E, in the area of maximum chlorophyll at the bottom of Figs. 2.1 and 2.3. The chlorophyll time series shows sporadic peaks with little clear annual cycle, but with a tendency for higher values in November to February. The normalized fluorescence time series (NFLH, dotted red) shows the reverse tendency, with a strong annual cycle, peaking near June, the season of lowest insolation. In fact, the normalization applied to fluorescence is responsible for almost the entire apparent annual cycle. With this normalization removed, the solid red line in Fig. 2.6 shows better, but still far from perfect agreement with the chlorophyll data.

Use of satellite fluorescence data has been discussed by Behrenfeld et al. (2009). It appears that fluorescence is fully stimulated by intensities of sunlight provided by all sun elevations above about 20°, which is the minimum elevation for which satellite results are computed. Behrenfeld et al. (2009) show a model with this saturation and with fluorescence quantum efficiency varying inversely with light intensity, as implied by full stimulation and as we find in these data. It is clear that FLH and not NFLH provides the better measure of chlorophyll. However, this has not yet been incorporated in NASA's satellite data processing.

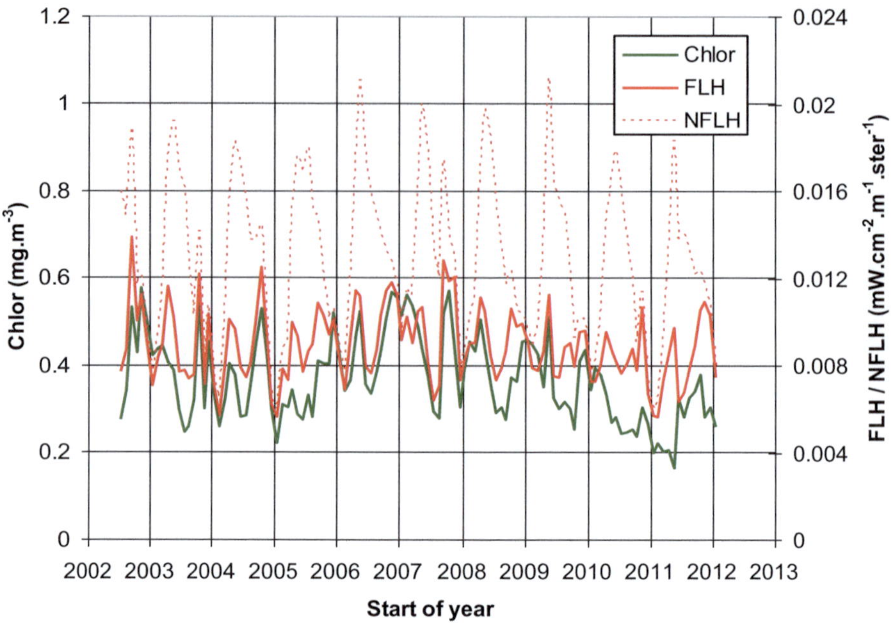

Fig. 2.6 Average surface chlorophyll concentrations (*green*) and fluorescence (*red*), plotted as normalized fluorescence (*dotted red*) and as-measured FLH (*solid red*), from MODIS Aqua data south of South Africa, from 39 to 41°S and 10 to 30°E. Relative scaling as in Fig. 2.2

2.6 Trichodesmium in the Red Sea

Figure 2.7 shows a pseudo-colour image of MCI signal in the Red Sea on 30 July 2005 near 23°N. The curved streaks are due to surface slicks of *Trichodesmium* that commonly blooms at this season. Small areas of high MCI signal near shore are due to coral reefs, whose zooanthellae also show the chlorophyll reflectance signature detected by this index.

The spectra in Figs. 2.8 and 2.9 show the "red edge" of vegetation near 700 nm with increased radiance at longer wavelengths consistent with the buoyant slicks formed by *Trichodesmium*. The spectra also show broad peaks between 500 and 620 nm which would be expected from detached trichomes in the water.

Figure 2.10 shows the monthly time series of total MCI signal for open water in the Red Sea. The monthly data show a peak near northern mid-summer in all years. The peak was large in 2005 at the time of Fig. 2.7, and even larger in 2006. On average the peaks are almost equal in June and July, but are down to 50 % of this level in August and 40 % in May. Amplitude is 15 % in April and small in other months. Peak MCI signal was reached in May in 2009 and 2010, in June in 2004, 2006, 2007 and 2011, in July in 2002, 2003 and 2005, and in August in 2008. Figure 2.11 shows the

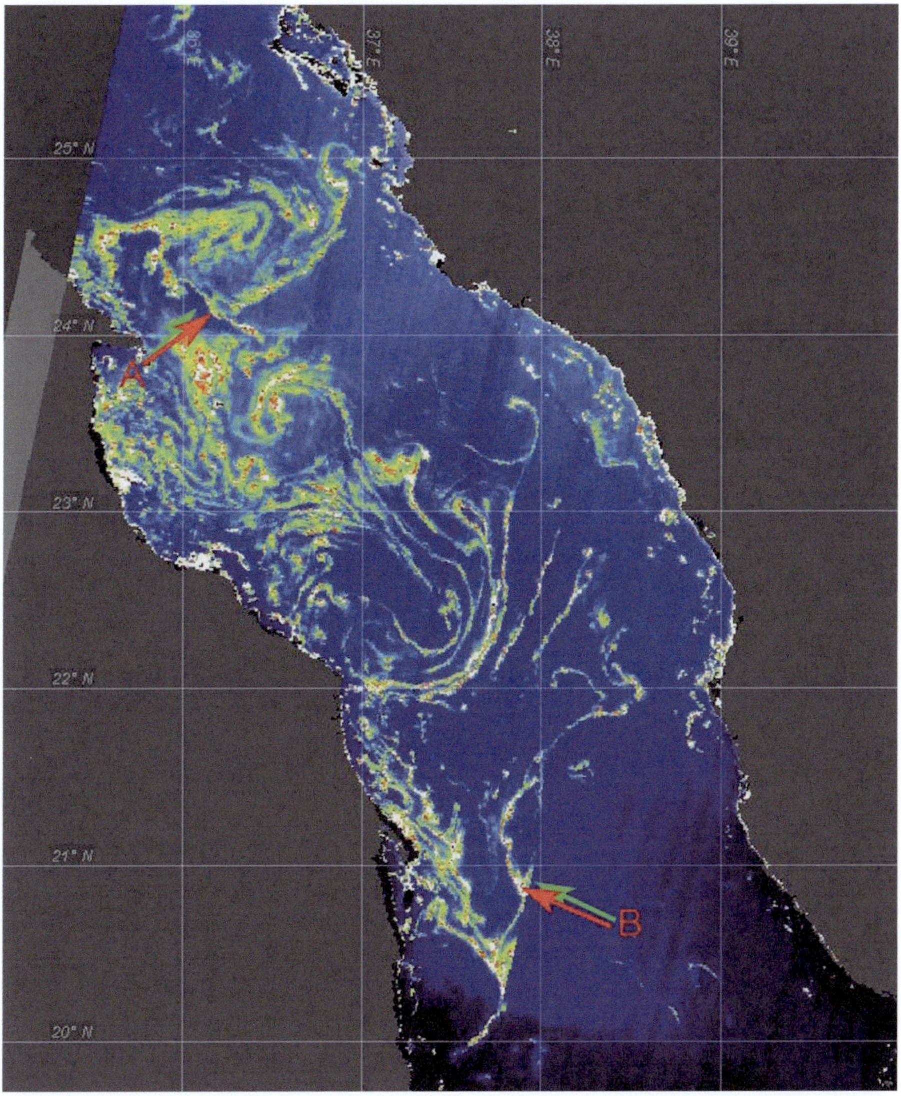

Fig. 2.7 Surface slicks of *Trichodesmium* in the Red Sea on 30 July 2005 between 20 and 25 N latitude show as areas of high MCI. This is a pseudo-colour image in which MCI radiance values are represented by a rainbow colour sequence with dark blue as the lowest signal level

average peak month for each year computed as in Eq. 2.3 using signal levels from April to August. The plot suggests a possible tendency for blooms to be happening earlier, in that the most recent three years show the three earliest computed average peak month values.

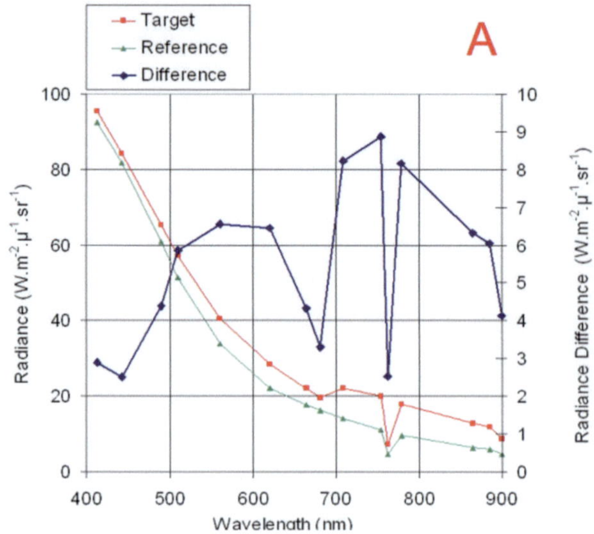

Fig. 2.8 Level 1 radiance spectra at points indicated by *arrows* (A) in Fig. 2.7. The red spectrum is of water containing *Trichodesmium*; the green spectrum of nearby "clear" water; the blue spectrum is the difference, showing the added radiance from the surface slick of *Trichodesmium* (*right axis on expanded scale*). The points indicate MERIS bands where radiances are measured

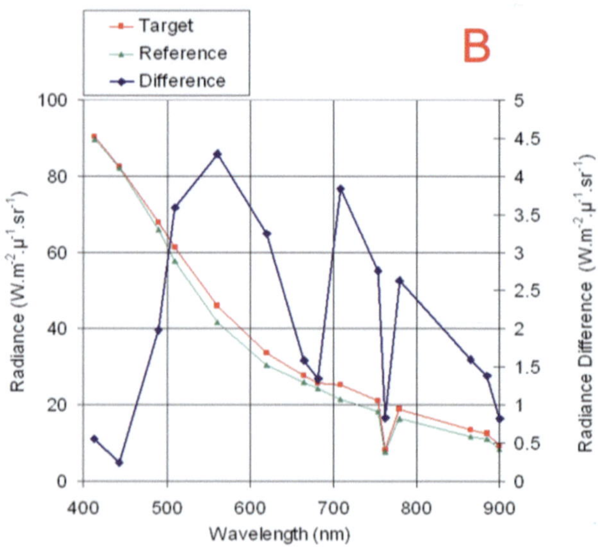

Fig. 2.9 As for Fig. 2.8, but for location B showing a similar *Trichodesmium* signature, here with relatively more detached trichomes causing a relatively larger peak at 500–620 nm

2.7 Trichodesmium Off Zanzibar

An intense bloom of *Trichodesmium* was reported in the channel between Zanzibar and mainland Tanzania, in the period from 3 to 15 January 2011. The monthly MERIS MCI time series for the area showed a significant MCI event in December 2010, but nothing in January 2011 (Fig. 2.12). There are extensive coral reefs in the area, but data for Fig. 2.12 were chosen to exclude these. Maximum bloom activity as shown

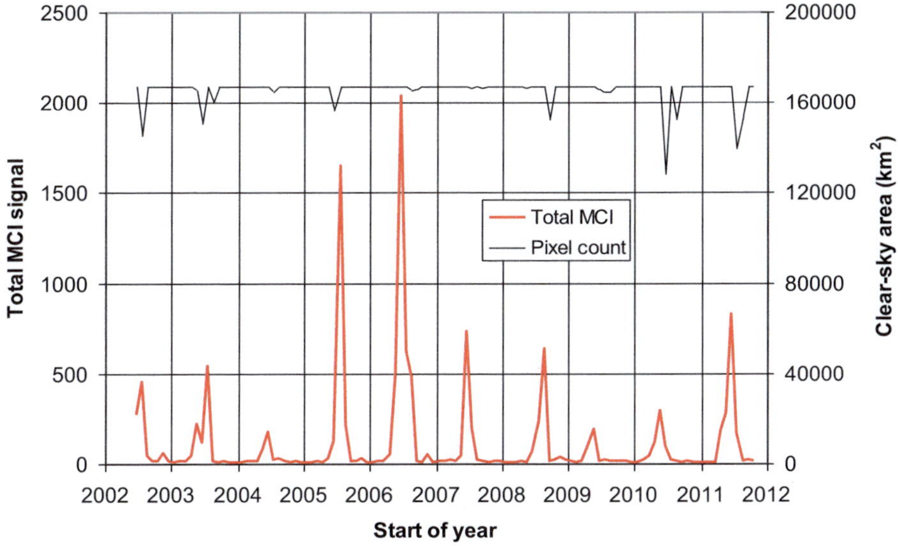

Fig. 2.10 Total MCI signal (*red*) and clear sky area (*black*) in a strip of water about 170 km wide and 1,000 km long, down the axis of the Red Sea from 27 N, diagonally across the area of Fig. 2.7, to about 17 N

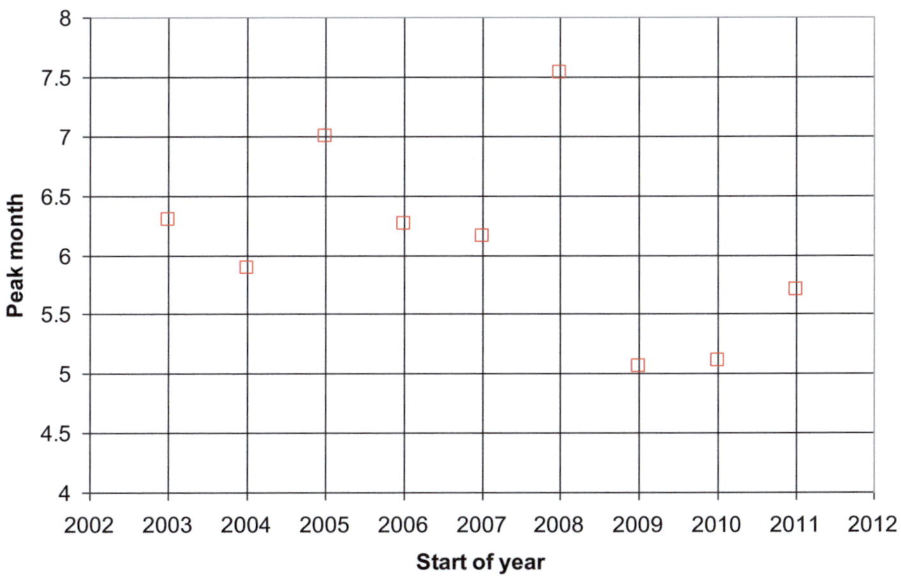

Fig. 2.11 Peak bloom month computed from the total MCI signal used in Fig. 2.10

by MCI occurs from October to January in all years since 2006, but in September 2005 and March 2003, with low signal in 2004. December 2010 stands out for the highest total MCI signal in the 10-year MERIS dataset, indicating intense, extensive

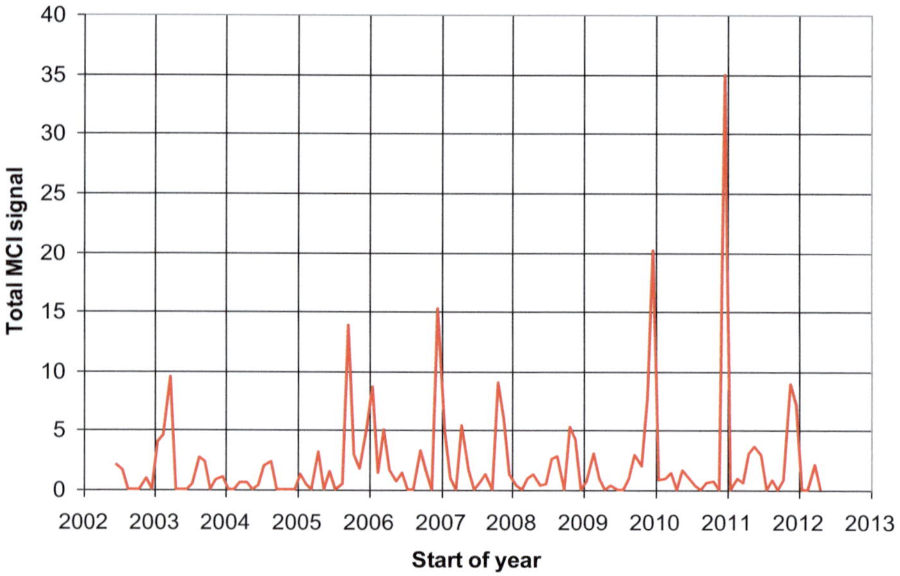

Fig. 2.12 Total MCI signal in a strip of water about 25 km wide, south of the island of Zanzibar off the coast of Tanzania, east Africa, and extending south-west about 100 km into the Indian ocean

blooms south of Zanzibar Island. Lesser events were recorded in Decembers of 2006, 2009 and 2011. The daily composite images show that the December event was imaged just on a single day, December 27. There are clear images on December 16 and 19, which show no event, and a partly cloudy image on December 30, which also shows nothing. Images of the area on January 15, 18 and 29 show small bloom activity outside the area used for Fig. 2.12.

Figure 2.13 shows MERIS data (at 1.2 km reduced resolution) of the event on December 27. Level 1 spectra are plotted from the observed radiances, with no atmospheric correction, for both a pixel with bloom (red arrow) and a nearby reference pixel with no apparent bloom indicated by MCI (green arrow). The difference spectrum (blue line, right axis) shows the bloom signature as a change in spectral radiance. MERIS shows bloom signals on both December 27 and January 15, but about ten times more strongly on December 27. The images suggest that in situ reports from Zanzibar missed the peak of the event. Sun glint, clouds and the relatively narrow MERIS swath reduce coverage to only 3–6 images of the area per month. The bloom might also have been dispersed by wind events. MERIS suggest an intense bloom period in late December with the bloom dispersed by December 30, with some residual bloom activity over the next 2 weeks.

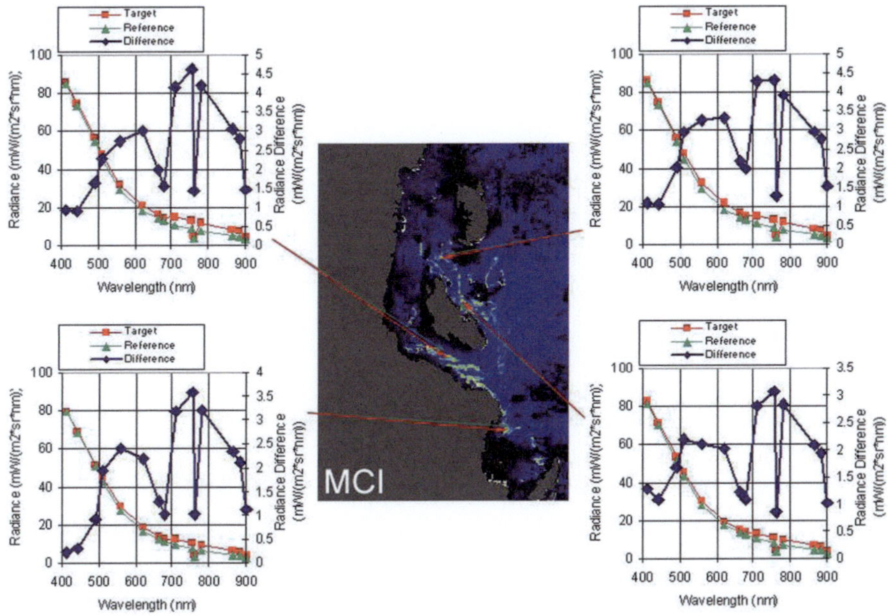

Fig. 2.13 MERIS RR level 1MCI image for the Tanzania coast with the island of Zanzibar (Unguja Island, *centre*) with the similar-sized island of Pemba to the north of it. Spectra for the four areas all show the chlorophyll red edge at 700 nm, indicating a slick of floating bloom or other vegetation. Figure 2.12 shows data for the area of the *top left* spectrum, south of Zanzibar

2.8 Trichodesmium Blooms Southwest of Madagascar

Global observations of MCI with MERIS show extensive linear slicks in the southern waters of the Mozambique Channel. These tend to occur in January and February (Fig. 2.14). Spectral properties of the MERIS data suggest that these slicks are most probably due to *Trichodesmium*, though there have been reports of pelagic *Sargassum* in the area. We have no in-situ reports of these events for confirmation. The observed time series has a maximum value in January 2009. In this year, signal was also present in December and February, amounting to 18 and 15 % respectively of the January peak. In 2006, a smaller peak was observed, also in January, and in 2007, 2010 and 2012 smaller peaks were observed in February. The plot shows some presence of slicks in the southern summer of all years.

Figure 2.15 shows an image with spectra from the pattern of slicks off the southwest coast of Madagascar, on 25 January 2009, covering a significant area. The time series plot suggests the event must have lasted several weeks, with peak extent in January, but significant signal in December and February. The difference spectrum shown is similar to those for *Trichodesmium* presented above. The MCI is responding to the presence of surface slicks in its detection of *Trichodesmium* and does not detect less intense events for which such slicks do not form.

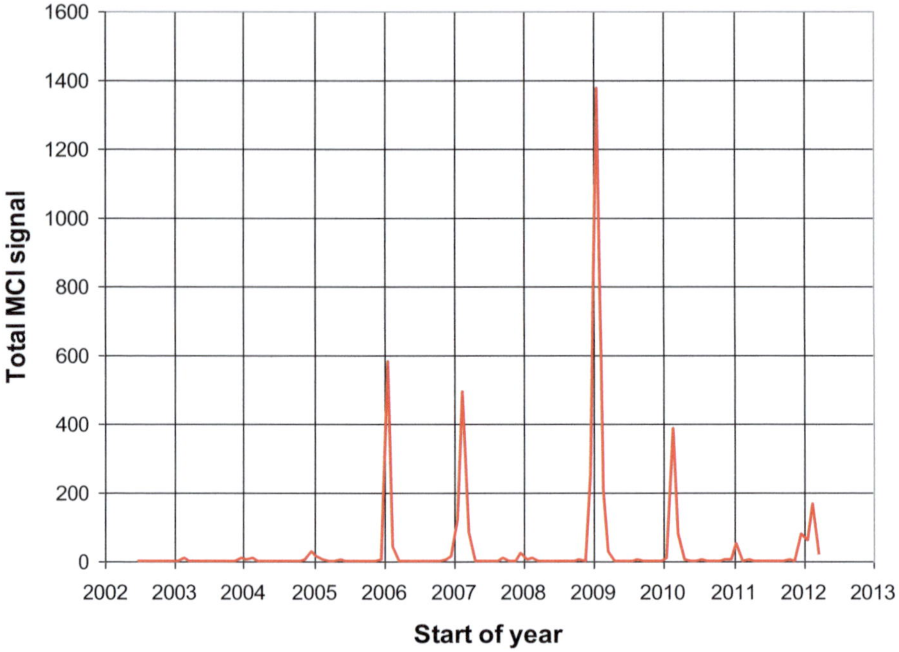

Fig. 2.14 The time series of total signal in the area 24–31°S, 36–43°E, for the waters of the Mozambique Channel, southwest of Madagascar

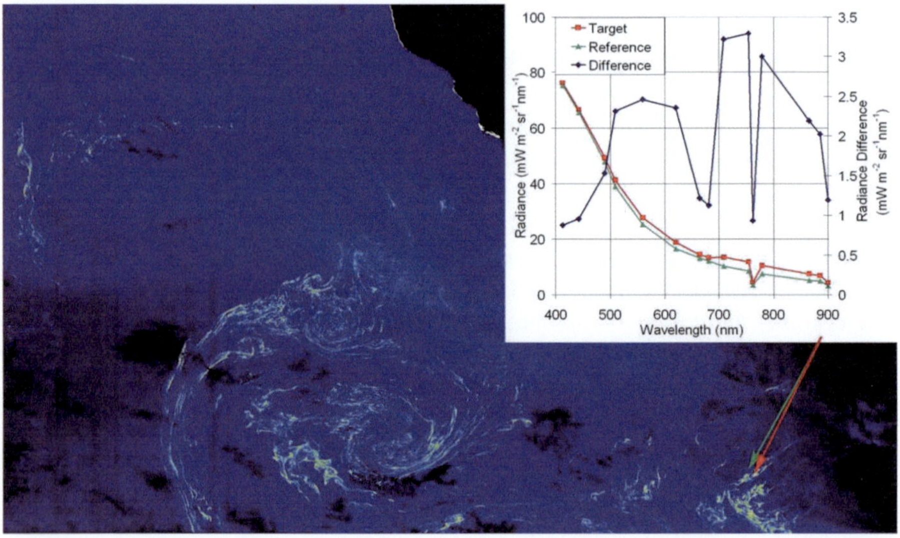

Fig. 2.15 MCI image (with spectra) of the slick pattern off Madagascar, 25 January 2009. The imaged area is about 500 km across, centred at about 26°S, 43°E. Cloud and land (*upper right*) and sun glint (*lower right*) are masked to black. Marine vegetation along the coast gives high MCI signal at upper right. The bloom spectrum is shown at top right in the same format as Figs. 2.8 and 2.9

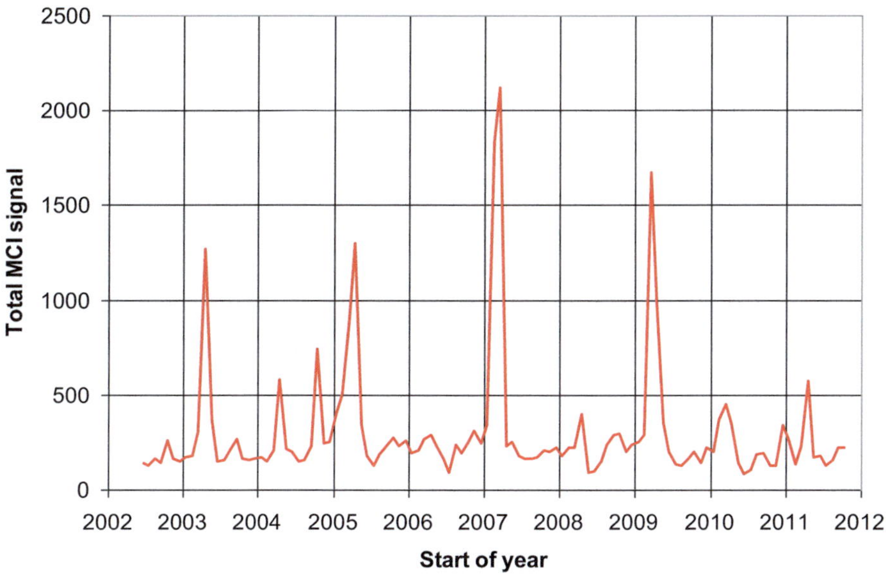

Fig. 2.16 The time series of total signal in the area 32–35°S, 15–20°E covering the waters off Cape Town, South Africa

2.9 Coastal Blooms Off South Africa

We present the statistics of blooms in two coastal areas of South Africa, one off Cape Town, and the other further east, between Cape Agulhas and Port Elizabeth. Figure 2.16 shows the time series of total signal in the Cape Town area, 32–35°S, 15–20°E, derived from the global monthly composite MCI images. This area includes coastal water and land. Because of this, the time series includes a variable background signal in all months from coastal vegetation, including coral reefs, if present. This signal is absent in Figs. 2.10 and 2.14 for the Red Sea and Mozambique Channel, which cover only open water.

Most of the peaks in Fig. 2.16 occur in April (large peaks in 2003, 2005 and March and April of 2009, and lesser peaks in April 2004, 2008 and 2011 and from February to April 2010). These are due to blooms close to shore in St Helena Bay, north of Cape Town. The largest peak shown, in February and March of 2007, is due to blooms close to the coast south of Cape Town and extending offshore to about 50 km off Cape Columbine, northwest of the city (Fig. 2.17). The only bloom outside the February to April period occurred in October 2004, mostly up to 50 km offshore of Cape Town, with some signal near shore in St Helena Bay.

The peaks in the time series in Fig. 2.16 are relatively broad, suggesting duration of bloom events of several weeks, so that signal appears in peak months and in months before and after. Figure 2.19 shows the time series for blooms in the area

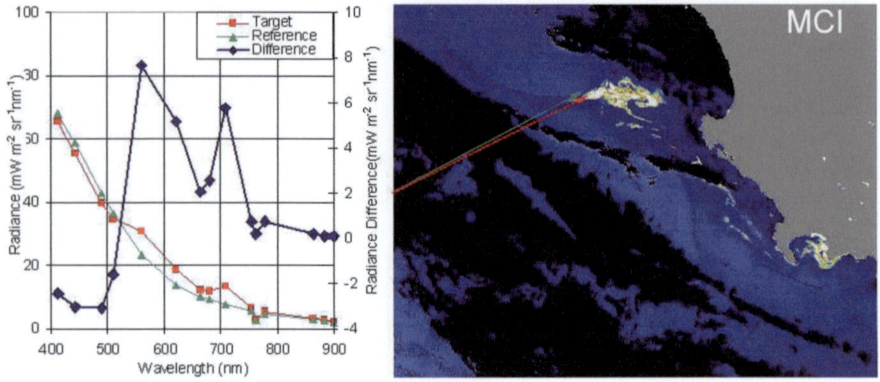

Fig. 2.17 MERIS image of high MCI signal from blooms near Cape Town, off Cape Columbine and in False Bay on 25 February 2007. The spectrum of the bloom off Cape Columbine is shown at left in the same format as Figs. 2.8 and 2.9

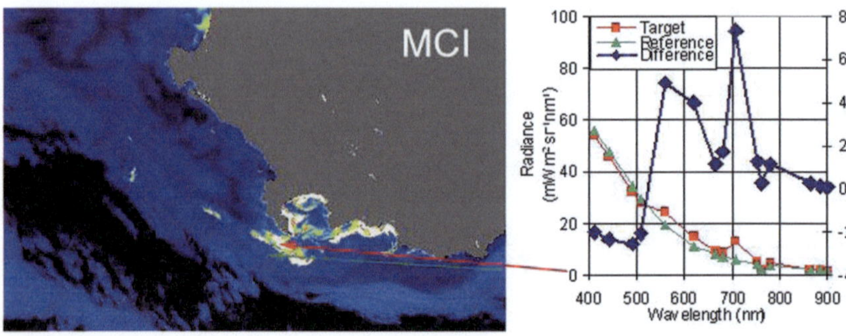

Fig. 2.18 MERIS image of the blooms on 13 March 2007 in the waters of False Bay, south east of Cape Town, South Africa. The spectrum of the bloom off False Bay is shown at right in the same format as Figs. 2.8 and 2.9

between Cape Agulhas, east of Cape Town, and East London. Here the time peaks are narrower, representing a high value in one month only, and suggesting events lasting for shorter periods. Of the three major peaks, two occur in April and one in November.

The high MCI signal indicates very high surface chlorophyll concentrations, probably at several hundreds of mg m^{-3}, due to intense blooms in surface waters on the South African coast. All events shown in Figs. 2.17, 2.18 and 2.20 show the peak at 709 nm as a major feature of the bloom spectra, with little or no added radiance at wavelengths longer than 720 nm. This indicates blooms in the near-surface waters, with no buoyant slicks covering parts of the water surface. This is in contrast to the spectra in Figs. 2.8, 2.9, 2.13 and 2.15, which show radiance increased also at longer wavelengths. This increase at longer wavelengths shows that all or most of the high-chlorophyll signal is from buoyant surface slicks of the type characteristic of *Trichodesmium*, or from floating *Sargassum*.

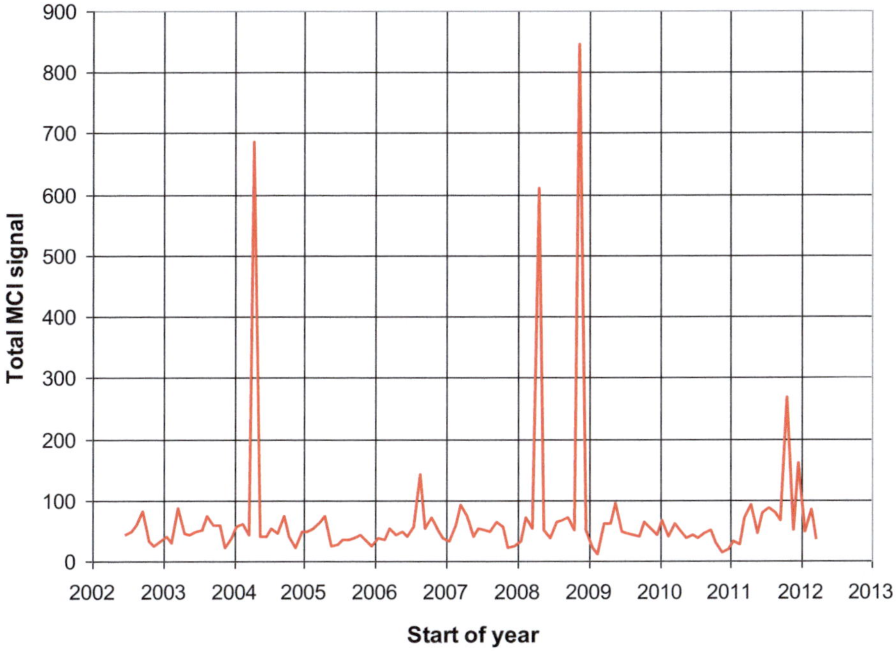

Fig. 2.19 The time series of total signal along the south coast of South Africa from east of Cape Town to East London, over the area 33–37°S, 20–28°E

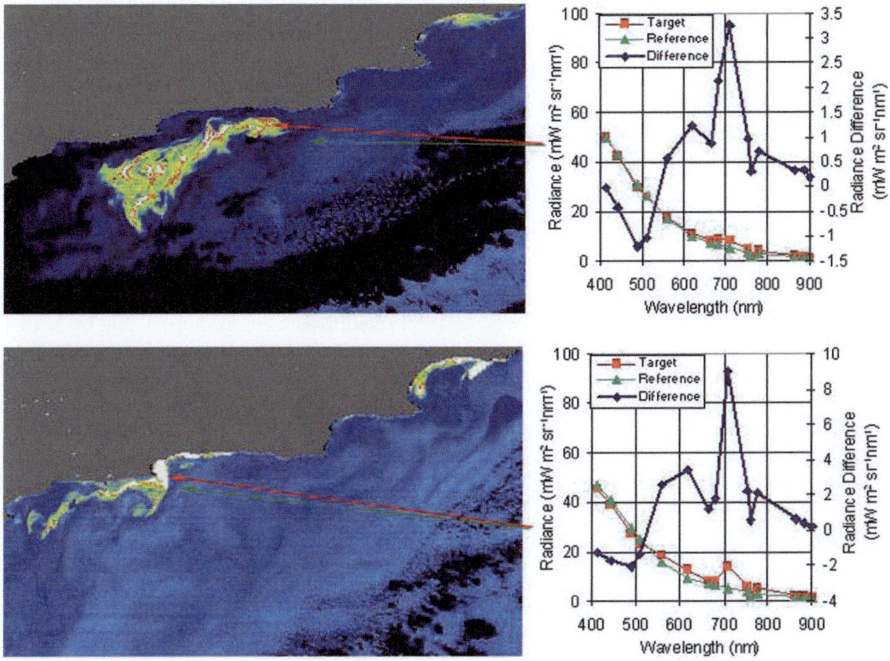

Fig. 2.20 MERIS MCI images of blooms in the waters west of Port Elizabeth, South Africa, top on 12 April 2004, bottom on 14 April 2008. In both cases, spectra show high peaks at 709 nm

We discuss above our reasons for believing that spectra shown in this chapter are from *Trichodesmium*. Pelagic *Sargassum* does not appear to be common in African waters. We have evidence for an occurrence in Sierra Leone following the equatorial Atlantic Sargassum event in the summer of 2011, but this appears to be unusual.

2.10 Summary and Conclusions

Ocean colour data from satellites now provide a valuable source of information on spatial and temporal patterns for all oceans and coastal waters of the world. Studies of African waters can be expected to especially benefit from satellite data, because of the lack of local observing facilities and supporting infra-structure, and the danger of piracy in some areas.

We hope that satellite water colour coverage will continue, though the recent loss of MERIS reminds us of the risk of a data gap in time. At present, images cover space scales from kilometers to global, and time scales from daily to the 15-year period for which data have been provided since 1997 by SeaWiFS, MODIS, MERIS and VIIRS. Other options, not considered here, include information on longer term trends that can be deduced by comparison of recent data with results from the CZCS (1978–1986), and the use of finer scale data from higher resolution, but less sensitive, imagers such as Landsat MSS or TM. Spatial resolution can be extended down to a scale of meters using commercial mapping sensors.

The fluorescence data from MODIS and MERIS are useful to confirm the chlorophyll values available through the green to blue ratio method. Examples are given in Figs. 2.2, 2.4, 2.5 and 2.6 above. Here we find that normalization of FLH as conventionally carried out, is not appropriate, in that un-normalized FLH provides a good measure of chlorophyll.

MERIS, through the spectral band at 709 nm, provided the added capability of detecting blooms using the MCI index. We would like to continue time series such as Figs. 2.10, 2.11, 2.12, 2.14, 2.16 and 2.19 indefinitely into the future, but there is now a data gap until the launch of ESA's Sentinel 3 satellite, planned for 2014.

Acknowledgments This work was supported by the Canadian federal Department of Fisheries and Oceans, and by the Canadian Space Agency. MODIS satellite data are made available by NASA, the US National Aeronautics and Space Administration. MERIS image data are made available by the European Space Agency, ESA. Time series analysis of MERIS data is made possible by the G-POD system of ESA. Some analyses and visualizations used in this study were produced with the Giovanni online data system, developed and maintained by the NASA GES DISC.

References

Abbott MR, Letelier RM (1999) Algorithm theoretical basis document chlorophyll fluorescence, MODIS product number 20, NASA. http://modis.gsfc.nasa.gov/data/atbd/atbd_mod22.pdf. Accessed 27 Dec 2013

Acker JG, Leptoukh G (2007) Online analysis enhances use of NASA earth science data. Eos Trans AGU 88(2):14–17

Behrenfeld MJ, Westberry TK, Boss ES, O'Malley RT, Siegel DA, Wiggert JD, Franz BA, McClain CR, Feldman GC, Doney SC, Moore JK, Dall'Olmo G, Milligan AJ, Lima I, Mahowald N (2009) Satellite-detected fluorescence reveals global physiology of ocean phytoplankton. Biogeosciences 6:779–794

Gitelson A (1992) The peak near 700 nm on radiance spectra of algae and water: relationships of its magnitude and position with chlorophyll concentration. Int J Remote Sens 13:3367–3373

Gower JFR, Borstad GA (1990) Mapping pf phytoplankton by solar-stimulated fluorescence using an imaging spectrometer. Int J Remote Sens 11:313–320

Gower JFR, Doerffer R, Borstad GA (1999) Interpretation of the 685 nm peak in water-leaving radiance spectra in terms of fluorescence, absorption and scattering, and its observation by MERIS. Int J Remote Sens 9:1771–1786

Gower JFR, Brown L, Borstad G (2004) Observation of chlorophyll fluorescence in west coast waters of Canada using the MODIS satellite sensor. Can J Remote Sens 30:17–25

Gower JFR, King S, Borstad G, Brown L (2005) Detection of intense plankton blooms using the 709 nm band of the MERIS imaging spectrometer. Int J Remote Sens 26:2005–2012

Gower JFR, Hu C, Borstad G, King S (2006) Ocean color satellites show extensive lines of floating *Sargassum* in the Gulf of Mexico. IEEE Trans Geosci Remote Sens 44:3619–3625

Gower JFR, King S (2007a) Validation of chlorophyll fluorescence derived from MERIS on the west coast of Canada. Int J Remote Sens 28:625–635

Gower JFR, King S (2007b) An Antarctic ice-related "superbloom" observed with the MERIS satellite imager. Geophys Res Lett 34. doi:10.1029/2007GL029638

Gower JFR, King S, Borstad G, Brown L (2008a) The importance of a band at 709 nm for interpreting water-leaving spectral radiance. Can J Remote Sens 34:287–295

Gower JFR, King SA, Goncalves P (2008b) Global monitoring of plankton blooms using MERIS MCI. Int J Remote Sens 29:6209–6216

Gower JFR, King S (2011) Distribution of floating *Sargassum* in the Gulf of Mexico and Atlantic Ocean mapped using MERIS. Int J Remote Sens 32:1917–1929

Morel A, Prieur L (1977) Analysis of variations in ocean color. Limnol Oceanogr 22:709–723

Moulin C, Gordon HR, Chomko RM, Banzon VF, Evans RH (2001) Atmospheric correction of ocean color imagery through thick layers of Saharan dust. Geophys Res Lett 28:5–8

Singh RP, Prasad AK, Kayetha VK, Kafatos M (2008) Enhancement of oceanic parameters associated with dust storms using satellite data. J Geophys Res 113(C11008). doi:10.1019/2008JC004815

Yacobi YZ, Gitelson A, Mayo M (1995) Remote sensing of chlorophyll in Lake Kinneret using high spectra resolution radiometer and Landsat TM: spectral features of reflectance and algorithm development. J Plankton Res 17:2155–2173

Chapter 3
Thermal Infrared Remote Sensing and Sea Surface Temperature of Marine and Coastal Waters Around Africa

Nicolas Hoepffner, Samuel Djavidnia, Leo Nykjaer and Pascal Derycke

Abstract Owing to satellite-derived Thermal Infrared (TIR) data, it is possible to derive large basin scale view of Sea Surface Temperature (SST) around the African continent over the last 2 or 3 decades. This may form the baseline against which future changes can be measured. Measurements of SST also serve as an entry point to characterize the water dynamics along the African coast and to understand where and when changes have taken place, such that possible short- to medium term scenarios can be developed and used as management tools for precautionary measures. Significant advances in TIR technology have been achieved to build the necessary confidence in satellite-based SST measurements for research, as well as for operational and commercial purposes. An effective dissemination of these EO value added products to African institutions and policy-makers is a key condition for long-term sustainability of coastal and marine resources. The chapter presents a variety of applications using TIR images and SST data that have contributed to develop the necessary knowledge and competences that are critical for the implementation of conservation and management strategies at continental, regional and national scales.

3.1 Introduction

With *ca.* 35,000 km of coastline, the African continent expands across 35° of latitude on either side of the equator and across similar distance in longitude from 50°E to 20°W. As a result of this geographical cover, Africa is surrounded by a complex system of ocean currents maintained by typical wind patterns, namely the north- and southeasterly trade winds, interacting with the shape of the coastline and bathymetry to create at regional and local scales some peculiar ocean surface dynamics that directly influence the distribution of the organisms and the ecosystem functioning at the coast. Among well-known features, the western African coast includes two

N. Hoepffner (✉) · L. Nykjaer · P. Derycke
European Commission, Joint Research Centre,
Institute for Environment and Sustainability, Ispra, Italy
e-mail: nicolas.hoepffner@jrc.ec.europa.eu

S. Djavidnia
European Maritime Safety Agency, Lisbon, Portugal

V. Barale, M. Gade (eds.), *Remote Sensing of the African Seas*,
DOI 10.1007/978-94-017-8008-7_3,

of the four major upwelling areas of the world ocean, the Benguela system in the south and the Morocco-Mauritanian system in the north (Hill et al. 1998). These are sites of economical importance due to high water productivity sustaining large biodiversity and valuable fish resources. Also, a number of mesoscale structures and oceanic fronts are recurrent e.g. along the eastern coast of Africa molding to a large extent the pelagic ecosystem and species distribution.

Due to this dynamic, coastal and marine environments play a vital role in the economy and society of many African countries (Odada 2010), contributing significantly to reducing the national balance of payments deficit, creating employment and meeting the protein needs of the local population. Fish products provide more than 60 % of the dietary proteins in many countries, such as Ghana and Gambia, stressing the need to develop fisheries management strategies to ensure efficient exploitation while keeping the stocks sustainable on the long-term. Given that marine physics can determine to a large extent the rate and distribution of the biological resources across a range of scales (Gargett 1997), monitoring the dynamics of water masses at appropriate resolution in the time-space domain will considerably facilitate the formulation and implementation of these strategies, as well as the mitigation of an increasing human impact on marine and coastal resources around Africa.

Earth Observations (EO) from satellite enable the provision of environmental information at low cost (to the users) and unprecedented time scales over large and distant areas of the oceans, complementing expensive, hence scarce, field campaigns using research vessels. Thermal Infrared (TIR) sensors on-board satellites, depicting changes in Sea Surface Temperature (SST) over large marine areas, are instrumental to monitor variations in the water masses at the ocean surface resulting from long or short-term atmospheric processes, as well as water column dynamics.

In the following, the applications of TIR to monitor and assess different marine issues around Africa are reviewed. A first section summarizes the fundamental basis to measure TIR signal over the ocean from space, listing the main sensors and databases available, and outlining the current status with the use of TIR data for accurate retrieval of SST. Other sections are referring to different applications of TIR over African coastal and marine waters, using SST as a single indicator of changes in water masses or in combination with other platforms to assess important processes such as fisheries. Finally, some concluding remarks are presented, providing a forward perspective in the use of TIR imagery to meet the societal needs of the African population in the context of international aid initiatives.

3.2 Sensors and Data Repositories

Quantitative information about the oceans is transferred from the sea surface to the satellite sensor through the atmosphere via electromagnetic radiation. This information transfer has some constraints, which are set by the relationship between the radiation emitted from the sea surface, and the atmosphere it travels through before reaching the satellite sensor. Thermal Infrared (TIR) sensors usually passively collect radiations at wavelengths between *ca.* 10 to 15 μm, although some instruments have additional bands in the visible and near-infrared parts of the spectrum for specific purposes, complementing information given by the far-infrared bands.

The strength of the infrared radiation emitted by the ocean is a function of the temperature at its surface: the higher the temperature, the greater is the radiant energy. Sea surface temperature is therefore the main product retrieved by these sensors, after correcting the water surface-emitted signal from the contribution due to the atmosphere. Note that a complete description of the retrieval of SST in terms of the physics of the signal and atmospheric effects is outside the scope of this paper. Details on these issues are available through a number of technical publications (e.g. Kilpatrick et al. 2001; Merchant et al. 2008) and books (Robinson 2004 and references therein).

The reliable calibration of the instruments, coupled with the regular validation exercises with field observations, allow SST measurements from TIR sensors to reach high levels of absolute accuracy, e.g. $< 0.2\,°C$ with the Advanced Along-Track Scanning Radiometer (AATSR) instrument (O'Carrol et al. 2008). In addition, polar-orbiting satellites can provide SST measurements twice daily at a spatial resolution of *ca.* 1 km. Alternatively, radiometers on-board geostationary satellites—e.g. the Meteosat Second Generation (MSG) satellites of the European Space Agency (ESA)—have coarser ground resolution (3–5 km) but much higher temporal sampling frequency (every 15 min or half-hourly) within its large but constant field of view.

Thermal sensors operating onboard satellite have been in orbit for the last 30 years, providing multi-decadal observations of the marine and ocean dynamics through SST measurements at global and regional scales. The Advanced Very High Resolution Radiometer (AVHRR) series operated by the US National Oceanic and Atmospheric Administration (NOAA) National Environmental Satellite, Data, and Information Service (NESDIS) represent undoubtedly the longest time series of SST measurements collected from space. The data are distributed operationally and globally through the AVHRR Pathfinder programme[1] (Kilpatrick et al. 2001) from 1981 to present, at spatial resolution of 4 km and daily daytime and nighttime fields. The processing and validation of Pathfinder SST are outside the scope of this work, but have been the focus of several publications (Kearns et al. 2000; Kumar et al. 2003; Marullo et al. 2007; Nykjaer 2009).

In addition to AVHRR, many other satellite missions equipped with thermal infrared bands to measure SST have been launched since 1999 in a variety of orbits (Donlon 2010). Such a multitude of available SST datasets derived from sensors with different characteristics and processed in slightly different ways called for an international effort to harmonize these data in order to provide users with consistent long-term data stream of SST products. As part of Global Ocean Data Assimilation Experiment (GODAE), the Group for High Resolution SST (GHRSST) provides a wide variety of SST data[2] (Donlon et al. 2007), including a multi-sensor merged product and uncertainty estimates for each of the products (e.g. Level 4 K10 SST

[1] The link http://www.nodc.noaa.gov/SatelliteData/pathfinder4km provides access to Pathfinder through the present, adding observations by the NOAA-19 sensor.

[2] Available at https://www.ghrsst.org.

data produced the Naval Oceanographic Office at 0.1° grid). Data is made available through both regional and global Data Assembly Centers, all of which offering the same interoperable metadata models, file format structures and access methods (Donlon et al. 2009).

In spite of the availability of multiple datasets from satellite, there are several obstacles that prevent African countries to make full use of this technology. These obstacles include the lack of appropriate infrastructure and data communication systems that would limit the diffusion of the data and subsequent use of these data by the African scientists. Also the level of technical expertise may not be sufficient or appropriately targeted to manipulate satellite data and promote their applications in the marine sector. A number of specific programs and projects and/or initiatives have been established to overcome these difficulties.

The information system GEONETCast is part of the Global Earth Observation System of Systems (GEOSS) established by the Group on Earth Observations (GEO) to provide environmental satellite data to users on a worldwide basis. It relies on the EUMETSAT's broadcast system for meteorological and environmental data, named EUMETCast, a multi-service dissemination system using standard Digital Video Broadcast technology and commercial telecommunication satellites (Jungbluth et al. 2011). Owing to a number of projects and European initiatives such as the Preparation for Use of MSG in Africa (PUMA) and its follow-up the African Monitoring of the Environment for Sustainable Development[3] (AMESD), more than 100 receiving stations have been deployed in sub-Saharan Africa countries collecting environmental information, including sea surface temperature from TIR sensors, through the EUMETCast system (see e.g. Clerici et al. 2009).

The Europe-Africa Marine EO Network[4] (EAMNet) has been established specifically in the marine and coastal domain to link Earth Observation (EO) data providers with centers of excellence in Europe and Africa to support sustainable development in Africa. Its data portal regroups several information systems providing the African user community with a set of EO-derived bio-physical variables, of importance to conduct water quality monitoring and resource assessment in their marine and coastal waters.

For example, the Global Marine Information System[5] (GMIS) has simple and easy-to-use mapping tool applications, created for the publication and dissemination of marine information via the web. GMIS relies on best quality EO data, including sea surface temperature from the Moderate Resolution Imaging Spectometer (MODIS)—available in two versions, the first onboard the Terra orbital platform (MODIS Terra), the second onboard the Aqua orbital platform (MODIS Aqua)—and AVHRR series, to generate indicators for regional diagnostic of the coastal state and analyses of changes in marine ecosystems. A specific window for Africa has been predefined in the system with five regional sub-windows corresponding to African Large Marine Ecosystems[6].

---

[3] See http://www.amesd.org for details.

[4] See http://www.eamnet.eu for details.

[5] Available at http://gmis.jrc.ec.europa.eu.

[6] See http://odin.ioc-africa.org/index.php/african-large-marine-eco-systems for details.

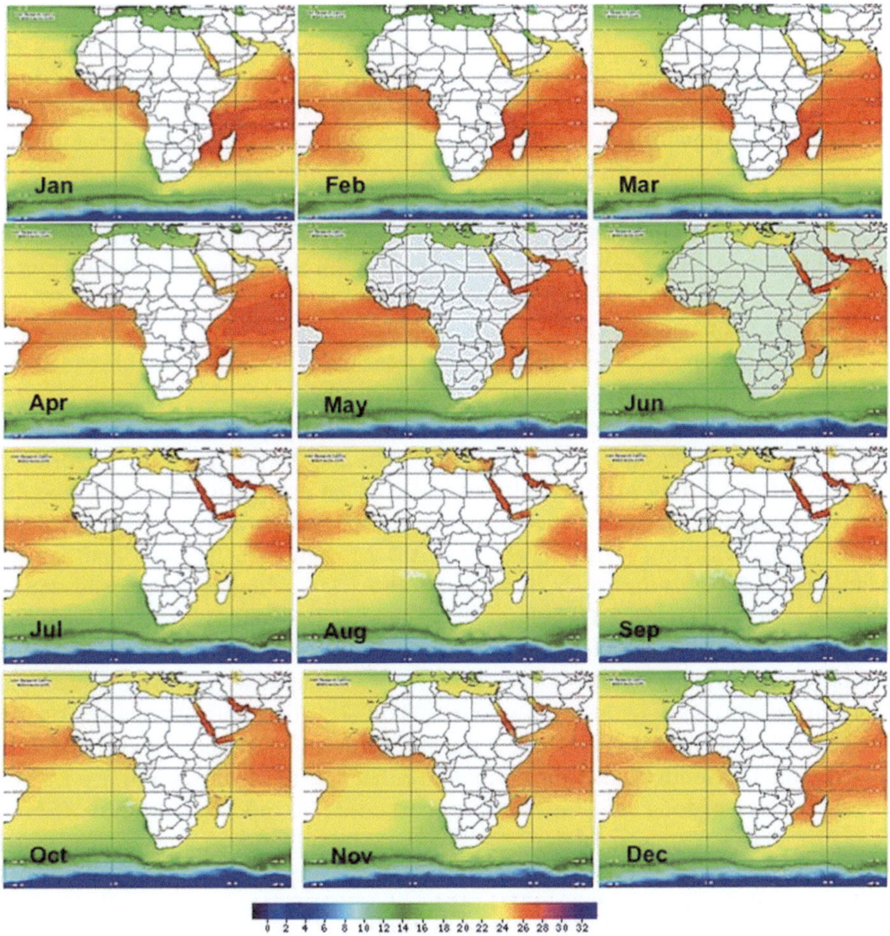

Fig. 3.1 Monthly climatologies of sea surface temperature (in°C) around Africa from MODIS Terra instrument (2000–2011). Images extracted from http://gmis.jrc.ec.europa.eu

3.3 Characterizing Water Masses Around Africa

Monthly SST climatology from MODIS (Fig. 3.1) illustrates the variety of the water masses around Africa under the influence of large-scale atmospheric processes such as the Trade Winds on the Atlantic Ocean side and the monsoon wind patterns on the Indian Ocean side. The oceanic circulation along the coast of Africa can be depicted from the SST images with the eastern Atlantic boundary currents on either side of the equator, i.e. southward Canary Current and northward Benguela Current feeding respectively the North and South Equatorial currents as part of the mid-Atlantic gyres.

All together, these currents contribute to a large extent to shape the global climate through exchange of heat/temperature and salt with adjacent waters (Tyson et al. 2002). Along the eastern coast of Africa, the seasonally-reversing monsoon winds over the Indian Ocean translate into the SST fields with a strong South Equatorial Current (22–23 °C) flowing westward from June to September (southwest monsoon) and feeding both the southward currents on either side of Madagascar and the northward Somali current. During the northeast monsoon (December–February), the flow reverse in the northern Indian Ocean carrying colder water westward and flowing south along the Somali and Kenya coast until it mixes with the warmer South Equatorial Current.

The benefit of satellite TIR data was exploited as early as the late 70's and early 80's to provide synoptic views of these strategic and dynamically-complex regions, and in particular the Aguhlas Current system (Lutjeharms 1981; Lutjeharms and Roberts 1988; see also references in Lutjeharms 2007). A large number of peer-reviewed publications and books (e.g. Lutjeharms 2006) were specifically dedicated to the Agulhas Current in which TIR images have an important role to describe the surface dynamics. Basically, the current flows southwestward along the southern African coast carrying warm waters from the Mozambique Channel and eastern Madagascar down to the southern tip of South Africa, where it get trapped within the Antarctic Circumpolar Current creating some spectacular mesoscale dynamics eastward along the Subtropical Convergence (Olson and Evans 1986). In Fig. 3.2, a monthly time series of MODIS SST data for 2005 has been analyzed using a threshold value in order to mask temperatures below 20 or 22 °C depending on the season. As a result, the sequence of images show the variability of the Agulhas flow, collecting waters from the South Equatorial Current on either side of Madagascar into a jet-like structure flowing along the South African coast until it retroflects to the east offshore Port Elizabeth, creating meanders and eddies as it collides with the south Atlantic waters.

Using TIR data from Meteosat II, Lutjeharms (1988) was also able to observe the shedding of a warm eddy detached from the Agulhas Current and crossing the Sub-Tropical Convergence and contributing to meridional heat transport at the border of the Southern Ocean. The TIR images also revealed some Agulhas eddies formed at the tip of South Africa and propagating into the Atlantic ocean. The variability of this so-called Agulhas leakage could have an important impact on the overall Atlantic overturning circulation (Whittle et al. 2008; Beal et al. 2011).

The stability of the Agulhas Current along the steep continental shelf is occasionally perturbed by transient events, so-called Natal pulses, generated by slight changes in the shelf width off Natal Bight (Lutjeharms and Roberts 1988; Lutjeharms et al. 2003; Bryden et al. 2005; Lutjeharms 2006). These instabilities are described as large solitary meanders associated with cold-water core readily observable within the warmer Agulhas Current from satellite TIR images. Using 6 years of data collected from the Spinning Enhanced Visible and Infra-Red Imager (SEVIRI), onboard the MSG geostationary satellite, Rouault and Penven (2011) provided a detailed analysis of these features occurring on average at a rate of 1.6 pulses per year, i.e. significantly less than previous estimates based on field observations and/or

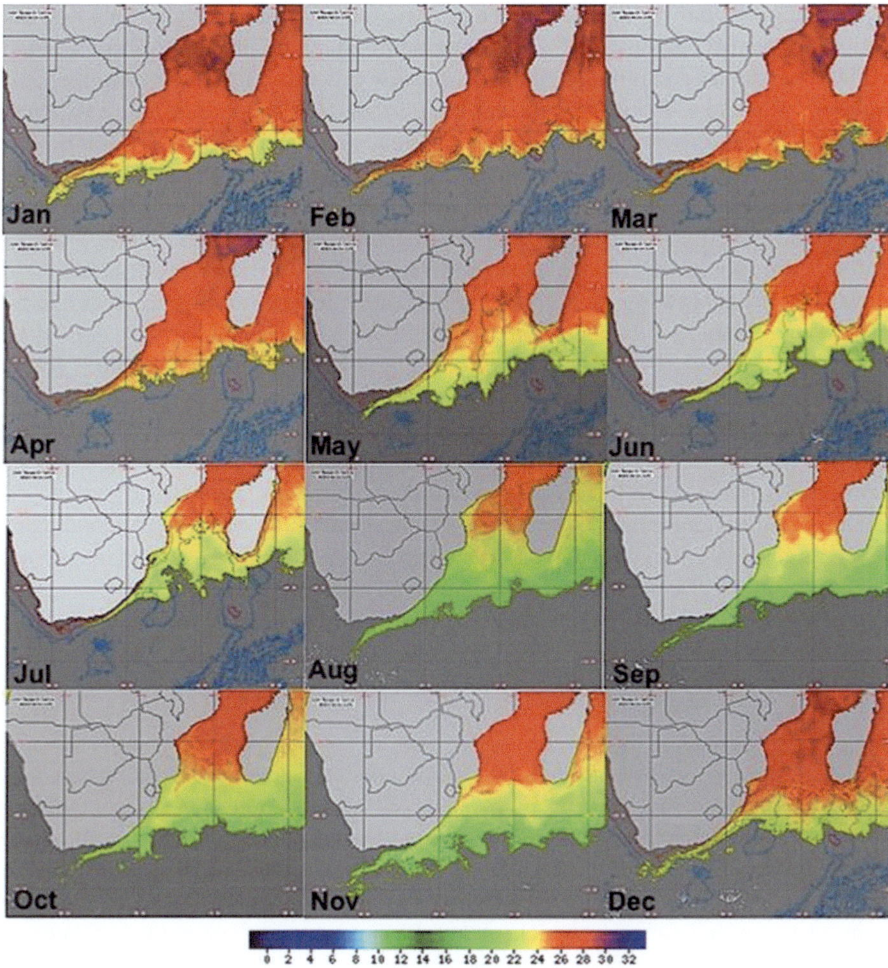

Fig. 3.2 SST monthly composites [°C] from MODIS Terra for 2005 in the Southern African area. SST lower than 22–23 °C have been masked out (dark grey) to highlight the extent of the Agulhas Current. Data extracted from http://gmis.jrc.ec.europa.eu using the 'threshold' analysis tool

numerical studies. Nevertheless, Natal pulses and associated eddies are critical in the life cycle of marine organisms and migration of commercial fishes (Fréon et al. 2010; Roberts et al. 2010), underlining the importance to monitoring these features on an operational basis using high-resolution TIR sensors.

North of the Agulhas Current, the hydrography in the Mozambique channel also show great spatio-temporal variability with a number of anti-cyclonic eddies propagating southward and directly affecting the Agulhas Current system and the variability of Natal pulses (De Ruijter et al. 2002; Schouten et al. 2003; Lutjeharms 2006; Swart et al. 2010).

Further north, during the southwest monsoon, the strong and northeasterly flowing Somali Current is responsible for high marine productivity along the Somali coast, while low biomass and phytoplankton production is commonly observed in the Gulf of Aden. The situation reverses during the northeast monsoon in winter (Veldhuis et al. 1997). Southwest winds during the boreal summer generate upwelling cells that represent important fishing zones for northeast African countries. MODIS image from June 2004 (Fig. 3.3) show typically two upwelling events respectively centered at 5°N and 10°N, starting in May and fading out late September. These features were already observed and analysed in the early 80's by Brown et al. (1980) and Evans and Brown (1980). At both locations, colder and nutrient-rich waters extend ca. 200 km offshore as filaments with a temperature difference of 4 to 5 °C compared to surrounding waters. A large anti-cyclonic ring known as the Great Whirl (Schott 1983) separates the upwelling cells, whereas other gyres can be observed southeast of the Socotra Island. Mafimbo and Reason (2010) used satellite data including thermal imagery to describe the mechanism of these upwelling events showing a strong covariability between SST and changes in mesoscale winds. This area represents an important fishing ground for the eastern African countries. A better understanding of its dynamic is essential to establish an ecosystem approach to fishery management in this part of the ocean currently under severe overfishing threats.

3.4 Upwelling Dynamics

Upwelling events are driven by alongshore wind stress inducing offshore transport of surface waters, replaced then by deeper cold water enriched in macronutrients, iron and inorganic carbon. This oceanographic feature yields optimal conditions to sustain growth of organisms at all level of the food chain (Cury et al. 2000). A satellite-based TIR can readily spot changes in the temperature signature resulting from an upwelling process, which makes SST product an optimum indicator to assess the variability and strength of such oceanographic phenomena. In particular, satellite TIR-derived SST has revealed the complexities and spatial extension of the mesoscale structures associated with upwelling, with off-shore streamers or narrow filaments that could reach few hundreds km offshore.

Van Camp et al. (1991) have described the upwelling variability off Northwest Africa using AVHRR data, observing filaments and meanders of cool coastal water extending well beyond the shelf break, and coupled with the large-scale seasonal variations of the trade winds (Fig. 3.4a). Hagen et al. (1996) and Barton et al. (1998) reported recurrent filaments from TIR imagery off Cape Ghir and north of Cape Bojador, respectively. These features resulting from a strong interactions between wind and local topography (Troupin et al. 2012) are quasi-permanent and extend few hundreds km offshore, significantly influencing the hydrodynamics and ecosystem behavior in the vicinity of the Canary Islands (Pelegri et al. 2005). An analysis of the temperature gradients on MODIS SST images through edge detection algorithm (Nieto et al. 2012) confirmed the intense mesoscale dynamics in this region with

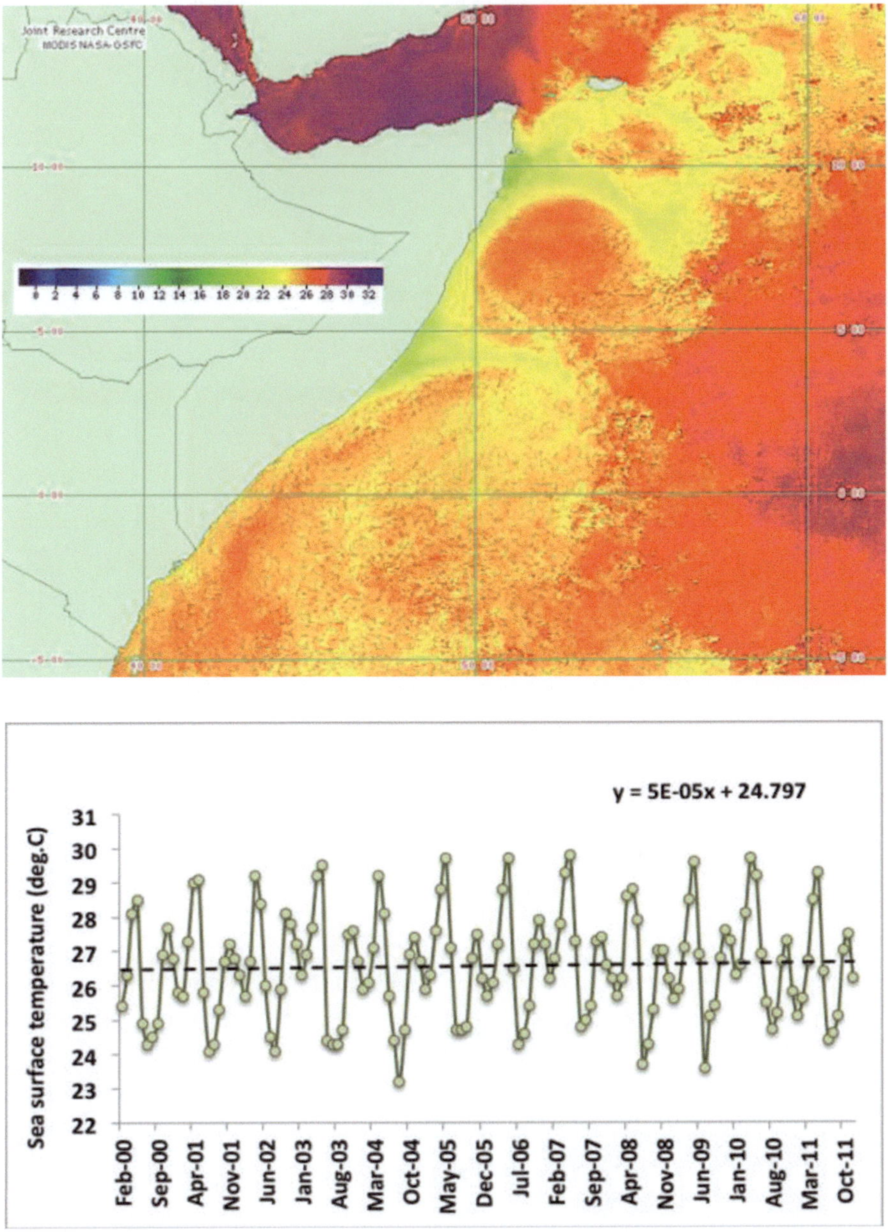

Fig. 3.3 Sea surface temperature (in°C) from MODIS Terra instrument for July 2004 showing filaments of upwelled waters along the coast of Somalia. The MODIS time series (January 2000–December 2011) reveals a clear seasonal cycle of the upwelling process with coldest temperature observed in July during the southwest monsoon. A second minimum temperature is observed in January as a result of the northeast monsoon and the reversal of the Somali current direction and cooler SST flowing south from the northern Indian Ocean. In spite of some inter-annual variability in the seasonal signal, no significant trend is observed over this MODIS decade. Image and plots are extracted from http://gmis.jrc.ec.europa.eu/using the 'timeseries' analysis tool

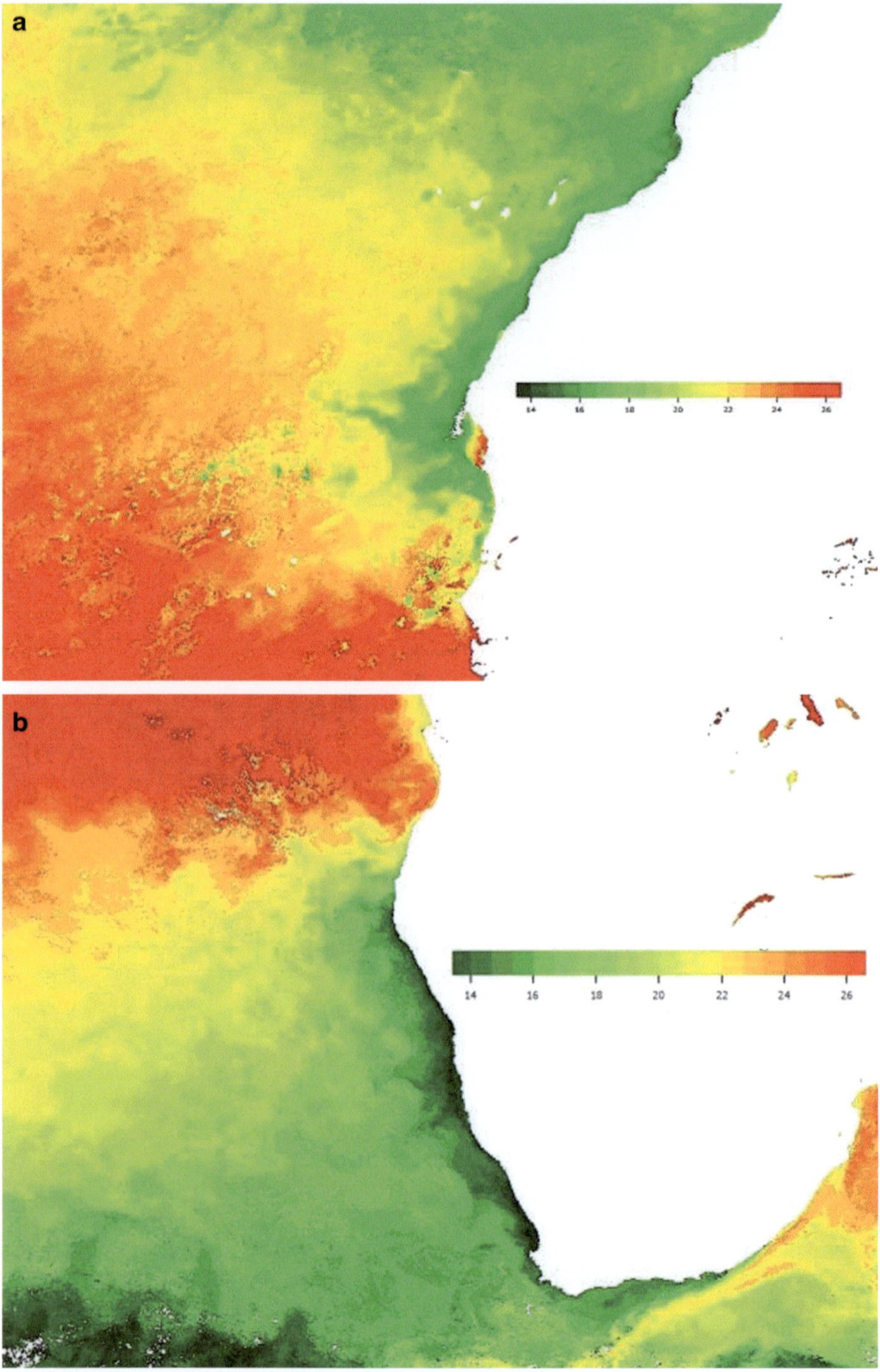

Fig. 3.4 SST (in°C) monthly composite (June 2004) from MODIS Terra instrument showing the extension and mesoscale features of coastal upwelling systems in northwest Africa (**a**) and the Benguela Current (**b**)

highest filament activities associated with Cape Bojador (26.1°N), Cape Ghir (30.5°N), and Cape Blanc (21°N).

In the south, Meeuwis and Lutjeharms (1990) also used AVHRR data set to investigate surface thermal characteristics of the Benguela upwelling system, showing similar permanent feature with seasonal variations depending on the location. The Benguela upwelling (Fig. 3.4b) extends from the southern tip of South Africa to the Angola front, with maximum intensity observed off Luderitz (Namibia). Owing to TIR satellite sensors, Lutjeharms et al. (1991) showed evidence of a complex structure of filaments exceeding 1000 km in length. Obviously, the spatial extent and time evolution of these filaments could not have been made using traditional ship surveys. Similar filaments and eddies are clearly observed in Fig. 3.4a (a MODIS SST monthly composite) along the coast of South Africa and Namibia with colder temperature extending several hundreds of km offshore.

In addition to map the spatio-temporal variability of upwelling events through sequential SST images, TIR data have been extensively used to estimate an upwelling index consisting in most cases of the temperature difference observed between coastal water and water further offshore at a given latitude (Nykjaer and Van Camp 1994). Figure 3.5 illustrates the seasonal cycle of the upwelling index in Northwest Africa as observed with AVHRR SST for the 1981–1989 period. Upwelling process is persistent throughout the year between 20°N and 26°N, with maximum intensity during the summer. South of 20°N, upwelling of lower intensity would mostly occur in winter, whereas north of 28°N, upwelling cells of significant intensity would take place during summer and autumn.

A similar scenario was observed using AVHRR SST data for the 1987–2006 period (Marcello et al. 2011), together with some evidence of an upwelling intensification during this period along the whole northwest African coast as a result of an increase in the water temperature offshore with respect to coastal waters. A direct link to meteorological forcing would expose upwelling systems to higher sensitivity to climate disturbances. Warming is associated with a stronger pressure gradient between land and ocean, which in turn, would reinforce alongshore geostrophic wind, hence coastal upwelling (Bakun 1990). The previous observation and this latter theory is somehow contrasting with another study showing a significant decrease in upwelling intensity within the Canary upwelling system between 1967 and 2006 as calculated from wind-derived Ekman transport (Gomez-Gesteira et al. 2008). A similar weakening of the upwelling intensity in the Iberian/Canary system is observed in the NCEP/NCAR time series resulting from hindcast modeling using observations from different platforms, including TIR satellite data, over the last 4 decades (Pardo et al. 2011). The same data, on the other hand, would show an enhancement of upwelling in the Benguela region. The impact of climate change on upwelling systems still relies on divergent indications. Santos et al. (2005), for example, argued that the variability of AVHRR-derived SST over 2 decades (1980s and 1990s) in the Canary Current is better explained by a breakpoint in the upwelling regime intensity as opposed to a linear long-term trend. Such a debate underlines the importance of maintaining a long-term continuity of satellite TIR data, and their subsequent combination with numerical models to get a better understanding of the processes in place.

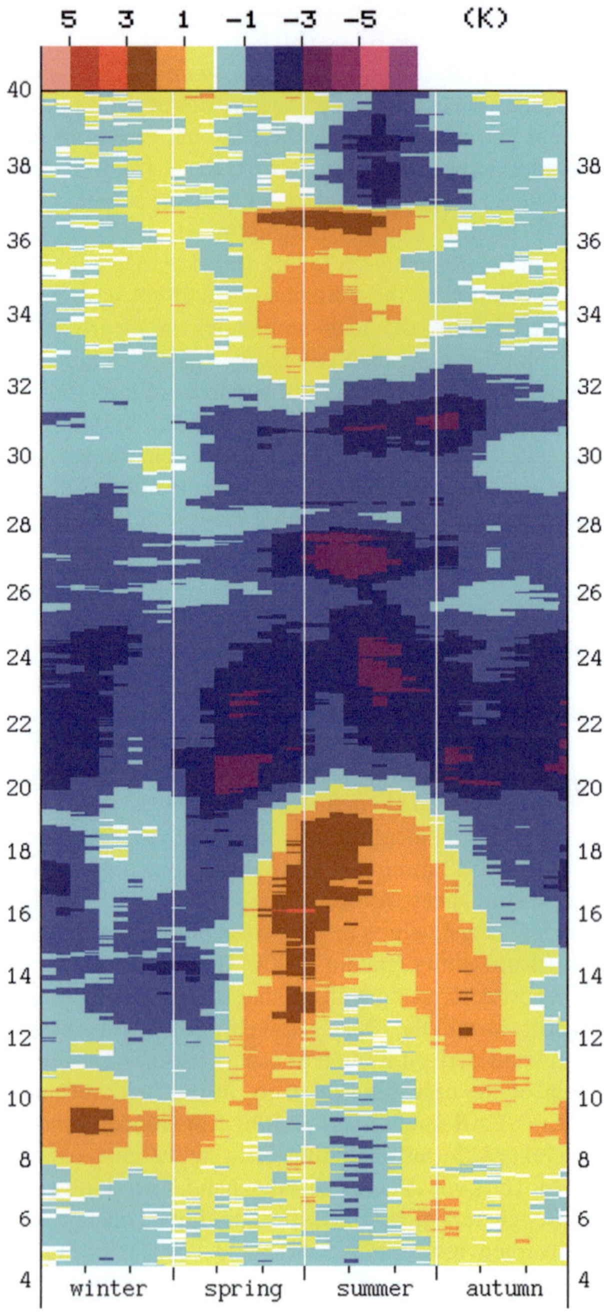

Fig. 3.5 Latitudinal-time section of SST upwelling index as defined by the zonal difference in TIR data for the period July 1981 to August 1989 (adapted from Nykjaer and van Camp 1994). Negative values refer to coastal temperature lower than 500 km offshore at given latitude

3.5 Applications to Fisheries & Marine Resources Management

Sustainable fisheries management practices are high priority on the development agenda for coastal African countries because they directly affect food security for the local population. Recent advances in data acquisition systems, analysis methods and communication technologies have considerably promoted the use of satellite data in the field of fisheries for both, near-real-time support of the day-to-day fishing activities and in the analysis of long-term variability in fish populations. Such synoptic observation of the sea surface is useful in several ways for the fisheries and aquaculture industries: for resource management, optimization of harvesting strategies, better understanding of the resource behavior and its spatial/temporal variability, site protection against recurrent phenomena (e.g. harmful algal blooms) and safety precautions. Chassot et al. (2011) have compiled a comprehensive review showing the importance of satellite remote sensing, including TIR images, to characterize the habitat and ecosystem properties that influence marine resources at large temporal and spatial scales, thus contributing to the development of an ecosystem approach to fisheries management.

As mentioned in the previous section, upwelling systems on the west coast of Africa support large populations of pelagic and demersal fish of considerable economic importance to regional and global fisheries. According to Pauly and Christensen (1995), coastal upwelling systems represents less than 1 % of the world oceanic area but accounts for ca. 20 % of the global fish catch.

In the Benguela-Angolan system, the catch of small pelagic fishes fluctuated greatly after 1980, including a dramatic collapse of the Namibian sardine from 1994 to 1996, despite the application of sound management principles. The observed variability was to a large extent due to the environment and not to fishing activities, as it is often the case in other modern fishery grounds. One of the key objectives of the EU ENVIFISH project (1998–2002; see Barange and Nykjaer 2003) was the identification and quantification of major environmental conditions that influence the recruitment and distribution of small pelagic fish stocks in the Benguela-Angola system. A substantial part of the scientific research in this project relied on TIR imagery derived for the period 1982–1999, starting with a synoptic-scale description of the South East Atlantic (Hardman-Moutford et al. 2003), and detailed SST pattern recognition using neural network method (Richardson et al. 2003).

Following the success of the ENVIFISH project, a similar analysis of the small pelagic fisheries was performed in the North West African upwelling system through the EU project NAT-FISH carried out between 2002 and 2005. As for its southern counterpart, NAT-FISH objectives was to identify and quantify environmental variability related to significant changes in abundance and distribution of small pelagic fish stocks, but also to investigate the potential of using models in combination with satellite SST data as a tool for suggesting precautionary measures to be implemented into responsible fishery management strategies. The project gave the opportunity to scientists in developing countries to gain insight into the use of long time-series of satellite-derived oceanographic variables otherwise not available at the level of their institutions.

Taking advantage of the high spatial and temporal resolution of thermal sensors, the detection and monitoring of thermal fronts from composite SST fields can contribute to develop forecasting capabilities of fish population and their movements (Agenbag et al. 2003), and subsequently, improve the harvest of the stocks in an effective and sustainable way. Temperature is indeed an important factor determining the distribution of fishes, as many of them tend to migrate along preferred temperature ranges. Lan et al. (2011) observed maximum aggregation of yellowfin tuna in equatorial Atlantic waters with temperature above 24–25 °C. According to Zainuddin et al. (2008), the distribution of albacore tuna in the northwestern Pacific is narrowly associated with a combination of environmental parameters, which can be described from thermal imagery, ocean colour, and altimetry. Specific isopleths in the sea surface temperature coincide with highest catch per unit efforts, underlining the benefit to monitor these temperature gradients to generate maps of potential fishing ground and predict spatial patterns of albacore habitats.

Also frontal structures associated with meanders and eddies are potential sites for fish aggregation due to increasing productivity at lower trophic levels (Bakun 2006). Consequently, TIR images are often used to identify so-called Potential Fishing Zones (PFZs) through the detection of SST gradients as generated by these oceanic features. The maps are then disseminated to fishermen in real time to help in reducing search time for fishing grounds, in saving on fuel and, thus, in reducing operational costs (Dwivedi et al. 2005). Moreover, the system could improve the quality of the catch and the sustainability of the stocks, targeting on adults populations rather than juvenile and nursery grounds, otherwise endangered by a blind and unsupervised harvesting process. This operational process to support fisheries is not yet very much used in Africa when compared to Asian countries where the benefit-to-cost ratio of fishing activity largely improves when using satellite data, combining thermal and visible imagery.

As part of the AMESD programme, a Thematic Action related to 'Coastal and Marine Management' was recently implemented through the Mauritius Oceanography Institute (MOI), thus acting as AMESD regional Implementation Centre. The main objective of this Action is to support the governments and institutions of the Indian Ocean Commission (IOC) and eastern African countries to make better use of Earth Observation data in their marine and coastal policies. Accordingly, following an agreement with EUMETSAT, MOI is regularly receiving daily chlorophyll and SST data at 1 km resolution to provide IOC users with oceanographic charts for the detection of Potential Fishing Zones. Figure 3.6 is an extract of such operational service showing sequential daily maps of MODIS SST in the Mozambique Channel. The data were further analysed to identify frontal structures favourable to aggregations of important commercial fishes. Using TIR data and other satellite sensors, Tew Kai and Marsac (2010) underlined the importance of eddy boundaries to monitor the distribution of top predators in this region.

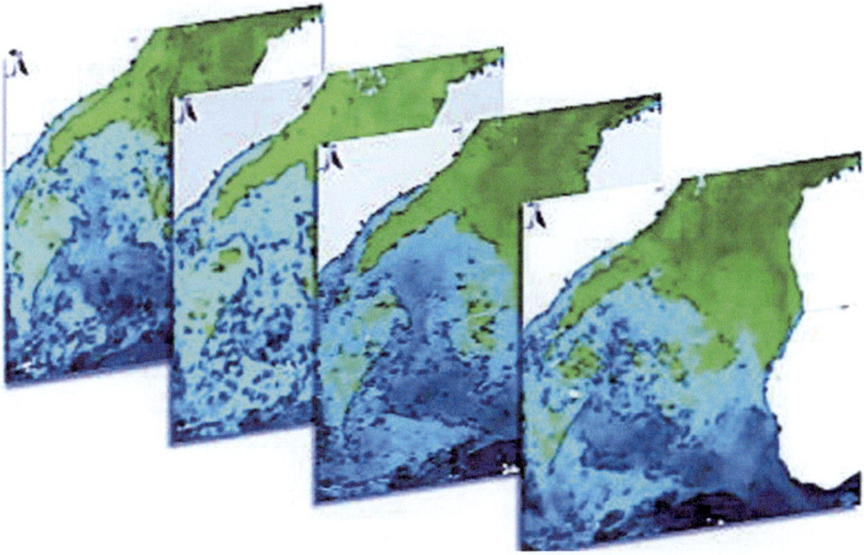

Fig. 3.6 Daily maps of SST (5 to 8 July 2012) as retrieved from MODIS Aqua instrument. Frontal features have been identified using an edge detection algorithm specifically adapted to SST images (Cayula and Cornillon (1992) further improved by Belkin and O'Reilly (2009). The images (courtesy: Vimal Ramchandur, MOI) are part of an operational service to be implemented within a sustainable management framework under the authority of Ministries of Fisheries of countries within the Indian Ocean Commission and nearby eastern African countries

3.6 Conclusion

Data from the thermal-infrared sensors based on satellite have been essential to characterize the complex and highly dynamic marine water properties around Africa. According to the Intergovernmental Panel on Climate Change (IPCC 2007), Africa is considered as one of the most vulnerable continents to climate change and climate variability with a large range of consequences for the environmental and social sectors. Over the last 2 or 3 decades, the continuing measurements of sea surface temperature from space represent by itself a simple and effective indicator of climate variability, identifying 'hot-spots' where SST changes are more pronounced than the average condition. Other derived information such as upwelling frequency and intensity, variability of mesoscale eddies, frontal structures are particularly important to assess the alterations of ecosystems and the impacts on biodiversity. Integrating TIR SST data in fisheries information system in relation to the spatial distribution of fishing grounds, spawning areas, nursery areas, represents thus an immediate step to take in order to move toward an ecosystem-based approach of ruling fisheries, with the possible identification and monitoring of additional Marine Protected Areas.

Acknowledgements Most of the data and maps were extracted from the Global Marine Information System web site, maintained as part of the institutional activities of the Water Resources Unit of the Institute for Environment and Sustainability (European Commission, Joint Research Centre).

We would like to thank the Ocean Biology Processing Group of National Aeronautics and Space Administration (NASA) for the distribution of MODIS SST products. Special thanks also directed to Vimal Ramchandur and the Mauritius Oceanographic Institute for providing regional maps of Potential Fishing Zones and comments on associated marine operational service.

References

Agenbag JJ, Richardson AJ, Demarcq H, Fréon P, Weeks S, Shillington FA (2003) Estimating environmental preferences of South African pelagic fish species using catch size- and remote sensing data. Prog Oceanogr 59: 275–300

Bakun A (1990) Global climate change and intensification of coastal ocean upwelling. Science 247(4939):198–201

Bakun A (2006) Fronts and eddies as key structures in the habitat of marine fish larvae: opportunity, adaptive response and competitive advantage. In: Olivar MP, Govoni JJ (eds) Recent advances in the study of fish eggs and larvae. Scientia Marina 70S2: 105–122

Barange M, Nykjaer L (2003) ENVIFISH: an EU/Southern African collaborative project investigating environmental causes of pelagic fisheries variability. Prog Oceanogr 59:177–179

Barton ED, Aristegui J, Tett P, Canton M, Garcia-Braun J, Hernandez-Leon S, Nykjaer L, Almeida C, Almunia J, Ballesteros S, Basterretxea G, Escanez J, Garcia-Weill L, Hernandez-Guerra A, Lopez-Laatzen F, Molina R, Montero MF, Navarro-Perez E, Rodriguez JM, van Lenning K, Velez H, Wild K (1998) The transition zone of the Canary Current upwelling region. Prog Oceanogr 41(4):455–505

Beal LM, De Ruijter WPM, Biastoch A, Zahn R, SCOR/WCRP/IAPSO WG136 (2011) On the role of the Agulhas system in ocean circulation and climate. Nature 472:429–436

Belkin IM, O'Reilly JE (2009) An algorithm for oceanic front detection in chlorophyll and SST satellite imagery. J Mar Syst 78:319–326

Brown OB, Bruce J, Evans RH (1980) Sea surface temperature evolution in the Somali basin during the southwest monsoon 1979. Science 209:595–597

Bryden HL, Beal LM, Duncan LM (2005) Structure and transport of the Agulhas current and its temporal variability. J Oceanogr 61:479–492

Cayula J-F, Cornillon P (1992) Edge detection algorithm for SST images. J Atmos Ocean Tech 9:67–80

Chassot E, Bonhommeau S, Reygondeau G, Nieto K, Polovina JJ, Huret M, Dulvy NK, Demarcq H (2011) Satellite remote sensing for an ecosystem approach to fisheries management. ICES J Mar Sci. doi:10.1093/icesjms/fsq195

Chen Z, Yan X-H, Jo Y-H, Jiang L, Jiang Y (2012) A study of Benguela upwelling system using different upwelling indices derived from remotely sensed data. Cont Shelf Res, in press

Clerici M, Hoepffner N, Diop M, Ka A, Kirugara D, Ndungu J (2009) SST derivation from MSG for PUMA Pilot Porject in Fisheries. Int J Remote Sens 30(8):1941–1959

Cury P, Bakun A, Crawford RJM, Jarre A, Quiones RA, Shannon LJ, Verheye HM (2000) Small pelagics in upwelling systems: patterns of interaction and structural changes in "wasp-waist" ecosystems. ICES J Mar Sci 57:603–618

De Ruijter WPM, Ridderinkhof H, Ltujeharms JRE, Schouten MW, Veth C (2002) Observations of the flow in the Mozambique channel. Geophys Res Let 29:1401–1403

Donlon CJ (2010) Sea surface temperature measurements from thermal infrared satellite instruments: status and outlook. In: Barale V, Gower JFR, Alberotanza L (eds) Oceanography from Space. doi:10.1007/978-90-481-8681-5_13. Springer, Science + Business Media B.V, pp 211–227

Donlon CJ, Casey KS, Robinson IS, Gentemann CL, Reynolds RW, Barton I, Arino O, Stark J, Rayner N, LeBorgne P, Poulter D, Vazquez-Cuervo J, Armstrong E, Beggs H, Llewellyn-Jones D, Minnett PJ, Merchant CJ, Evans R (2009) The GODAE high resolution sea surface temperature pilot project (GHRSST-PP). Oceanography 22(3):34–45

Donlon CJ, Robinson IS, Casey KS, Vazquez-Cuervo J, Armstrong E, Arino O, Gentemann C, May D, LeBorgne P, Piollé J, Barton I, Beggs H, Poulter DJS, Merchant CJ, Bingham A, Heinz S, Harris A, Wick G, Emery B, Minnett P, Evans R, Llewellyn-Jones D, Mutlow C, Reynolds R, Kawamura H, Rayner N (2007) The global Ocean data Assimilation Experiment (GODAE) high resolution sea surface temperature pilot project (GHRSST-PP). Bull Amer Meteor Soc 88(8):1197–1213

Dwivedi RM, Solanki HU, Nayak S, Gulati D, Sonvanshi VS (2005) Exploration of fishery resources through integration of ocean colour and sea surface temperature. Indian J Mar Sci 34(4):430–440

Evans RH, Brown OB (1980) Propagation of thermal fronts in the Somali current system. Deep-Sea Res 28A(5):521–527

Fréon P, Coetzee JC, Van der Lingen CD, Connell AD, O'Donoghue SH, Roberts MJ, Demarcq H, Attwood CG, Lamberth SJ, Hutchings L (2010) A review and tests of hypotheses about causes of the Kwazulu-Natal sardine run. Afr J Mar Sci 32(2):449–479

Gargett AE (1997) Physics to fish: interactions between physics and biology on a variety of scales. Oceanography Mag 10(3):128–131

Gomez-Gesteira M, de Castro M, Alvarez I, Lorenzo MN, Gesteira JLG, Crespo AJC (2008) Spatio-temporal upwelling trends along the Canary Upwelling System (1967–2006). Trends Dir Clim Res: Ann NY Acad Sci 1146:320–337

Hagen E, Zulicke C, Feistel R (1996) Near-surface structures in the Cape Ghir filament off Morocco. Oceanol Acta 19(6):577–598

Hardman-Mountford NJ, Richardson AJ, Agenbag JJ, Hagen E, Nykjaer L, Shillington FA, Villacastin C (2003) Ocean climate of the South East Atlantic observed from satellite data and wind models. Prog Oceanogr 59(2–3):181–223

Hill AE, Hickey BM, Shillington FA, Strub PT, Brink KH, Barton ED, Thomas AC (1998) Eastern ocean boundaries coastal segment (E). In: Robinson AR, Brink KH (eds) The global coastal ocean, regional studies and syntheses, the sea, Vol 11. Wiley, New York, pp 29–67

Jungbluth G, Fulton R, Moodie L, Seymour P, Williams M, Wolf L, Zhang J (2011) GEONETCast: global satellite data dissemination and the technical and social challenges. GSA Spec Pap 482:77–85

Kearns EJ, Hanafin JA, Evans RH, Minnett PJ, Brown OB (2000) An independent assessment of pathfinder AVHRR sea surface temperature accuracy using the marine atmosphere emitted radiance interferometer (M-AERI). Bull Am Meteorol Soc 81:1525–1536

Kilpatrick KA, Podesta GP, Evans R (2001) Overview of the NOAA/NASA advanced very high resolution radiometer Pathfinder algorithm for sea surface temperature and associated matchup database. J Geophys Res C 106:9179–9197

Kumar A, Minnett PJ, Podesta GP, Evans RH (2003) Error characteristics of the atmospheric corrections algorithms used in retrieval of sea surface temperatures from infrared satellite measurements: global and regional aspects. J Atmos Sci 60:575–585

Lan K-W, Lee M-A, Lu H-J, Shieh W-J, Lin W-K, Kao S-C (2011) Ocean variations associated with fishing conditions for yellowfin tuna (*Thunnus albacares*) in the equatorial Atlantic Ocean. ICES J Mar Sci 68(6):1063–1071

Lutjeharms JRE (1981) Features of the southern Agulhas current circulation from satellite remote sensing. S Afr J Sci 77:231–236

Lutjeharms JRE (1988) Meridional heat transport across the sub-tropical convergence by a warm eddy. Nature 331:251–254

Lutjeharms JRE (2006) The Agulhas current. Springer, Berlin

Lutjeharms JRE (2007) Three decades of research on the greater Agulhas Current. Ocean Sci 3:129–147

Lutjeharms JRE, Roberts HR (1988) The natal pulse: an extreme transient on the Agulhas current. J Geophys Res C1(93):631–645

Lutjeharms JRE, Boebel O, Rossby HT (2003) Agulhas cyclones. Deep Sea Res II 50:13–34

Lutjeharms JRE, Shillington FA, Duncombe Rae CM (1991) Observations of extreme upwelling filaments in the South East Atlantic Ocean. Science 253:774–776

Mafimbo AJ, Reason CJC (2010) Air-sea interaction over the upwelling region of the Somali coast. J Geophys Res 115:1–11

Marcello J, Hernandez-Guerra A, Eugenio F, Fonte A (2011) Seasonal and temporal study of the northwest African upwelling system. Int J Remote Sens 32(7):1843–1859

Marullo S, Nardelli BB, Guarracino M, Santoleri R (2007) Observing the Mediterranean Sea from space: 21 years of Pathfinder-AVHRR sea surface temperature (1985 to 2005): re-analysis and validation. Ocean Sci 3:229–310

Meeuwis JM, Lutjeharms JRE (1990) Surface thermal characteristics of the Angola-Benguela front. S Afr J Mar Sci 9:261–280

Merchant CJ, Le Borgne P, Marsouin A, Roquet H (2008) Optimal estimation of sea surface temperature from split-window observations. Rem Sens Env 112(5):2469–2484

Nieto K, Demarcq H, McClatchie S (2012) Mesoscale frontal structures in the Canary Upwelling System: new front and filament detection algorithms applied to spatial and temporal patterns. Rem Sens Env 123:339–346

Nykjaer L (2009) Mediterranean Sea surface warming 1985–2006. Clim Res 30:11–17

Nykjaer L, Van Camp L (1994) Seasonal and interannual variability of coastal upwelling along northwest Africa and Portugal from 1981 to 1991. J Geophys Res 99:14197–14207

O'Carrol AG, Eyre JR, Saunders RW (2008) Three-way error analysis between AATSR, AMSR-E and in situ sea surface temperature observations. J Atmos Oceanic Technol 25:1197–1207

Odada EO (2010) Integration of coastal and marine areas into sustainable development strategies: A case study of Africa. J Oceanogr Mar Sci 1(3):40–52

Olson DB, Evans RH (1986) Rings of the Agulhas. Deep-Sea Res 33(1):27–42

Pardo PC, Padín XA, Gilcoto M, Farina-Busto L, Pérez FF (2011) Evolution of upwelling systems coupled to the long-term variability in sea surface temperature and Ekman transport. Clim Res 48:231–246

Pauly D, Christensen V (1995) primary production required to sustain global fisheries. Nature 374:255–257

Pelegri JL, Marrero-Diaz A, Ratsimandresy A, Antoranz A, Cisneros-Aguirre J, Gordo C, Grisolia D, Hernandez-Guerra A, Laiz I, Martinez A, Parrilla G, Perez-Rodriguez P, Rodriguez-Santana A, Sangra P (2005) Hydrographic cruises off northwest Africa: the Canary current and cape Ghir region. J Mar Syst 54:39–63

Richardson AJ, Risien C, Shillington FA (2003) Using self-organizing maps to identify patterns in satellite imagery. Prog Oceanogr 59:223–241

Roberts MJ, Van der Lingen CD, Whittle C, Van den Berg M (2010) Shelf currents, lee-trapped and transient eddies on the inshore boundary of the Agulhas Current, South Africa: their relevance to the Kwazulu-Natal sardine run. Afr J Mar Sci 32:423–447

Robinson I (2004) Measuring the oceans from space: the principles and methods of satellite oceanography. Springer/Praxis, Berlin. ISBN 3-540-42647-7, p 670

Rouault MJ, Penven P (2011) New perspectives on Natal Pulses from satellite observations. J Geophys Res 116:C07013. doi:10.1029/2010JC006866

Santos AMP, Kazmin AS, Peliz A (2005) Decadal changes in the Canary upwelling system as revealed by stallite observations: their impact on productivity. J Mar Res 63:359–379

Schott F (1983) Monsoon response of the Somali current and associated upwelling. Prog Oceanogr 12:357–381

Schouten MW, de Ruijter WPM, van Leeuwen PJ, Ridderinkhof H (2003) Eddies and variability in the Mozambique channel. Deep Sea Res II 50:1987–2003

Swart NC, Lutjeharms JRE, Ridderinkhof H, de Ruijter WPM (2010) Observed characteristics of Mozambique channel eddies. J Geophys Res 115:C09006. doi:10.1029/2009JC005875

Tew Kai E, Marsac F (2010) Influence of mesoscale eddies on spatial structuring of top predators' communities in the Mozambique channel. Prog Oceanogr 86:214–223

Troupin C, Mason E, Beckers J-M, Sangrà P (2012) Generation of the Cape Ghir upwelling filament: a numerical study. Ocean Model 41:1–15

Tyson P, Odada E, Schulze R, Vogel C (2002) Regional-global change linkages: Southern Africa. In: Tyson P, Fuchs R, Fu C, Lebel L, Mitra AP, Odada E, Perry J, Steffen W, Virji H (eds) Global-regional linkages in the Earth system. IGPB Series, Springer, Chapter 2 pp 3–64

Van Camp L, Nykjaer L, Mittelstaedt E, Schlittenhardt P (1991) Upwelling and boundary circulation off Northwest Africa as depicted by infrared and visible satellite observations. Prog Oceanogr 26(4):357–402

Veldhuis MJW, Kraay GW, Van Bleijswijk JDL, Baars MA (1997) Seasonal and spatial variability in phytoplankton biomass and growth in the northwestern Indian Ocean: the southwest and northeast monsoon, 1992–1993. Deep Sea Res I 44(3):425–449

Whittle C, Lutjeharms JRE, Duncombe Rae CM, Shillington FA (2008) Interaction of Agulhas filaments with mesoscale turbulence: a case study. S Afr J Sci 104:135–139

Zainuddin M, Saitoh K, Saitoh SI (2008) Albacore (*Thunnus alalunga*) fishing ground in relation to oceanographic conditions in the western north Pacific Ocean using remotely sensed satellite data. Fish Oceanogr 17:61–73

Chapter 4
Remote Sensing of African Coastal Waters Using Active Microwaves Instrument

Werner Alpers

Abstract Active microwave instruments flown on satellites to remotely sense the ocean are the radar altimeter, the scatterometer, and the Synthetic Aperture Radar (SAR). While radar altimeter data are mainly used as input for general ocean circulation models and for wind and global wave forecast, scatterometer and SAR data are well suited to investigate singular events in the marine boundary layer and in the ocean. Examples of atmospheric and oceanic phenomena observed by scatterometers and SARs over African coastal waters are presented. Concerning atmospheric phenomena, first near-surface wind fields derived from scatterometer data acquired over a cylone in the Mozambique Channel and over a wind front west of Morocco are presented. Then SAR images showing sea surface signatures of an atmospheric front, atmospheric gravity waves and a wind jet are presented. Quantitative information on the near-surface wind field is derived from two of the four SAR images shown. Concerning oceanic phenomena, three SAR images are presented. The first two SAR images show sea surface signatures of internal waves generated in the Strait of Gibraltar and at the Atlantic shelf west of Morocco and the third one shows sea surface signatures of biogenic surface films in the upwelling region off the coast of Senegal.

4.1 Introduction

Microwaves are electromagnetic waves with wavelengths between one millimeter and 1 m, or equivalently, with frequencies between 300 MHz (0.3 GHz) and 300 GHz. However, microwave instruments used for remote sensing of the ocean from satellites use only microwaves with wavelengths above 2 cm (below 15 GHz) because above this wavelength the microwaves are very little affected by clouds and (ligth) rain. Thus microwave instruments can provide information on the ocean (almost) independent of weather conditions. Depending on their mission objective, microwave instruments use different frequency bands. Their names are L-band (1–2 GHz, 30–15 cm), S-band (2–4 GHz, 15–7.5 cm), C-band (4–8 GHz, 7.5–3.8 cm), X-band (8–12 GHz,

W. Alpers (✉)
Institut für Meereskunde, Universität Hamburg,
Hamburg, Germany
e-mail: werner.alpers@zmaw.de

V. Barale, M. Gade (eds.), *Remote Sensing of the African Seas*,
DOI 10.1007/978-94-017-8008-7_4,

3.8–2.5 cm), and Ku-band (12–18 GHz, 2.5–1.7 cm). These strange names for the frequency bands are leftovers of World War 2, when microwave technology was developed in different countries and when frequency bands were top-secret.

There exist two kinds of microwave instruments used for remote sensing: passive and active instruments. Passive instruments, called microwave radiometers, measure the emitted microwave radiation from an object at the Earth surface, while active instruments emit microwave pulses and receive them after backscattering from the object. The active microwave instruments are called radars, an abbreviation for Radio Detection and Ranging. Three different types of radars are employed from space for remote sensing of the ocean: (1) scatterometers for measuring near-surface wind vectors over the ocean on a global scale; (2) Synthetic Aperture Radars (SARs) for measuring ocean surface waves, meso-scale phenomena in the ocean and in the marine boundary layer, near-surface ocean winds, and surface currents; and (3) radar altimeters for measuring geostrophic ocean currents, wave heights, and near near-surface wind speeds (not wind vectors).

The microwaves do not penetrate into the water body and thus microwave instruments can only provide information on the ocean and the marine boundary layer via sea surface characteristics. In the case of active microwave instruments, all information is extracted from the sea surface roughness and from the time a radar pulse needs to travel the distance radar- sea surface and back. In the case of passive microwave instruments (radiometers), all information is extracted from the microwave radiation emitted from the sea surface at different frequencies and polarizations. The emitted radiation (or the brightness temperature) contains information on sea surface roughness (and thus the near-surface wind speed), Sea Surface Temperature (SST) and Sea Surface Salinity (SSS).

Remote sensing of the ocean became a large boost from the American Seasat mission in 1978 (Evans et al. 2005). This satellite carried a scatterometer, a SAR, a radar altimeter, and a microwave radiometer. Although this mission only lasted for about three months, it provided the oceanographic community with a wealth of data that motivated the European Space Agency (ESA) and several other space agencies to launch satellites carrying similar active microwave sensors, among them (1) scatterometers onboard the European ERS-1, ERS-2 and MetOp satellites, the Japanese ADEOS satellite, the American Quikscat satellite, the Indian OceanSat-2 satellite, and the Chinese HY-2A satellite; (2) SARs onboard the European ERS-1, ERS-2, and Envisat satellites, the Russian Almaz-1 satellite, the Canadian Radarsat-1 and Radarsat-2 satellites, the Japanese JERS-1 and ALOS satellites, the German TerraSAR-X and TanDEM X satellites, the Italian COSMO-Skymed satellites, the Indian RISAT-1, and Risat-2 satellites, and the Chinese HJ-1C satellite; and (3) radar altimeters onboard the European ERS-1, ERS-2, Envisat, the French-American Topex-Poseidon, Jason-1 and Jason-2 satellites, and the Indian-French SARAL/AltiKa satellite.

In the following, we shall present only scatterometer and SAR data that were acquired over coastal areas of the African continent. Radar altimeter data are less suited for investigations of coastal areas: they unfold their full value only when assimilated in general ocean circulation models and wind/wave forecast models.

4.2 Scatterometer

A scatterometer is an all-weather instrument which is used from satellites to measure two-dimensional near-surface wind fields (wind vectors). It measures the wind vectors indirectly via the short-scale sea surface roughness. Short-scale roughness means in this context waves having wavelengths of the order of centimeters to decimeters. A scatterometer emits radar pulses from several antennas or from a rotating antenna and records the backscattered radar power or normalized radar cross section (NRCS) from the wind-roughened sea surface. The higher the wind speed, the rougher the sea surface, and the higher is the backscattered radar power or the normalized radar cross section (NRCS). However, the backscattered radar power depends on not only on wind speed, but also on incidence angle and wind direction relative to the antenna look direction (i.e., on azimuth angle). Thus, in order to measure wind speed and direction with a scatterometer, the backscattered radar power from a resolution cell must be measured from different azimuth directions. This is achieved by using several antennas or a rotating antenna.

The scatterometer onboard the American Seasat satellite, called SASS (launched in 1978), and the one on the Japanese satellite ADEOS, called NSCAT (launched in 1996) had 6 antennas, three antennas looking to the right and three to the left of the satellite track. The C-band scatterometer Active Microwave Instrument (AMI) onboard the European satellites ERS-1 and ERS-2, which were launched in 1991 and 1995, respectively, had three antennas pointing only to the one side of the satellite track (to the right side). On the other hand, the Ku-band (13.4 GHz) scatterometer SeaWinds, later dubbed QuikSCAT, onboard the American Quikscat satellite, which was launched in June 1996 and whose mission ended in November 2009, used a rotating antenna with two spot beams that swept in a circular pattern. The instrument collected data in a continuous, 1800-kilometer-wide swath covering 90 % of Earth's surface in one day. Standard Quikscat wind fields have a resolution of 25 km (12.5 km pixel spacing) and are available on-line[1] until 19 November 2009.

At present, two scatterometers are flying in space, the C-band ASCAT (Advanced Scatterometer) onboard the European MetOp satellite (launched in 2006) and the Ku-band OSCAT (Oceansat-2 Scatterometer) onboard the Indian Oceansat-2 satellite (launched in 2009). While ASCAT uses the multiple antenna principle (three antennas looking to each side of the satellite track), OSCAT uses, like SeaWinds, the rotating antenna principle. ASCAT wind fields[2] have a resolution of 25 km (12.5 km pixel spacing) and are measured on two sides of the satellite track on two 500 km wide swathes (strips). OSAT wind fields[3] have a resolution of 50 km and are measured on a 1420 km wide swath.

---

[1] For details, see the websites: http://www.remss.com/qscat/qscat_browse.html, http://www.knmi.nl/scatterometer/qscat_prod/.

[2] See http://www.knmi.nl/scatterometer/ascat_ear_25_prod/ascat_app.cgi.

[3] See http://www.knmi.nl/scatterometer/oscat_50_prod/oscat_app.cgi.

In order to retrieve wind fields from the NRCS values measured by scatterometers (and SARs), one needs a transfer function, usually called Geophysical Model Function (GMF) or wind scatterometer model functions, which relates NRCS values to wind vectors for a neutrally stable atmosphere at a height of 10 m above the sea surface. Several such functions are in use for different scatterometers operating at different radar frequencies. For C-band scatterometers, the most popular GMFs are CMOD4 (Stoffelen and Anderson 1997), CMOD_IFR2 (Quilfen et al. 1998) CMOD5 (Hersbach et al. 2007), and the CMOD5n (Hersbach 2008). The latter one is presently used to retrieve neutral winds (not real 10 m winds) from ASCAT data.

All GMFs saturate at wind speeds around 25 m/s, independent of the radar frequency. Thus it seems impossible to retrieve winds higher than 25 m/s from scatterometer data. However, recently an advanced Ku-band GMF has been developed by Remote Sensing Systems (RSS) (Ricciardulli and Wentz 2011) that seem to allow retrieving also higher winds from Quikscat data. A disadvantage of Ku-band scatterometers compared to C-band scatterometers is that Ku-band microwaves interact strongly with rain drops in the atmosphere (volume scattering and, attenuation) causing biases in the wind retrieval when rain is present.

4.2.1 Tropical Cyclone

Examples of near surface wind fields retrieved from Quikscat data are depicted in Fig. 4.1. These wind maps have been obtained from the website of Remote Sensing Systems[4]. The four wind maps show the passage of the tropical cyclone Flavio from the Indian Ocean towards to coast of Mozambique, East Africa. Flavio made landfall on 22 February 2007 at the Mozambican coast near the city of Inhambane, where it caused considerable damage. Note that the direction of rotation in the cyclone is clockwise, which is typical for a southern-hemisphere cyclone. The scatterometer data were acquired at 03:12 UTC on 18 February 2007 (upper left panel) at 0400 UTC on 20 February (upper right panel), at 1548 UTC on 20 February (lower left panel), and at 1544 UTC on 21 February (lower right panel). The data shown in panels a and b were obtained during descending (morning) satellite passes of the satellite and the one shown in Panels c and d during ascending (evening) passes. Such wind maps are very useful for tracking the position of cyclones. While, in general, the wind speeds and directions derived from Quickscat are quite reliable, this does not apply for the inner regions of a tropical cyclone (or typhoon/hurricane), where the winds can be quite high (above 25 m/s). As mentioned before, retrieving near-surface wind fields with speeds above 25 m/s is questionable at all radar bands because above this value the GMF saturates. At Ku band, the retrieval of typhoon/hurricane winds from scatterometer data is further aggravated by the fact that typhoons/hurricanes are associated with heavy rain. Raindrops in the atmosphere both scatter and attenuate Ku-band microwaves and cause errors in the wind retrieval algorithm.

---

[4] At http://www.ssmi.com/qscat/scatterometer_data_daily.html?rgn=indian_west&size=large.

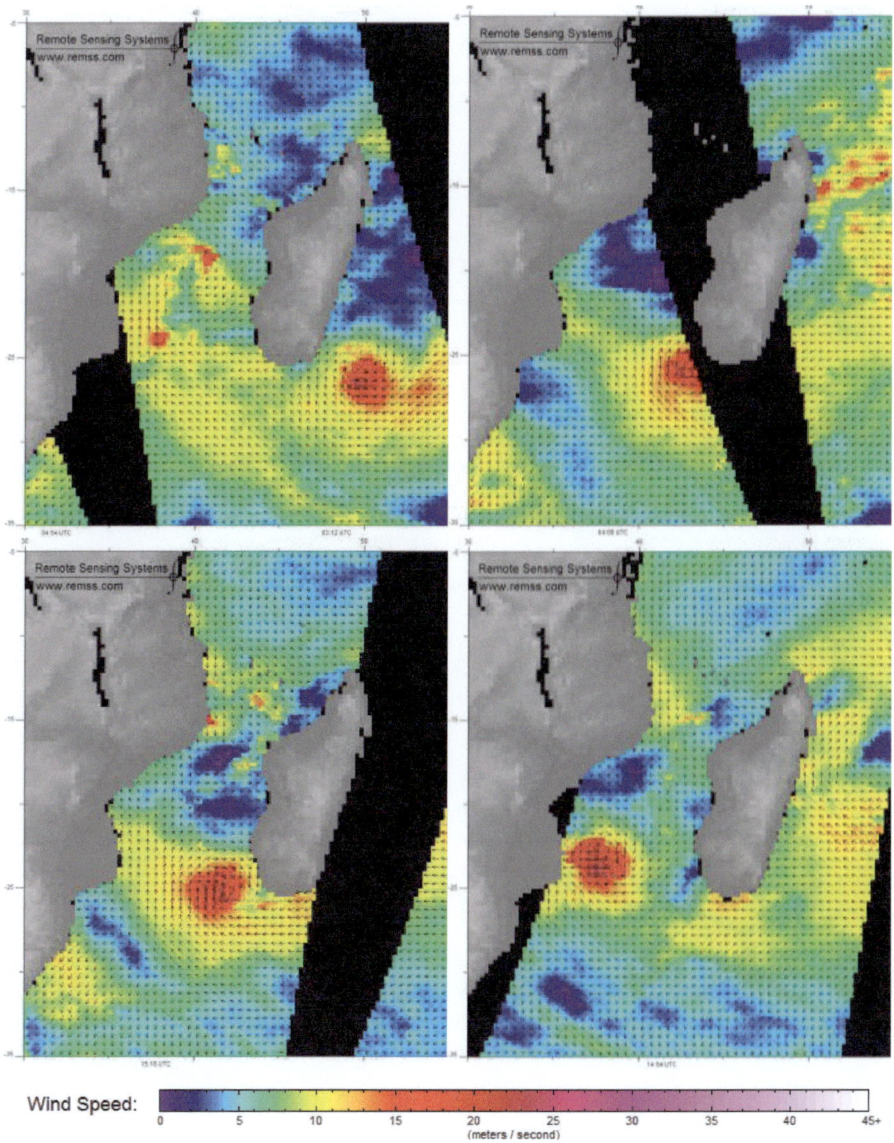

Fig. 4.1 Wind fields measured by Quikscat (*upper left panel*) at 0312 UTC on 18 February and (*upper right panel*) at 0400 UTC on 20 February. 2007 during morning passes, (*lower left panel*) at 1548 UTC on 20 February and (lower right panel) at 1454 UTC on 21 February 2007 during evening passes. They show the wind pattern of the tropical cyclone Flavio propagating eastward from the Indian Ocean, past the southern tip of Madagascar, towards the coast of Mozambique. The grey areas are land. © Remote Sensing Systems

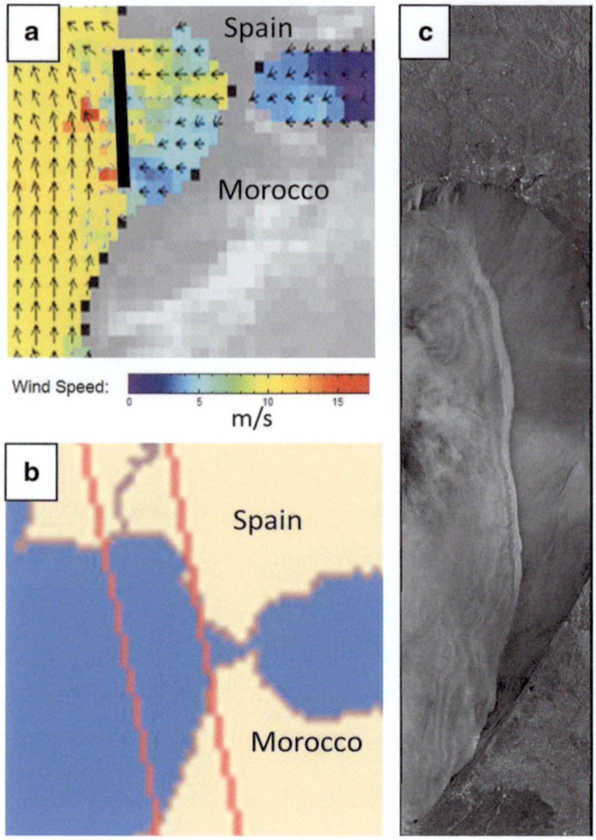

Fig. 4.2 **a** Wind field measured by Quikscat (resolution: 25 km) at 1919 UTC on 9 February 2006 over the Atlantic Ocean off the coast of Morocco and Spain. The inserted thick black line denotes the approximate position of the atmospheric front, where the wind direction and speed change. **b** Location of the SAR swath (width: 100 Km) of Envisat ASAR, which imaged this wind front at 2214 UTC on 9 February 2008. **c** Section of a SAR image acquired by Envisat ASAR showing the atmospheric front at high resolution (25 m). © Remote Sensing Systems, ESA

4.2.2 Wind Front

Wind fronts, defined as fronts where wind direction and speed change, are also detectable by scatterometers. An example is depicted in Fig. 4.2a, which shows the near-surface wind field over the Atlantic Ocean off the west coast of Morocco, North Africa, which was retrieved from Quikscat data acquired at 1919 UTC on 9 February 2006. The poor spatial resolution of 25 km of Quikscat allows only an approximate determination of the position of the atmospheric front, see the black line inserted in the wind map (Fig. 4.2a). However, atmospheric fronts and other atmospheric phenomena in the marine boundary layer are resolved much better by SAR as will be shown in the Sect. 4.3.1. Here we show for comparison a section SAR image which was acquired by the SAR onboard the Envisat satellite, called Advanced Synthetic Aperture Radar (ASAR), 2 h and 55 min later (at 2214 UTC) than the Quikscat data. The SAR image was acquired in the Image Mode (IM) along the swath depicted in Fig. 4.2b (swath width: 100 km) during an asceding pass of the satellite. A section of this SAR is depicted in Fig. 4.2c, which shows the same atmospheric front with a resolution of 25 m. This SAR image shows the position of the wind front much more accurately than the Quikscat wind map and reveals fine-scale structures of the front. Note the wave-like structure of the front (in particular in the northern section), which suggests that the front generated atmospheric gravity waves (see also next section).

4.3 Synthetic Aperture Radar

A SAR is also an active microwave instrument, but can image the sea surface with a much higher resolution than a scatterometer. SAR images acquired over the ocean from satellites have spatial resolutions ranging from few meters to several 100 m depending on the type of SAR and on the data processing. The high resolution is achieved by exploiting the phase history of the backscattered signals and then using a rather complex data processing algorithm to generate the SAR image. The SAR principle works well when stationary targets are imaged, but when the targets are moving, the phase history is distorted, which leads to artifacts in the SAR image. For example, when SAR images a moving train, then, on the SAR image, the train is displaced from the railroad track ("train- off- the track effect"). Since in the presence of ocean surface waves the ocean surface is moving, the SAR image provides a distorted image of the wave field. This is a severe problem, which limits the ability of SAR to provide reliable information on ocean waves (Alpers et al. 1981).

Originally, the main motivation to fly SARs onboard ocean observing satellites was to measure wind generated ocean waves and use the data to improve ocean wave forecast. But even until today, it is still controversial whether assimilation of SAR-derived ocean wave products has a positive impact on wave prediction (Abdallah et al. 2010) because of the motion-induced distortions. However, in the case of swell imaging, the distortion is small when the swell has long wavelength and small amplitude. In this case, wave information can be retrieved from SAR images. Contrasting the poor performance of SAR for ocean wave imaging, SAR has turned out to be a very valuable instrument to image other oceanic phenomena, like internal waves, underwater bottom topography, oceanic fronts, oceanic eddies, and surface film coverage. Furthermore, it is also a very valuable instrument to image atmospheric phenomena in the marine boundary layer, like atmospheric gravity waves, atmospheric fronts, atmospheric eddies/cyclones, and coastal wind fields.

In this paper we present only SAR images acquired by the C-band SARs onboard the ERS-1, ERS-2, and Envisat satellites. The ERS SAR images have always a resolution of 25 m and a swath width of 100 km, while the Envisat SAR images, called ASAR (Advanced SAR) images, have different resolutions: 25 m in the Image Mode (IMM), 150 m in the Wide Swath Mode (WSM) and 1 km in the Global Mode (GM). The swath widths are 100 km for the Image Mode and about 400 km for the Wide Swath Mode and the Global Mode.

4.3.1 Atmospheric Phenomena

Atmospheric phenomena in the marine boundary layer become visible on SAR images of the sea surface by the variations of the short-scale sea surface roughness. The near-surface wind modulates the short-scale sea surface roughness and thus, via Bragg scattering, also the backscattered radar power or the Normalized Radar Cross Section (NRCS) (see e.g. Valenzuela 1978).

On SAR images, the modulation of the NRCS becomes visible as image intensity or image brightness modulation. The high resolution SAR already gives valuable information on the location and the spatial structure of atmospheric phenomena, but more information is extracted from them when the NRCS values are converted into near-surface wind vectors. Such conversion is achieved by applying Geophysical Model Functions (GMFs) or wind scatterometer model functions used in scatterometry (see Sect. 4.2). However, quantitative information on the sea surface wind speed can only be extracted from SAR images when the wind direction near the sea surface is known.

The wind direction is sometimes inferred from the SAR images itself, e.g., from the direction of wind streaks or from the direction of wind shadows behind coastal mountains or mountainous islands (Monaldo et al. 2001, Monaldo and Karbaol 2003; Horstmann and Koch 2005). Another way to obtain wind direction is from atmospheric models, like the one developed by the National Center for Environmental Prediction (NCEP) in the US. Recently, a new wind retrieval has been developed in which also Doppler shifts induced by motions of the sea surface is included (Mouche et al. 2012). This new wind retrieval algorithm uses all three sources of information on wind direction (model, linear features, and Doppler shift) and combines them using the Bayesian method.

However, near-surface wind fields cannot be retrieved from SAR images at the spatial resolution of the SAR because of the speckle inherent in all SAR images. In order to reduce the speckle, one has to carry out averaging over pixels which implies reduction of the spatial resolution. It requires a trade-off between accuracy of the wind field and spatial resolution. Typically, wind fields are retrieved from SAR images with a spatial resolution of 200–500 m.

4.3.1.1 Atmospheric Gravity Waves

Atmospheric Gravity Waves (AGWs), which sometimes are also called atmospheric internal waves, are encountered in the atmosphere when it consists of layers of different densities (potential temperature). They either occur as quasi-periodic waves or as strongly nonlinear waves in the form of solitary waves or wave packets. AGWs are often generated behind mountain ranges in which case they are called lee waves. Usually lee waves are stationary with respect to the terrain feature, but they are propagating relative to the mean air flow above the earth surface.

In Fig. 4.3 an ERS-1 SAR image is depicted which was acquired at 2239 UTC on 3 September 1993 during an ascending satellite pass. It shows in the lower section sea

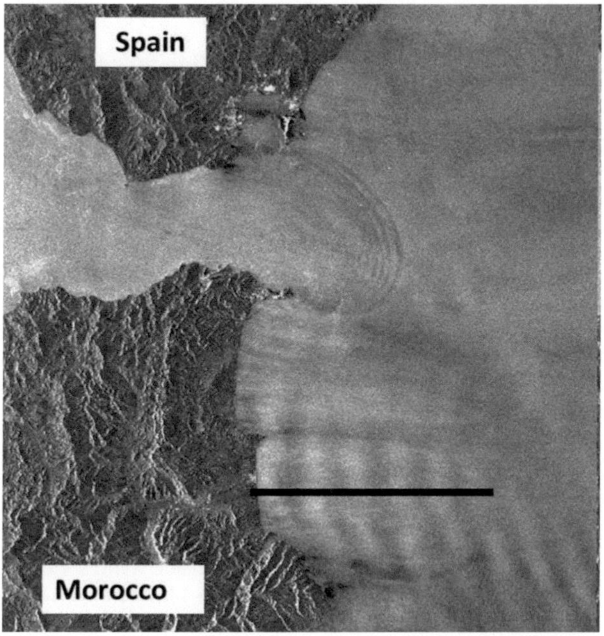

Fig. 4.3 Section of a SAR image acquired by ESR-1 at 2239 UTC on 3 September 1993 showing the area around the Strait of Gibraltar. Visible are in the lower section sea surface signatures of atmospheric gravity waves (lee waves) generated by an eastward wind blowing over the eastern coastal mountain range of Morocco. The inserted black line denotes the transect along which the variation of the normalized radar cross section (NRCS) was determined (see Fig. 4.5). In the upper section also sea surface signatures of oceanic internal waves are visible which were generated in the Strait of Gibraltar. The Imaged area is 100 km x 140 km. © ESA

surface signatures of AGWs. They are lee waves which were generated by a westerly wind blowing with 7 m/s over the coastal mountain range (Er Rif) of eastern Morocco. The inserted black line denotes the transect along which the variation of the NRCS was determined (see Fig. 4.5). In the upper section of the image also sea surface signatures of oceanic internal waves are visible, which propagate out of the Strait of Gibraltar into the Mediterranean Sea.

AGWs become visible on SAR images because they are associated with a varying wind speed at the sea surface which modulates the short-scale sea surface roughness (see e.g. Alpers and Stilke 1996; Cheng and Alpers 2010). This is shown schematically in Fig. 4.4 (adapted from Doyle and Durran 2002): The wavy lines denote streamlines associated with a nonlinear AGW behind a mountain (lee wave). In the shadowed region, the airflow associated with the AGW is directed opposite to the ambient airflow. Note, that in this figure the wind direction is defined as the direction into which the wind is blowing.

The variation of NRCS along the transect is plotted in Fig. 4.5. The left vertical coordinate axis shows the NRCS values and the right axis the near-surface wind

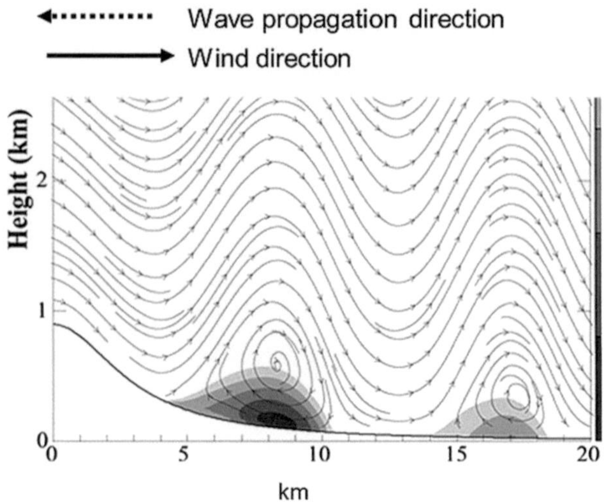

Fig. 4.4 Streamlines associated with a nonlinear atmospheric gravity wave behind a mountain range (lee wave). In the shadowed region, the airflow associated with the wave is opposite to the ambient wind direction (adapted from Doyle and Durran 2002). Here the wind direction is defined as the direction into which the wind is blowing

speed. For converting NRCS values in into near-surface wind speeds, we have applied the wind scatterometer model function CMOD4 and inferred the wind direction from the wind streaks visible on the SAR image. Since for a homogeneous roughness pattern the NRCS decreases with incidence angle and since the incidence angle is larger on the right-hand side of the image than on the left-hand side, the NRCS decreases from left to right on this NRCS plot.

But not all AGWs are lee waves. They also can be generated by horizontal wind shear as visible on the ERS-2 SAR image depicted in Fig. 4.6, which was acquired at 1105 UTC on 6 March 2000. The wind map derived from Quikscat data acquired at 0606 UTC on 6 March 2000 (not shown here) shows an easterly wind of 16 m/s blowing through the Strait of Gibraltar. It was the strong wind which generated

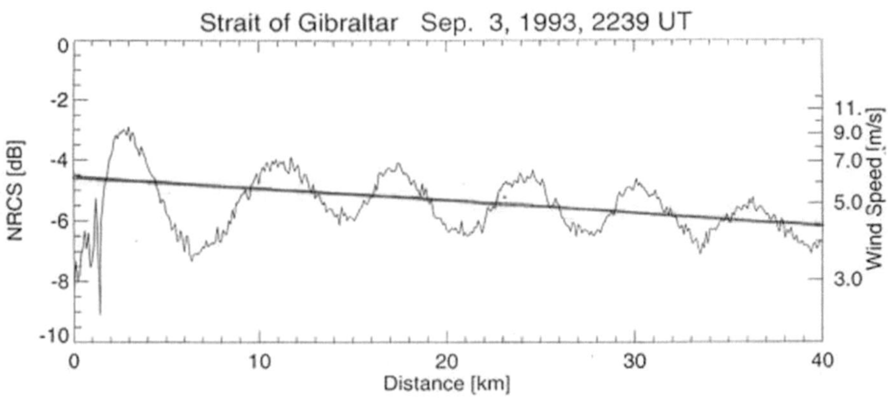

Fig. 4.5 Scan of the normalized radar cross section (NRCS) along the transect (from west to east) inserted in Fig. 4.3. The left vertical coordinate axis shows the NRCS values and the right axis the wind speed at a height of 10 m above sea level as calculated from the wind scatterometer model function CMOD4

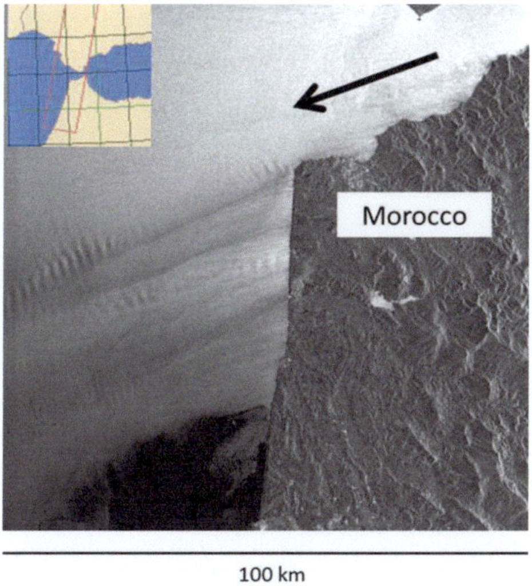

Fig. 4.6 Section of the SAR image acquired by ERS-2 at 1105 UTC on 6 March 2000 over the sea area west of the Strait of Gibraltar. Visible are sea surface signatures of atmospheric gravity waves generated by an easterly wind blowing through the Strait of Gibraltar and over the coastal mountains of Morocco. The arrow denotes the direction into which the wind is blowing, The wave train to the north is generated by a horizontal shear in the wind field and the other wave trains are typically lee waves generated by the interaction of the wind with topographic features at the Moroccan Atlantic coast. The inset shows the SAR swath © ESA

AGWs by horizontal wind shear at the northern coast of Morocco. Note also train of AGWs further south which are lee waves generated by interaction of the wind with topographic features at the Atlantic coast of Morocco.

4.3.1.2 Coastal Wind Jet

In coastal areas often highly variable winds are encountered that cannot be resolved by scatterometers, but can be resolved by SARs. Coastal winds include land/sea breeze, katabatic winds, gap winds, bora winds, and wind jets. Here we present as an example a SAR image that shows a wind jet blowing from the Strait of Gibraltar into in the Mediterranean Sea. The SAR image (Fig. 4.7a) was acquired by Envisat ASAR at 1027 UTC on 3 February 2011 in Wide Swath Mode (swath width: 400 km). Figure 4.7b shows the near-surface wind field retrieved from this image by using the CMOD 4 wind scatterometer model and by taking the wind direction from the NCEP model valid for 0900 UTC, i.e. 1 h and 27 min before the SAR image was acquired. (courtesy NERSC, Bergen). The wind jet originates in the Strait of Gibraltar and blows into the Mediterranean Sea with a speed of up to 13 m/s. It is generated by a confluence of air flow from the north and the south and funneled through the Strait of Gibraltar giving gives rise to quite high winds east of the strait.

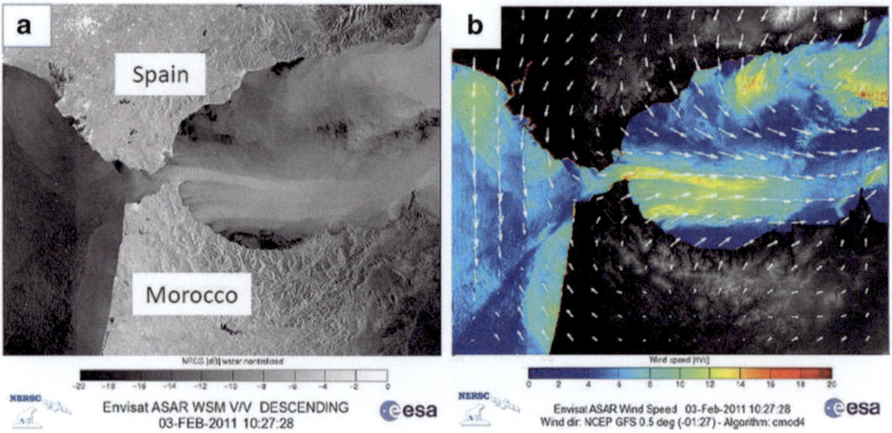

Fig. 4.7 **a** Section of the SAR image acquired by Envisat ASAR at 1027 UTC on 3 February 2011 over the Strait of Gibraltar. **b** Near-surface wind field derived from the SAR image using the wind scatterometer model CMOD4 and the wind direction from the NCEP model for 0900 UTC. It shows a wind jet originating in the Strait of Gibraltar and blowing with a speed of 13 m/s into the Mediterranean Sea. © ESA, NERSC

4.3.2 Oceanic Phenomena

Oceanic phenomena are also often be delineated on SAR images of the sea surface because they modulate the sea surface roughness. The fact that sea surface roughness variations can be indicators of sub-surface phenomena, is known since centuries. For example, mariners used often sea surface roughness variations as an indicator for shallow underwater features. Since SAR is a very sensitive roughness sensor and has all-weather capability, it is a valuable instrument to observe oceanic phenomena. Oceanic phenomena detectable by SAR include ocean surface waves, internal waves, eddies, oceanic fronts, underwater bottom topography, ship wakes, oceanic wakes behind islands, oil slicks, river plumes and upwelling areas. The variation of the sea surface roughness associated with oceanic phenomena can have different causes. It can be caused *e.g* by (1) variable surface currents associated with oceanic internal waves or with tidal flow over shallow underwater bottom topography (Alpers and Hennings 1984; Alpers 1985); (2) mineral oil or surface-active substances floating on the sea surface (Alpers and Espedal 2004); or (3) cold surface water surrounded by warmer ambient water (Koslov et al. 2012).

4.3.2.1 Oceanic Internal Waves

As mentioned before, Oceanic Internal Waves (OIWs) become visible on SAR images because they modulate the sea surface roughness caused by variations of the surface current associated with these waves. Usually, the OIWs visible on SAR images are strongly nonlinear and are organized in wave packets. An example of an ERS-1

Fig. 4.8 ERS-1 SAR image acquired at 2239 UTC on 1 January 1993 over the Mediterranean Sea east of the Strait of Gibraltar. It shows sea surface signatures of a packet of internal solitary waves generated in the Strait of Gibraltar propagating eastward into the Mediterranean Sea. The dark line intersecting the packet results very likely from a discharge of oily substances from a moving ship f (sludge oil from the ship's engine room). Imaged area: 100 km x 50 km. © ESA

SAR image showing sea surface signatures of an OIW packet is depicted in Fig. 4.8, which was acquired at 2239 UTC on 1 January 1993 over the Mediterranean during an ascending satellite pass. It shows sea surface signatures of a packet of OIWs generated in the Strait of Gibraltar and propagating eastward into the Mediterranean Sea. The dark wriggled line intersecting the packet results from surface active material floating on the sea surface. From the linear shape we conclude that it was discharged from a moving ship. The discharged material is very likely a mixture of diesel oil and water (sludge oil) pumped into sea from the ship's engine room during cleaning operation.

Oily substances floating on the sea surface attenuate the short sea surface waves which are responsible for the radar backscattering. Thus sea areas covered with films consisting of oily substances are areas of strongly reduced NRCS values and thus become visible on SAR images as dark areas.

OIWs are encountered in the ocean only when the water body consists of water layers of different densities. This is always the case in the Strait of Gibraltar, where the water body consists of two layers: a heavy bottom layer of salty Mediterranean water and a lighter upper layer of less salty Atlantic water. The boundary between these two layers is called pycnocline or halocline (from Greek pyknos = dense, halos = salt, klinein = to lean). When the water body is pushed by tidal forcing over the shallow sills in the Strait Gibraltar, the pycnocline gets distorted and strongly nonlinear OIWs evolve (Brandt et al. 1996). Observations and model calculations have shown that the nonlinear OIWs are here always waves of depression, i.e., the pycnocline is distorted in the way shown in Fig. 4.9. Theoretical models describing the modulation of the sea surface roughness by the variable surface currents have been developed previously in the framework of weak hydrodynamic interaction theory (Alpers 1985). The larger the gradient of the surface current in look direction of the antenna, the larger is the

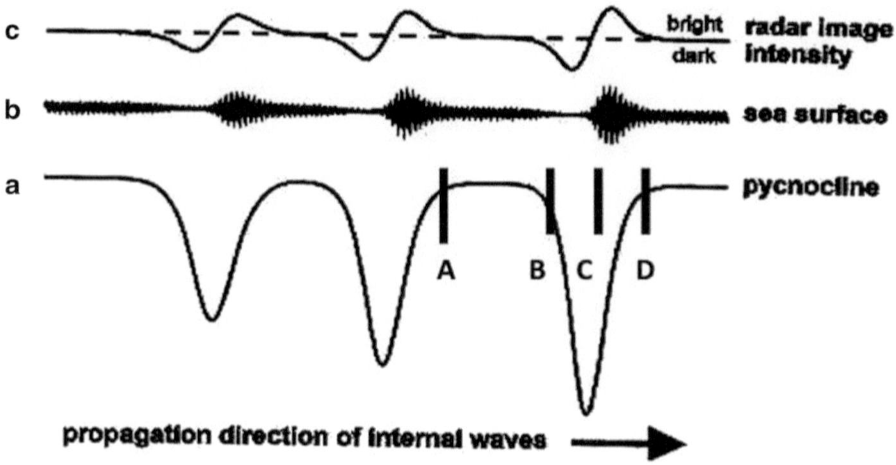

Fig. 4.9 Schematic diagram illustrating the generation of the radar signature of a non-linear oceanic internal wave packet consisting of three internal waves (solitons) of depression. **a** Shape of the pycnocline in a 2-layer model, **b** variation of the short scale sea surface roughness, and **c** variation of the radar image intensity (image brightness), The distances B-C and C-D give the widths of the bands with decreased and increased radar image intensity, respectively, and A-B the width of the band with background image intensity

modulation of the short-scale roughness, and the stronger is the NRCS modulation. The gradient of the surface current is related to OIW parameters, like wave amplitude and depth and the variation of water density across the pycnocline.

The relationship between the distortion of the pycnocline caused by a non-linear oceanic internal wave packet consisting of three internal waves (solitons) of depression. and the variation of the NRCS or the SAR image intensity is shown schematically in Fig. 4.9. Such relationship is predicted by theory and has also been confirmed by observations (Alpers and Huang 2011). Note that in the case of OIWs of depression, the front of the wave packet is always imaged as a bright band.

OIWs are not only generated by tidal forcing in straits with shallow sills, like in the Strait of Gibraltar or the Strait of Messina, but also at shelf breaks. An ERS-1 SAR image showing internal wave packets generated at successive tidal cycles at the shelf break on the Moroccan Atlantic coast is depicted in Fig. 4.10. The shape of the wave patterns is closely related to the bathymetry. Another area around Africa where frequently oceanic internal waves are encountered is the Mozambique Channel. Here a "hot spot" for the generation of internal waves is the Sofala shelf (20°S, 36°E) (da Silva et al. 2009).

4.3.2.2 Biogenic Surface Films

It is well known that oily substances oil floating on the sea surface attenuate the short surface waves and thus cause a reduction of the backscattered radar power or NRCS

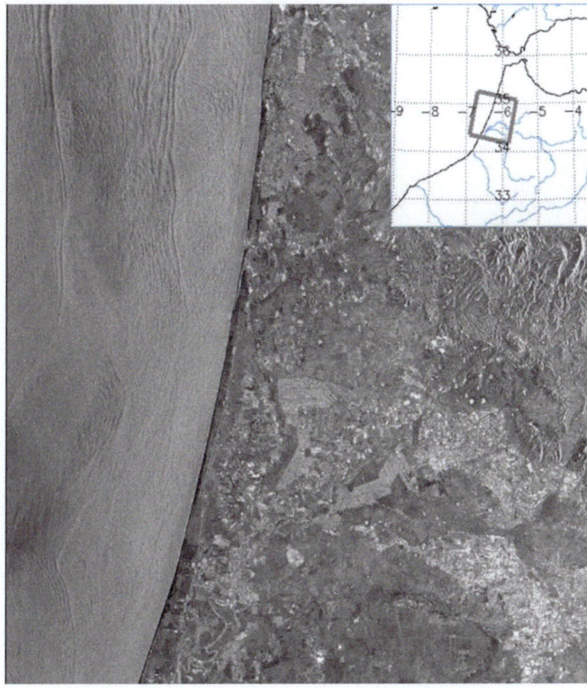

Fig. 4.10 ERS-1 SAR image acquired at 1105 UTC on 25 September 1992 over the west coast of Morocco, see inset. It shows sea surface signatures of oceanic internal wave packets generated during successive tidal cycles at the shelf break on the Atlantic coast of Morocco. Imaged area: 100 km x 105 km. © ESA

as seen on the ERS-1 SAR image depicted in Fig. 4.8. Dark (wriggled) lines are often observed on SAR images of the sea, in particular along shipping lines. They usually result from discharges of crude oil pumped into the sea from tankers during tanker washing or from discharges of a mixture of diesel oil and water (sludge oil) from the ship's engine room.

But not only anthropogenic pollution causes a reduction of the NRCS, but also surface films of natural origin, so called biogenic surface films. They are encountered usually in areas of high biological productivity with high fish population. This implies that in these regions the amount of surface active substances secreted by marine plants and animals is greatly enhanced. These substances rise to the sea surface and can cover large areas. Although these biogenic surface films are only one molecular layer thick, they can attenuate the short ocean waves, which are responsible for the radar backscattering, as strongly as mineral oil films. Thus biogenic surface films also give rise to dark areas on SAR. However, the dark areas do not have the form of a line, but usually of a (large) connected patch.

Since upwelling regions are regions of high biological activity, we expect that in these regions frequently biogenic surface films are encountered which give rise to dark patches on SAR images. In upwelling regions not only biogenic surface films can give rise to a reduction of the NRCS, but also the upwelled cold water. Cold water changes the stability of the marine boundary layer from (usually) neutrally stable to stable which causes a reduction of the wind stress (Koslov et al. 2012) and thus of the

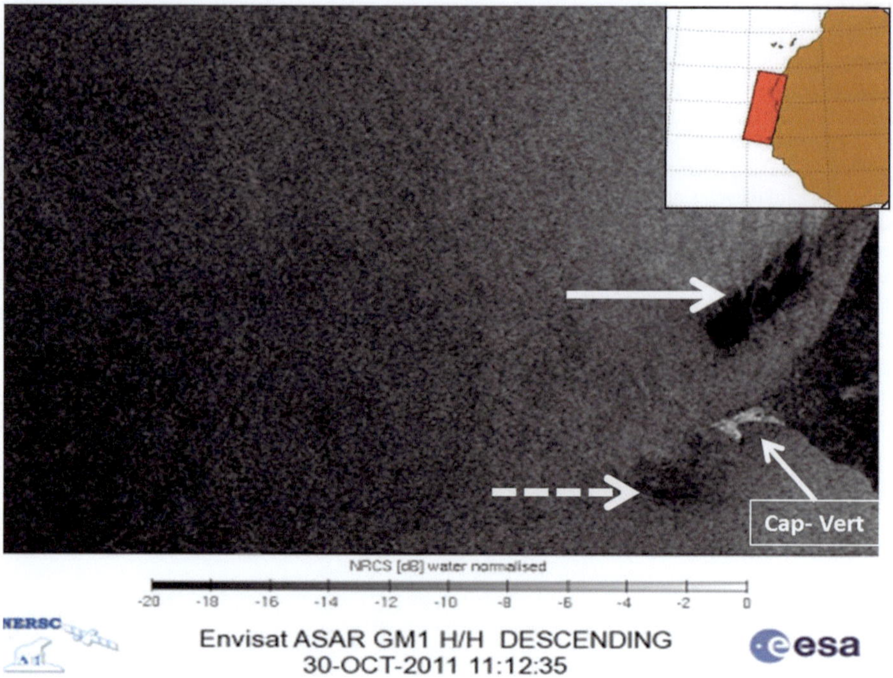

Fig. 4.11 Section of the SAR image acquired by Envisat at 1112 UTC on 30 October 2011 over the coastal waters off the coast of Senegal (West Africa). The dark area to the north (marked by a solid arrow} is located in an upwelling region with high concentration of biota generating surface active material that ascends to the sea surface and forms there a biogenic (monomolecular) surface film that damps the short-scale sea surface waves. The dark area further south (marked by a *dashed arrow*) is the sea surface signature of a small-scale oceanic cold eddy generated at the headland of Cap-Vert (see Fig. 4.12 and text). The inset shows the SAR swath. © ESA

NRCS. This is a well-known phenomenon and has been confirmed, among others, by radar backscatter measurements carried out from a platform in the North Sea, see Keller et al. (1989). Although upwelling areas are often detected on SAR images as areas of reduced NRCS values, it is often not clear, what has caused this reduction: surface films or cold surface water. However, the amount of the reduction can give an answer: cold water causes only a small reduction (of the order of 3 dB or less), while biogenic surface films can cause much higher reductions (of the order of 10 dB).

An example of such a SAR image acquired over a well-known upwelling area, the coast of Senegal (West Africa), is shown in Fig. 4.11. This is an Envisat ASAR image which was acquired in the Global Mode (GM, resolution: 1 km) at 1112 UTC on 30 October 2011. It shows two darkish patches, one deep dark (marked by a solid arrow) and one light dark patch (marked by a dashed arrow). We have measured the reduction of the NRCS relative to the background of these two darkish patches and found 10–12 dB for the deep dark patch and 3 dB for the light dark patch

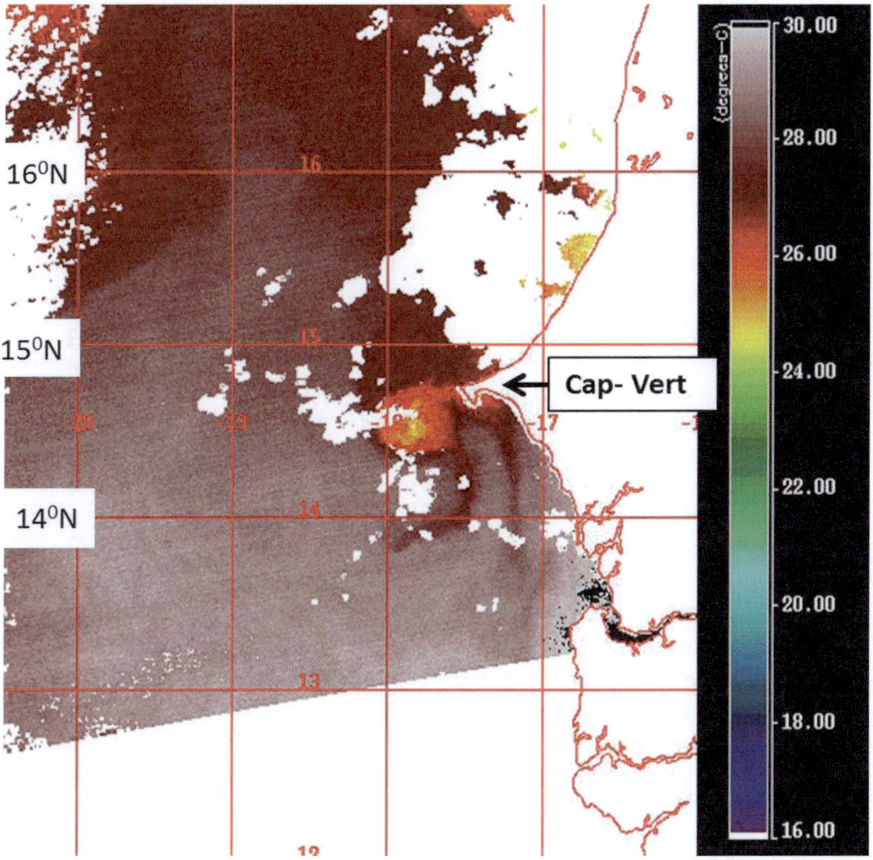

Fig. 4.12 SST in°C retrieved from MODIS data acquired at 14;30 UTC on 31 October 2011. The white areas are land (*on the right*) or clouds where no SST values could be retrieved. © NASA

(Alpers et al. 2013). From these values we conclude that the deep dark patch cannot be caused (alone) by cold water, but must be caused primarily by surface films. On the other hand, the light dark patch can be caused by partial coverage with surface films or by the change in the stability of the air-sea interface due to cold upwelled water or by both. Since in the first case the dark patch is located in the upwelling at the coast of Senegal, we conclude that it is caused by biogenic surface films, although we have no direct proof.

However, In the second case we have for comparison a sea surface temperature (SST) map derived from MODIS data acquired one day later (at 1430 UTC on 31 October 2011), see Fig. 4.12. It shows the sea surface signature of a cold (cyclonic) eddy off the headland Cap– Vert. For details the reader is referred to Chapter 11 entitled "Satellite observations of oceanic eddies around Africa". From the value of the NRCS reduction of 3 dB we conclude that it is caused by the cold upwelled water.

At this stage of the development of the eddy, the biological activity was still low such that no or very little surface active material was secreted by the biota within the eddy which could give rise to biogenic surface films.

4.4 Summary and Conclusions

Scatterometers and SARs are active microwave instruments which have been flown since 1978 on several satellites. Both instruments measure over the ocean the small-scale roughness, which can be used to measure near-surface winds. While the near-surface wind fields derived from scatterometer data have a resolution of typically 25 km, the ones derived from SAR data have a resolution of typically 200–500 m. Examples of SAR images showing an atmospheric front, atmospheric gravity waves, and a wind jet over African coastal waters have been presented. While the SAR can resolve the near-surface wind field with a much higher spatial high resolution than the scatterometer, it has the disadvantage that, from a single satellite, it can provide near-surface wind information from a given area only with a time separation of several days. On the other hand, one can obtain wind field maps of most areas of the World's ocean from scatterometer data every day or every second day.

Not only atmospheric phenomena in the marine boundary layer are detectable by SAR, but also oceanic phenomena, like internal waves, tidal flow over shallow underwater topography, oceanic fronts, upwelling, oceanic eddies, and oil or biogenic surface films. These oceanic phenomena are primarily encountered in coastal areas. The atmospheric phenomena most often observed on SAR images of African coastal waters are land/sea breeze, katabatic winds, gap winds, wind jets, and atmospheric gravity waves. The oceanic phenomena observed most often are upwelling phenomena, internal waves, oceanic eddies, and surface film coverage (mineral oil films or biogenic surface films). The interpretation of features visible on SAR images of the sea surface is not always straightforward and requires some knowledge of the SAR imaging mechanism. E.g., it is often not easy to decide whether a features visible on a SAR image results from an atmospheric or an oceanic phenomenon. This applies in particular to atmospheric/oceanic fronts and atmospheric/oceanic gravity waves (Alpers and Huang 2011). Therefore, in order make optimum use of SAR data acquired over the ocean, one should combine them with other satellite data acquired in the infrared, visible or ultraviolet bands, with in -situ data, and with results of model calculation.

References

Abdalla S, Janssen PAFM, Bidlot JR (2010) Status of Envisat fast delivery wind wave products. Proceedings of the ESA Living Planet Symposium held in Bergen, Norway from 28 June to 2 July 2010, ESA publication SP-686

Alpers W (1985) Theory of radar imaging of internal waves. Nature 314:245–247

Alpers W, Hennings I (1984) A theory of the imaging mechanism of underwater bottom topography by real and synthetic aperture radar. J Geophys Res 89:10529–10546

Alpers W, Espedal H (2004) Oils and Surfactants. Chapter 11 In: Jackson ChR, Apel JR (eds) Synthetic Aperture Radar Marine User's Manual. National oceanic and atmospheric administration, center for satellite application and research, NOAA/NESDIS, Washington, DC, USA, ISBN 0–16-073214-X. Chapter 11:263–275

Alpers W, Huang W (2011) On the discrimination of radar signatures of atmospheric gravity waves and oceanic internal waves on synthetic aperture radar images of the sea surface. IEEE T Geosci Remote 49:1114–1126

Alpers W, Stilke G (1996) Observation of a nonlinear wave disturbance in the marine atmosphere by the synthetic aperture radar aboard the ERS-1 satellite. J Geophys Res 101:6513–6525

Alpers W, Ross DB, Rufenach CL (1981) On the detectability of ocean surface waves by real and synthetic aperture radar. J Geophys Res 86:6481–6498

Alpers W, Chen JP, Pi CJ, Lin II (2010) On the origin of atmospheric frontal lines off the east coast of Taiwan observed on space-borne synthetic aperture radar images. Mon Weather Rev 138:475–496. doi:10.1175/2009MWR2987.1

Alpers W, Brandt P, Lazar A, Dagorne D, Sow B, Faye S, Hansen MW, Rubino A, Poulain PM, Brehmer P (2013) A small-scale oceanic eddy off the coast of West Africa studied by multi-sensor satellite and surface drifter data, Remote Sens Environ 129:132–143. doi 10.1016/j.rse.2012.10.032

Brandt P, Alpers W, Backhaus JO (1996) Study of the generation and propagation of internal waves in the Strait of Gibraltar using a numerical model and synthetic aperture radar images of the European ERS satellite. J Geophys Res 101:14237–14252

Cheng CM, Alpers W (2010) Investigation of trapped atmospheric gravity waves over the South China Sea using Envisat synthetic aperture radar images. Int J Remote Sens 31:4725–4743

Da Silva JCB, New AL, Magalhaes JM (2009) Internal solitary waves in the Mozambique Channel: Observations and interpretation. J Geophys Res 114(C05001). doi:10.1029/2008JC005125

Doyle JD, Durran DR (2002) The dynamics of mountain-wave induced rotors. J Atmos Sci 59:186–201

Evans DL, Alpers W, Cazenave A, Elachi C, Farr T, Glackin D, Holt B, Jones L, Liu WT, McCandless W, Menard Y, Moore R, Njokua E (2005) Seasat—A 25-year legacy of success. Remote Sensing of. Environment 94:384–404. doi:10.1016/j.rse.2004.09.011

Hersbach H (2008) CMOD5.N: A C-band geophysical model function for equivalent neutral wind. ECMWF Technical Memorandum No 554. http://www.ecmwf.int/publications/library/ecpublications/_pdf/tm/501-600/tm554.pdf

Hersbach H, Stoffelen A, de Haan S (2007) Ocean wind field and their variability derived from SAR. Earth Obs Quart 59:8–12

Horstmann J, Koch W (2005) Comparison of SAR wind field retrieval algorithms to a numerical model utilizing ENVISAT ASAR data. IEEE J Oceanic Eng 30:508–515

Keller WC, Wismann V, Alpers W (1989) Tower-based measurements of the ocean C-band radar backscattering cross section. J Geophys Res 94:924–930

Kozlov IE, Kudryavtsev VN, Johannessen JA, Chapron B, Dailidiene I, Myasoedov AG (2012) ASAR imaging for coastal upwelling in the Baltic Sea. Adv Space Res. doi:10.1016/j.asr.2011.08

Monaldo F, Kerbaol V (2003) The SAR measurement of ocean surface winds: an overview. Proc. 2nd workshop on Coastal and Marine applications of SAR, Svalbard (Norway), 8–12 September 2003, 15–32

Monaldo FM, Thompson DR, Beal RC, Pichel WG, Clemente-Colon P (2001) Comparison of SAR derived wind speed with model predictions and ocean buoy measurements. IEEE T Geosci Remote 39:2587–2600

Mouche A, Dagestad KF, Collard F, Guitton G, Chapron B, Johannessen J, Kerbaol V, Hansen MW (2012) On the use of Doppler shift for sea surface wind retrieval from SAR. IEEE T Geosci Remote 50:2901–2909

Quilfen Y, Chapron B, Elfouhaily T, Katsaros K, Tournadre J (1998) Observation of tropical cyclones by high-resolution scatterometry. J Geophys Res 103:7767–7786

Ricciardulli L, Wentz F (2011) Reprocessed QuikSCAT (V04) wind vectors with Ku-2011 Geophysical model function. Remote Sensing Systems Technical Report 043011[2]. http://www.ssmi.com/qscat/qscat_Ku2011_tech_report.pdf
Stoffelen A, Anderson D (1997) Scatterometer data interpretation: Estimation and validation of the transfer function CMOD4. J Geophys Res 102:5767–5780
Valenzuela GR (1978) Theories for the interaction of electromagnetic and oceanic waves: a review. Bound Lay Meteorol 13:61–85

Part II
Western Africa Near-Coastal Waters

Chapter 5
Mesoscale Dynamics in the Canary Islands Area as Observed Through Complementary Remote Sensing Techniques

Luis García-Weil, Antonio G. Ramos, Josep Coca and Alexandre Redondo

Abstract The combined use of data collected by multi-spectral radiometers, operating in the visible and infrared spectrum, and altimeter radars, operating in the microwave spectrum, allowed the detailed assessment of oceanic mesoscale dynamics in the Canary Islands area. Island-induced eddies are recurrent oceanographic structures of the region and contribute, together with other features originating from the African near-coastal zone, like upwelling filaments, to the variability of the physical and biological field of the Canarian Archipelago. The analysis of satellite-measured parameters with complementary specificities—namely brightness temperature and sea surface temperature, chlorophyll-a, as well as sea level anomaly—provided a description of the mesoscale variability around the Islands, covering in detail an entire seasonal cycle in the period 1998–1999. An overview is given of the spatial and temporal characteristics of all mesoscale features observed in the satellite imagery, including their origin, evolution and interaction with each other and their environmental surroundings.

5.1 Introduction

In the last 2 decades, different studies using in-situ and Remote Sensing (RS) data have allowed to portray the Canary Islands area—through which the main current, known as Canary Current, flows in a southwestward direction—as a region showing a complex pattern of oceanic mesoscale variability, consisting primarily of eddies and mesoscale meanders or filaments (Barton et al. 1998). Similar patterns have

L. G.-Weil (✉)
Departamento de Física, Universidad de Las Palmas de Gran Canaria (ULPGC), Las Palmas, Spain
e-mail: lgarcia@dfis.ulpgc.es

A. G. Ramos · J. Coca · A. Redondo
SeaSNET Canarias ground station. Departamento de Biología, Universidad de Las Palmas de Gran Canaria (ULPGC), Las Palmas, Spain

V. Barale, M. Gade (eds.), *Remote Sensing of the African Seas*,
DOI 10.1007/978-94-017-8008-7_5,

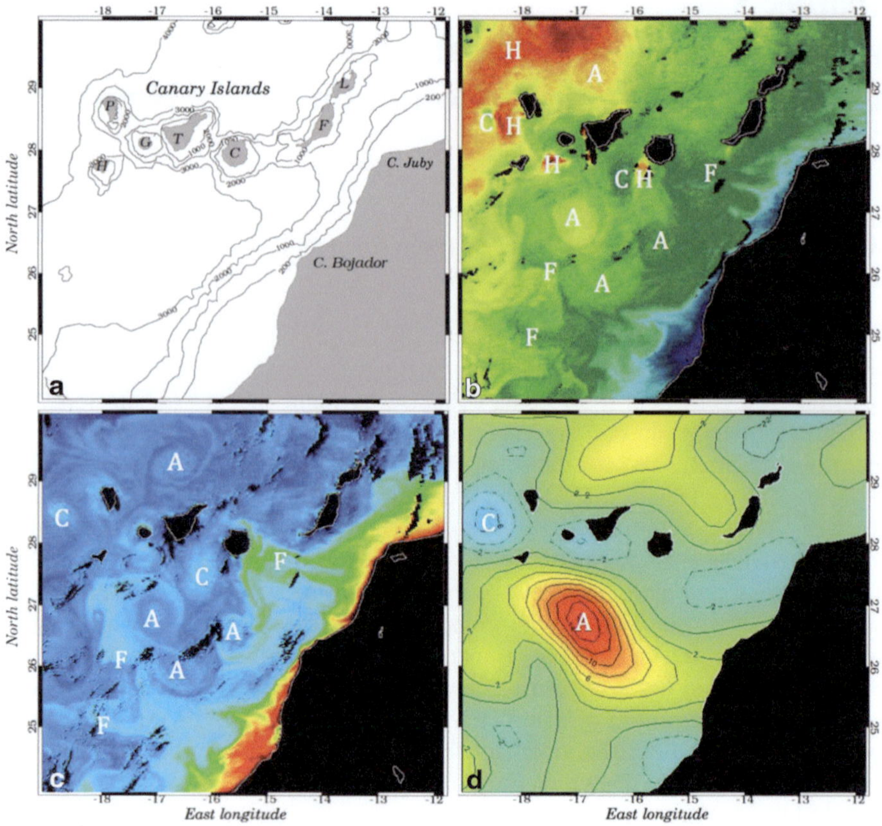

Fig. 5.1 **a** Map of the study area. The letters inside the Canary Islands maps stand for La Palma (*P*), El Hierro (*H*), Gomera (*G*), Tenerife (*T*), Gran Canaria (*C*), Fuerteventura (*F*) and Lanzarote (*L*). The 200, 1000, 2000, 3000 and 4000 m isobaths are shown. **b** Sea surface temperature (*SST*) from MODIS data collected on 7 September 2007. **c** Chlorophyll a (*Chl-a*) concentration from MODIS data collected on 7 September 2007. **d** Sea Level Anomaly (*SLA*) map from altimeter data collected on 5 September 2007. The letters inside the satellite-derived maps stand for anticyclonic eddy (*A*), cyclonic eddy (*C*), filament (*F*) and *SST* hot spot (*H*)

been observed also in other eastern boundary current regions, where they are found to be rather common features. The Canary Islands area, however, does have a distinctive topographic peculiarity, which exerts an important influence on the observed mesoscale variability (Barton et al. 1998), i.e. the Canarian Archipelago itself. As outlined in Fig. 5.1a, this consists of seven main islands, separated by deep channels and distributed zonally around 28° N, at approximately 100 km to 600 km from the African coastline.

The Canarian Archipelago spans the transition zone between upwelling-dominated, eutrophic inshore waters and oligotrophic offshore waters, determining a perturbation of the Canary Current equatoward flow and of the Trade winds—thus generating mesoscale variability downstream of the islands in the form of warm wakes and of both cyclonic and anticyclonic eddies. RS images derived from sensors operating in the visible and thermal infrared parts of the spectrum provide indications of the systematic presence of mesoscale eddies and warm regions in the lee of nearly all islands (Van Camp et al. 1991; Pacheco and Hernández-Guerra 1999; Barton et al. 2000), but so far the most studied and best documented features, by means of both in-situ and RS observations, are those found south of Gran Canaria island (Arístegui et al. 1997; Barton et al. 1998; Basterretxea et al. 2002; Sangrà et al. 2005).

Previous investigations such as those mentioned above, focusing mainly on the analysis of the mesoscale features found south of Gran Canaria, concluded that the island-induced eddies of this region are recurrent oceanographic structures, which—in combination with other features originating from the coastal zone, like upwelling filaments and eddies—contribute significantly and in different ways to the variability of the observed physical and biological field south of the Archipelago (Arístegui et al. 1997). The main purpose of this work is to provide an additional description of the mesoscale variability around the Canary Islands, covering in detail an entire seasonal cycle.

This will be done using a combination of RS data with complementary specificities, taking into account that earlier work in the region has already shown the correlation existing between the signature of mesoscale features in satellite data and concurrent in situ observations, and validating their use to assess oceanographic processes of the whole area (Arístegui et al. 1994; Barton et al. 1998; Tejera et al. 2002). The main advantages and drawbacks of the RS techniques used in the present work will be briefly reviewed. Further, the results of an altimeter data time series analysis will be given, together with a detailed account of the mesoscale variability in the region, as obtained through the combined analysis of altimeter and multi-spectral radiometer data.

5.2 Remote Sensing Data and Mesoscale Variability

Chlorophyll-a (Chl-a), Brightness Temperature (BT) and Sea Surface Temperature (SST) data, gathered by multi-spectral radiometers operating in the visible and infrared spectrum, have all been used in this work, in combination with Sea Level Anomaly (SLA) data from altimeter radars operating in the microwave spectrum. Both BT and SST can be obtained from thermal infrared sensors, but unlike for SST, the atmospheric signal is not removed for BT.

Algorithms used to remove atmospheric effects increase the noise level and reduce the temperature gradients in the data (La Violette and Holyer 1988). This sometimes renders BT images preferable to SST images, when observing mesoscale features. A description of how these geophysical parameters are derived from RS measurements is out of the scope of this work and can be found in a vast literature on the subject (see e.g. Robinson 2004). However, in relation to the retrieval of information on mesoscale phenomena in the Canary Islands area, it is worth examining the sampling performance of different sensors and the main characteristics of their derived parameters.

5.2.1 Sampling Properties of Remote Sensors

There are important differences between the sampling scheme of multi-spectral radiometers and radar altimeters. The first are imaging systems providing a synoptic view of the ocean surface. Those used in this work (e.g. the Advanced Very High Resolution Radiometer, AVHRR; the Moderate Resolution Imaging Spectroradiometer, MODIS; and the Sea-viewing Wide Field-of-view Sensor, SeaWiFS) have a spatial resolution on the order of ~ 1 km at nadir, and a viewing swath width larger than 1500 km. The spatial resolution degrades toward the edge of the swath, such that in many applications, as in this study, these parts of the scene are not used. The second, being non-imaging radars, only provide a profile of sea surface height below the satellite along their ground tracks. Then, interpolation is needed to build SLA maps. In the present case, SLA maps were generated by the French Archiving, Validation and Interpretation of Satellite Oceanographic (AVISO) data center, combining the measurements of several radar altimeters in order to minimize errors due to the paucity of data of one single altimeter missions in the composition of SLA maps (Le Traon and Dibarboure 1999). Specifically, in this work it was used the reference delayed time product (SSALTO/DUACS User Handbook 2012). This merges in one map the measurements of two altimeter missions, on the same two orbits (first Topex/Poseidon and ERS followed by Jason-1 and Envisat). Thus it ensures a homogeneous and stable sampling along the available time period. It has been shown that Gaussian-shaped eddies with e-folding scales of at least 0.4° are resolved in these maps (Chelton et al. 2011).

With respect to temporal sampling, during the years analyzed in more detail in this work (i.e. the 1998–1999 period), there were usually 4 AVHRR-recorded SST (or BT) scenes available per day. This high-frequency rate is due to the fact that infrared sensors can observe the Earth during day and night, and that in 1998–1999 there were two satellites, with an AVHRR on board, observing the area of study at different times. Conversely, just 1 Chl-a scene per day was available, because visible sensors operate during daytime only, and in 1998–1999 SeaWiFS was the only one available to observe ocean colour from space. The actual frequency of useful SST and Chl-a images is lower than the nominal one mentioned above, because no information can be retrieved by infrared and visible radiometers when clouds cover the ocean, a limitation that renders the time interval between consecutive useful SST or Chl-a scenes highly irregular. On the other hand, radar altimeters measure a microwave signal going through clouds, but this signal is degraded if rain is present. The Canary Islands area is placed in a dry subtropical zone and therefore altimeter data from this region are not seriously affected by rain (Tran et al. 2005). Although atmospheric effects are less a problem in the microwave region, the altimeter sampling characteristics generate a very poor temporal resolution (in excess of 10 days). By merging data from more than one altimeter in the same data product, AVISO produces regularly one SLA map of any given area every 7 days.

5.2.2 *Properties of Retrieved Geophysical Parameters*

The objective of this subsection is to provide a brief overview of how the geophysical parameters retrieved from RS measurements are related to the observation of mesoscale features in the Canary Islands area. The interested reader is referred to Robinson (2010) for a more detailed review of the relationship between the knowledge of mesoscale ocean processes and RS measurements.

SST and Chl-a are frequently used to observe ocean processes. The contrast between cold and productive waters, normally associated to coastal upwelling areas and cyclonic eddies, and warm and poorer waters, related to oceanic areas and anticyclonic eddies, frequently makes both these parameters useful to visualize ocean mesoscale features. As for SLA, it is directly related to ocean dynamics. This connection is set through geostrophic balance, which is expressed as:

$$u_g = -\frac{g}{f}\frac{\Delta\eta}{\Delta y}, v_g = \frac{g}{f}\frac{\Delta\eta}{\Delta x} \tag{5.1}$$

where η is the SLA, f is the Coriolis parameter (that is positive in the Northern Hemisphere), and u_g and v_g are the geostrophic velocity anomalies (Ducet et al. 2000). Then, in Northern Hemisphere, anticyclonic eddies, in which water rotates in a clockwise sense around the eddy center, appear as concave downward in SLA anomaly maps, while cyclonic eddies, in which water rotates in anticlockwise sense, appear as concave upward.

The basic correlation between variability of these satellite-derived geophysical parameters and mesoscale dynamics is not always evident. In the case of infrared radiometers, the electromagnetic radiation recorded by the sensor comes from a layer adjacent to the surface that is only 10-20 μm thick. Under strong insolation and light wind conditions, a highly stable diurnal thermocline is formed, and satellite-derived SST does not correspond to mixed layer temperature, which is related to ocean dynamics. Figure 5.1b shows a diurnal MODIS SST image, with warm wakes downwind of some of the islands and with a high-temperature feature in the top left part of the image. These patterns are not observed in the coincident MODIS Chl-a image shown in Fig. 5.1c, because the signal recorded in the visible spectral interval is integrated over a deeper layer of water.

Mesoscale features like eddies can be detected in SST (or BT) and Chl-a scenes, because they can produce strong vertical isopycnal displacements that change the surface value of these parameters. In a stable eddy in geostrophic balance, this displacement requires isopycnals (and isotherms) to dip down in the center of an anticyclonic eddy and upwards in a cyclonic eddy. In order to reach this dynamical balance, there is initially a downwelling at the center of anticyclonic eddies which draws down the isotherms, so warmer and nutrient depleted water penetrates deeper than normal. Subsequently warm water is dynamically maintained in this position. The converse situation occurs in cyclonic eddies. So anticyclonic (cyclonic) eddies can show in the center high (low) values of SST and low (high) values of Chl-a as observed in satellite images.

But this is not the only way that eddies are detected in SST and Chl-a scenes. Another possibility is that water with different surface SST and/or Chl-a properties can be wrapped around the eddies' core. Then, SST and Chl-a can be considered as tracers of the water flow. This mechanism of eddy observation is useful when large scale gradients of SST an Chl-a, like those observed frequently in the coastal transition zone, are present. As it will be shown in next section, the signal of ocean eddies in the SST and Chl-a fields disappears, albeit not at the same time, when they move away from the coastal transition zone and into offshore areas. This is produced because of the intensity decay of eddies and the simultaneous weakening of the surface gradients of SST and Chl-a fields. The signal of the eddies usually lasts for a longer time in SLA maps, as in Fig. 5.1d, again this being a property directly related to ocean dynamics.

5.3 Mesoscale Variability in the Canary Islands Area

5.3.1 Annual and Seasonal Characteristics of Mesoscale Activity

The eddy kinetic energy, drawn from the SLA-derived geostrophic velocity anomalies, i.e.

$$EKE = \frac{1}{2}(u_g^2 + v_g^2) \tag{5.2}$$

is a quite often used measurement of mesoscale variability (Ladd 2007). By means this parameter, computed on the basis of an altimeter data record that spans over more than a decade, it is possible to show which are the areas in the Northwest African ocean region that show distinctive high levels of mesoscale variability, and also to detect seasonal patterns and possible interannual variations in the strength of this mesoscale activity for specific areas.

As observed in the mean EKE map shown in Fig. 5.2a, there is a local maximum ($> 110\,cm^2s^{-2}$) south of the Canary Archipelago. This region, centered near 26.5° N and 17.5° W, does not correspond to the exact location where mesoscale features are generated (which is actually closer to the Canary Islands and the African shelf; Arístegui et al. 1997; Barton et al. 2004; Sangrà et al. 2005), but does indicate that the Canary Islands correspond to an important eddy formation area. A band of moderately high values ($> 75\,cm^2s^{-2}$) extends southward and westward from the local maximum position. As will be shown later, this indicates the pathway of eddies into the open ocean.

A clear annual climatological cycle appears in the 1993–2008 time series of EKE values shown in Fig. 5.2b, with a maximum in October ($\sim 200\,cm^2s^{-2}$) and a minimum in April ($\sim 50\,cm^2s^{-2}$). This monthly climatology is obtained computing, over the area shown in Fig. 5.2a, the mean value for each month over the time range 1993–2008. EKE maxima are reached several months after the normal eddy generation period. Eddy genesis is more frequent in summer (Piedeleu et al. 2009),

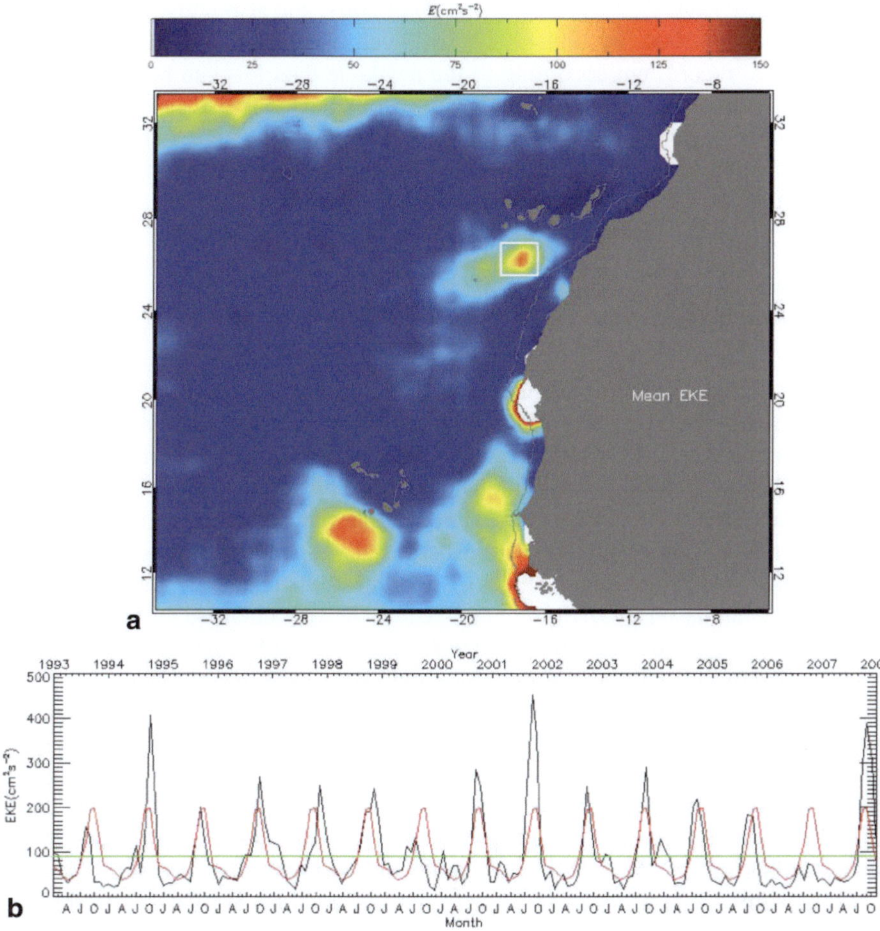

Fig. 5.2 **a** Mean *EKE* values [cm^2s^{-2}] off northwest Africa for the 1993–2008 period. **b** 1993–2008 time series of monthly (*black line*), monthly climatological (*red line*) and average (*green line*) *EKE* values [cm^2s^{-2}], as computed in the *white-line box* drawn around the local maximum, south of the Canary Islands, of Fig. 5.2a

when currents and winds are stronger in this area[1]. The time and space location of EKE maxima indicate that the most energetic features are not observed immediately after their generation. After reaching a maximum, EKE values decay as mesoscale features progress in their life cycle, following a southwestward path.

In general, the time series of monthly EKE data correlate quite well with the climatological cycle, as seen in Fig. 5.2b, also showing an analogous seasonal pattern. The most important deviations, for the time series used here, seem to have occurred

[1] See the website (http://cioss.coas.oregonstate.edu/scow/index.html) for a scatterometer climatology of ocean winds.

with the highest EKE values ($> 400\,cm^2s^{-2}$) of 2001, and also with the lowest ones ($\sim 80\,cm^2s^{-2}$) of 2006. In 1998–1999, EKE values were slightly higher than the climatological record, but according to the available data this can be considered a "normal" period. The important peculiarity of this period is that it is the first one when altimeter, ocean color an SST data became available together. In the following, the mesoscale variability in the Canary Islands area will be detailed during these years, through the combined use of different RS data.

5.3.2 *Observations of Mesoscale Features*

Taking advantage on the complementarity of the sampling scheme offered by different RS data, the main mesoscale features generated during 1998–1999 were studied in some detail. AVHRR-derived BT data were analyzed in combination with SeaWiFS-derived Chl-a data and SLA data recorded by various altimeters during the same period. Both BT and Chl-a data show similar patterns next to the area where mesoscale features are generated (i.e. mesoscale eddies, fronts and coastal upwelling can be seen in color and infrared images). An initial assessment of dynamical features was based primarily on AVHRR images, because they are obtained more frequently. SeaWiFS images were used only when able to add new information, in addition to that provided by BT patterns. Further, SLA data were used in combination with the Chl-a and BT data to provide a direct link to ocean dynamics and a better feature resolution. Mesoscale features are observed in altimeter maps when they reach suitable size and strength, and can be detected even when Chl-a and BT gradients are too weak to be recorded by a radiometer.

There is ample evidence, including that shown in the present work, that Canary Islands eddies are generated all year around, which suggests that current flow disturbances are strong enough to spawn these features at any time. Similarly, disturbances in the form of frontal instabilities and upwelling filaments, usually associated with mesoscale eddies, are regularly observed next to the African coast.

A complete census of all these mesoscale features is not possible using the chosen RS data, due to implicit limitations in their sampling characteristics. Thus, the present study is focused on the most prominent mesoscale features that could be observed in the satellite data for a period longer than 2 months, and of which both generation and life history characteristics could be unambiguously established. These coherent, long-lived features, in fact, are those that have more implications in the transport of heat, nutrients, carbon and other properties from the shelf to the coastal transition zone, and from this region to the open ocean.

5.3.2.1 Description of Variability in the Canary Islands Area

Several eddy-like structures appear in the series of BT (preferred here to SST for the reason explained earlier), Chl-a and SLA images, shown in Fig. 5.3 and following

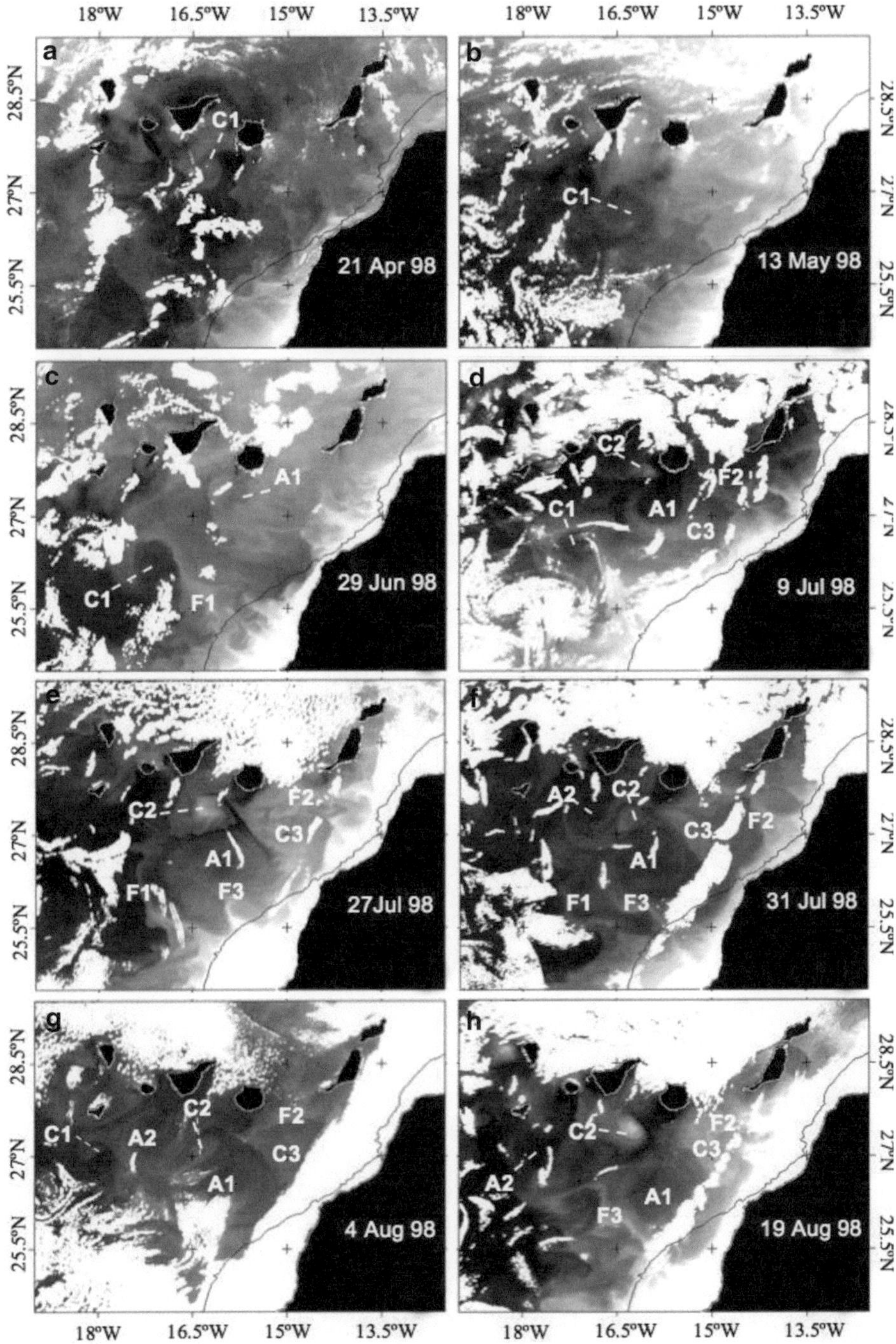

Fig. 5.3 BT images from AVHRR channel 4 (10.3–11.3 μm) over the period from April to August 1998 **(a–h)**. Each image has been specifically enhanced to emphasize the features. Brightness temperature increase from light to dark shades. The black line indicates the 200 m depth contour

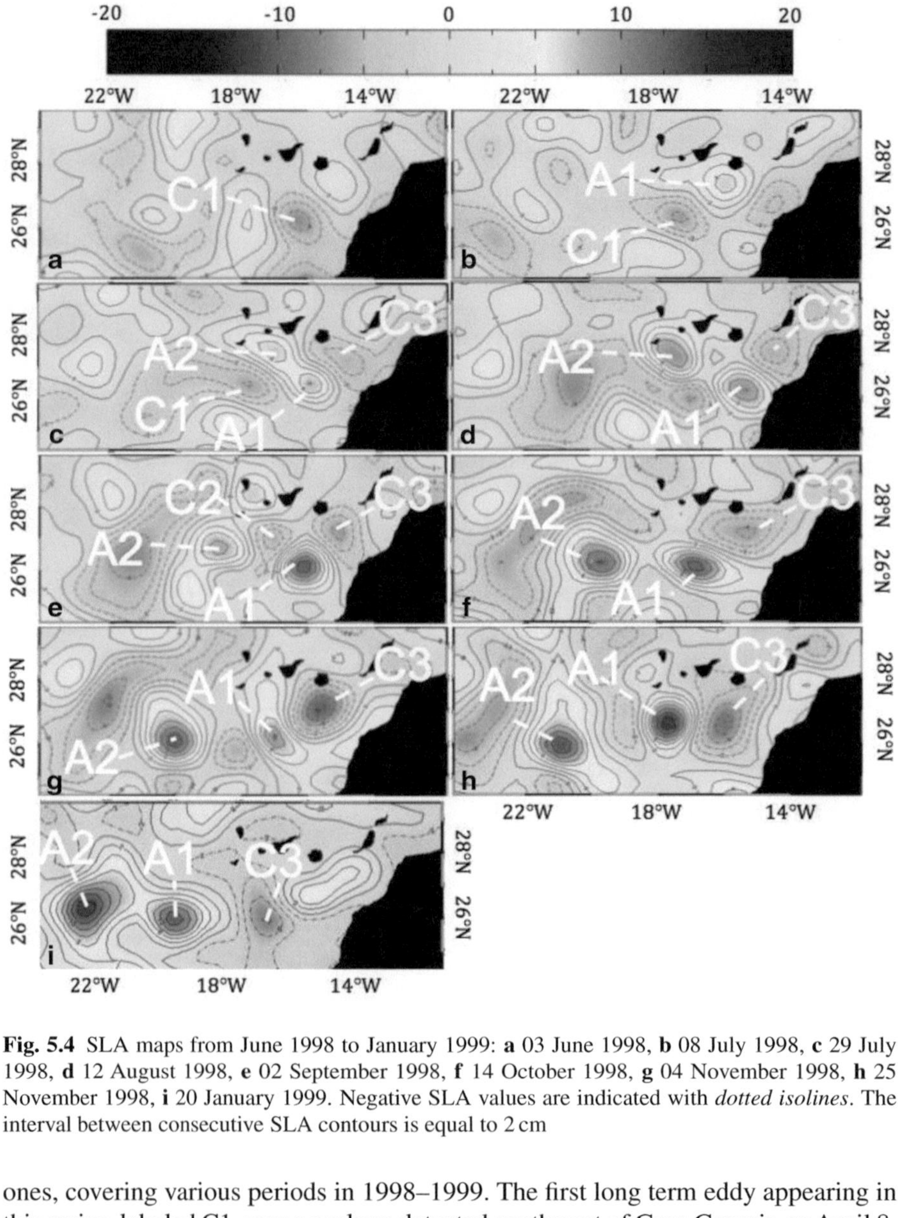

Fig. 5.4 SLA maps from June 1998 to January 1999: **a** 03 June 1998, **b** 08 July 1998, **c** 29 July 1998, **d** 12 August 1998, **e** 02 September 1998, **f** 14 October 1998, **g** 04 November 1998, **h** 25 November 1998, **i** 20 January 1999. Negative SLA values are indicated with *dotted isolines*. The interval between consecutive SLA contours is equal to 2 cm

ones, covering various periods in 1998–1999. The first long term eddy appearing in this series, labeled C1, was a cyclone detected southwest of Gran Canaria on April 8, 1998. This near-circular eddy (Fig. 5.3a) was advected southward until the beginning of June 1998, at an estimated mean speed of $\sim 3\,\text{cm s}^{-1}$, nearly doubling its initial size to ∼75 km (Fig. 5.3b). This made it observable also in SLA data (Fig. 5.4a). When the eddy approached 26.5° N, it started to progress westward, from 16.5° W to 18° W, at an estimated mean speed of $\sim 2.8\,\text{cm s}^{-1}$, increasing its size to ∼100 km

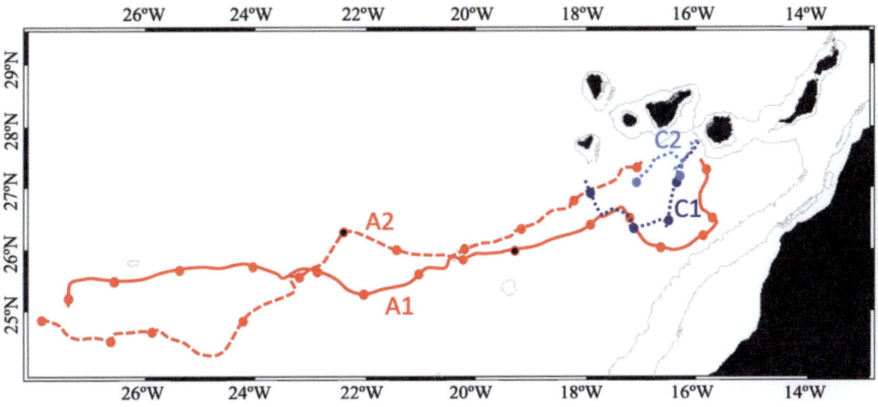

Fig. 5.5 Trajectories of island-generated eddies *A1* (*red solid line*), *A2* (*red dashed line*), *C1* (*dark blue dotted line*) and *C2* (*light blue dotted line*). The first observation date of these eddies is indicated in Table 5.1. The *colour circles* show the approximate eddies positions at the beginning of each month, and the *black circles* show their positions starting 1999

(Figs. 5.3c, d and 5.4b, c). Then, the eddy appeared to weaken, reducing its size to ~ 30 km and turning northwestward at a higher speed (~ 6 cm s^{-1}; Fig. 5.3g). This eddy trajectory and other ones described next are shown in Fig. 5.5. First, eddies center locations were derived from ellipse fitting to RS images and then the positions at the beginning of each month were computed by simple interpolation. This change in orientation and velocity of C1 shown in Fig. 5.5 could have been due to the proximity of a bigger and stronger anticyclonic eddy (labeled A2). In fact, in case of interaction of 2 vortices of different strength and opposite rotation, the weaker is driven around the stronger vortex in a parabolic path (Simpson and Lynn 1990).

On June 26, 1998, an anticyclonic eddy, A1, was detected south of Gran Canaria in BT (Fig. 5.3c) and Chl-a images and nearly at the same time in SLA maps (Fig. 5.4b). In a manner similar to C1, this eddy was initially advected to the south at an estimated mean speed of ~ 3.5 cm s^{-1} (Figs. 5.3d–f, 5.4c and 5.5). Then, starting in August, it moved ~ 35 km eastward instead of westward, possibly due to the presence of anticyclonic eddy A2, located further offshore (Figs. 5.3g, 5.4d and 5.5), approaching the African continental shelf and remaining nearly stationary around 26° N (Fig. 5.5 and 5.6a). Taking into account that Canary Islands eddies usually extend to several hundred meters of depth (Barton et al. 1998), and that in addition A1 had an initial core diameter of ~ 100–120 km, starting in September, it is likely that A1 might have been slowed down or even arrested by bottom topography, in agreement with what observed in the imagery. There is in fact previous evidence of *quasi* stationary anticyclones at this location, as in the case observed by García-Weil (1998) in AVHRR scenes dated from August to November 1992.

In September 1998, A1 started to move westward again, at an estimated mean speed of ~ 3 cm s^{-1}, in correspondence with the displacement of A2 further offshore, additionally increasing its size to ~ 160 km (Figs. 5.4e, f, 5.5 and 5.6b, c). Then, A1 apparently collided with yet another anticyclonic eddy located just

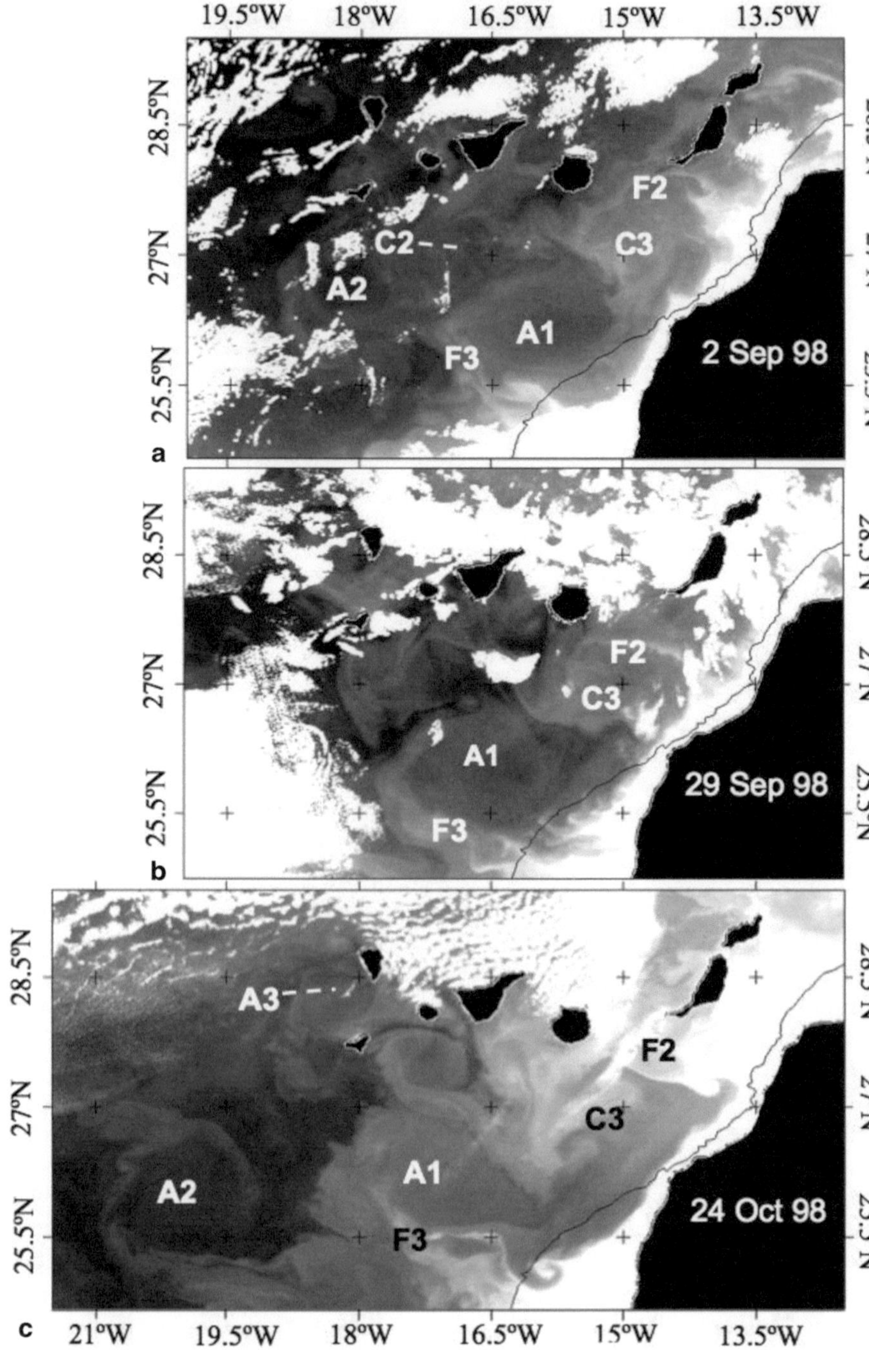

Fig. 5.6 BT images from AVHRR channel 4 (10.3–11.3 μm) over the period from September to October 1998 **(a–c)**

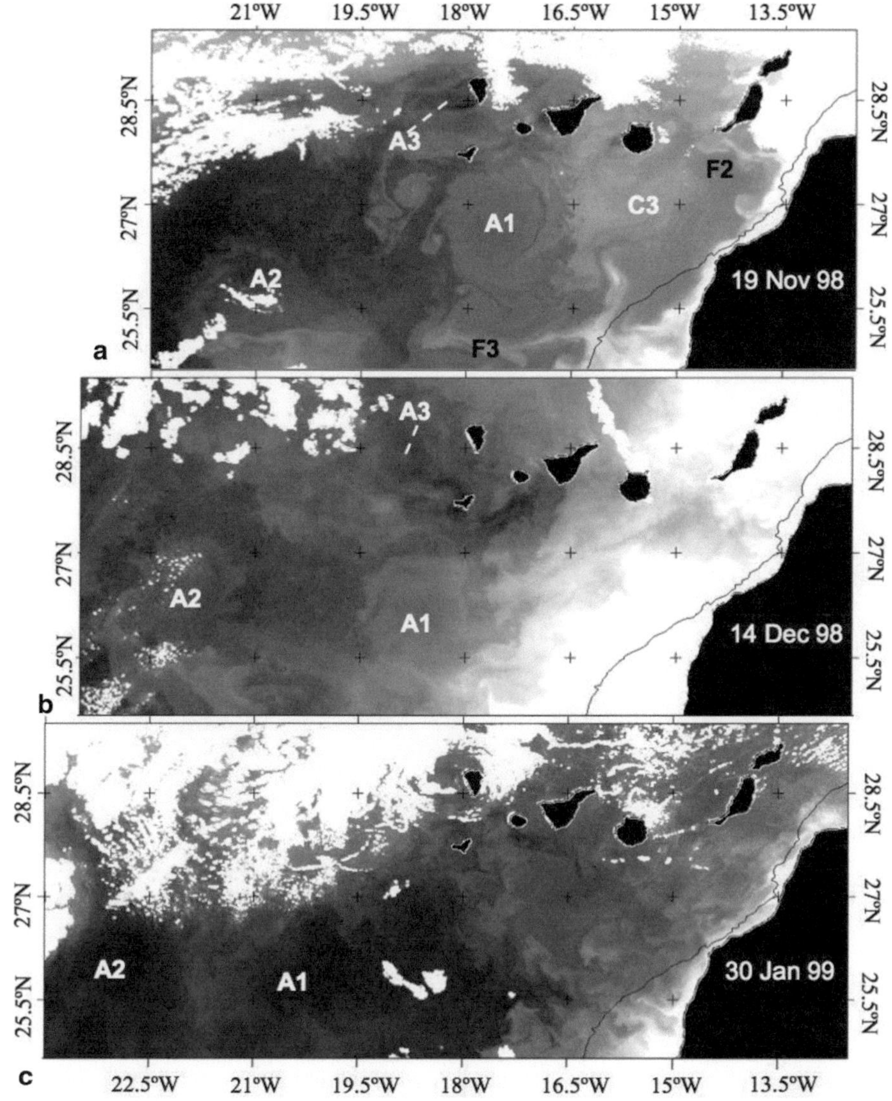

Fig. 5.7 BT images from AVHRR channel 4 (10.3–11.3 μm) over the period from November 1998 to January 1999 **(a–c)**

south of Tenerife (Figs. 5.4g and 5.6c). Once the coalescence process was completed, A1 resumed its southwestward displacement (Figs. 5.4h, 5.5 and 5.7a) at an estimated mean speed of $\sim 4.6\,\text{cm s}^{-1}$ (35 days average). Successively, in the middle of December (Fig. 5.7b), when the eddy reached a latitude next to 26° N, it advanced westward (Figs. 5.4i, 5.5 and 5.7c) at an estimated mean speed of $\sim 3.5\,\text{cm s}^{-1}$. A1 was clearly identified in AVHRR scenes only until January 31, 1999. After

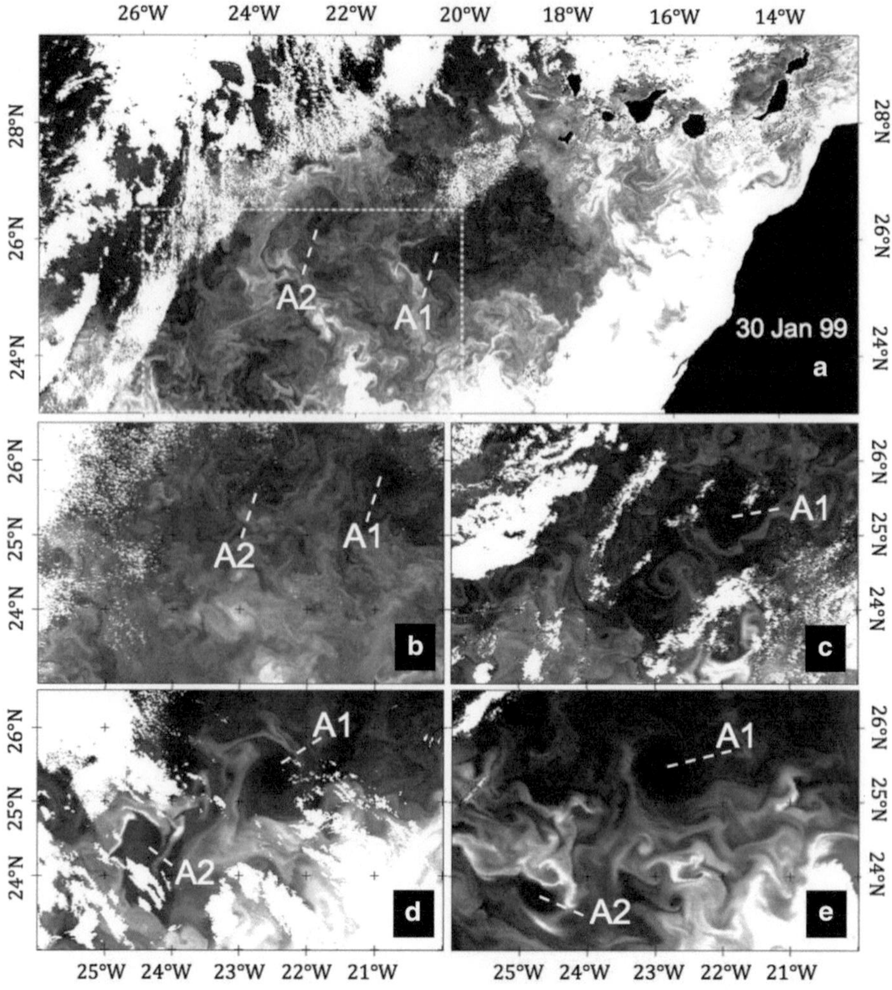

Fig. 5.8 Chl-a images from SeaWiFS over the period from January to April 1999. **a** 30 January 1999, **b** 21 February 1999, **c** 31 March 1999, **d** 11 April 1999, **e** 24 April 1999. Each image has been specifically enhanced to emphasize the features. Chlorophyll-like pigment concentrations increase from *dark* to *light* shades. The area inside the *dotted white line* in Fig. 5.8a is the region displayed in the rest of figures

that, the tracking of the same feature was done combining Chl-a (Fig. 5.8) and SLA data (Fig. 5.9). In both cases, A1 was observed moving westward. Its signal was last detected in the Chl-a data dated May 1999, but remained in SLA maps until August 18, 1999, more than 1 year after its first appearance in the satellite record.

Another cyclonic eddy located southwest of Gran Canaria, C2, was detected on July 3, 1998 (Fig. 5.3d) in approximately the same position than C1 (Fig. 5.5). It showed a mean diameter of ~ 50 km and a general westward motion, with important

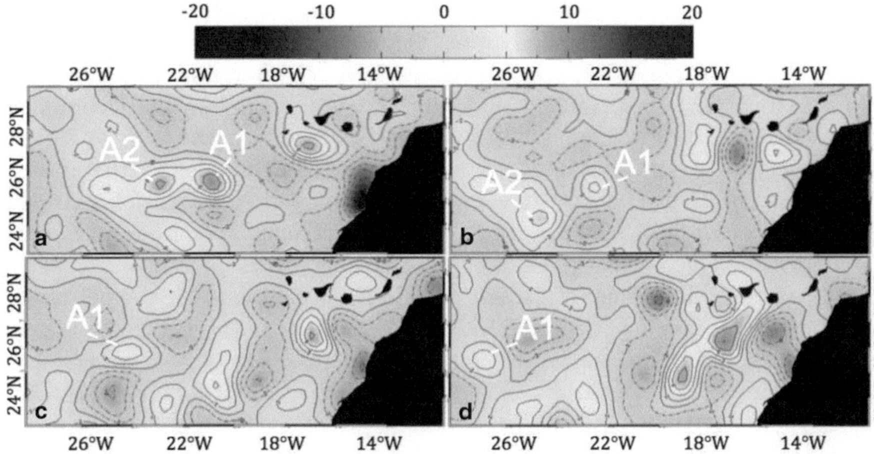

Fig. 5.9 SLA maps from February 1999 to August 1999: **a** 17 February 1999, **b** 14 April 1999, **c** 09 June 1999, **d** 11 August 1999

south–north meanderings (Fig. 5.5). In its early development, it advanced with a predominantly southward component and an estimated mean speed of $\sim 3\,cm\,s^{-1}$. After it detached from Gran Canaria, the eddy shape changed from circular (Fig. 5.3e) to elliptical, while its surface signal in the AVHRR scenes appeared to weaken (Fig. 5.3f, g). The apparent cause of the C2 shape and strength modifications was the squeeze of two anticyclones, A1 and A2 (Fig. 5.3f, g).

Once A2 moved away, on August 20, the surface signal of C2 intensified in both AVHRR and SeawiFS images, its shape returned circular (Fig. 5.3h), and it appeared also in the SLA maps (Fig. 5.4e). Next, C2 moved first to the north, probably under the influence of A1, and then to the west, at an estimated mean speed of $\sim 9\,cm\,s^{-1}$, for 6 days (Fig. 5.5). Finally, C2 turned to the south and propagated in a southwestward direction, being identified the last time on September 2 in both BT and Chl-a images (Fig. 5.6a). Negative values of SLA associated to C2 are observed in the altimeter maps until the end of September (Fig. 5.4e).

The long-term anticyclonic eddy labeled A2 was detected simultaneously in passive radiometer (Fig. 5.3e, f) and altimeter data (Fig. 5.4c) south of Tenerife on July 27, 1998. This eddy reached a diameter of $\sim 150\,km$ in only 1 month (Fig. 5.6a). It travelled southwestward at an estimated mean speed of $\sim 4.6\,cm\,s^{-1}$ (111 day average), but showing a predominant western component (Figs. 5.4e–g, 5.5 and 5.6c). The same size and direction were maintained until the second half of November, when the eddy center reached a position next to 25.8° N and 21° W (Figs. 5.4h, 5.5, 5.6c and 5.7a). Afterwards, the eddy continued progressing westward, meandering first north until the end of January 1999 (Figs. 5.4i, 5.5, 5.7b, c and 5.8a) and then south (Figs. 5.5, 5.8b–e and 5.9a, b). A2 was last seen in the AVHRR scene dated January 30, 1999, and in the SeawiFS scene dated April 24, 1999. In the SLA maps, the A2 signal could be detected until June 2, 1999. In total, A2 was followed for nearly 1 year by means of satellite observations.

An interesting aspect of the observations above is the apparent change in shape of the long-lived eddies A1 and A2, when they moved away from their generation area. Even if anticyclones are generally recognized as high-termperature (and low-chlorophyll) eddies, on occasion the converse also appears to be true. Accordingly, both A1 and A2 seem to exhibit a cold core in Fig. 5.7a, b. This is due to the fact that at this time the high insolation and the low winds render offshore surface waters warmer than that transported by the eddies. Indeed, the same eddies displayed again a high-temperature (and low-chlorophyll) core 1 month later (Figs. 5.7c and 5.8a–e), when convection produced by surface cooling, aided by wind stirring, weakened the seasonal stratification. This resulted in colder offshore waters, with higher phytoplankton content than that observed in A1 and A2. These changing environmental conditions could be also related to modifications of the sub-mesoscale regime in which the mesoscale features, A1 and A2, are embedded. The Chl-a images of January and February 1999 (Fig. 5.8a, b) show a sub-mesoscale field dominated by short wavelength features that complicate even the mere identification of A1 and A2 in the satellite images. The spatial scales of this sub-mesoscale field increased gradually with time, as shown in the SeaWiFS scenes of April 1999 (Fig. 5.8d, e), in which A1 and A2 are easily observed.

A third anticyclonic eddy, labeled A3, was observed southwest of La Palma island on September 16, 1998. Opposite to the previously described eddies, which separate from their generation area in approximately 2 weeks, A3 was retained in the same position close to La Palma island until the end of November (i.e. more than 70 days), displaying a constant diameter of ~ 80 km (Figs. 5.6c and 5.7a). Afterwards, it shifted quickly to the west (Fig. 5.7b), at an estimated mean speed of $\sim 6.5\ \mathrm{cm\ s^{-1}}$ (17 day average), slowing down subsequently, and being observed last on December 26, 1998. This anticyclonic eddy was never coupled to a counter-rotating cyclonic eddy, for the whole period in which it was attached to La Palma island (unlike what occurs in the leeway of an isolated obstacle when leeside attached eddies appear). Other works reporting the presence of trapped eddies, downstream of isolated islands and seamounts, argue that sometimes incident flows are not strong enough to generate a vortex street of eddies (Bograd et al. 1997). A similar explanation could be given for the prolonged (September to November) attachment of A3 southwest of La Palma.

Another long-lived cyclonic eddy, labeled C3 and located over the bathymetric trough downstream of the shallower channel between Fuerteventura and Africa, was observed for the first time on June 27, 1998 (Fig. 5.1a). Unlike the other eddies described earlier, C3 was not generated and shed from one of the islands. Its origin is thought to be related to potential vorticity conservation, as the alongshore flow encounters the deeper water of the trough, producing such an eddy as long as the flow through the zone is equatorward (Barton et al. 1998, 2004). C3, detected in Chl-a, BT and SLA observations, showed a mean diameter of ~ 90 km and initially remained stationary around 27.2° N and 15° W (Figs. 5.3d–h, 5.4c–e and 5.6a). In September, it started to shift westward, approximately at the same time as A1 began to move offshore (Figs. 5.4f and 5.6b, c). One month later, C3 was located south of Gran Canaria (Figs. 5.4g, h and 5.7a). The last evidence of C3 in passive radiometer data occurred on November 19. It was still observed, moving southwestward (Fig. 5.4i), in SLA maps until February 10, 1998.

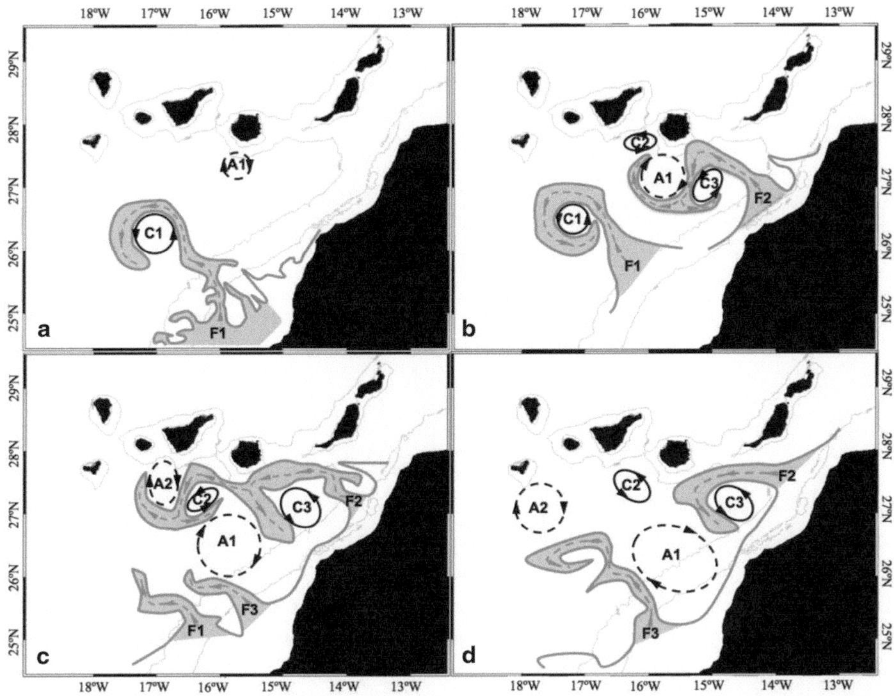

Fig. 5.10 Sketches of eddy-filament and eddy–eddy interactions for different dates in 1998: **a** 13 May 1998, **b** 9 July 1998, **c** 31 July 1998, and **d** 19 August 1998

Mesoscale eddies are not the only long-lived features appearing in the area of interest. Some upwelling filaments, driving cold and chlorophyll-rich waters from the African coast to the open ocean, were observed for several months in RS data generated by passive radiometers (Fig. 5.10). The first one, labeled F2, was first detected in a Chl-a image on June 27, 1998, extending offshore between Cape Juby and Cape Bojador. It then turned southward and approached Gran Canaria, reaching a length of more than 250 km (Figs. 5.3d–h, 5.6a–c and 5.10). This narrow feature appeared shortened on November 19 (Fig. 5.7a) and was observed until December 10, 1998 (Fig. 5.7a). A second filament, labeled F3, was present on July 27, 1998, in a BT image south of Cape Bojador (Figs. 5.3e and 5.10). F3, also with a length greater than 250 km, invaded open ocean waters deflecting its trajectory northward (Fig. 5.6a–c). This filament was observed for the last time on November 19, 1998 (Fig. 5.7a).

Table 5.1 summarizes the spatial and temporal characteristics of all mesoscale features described above, providing the coordinates of the first location place, the first observation date and the period of time, in days, during which each feature remained visible.

Table 5.1 Long-term mesoscale features generated in the Canary Islands area during 1998

Feature	First location place	First observation date	Days
C1	SW Gran Canaria (27.55° N, – 16.18° W)	08/04/98	124
C2	SW Gran Canaria (27.80° N, – 16.05° W)	03/07/98	83
C3	Fuerteventura-Africa (27.54° N, – 14.87° W)	27/06/98	230
A1	S Gran Canaria (27.37° N, – 15.81° W)	26/06/98	419
A2	S Tenerife (27.56° N, – 17.02° W)	27/07/98	311
A3	SW La Palma (28.28° N, – 18.30° W)	16/09/98	102
F2	S Cape Juby (27.60° N, – 13.71° W)	27/06/98	167
F3	S Cape Bojador (25.51° N, – 15.80° W)	27/07/98	87

5.3.2.2 Eddies and Cross Shelf Transport

The creation of filaments in the Canary Islands area is strongly promoted by mesoscale eddies. In fact, eddies C1 and A1, both of which detached from Gran Canaria island, generated upwelling filaments F1 (Fig. 5.3c–e) and F3 (Fig. 5.3e–h) when moving southward and approaching the African continental shelf. Starting in June 1998, C1 reached its most southern position at 26.2° N and 16.5° W, just south of Cape Bojador and next to the latitude where the coastal shelf gets wider (see Fig. 5.1a). At this time, a fully developed and irregular coastal upwelling front, showing wavelike disturbances (some of them evolving into small filaments), was observed. Then, the interaction between the eddy and one of these disturbances gave place to a narrow and long tongue, about 10–15 km wide and $\sim$200 km long (from the coastal upwelling front) labeled F1. This is shown in Figs. 5.3c, d and 5.10a, b, in which C1 appears to have reached its final position.

At the end of July 1998, after moving southward, A1 was also implicated in the formation of filament F3, located south of Cape Bojador. This process started when the upwelled waters situated south of 26° N (again the location where the continental shelf expands) felt the shear induced by the southern rim of the eddy (Fig. 5.3e). The upwelled waters were expelled away from the coast, forming a small plume that quickly advanced offshore, around eddy A1 (Figs. 5.3f and 5.10c). Then, in 3 weeks, a long and narrow filament was clearly seen extending offshore, as A1 moved westward (Figs. 5.6a–c and 5.10d). F3 persisted, in association with A1, until the end of October, when A1 collided with another anticyclonic eddy (Fig. 5.7a).

In early June 1998, a small disturbance was observed in the coastal upwelling front close to Cape Juby, in both the Chl-a and BT images. In successive images, this small-scale feature was seen growing, until finally, on June 27, 1998, the long filament F2 appeared, as entrained around the nearly stationary eddy C3. This filament stayed associated with C3 (Fig. 5.10b–d) until the end of October 1998, when the eddy was placed south of Gran Canaria.

Both filaments F2 and F3 seem to be recursive oceanographic features linked to eddies in the Canary Islands area, as analogous features have been observed jointly in other occasions (see for example Fig. 11c in Barton et al. 1998). In addition, not only the eddies participate in filament generation, but they can also operate in combination with the same filaments to export the coastal upwelled waters at great distances. RS images show often anticyclonic eddies shedded by Gran Canaria interacting with a filament similar to F2, entraining cold and high-chlorophyll waters in their peripheries (Arístegui et al. 1997; Pacheco and Hernández-Guerra 1999). This situation repeated again in July, 1998 (Figs. 5.3d and 5.10d) and was produced by the newly generated anticyclonic eddy A1. Occasionally, various eddies may be involved at the same time in the offshore export of upwelled water. This can be observed starting in August 1998 (Figs. 5.3f, g and 5.10c), when the eddies A2 and C2 worked in combination as a gear-like system, spreading coastal water transported by the filament at a distance of ~350 km from the coast.

5.3.2.3 Eddy–Eddy Interactions

South of the Canary Islands, eddy motion is not only influenced by advection due to the Canary Current, but also through the interaction with other vortices. Relatively small cyclonic eddies, like C1 and C2, are driven around stronger anticyclonic vortices, and in the case of C2 even squeezed and distorted, when entrained between A1 and A2, as already seen earlier (Fig. 5.10c).

Similarly, the motion of bigger eddies is also influenced by other factors besides the Canary Current. Thus, the path of A1 next to its generation area was determined by the presence offshore of A2 and by the proximity of the African coast. This produced the arresting or slowing down of eddy A1 (Figs. 5.6a and 5.10c, d). In a similar way, the free movement of C3 was impeded by the coastal topography and the blocking effect of A1, located south of C3 (Figs. 5.3e–h and 5.10c, d). When A2 moved westward, at the beginning of September 1998, A1 and C3 started to drift also from their nearly stationary positions (Fig. 5.6b, c). The movement of these larger vortices is coupled next to their generation area. Far from the formation region, eddies move to the west under the combined influence of the Canary Current and the beta effect, covering about 100 km *per* month (Figs. 5.5, 5.7, 5.8 and 5.9).

The larger size and longer time span of anticyclonic eddies contribute to the collision and possible merging of these mesoscale features in Canarian waters. In October 1998, mature eddy A1 interacted with a new anticyclonic eddy located south of Tenerife. On October 24, A1, having a mean size of ~135 km, and the younger eddy, having a mean size of ~90 km, adopted a figure 8 structure (Fig. 5.6c), with the centres of the eddies separated by 130 km (less than three times the mean radius of the vortices, which was ~55 km). From October 26 to November 8 there were no clear AVHRR or SeaWiFS images available. Afterwards, a bigger, reinforced anticyclonic eddy, with a diameter of ~160 km, was observed, as the result of the coalescence between the two eddies (Figs. 5.4h and 5.7a). This suggests that vortex merging is a quick process, which can last less than 10 days. In spite of its short duration, the process could be detected also in the weekly SLA maps, where the two approaching eddies appear as an elongated feature (Fig. 5.4g).

5.4 Concluding Remarks

The combination of different RS data types, with different spatial and temporal resolutions, and obtained in different regions of the electromagnetic spectrum, allows for the effective monitoring of a region with complex mesoscale variability patterns such as the Canary Islands area. It has been shown that this area has a higher mesoscale activity than other parts of the Northwest African coastal transition zone. Its distinctive mesoscale activity has a clear seasonality, with a maximum in October and a minimum in April, and is strongly related to the eddy generation processes typical of the Canarian Archipelago.

The detailed analysis of the 1998–1999 historical data record, combining the complementary characteristics of various RS data sources, shows that most of the the long-lived mesoscale eddies develops during the second half of the year. A similar number of cyclonic and anticyclonic eddies was generated, in the period of observation, but over extended periods the mesoscale dynamics seemed to be dominated by anticyclones. These displayed a bigger size (~ 150 km) and were observed during a longer time than cyclones.

Eddy trajectories and velocities revealed that the actual motion observed depended on advection by ocean currents; on the influence of other eddies; on the arresting effect exerted by the coast, when eddies were close to the upwelling area; and also on the own drift of eddies, produced by the planetary beta effect. Related to this, a common characteristic of the path of eddies was the westward movement of the vortices, certainly conditioned by the combined influence of the southwestward Canary Current and the westward drift of the beta effect. Consequently, it is possible to find long-lived anticyclonic eddies more than 1000 km west of the island where they were generated, at approximately 26°N.

Finally, the available data set has revealed significant interactions between eddies generated in the Canary Islands area. Drastic structural changes could be observed when anticyclonic eddies became quite close, giving place to a new eddy resultant by their coalescence. The observations suggest that this kind of interaction can occur commonly south of the Canarian Archipelago. Island-shed eddies interact also with waters of the upwelling region, and can operate jointly with upwelling filaments to export effectively coastal waters far offshore. At the same time, if the eddies get closer to the continental shelf, they can induce the formation of upwelling filaments, as observed for the cyclonic and anticyclonic eddy pair, observed south of Cape Juby and Cape Bojador.

Acknowledgments The authors would like to thank AVISO Altimetry for providing the altimeter data, and also CREPAD (INTA) for supplying the Level-1b ESA-Sharp AVHRR data. SeaWiFS Level-2 data were provided by the NASA Ocean Biology Processing Group (Feldman and McClain 2007). This work was supported by Spanish *Ministerio de Ciencia e Innovación* through the INTEGCAN (MCINN, CGL2004 02235) project and the MOMAC (MCINN, CTM2008-05914/MAR) project.

References

Arístegui J, Sangrá P, Hernández-León S, Cantón M, Hernández-Guerra A, Kerling JL (1994) Island-induced eddies in the Canary Islands. Deep-Sea Res 41(1):1509–1525

Arístegui J, Tett P, Hernández-Guerra A, Basterretxea G, Montero MF, Wild K, Sangrá P, Hernández-León S, Cantón M, García-Braun JA, Pacheco M (1997) The influence of island-generated eddies on chlorophyll distribution: a study of mesoscale variation around Gran Canaria. Deep-Sea Res 44(1):71–95

Barton ED, Arístegui J, Tett P, Cantón M, García-Braun J, Hernández-León S, Nykjaer L, Almeida C, Almunia J, Ballesteros S, Basterretxea G, Escánez J, García-Weil L, Hernández-Guerra A, López-Laatzen F, Molina R, Montero MF, Navarro-Pérez E, Rodríguez-Pérez JM, Van Lenning K, Vélez H, Wild K (1998) The transition zone of the Canary Current. Prog Oceanogr 41:455–504

Barton ED, Basterretxea G, Flament P, Mitchelson-Jacob EG, Jones B, Arístegui J, Herrera F (2000) Lee region of Gran Canaria. J Geophys Res 105:17173–17193

Barton ED, Arístegui J, Tett P, Navarro-Pérez E (2004) Variability in the Canary Islands area of filament-eddy exchanges. Progr Oceanogr 62(2004):71–94

Basterretxea G, Barton ED, Tett P, Sangrá P, Navarro-Pérez E, Arístegui J (2002) Eddy and deep chlorophyll maximum response to wind-shear in the lee of Gran Canaria. Deep-Sea Res 49(1):1087–1101

Bograd SJ, Rabinovich AB, LeBlond PH, Shore JA (1997) Observations of seamount-attached eddies in the North Pacific. J Geophys Res 102:12441–12456

Chelton DB, Schlax MG, Samelson RM (2011) Global observations of nonlinear mesoscale eddies. Progr Oceanogr 91:167–216

Ducet N, Le Traon PY, Reverdin G (2000) Global high-resolution mapping of ocean circulation from TOPEX/Poseidon and ERS-1 and -2. J Geophys Res 105:19477–19498

Feldman GC, McClain CR (2007) Ocean Color Web, SeaWiFS Reprocessing R2010.0, NASA Goddard Space Flight Center. Kuring N, Bailey SW (eds) http://oceancolor.gsfc.nasa.gov. Accessed 10 August 2011

García-Weil L (1998) Descripción y análisis cuantitativo mediante series de imágenes de satélite de la dinámica de las aguas superficiales del noroeste de África. Ph D Thesis, Univ de Las Palmas de Gran Canaria. http://acceda.ulpgc.es/handle/10553/1996

La Violette PE, Holyer RJ (1988) Noise and temperature gradients in multichannel sea surface temperature of the ocean. Remote Sens Environ 25:231–241

Ladd C (2007) Interannual variability of the Gulf of Alaska eddy field. Geophys Res Lett 34:L11605. doi:10.1029/2007GL029478

Le Traon PY, Dibarboure G (1999) Mesoscale mapping capabilities od multiple-satellite altimeter missions. J Atmos Ocean Tech 16:1208–1223

Pacheco MM, Hernández-Guerra A (1999) Seasonal variability of recurrent phytoplankton patterns in the Canary Island area. Int J Remote Sens 20:1405–1418

Piedeleu M, Sangrà P, Sánchez-Vidal A, Fabrés J, Gordo C, Calafat A (2009) An observational study of oceanic eddy generation mechanisms by tall deep-water islands (Gran Canaria). Geophys Res Lett 36:L14605. doi:10.1029/2008GL037010

Robinson IS (2004) Measuring the oceans from space: The principles and methods of satellite oceanography. Springer-Praxis, Chichester

Robinson IS (2010) Discovering the oceans from space: The unique applications of satellite oceanography. Springer-Praxis, Chichester

Sangrà P, Pelegrí JL, Hernández-Guerra A, Arregui I, Martín JM, Marrero-Díaz A, Martínez A, Ratsimandresy AW, Rodríguez-Santana A (2005) Life history of an anticyclonic eddy. J Geophys Res 110:C03021. doi:10.1029/2004JC002526

Simpson JJ, Lynn RJ (1990) A mesoscale eddy dipole in the offshore California Current. J Geophys Res 95:13009–13022

SSALTO/DUACS user handbook (2012), (M)SLA and (M)ADT near-realtime and delayed time products, CLS-DOS-NT-06–034, Toulouse, France

Tejera A, García-Weil L, Heywood KJ, Cantón-Garbín M (2002) Observations of oceanic mesoscale features and variability in the Canary Islands area from ERS-1 altimeter data, satellite infrared imagery and hydrographic measurements. Int J Remote Sens 23:4897–4916

Tran N, Obligis E, Ferreira F (2005) Comparison of two Jason-1 altimeter precipitation detection algorithms with rain estimates from the TRMM microwave imager. J Atmos Ocean Tech 22:782–794

Van Camp L, Nykjaer L, Mittelstaedt E, Schlittenhardt P (1991) Upwelling and boundary circulation off northwest Africa as depicted by infrared and visible satellite observations. Prog Oceanogr 26:357–402

Chapter 6
Impacts of Saharan Dust on the Marine Environment in the Area off Northwest Africa

Thomas Ohde and Herbert Siegel

Abstract The main sources of nutrient supply in the area off Northwest Africa are upwelling and Saharan dust deposition. Their influence on phytoplankton development was evaluated by different methods. Statistical analyses showed that alongshore wind stress and induced upwelling were most significantly responsible for surface Chl-a variability mainly in winter and spring with delay of up to 16 days. Only 5 % of the Chl-a variability was related to the dust input. All strong Saharan dust storms between 2000 and 2008 were identified by remote sensed dust aerosol optical depth. Only some of the events caused a biological response expressed by an increase of surface chlorophyll-a. Time lags between dust deposition and enhancement of chlorophyll-a between 8 and 16 days were determined. The chlorophyll-a concentration was increased up to 2.4 mg m^{-3}. Atmospheric dust modifies the amount and spectral distribution of the water light field. In the applied optical model the dust effect on incident solar radiation was parameterized by radiation measurements. The photosynthetically active radiation was reduced up to 19.3 % in the upper water column. The impacts of spectral effects on photosynthetically active radiation were different in oceanic and coastal regions. A compensation of the spectral effects at water depths of about 7 m was only observed in ocean regions because of different light attenuation in coastal areas. Saharan dust also impacts satellite derived ocean surface wind speeds. The influence depended mainly on the strength of dust storms, the microwave frequency and the wind speed. The impact was higher for stronger dust storms, higher frequencies and lower wind speeds.

6.1 Introduction

This chapter summarises satellite based investigations in the upwelling area off Northwest (NW) Africa focussed on the biological response to upwelling and dust deposition, as well as on the quantification of atmospheric dust impact on

T. Ohde (✉) · H. Siegel
Leibniz Institut für Ostseeforschung Warnemünde,
Rostock, Germany
e-mail: thomas.ohde@io-warnemuende.de

V. Barale, M. Gade (eds.), *Remote Sensing of the African Seas*,
DOI 10.1007/978-94-017-8008-7_6,

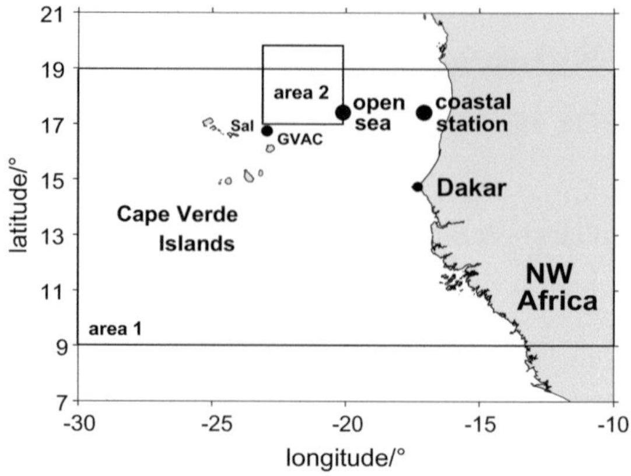

Fig. 6.1 The study area for statistical analyses enclosed the large box (area 1: 10° to 30°W, 9° to 19°N). The small box near the Cape Verde Island Sal was selected in order to study the influence of Saharan dust on satellite-derived ocean surface wind speeds (area 2: 20° to 23°W, 17° to 20°N). The map shows the location of the airport weather station on the island Sal (GVAC). The *in situ* optical measurements at the open sea station and at the coastal station are marked with *black dots*

Photosynthetically Available Radiation (PAR) below the water surface and on microwave satellite derived surface wind speed.

Saharan dust can act as a fertilizer for phytoplankton and may influence its development and composition (e.g. Young et al. 1991; Meskhidze et al. 2005; Duarte et al. 2006). Nutrient supply in the study area (Fig. 6.1) is mainly regulated by coastal upwelling and dust deposition (e.g. Baker et al. 2003; Duarte et al. 2006). The strength of coastal upwelling and the nutrient supply by dust depend on local wind conditions (e.g. Van Camp 1991; Pradhan et al. 2006). The dust input is strongly variable and depends on the season (e.g. Prospero and Carlson 1972; Gao et al. 2001). Differences concerning the biological response to dust input were determined in different regions of the oceans. Dust input into the Pacific Ocean did not always lead to phytoplankton blooms (Meskhidze et al. 2005), but examples showed an increase in primary production of more than 60 % (Young et al. 1991). In the eastern subtropical North Atlantic a ten-fold increase of phytoplankton production was observed within 4 days (Duarte et al. 2006).

Saharan dust in the atmosphere affects the amount and the spectral distribution of the incident radiation (e.g. Frederick and Erlick 1997; Ohde and Siegel 2013) and influences the PAR in the water column. PAR is an important quantity in marine biology because it is the solar radiation in the visible spectrum from 400 to 700 nm that phytoplankton uses for photosynthesis. The incident solar radiation above the water surface can be attenuated by dust up to 22 % (Otto et al. 2009; Ohde and Siegel 2013). Dust impacts above all the spectral distribution in the blue spectral range. Reductions up to 19.7 % at 400 nm were observed (Ohde and Siegel 2012a).

The deposition of atmospheric dust in the water column can affect the seawater optical properties (Kopelevich 2012). The absorption at 440 nm of supplied dust can be in the order of pure water absorption and the scattering can be twenty times higher than the pure seawater scattering (Stramski et al. 2004).

Atmospheric dust can disturb satellite measurements of surface wind speeds because of the attenuation of the microwave signal. Recent investigations established that the attenuation depends on the operating satellite frequency (Comparetto 1993), the dust particle radius, the density of dust particles, the permittivity (Altshuler 1983) and the moisture content (e.g. Ansari and Evans 1982).

One goal of the present investigation was the study of the relationships between Saharan dust deposition, upwelling process and phytoplankton on the basis of statistical analysis of remotely sensed data and event statistics of dust storms. Another aim was the determination of the influence of atmospheric dust on PAR below the water surface on the basis of an optical model along with parameterized spectral effects. These studies were restricted to the dust effect on PAR if the dust is in the atmosphere and not in the water column. The present investigation also contributed to the validation of satellite microwave sensors by the quantification of the impact of Saharan dust on satellite derived ocean surface wind speeds.

6.2 Methodology

The investigations in this chapter based on satellite observations of different sensors, radiation measurements below and above the water surface in the area of NW Africa and modelling of optical processes in the water column. The following remarks of the used methods were given in detail in different papers (Ohde and Siegel 2010, 2012a, b, 2013).

6.2.1 Methods in Relation to Dust Impact on Phytoplankton

A time series of 8-day (8d) means of geophysical data of Sea-Wide Field-of-view Sensor (SeaWiFS), microwave scatterometer SeaWinds (QuikScat) and Moderate Resolution Imaging Spectroradiometer (MODIS) were the basis of the investigations of the dust impact on phytoplankton, in the investigation period from 2000 to 2008 (Ohde and Siegel 2010).

Remotely sensed chlorophyll-a (Chl-*a*) concentration[1] was used as a proxy for phytoplankton biomass. As a proxy for the strength of the coastal upwelling, the satellite-derived alongshore wind stress (τ_v) was determined according to Yelland and Taylor (1996) based on satellite wind data[2]. The dust component of the Aerosol Optical Depth (AOD_{dust}) was used as a proxy for the atmospheric dust column load[3].

[1] From http://daac.gsfc.nasa.gov.

[2] From http://www.remss.com.

[3] From http://daac.gsfc.nasa.gov.

The AOD_{dust} data were derived from the aerosol optical depths at 550 nm of MODIS using the approach of Kaufman et al. (2005). A linear approach between AOD_{dust} and the surface deposition was used (see e.g. Kaufman et al. 2005).

Area-averaged means were calculated from 8d means of the geophysical variables AOD_{dust}, Chl-*a* and τ_v. The signals of yearly cycle of AOD_{dust}, Chl-*a* and τ_v were eliminated by the determination of anomalies. The seasonal means of time period from 2000 to 2008 were subtracted from each area-averaged 8d mean data point. The anomalies were averaged over three month to eliminate fluctuations.

The generated time series of anomalies were analyzed by cross- and multiple correlations to get potential relations between the geophysical variables AOD_{dust}, Chl-*a,* and $\tau_{v.}$ The cross-correlation establishes the degree to which two time series are correlated and the potential time lag between cause and effect. Multiple correlation evaluates the contribution of each independent process to the variability of the dependent variable. Detailed information's about the used methods were given in Ohde and Siegel (2010).

The derived time series of area-averaged means were used to determine all strong Saharan dust storms in the time period from 2000 to 2008. The identified storms were investigated whether there was an increase in the Chl-*a* concentrations that could have been caused by the dust storm. The Chl-*a* concentration could also be influenced by intensified coastal upwelling because of the increased alongshore wind stress during storms. Therefore, the dataset of τ_v was searched for constant or decreased values during dust storms. The increase in the Chl-*a* concentration of such events should be related to nutrient supply by dust input, but not by coastal upwelling.

6.2.2 Methods in Relation to Dust Impact on PAR

6.2.2.1 Optical Model Description

The influence of Saharan dust on PAR in the water column was investigated with an optical model which separated the effects of dust on the amount and the spectral distribution of the incident radiation. This optical model was introduced here shortly because the detailed description was given in Ohde and Siegel (2012b). The influence of dust on the magnitude of PAR was studied with the relation

$$\frac{\Delta_m Q_{PAR}(z)}{Q_{PAR}(z, N=1, S=1)} = \frac{Q_{PAR}(z) - Q_{PAR}(z, N=1, S=1)}{Q_{PAR}(z, N=1, S=1)} * 100\% \qquad (6.1)$$

The term $\Delta_m Q_{PAR}(z)/Q_{PAR}(z, N = 1, S = 1)$ described the relative deviation of PAR of the dust case in relation to the non-dust case where $N = 1$ and $S = 1$. $Q_{PAR}(z)$ was the PAR including the dust effect and $Q_{PAR}(z, N = 1, S = 1)$ was the PAR without the dust impact. On the same way, the influence of the spectral effect of dust on PAR was investigated with the relative deviation given by

$$\frac{\Delta_s Q_{PAR}(z)}{Q_{PAR}(z, S=1)} = \frac{Q_{PAR}(z) - Q_{PAR}(z, S=1)}{Q_{PAR}(z, S=1)} * 100\% \qquad (6.2)$$

where $Q_{PAR}(z)$ included the spectral effect and $Q_{PAR}(z, S = 1)$ excluded this effect. The PAR was defined by the equation

$$Q_{PAR}(z) = \frac{1}{hc} * \int_{400}^{700} \lambda * E_d(z, \lambda) d\lambda. \tag{6.3}$$

The quantity h was the Planck's constant, c was the velocity of light in vacuum, λ was the wavelength and z was the water depth. The quantity $E_d(z, \lambda)$ was the downward irradiance below the water surface determined by

$$E_d(z, \lambda) = E_d(0^-, \lambda) \exp\left[- \int_0^z K_d(z, \lambda) dz \right] \tag{6.4}$$

The exponential described the change of downward irradiance in the water column by the light attenuation expressed in terms of the diffuse attenuation coefficient $K_d(z, \lambda)$. The downward irradiance just below the water surface was given by

$$E_d(0^-, \lambda) = (1 - \sigma + \overline{\sigma} * r(\lambda)) * E_d(0^+, \lambda) \tag{6.5}$$

The first term defined the modification of the downward irradiance by reflection processes at the interface between air and water (σ- reflection coefficient for downward irradiance from air to water, $\overline{\sigma}$- reflection coefficient for upward irradiance from water to air, $r(\lambda)$- total internal reflectance given by $r(\lambda) = \pi L_u(0^-, \lambda)/E_d(0^+, \lambda)$ with the upward radiance just below the water surface $L_u(0^-, \lambda)$ and the above water downward irradiance $E_d(0^+, \lambda)$). The second term was the downward irradiance just above the water surface given by

$$E_d(0^+, \lambda) = E_{d,cs}^{gc}(0^+, \lambda) * M(0^+, \lambda) * N(0^+, 490\ nm) * S(0^+, \lambda, N) \tag{6.6}$$

The first term described the downward irradiance of clear skies using the model of Gregg and Carder (1990) and the second one the correction of instrumental errors or inaccurate modeling. The third term is a normalization factor and determined the influence of dust on magnitude of incident radiation. The quantity $S(0^+, \lambda, N)$ described the effect of dust on the spectral distribution of the incident radiation (Bartlett et al. 1998; Ohde and Siegel 2012b).

6.2.2.2 Optical Model Input

The application of the optical model required the knowledge of the clear sky downward irradiance $E_{d,cs}^{gc}$, the correction term M, the normalization factor N, the spectral effect S, the total internal reflectance r, the diffuse attenuation coefficient K_d as well as the reflection coefficients σ and $\overline{\sigma}$. The detailed description was given in Ohde and Siegel (2012b).

The clear sky downward irradiance $E^{gc}_{d,cs}(0^+,\lambda)$ was determined by the model of Gregg and Carder (1990) usable for cloudless and dustless maritime atmospheres. The model delivered $E^{gc}_{d,cs}(0^+,\lambda)$ from 400 to 700 nm with a spectral resolution of 1 nm and a root-mean-squared error of only 6.2 %. Meteorological data of different sources were used for the calculation of $E^{gc}_{d,cs}(0^+,\lambda)$. The atmospheric sea level pressure and the relative humidity were taken from NCEP (National Centers of Environmental Prediction). Daily wind speeds and daily total precipitable water of AMSR-E (Advanced Microwave Scanning Radiometer) were obtained from the Remote Sensing System, Santa Rosa. The total ozone was extracted from the MSG (Meteosat Second Generation) product of EUMETSAT (European Organisation for the Exploitation of Meteorological Satellites). The METAR (Aviation Routine Weather Reports) dataset provided by NCAR/EOL (National Center for Atmospheric Research/Earth Observation Laboratory) under sponsorship of the NSF (National Science Foundation) was the source of the visibility of the airport station GVAC (Fig. 6.1).

Measurements of downward irradiance $E_d(0^+,\lambda)$ at sea level in February 2008 were used to determine the correction term M, to derive the range of the normalization factor N and to parameterize the spectral effect of dust S. Measurements during clear skies and Sahara dust storms were identified by study of the corresponding meteorological dataset, by analysis of the irradiance spectra and by comparison to the dust aerosol optical depth. The correction term $M(0^+,\lambda)$ considered instrumental errors and inaccurate modeling which caused deviations of the observed clear sky irradiances from the modeled ones (Ohde and Siegel 2013). The normalization factor $N(0^+,490nm)$ was the ratio of observed downward irradiance to the modelled clear sky irradiance at the wavelength of 490 nm. The normalization factor was equal to 1 for the special case of clear sky. Minimum ($N^{\min}=0.81$) and maximum normalization factors ($N^{\max}=0.93$) were determined from measurements of downward irradiance at sea level during dusty skies. The parameterized spectral effect of dust $S(0^+,\lambda,N)$ was derived from the classified downward irradiances according to the method of Bartlett et al. (1998). The following power function was delivered (Ohde and Siegel 2013)

$$S(0^+,\lambda,N) = 0.82 + 0.31 * N + (0.18 - 0.31 * N) * (\lambda/490)^{-2} \tag{6.7}$$

Measurements of downward irradiance $E_d(0^-,\lambda)$ in the water column in February 2008 were used to derive the total internal reflectance $r(\lambda)$ as well as the diffuse attenuation coefficient $K_d(z,\lambda)$. Only stations with no atmospheric dust were selected. Both quantities are shown in Fig. 6.2.

The total internal reflectance varied with the distance to the coast. The water was green at the coastal station and blue at the open ocean station (see Fig. 6.1). The diffuse attenuation coefficients decreased in the whole spectral range with increasing distance to the coast. The optical water properties of ocean water station were in the range of ocean types II and III in relation to the classification of Jerlov (1976). The coastal station belonged to coastal waters of types 3 to 5. The reflection properties at the water surface have to be known for the optical model. The coefficients σ and $\overline{\sigma}$ were taken from literature (e.g. Jerlov 1976; Mobley 1999; Lee et al. 2010).

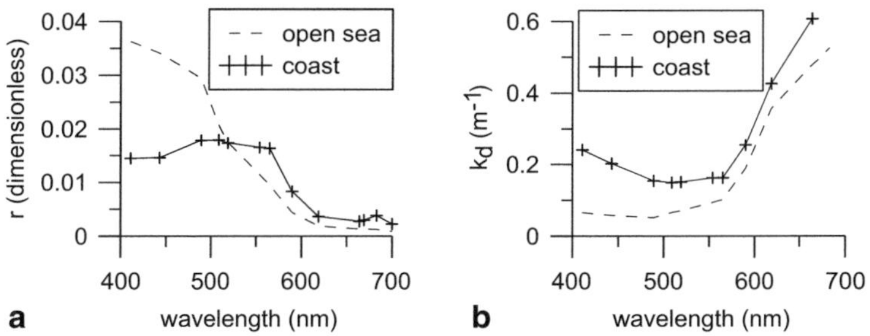

Fig. 6.2 The total internal reflectance (**2a**) and the diffuse attenuation coefficient (**2b**) are shown for the open sea station and the coastal station (according to Ohde and Siegel 2013)

6.2.3 Methods in Relation to Dust Impact on Satellite Wind speed

The study area 2 in Fig. 6.1 was selected to investigate the influence of Saharan dust on satellite derived ocean surface wind speeds (details in Ohde 2010). The area was located close to the airport on the island Sal in order to use ground truth wind speeds (w_{METAR}) from the METAR weather station GVAC[4]. The wind speed products $w_{QuikSCAT}$ and w_{TMI} of QuikSCAT and TMI (TRMM Microwave Imager) were downloaded from http://www.remss.com. QuikSCAT and TMI measured wind speeds at a height of 10 m above the ocean surface using frequency of 13.4, and 10.7 and 37 GHz, respectively. Data of aerosol optical depth of MODIS were used to approximate the optical depth of the dust component (AOD_{dust}). The data were downloaded from the Giovanni archive (http://daac.gsfc.nasa.gov). The aerosol dust components were derived from the aerosol optical depths at 550 nm of MODIS using the method of Kaufman et al. (2005). Quasi true-colour images of the MODIS sensor were used to verify Saharan dust storms[5].

The ground truth wind speeds of the GVAC METAR station were converted to the same height of 10 m as the satellite derived wind speeds using the atmospheric dispersion relation (Prandtl 1904). The wind speed products of QuikSCAT and TMI as well as the dust aerosol optical depths were area-averaged in the study area 2 (Fig. 6.1). 3d means and seasonal cycles of monthly means of w_{METAR}, $w_{QuikSCAT}$, w_{TMI} and AOD_{dust} as well as seasonal cycles of monthly mean differences $w_{QuikSCAT-}w_{METAR}$ of years 2001 to 2008 were determined. The relative errors of satellite derived wind speeds were determined using the relation $\Delta w = ((w_{sat,uncorr}-w_{sat,corr})/w_{sat,corr})*100\,\%$ where $w_{sat,uncorr}$ was the uncorrected satellite derived wind speed and $w_{sat,corr}$ was the corrected ones (sat = QuikSCAT, TMI). The corrected satellite derived wind speed $w_{sat,corr}$ was given by $w_{sat,corr} = w_{sat,uncorr} - U*AOD_{dust.}$ The term U was the slope of the linear fit of seasonal cycles of monthly mean differences. A detailed description of the methods was given in Ohde 2010.

[4] From http://data.eol.ucar.edu.

[5] From http://rapidfire.sci.gsfc.nasa.gov.

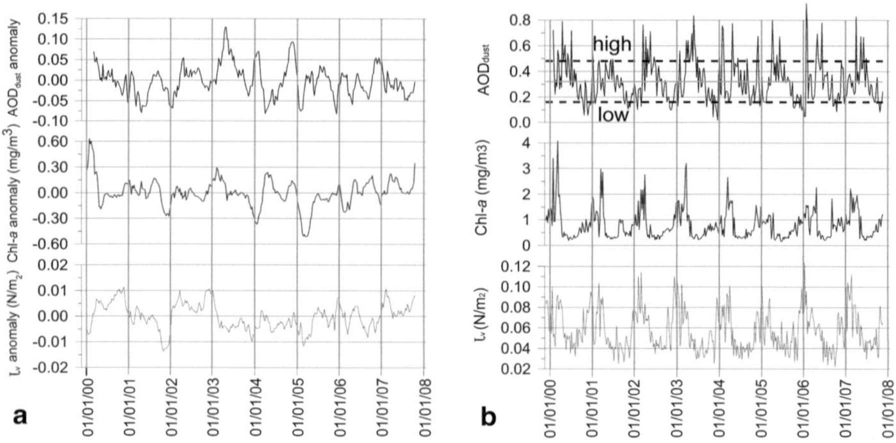

Fig. 6.3 The anomalies of area-averaged AOD$_{dust}$, Chl-*a* and τ_v for the period 2000 to 2008 are shown (Fig. 6.3a, Ohde and Siegel 2010). The anomalies were smoothed using a 3-month running average. The inter-annual variability of area averaged 8d means of AOD$_{dust}$, Chl-*a* and τ_v for the same time period are given in Fig. 6.3b

6.3 Results and Discussion

6.3.1 Dust Impact on Phytoplankton

The dust impact on phytoplankton was investigated on the basis of AOD$_{dust}$, Chl-*a*- and τ_v-anomalies for the time period 2000 to 2008 (details in Ohde and Siegel 2010). The anomalies which do not contain signals from the yearly cycle are presented in Fig. 6.3a. The generated time series were analyzed using the statistical methods of cross- and multiple correlation.

The highest correlations between Chl-*a* and τ_v were observed in the period of northern winter and spring ($0.41 \leq |r| \leq 0.49$, 99.9 % confidence limit, r correlation coefficient). Time lags up to16 days between the anomalies of Chl-*a* and τ_v were determined. Only a small significant correlation between the anomalies of Chl-*a* and AOD$_{dust}$ without a time lag was detected for the northern winter. The alongshore wind stress τ_v and induced upwelling were most significantly responsible for surface Chl-a variability with 24 % mainly in winter and spring with delay of up to 16 days. AOD$_{dust}$ contributed only about 5 % to the variability of the surface Chl-*a* anomaly in the winter period when Saharan dust is transported to the study area by strong westward winds in the lower atmosphere (e.g. Schepanski et al. 2009). The small statistical relationship can be explained. Different residence times of 6 to 62 days of dissolved iron in surface waters, a potentially usable nutrient for phytoplankton, were determined (e.g. Croot et al. 2004). The iron could be rapidly removed from the upper water column and could be not available for phytoplankton in the potential case of lower limit of residence time. The iron pool could be rebuilt in the case of long

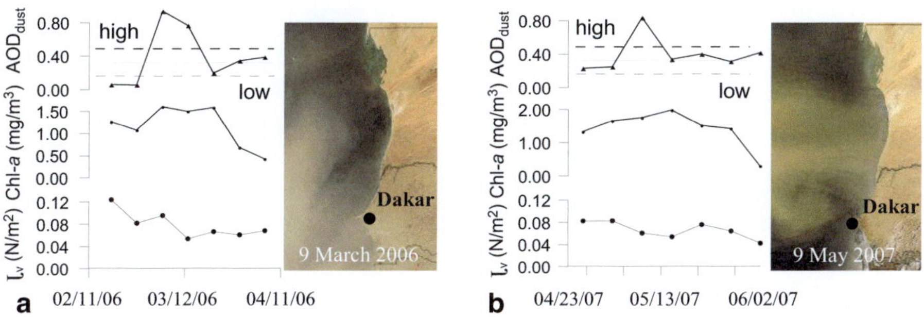

Fig. 6.4 The 8d means of AOD$_{dust}$, Chl-*a* and τ_v during the Saharan dust storms around 9 March 2006 (Fig. 6.4a) and 9 May 2007 (Fig. 6.4b) are shown. The RGB-MODIS images of the dust storms were downloaded from rapidefire.sci.gsfc.nasa.gov

residence times. No impact on phytoplankton could be expected and the statistical relationship would be small. Another reason could be the use of the quantity AOD$_{dust}$. This could be a poor proxy for dust deposition and the linear approach could be invalid (Kaufman et al. 2005; Schepanski et al. 2009). Further details were given in Ohde and Siegel (2010) where even the Sea Surface Temperature (SST) was included in the statistical analysis. A correlation coefficient of up to–0.75 (99.9 % confidence limit) with a time lag of 24 days was determined for the yearly time series of anomalies of τ_v and SST.

Another investigation concerning the biological response to dust supply was carried out because the statistical approach delivered no sufficient answers. All strong dust storms between 2000 and 2008 were identified in the study area (Fig. 6.1) according to the definitions of low (AOD$_{dust}$ values smaller than $< AOD_{dust} > -\sigma$), moderate (AOD$_{dust}$ values between $< AOD_{dust} > -\sigma$ and $< AOD_{dust} > +\sigma$) and high amount (AOD$_{dust}$ values greater than $< AOD_{dust} > +\sigma$) of atmospheric dust. A mean dust optical depth $< AOD_{dust} >$ of 0.32 and a standard deviation σ of 0.16 were observed in the investigation period of 2000 to 2008. The dashed lines in Fig. 6.3b represent the limits of low and high amount of atmospheric dust. The area-averaged 8d means of AOD$_{dust}$ values were 53 times in the range of strong Saharan dust storms.

Between 2000 and 2008 six strong dust storms were observed on 10 February 2001, 30 April 2003, 10 February 2004, 16 May 2005, 9 March 2006 and 9 May 2007 which fulfilled the conditions of increased AOD$_{dust}$ and Chl-*a*, respectively, and constant or decreased τ_v values. The Chl-*a* concentrations increased between 0.3 and 2.4 mg m^{-3} within 8 to 16 days after the dust storms with AOD$_{dust}$ values of up to 0.9.

Two examples of Saharan dust storms around 9 March 2006 and 9 May 2007 are given in Fig. 6.4. During the dust storms the AOD$_{dust}$ values increased by 0.9 (0.6) and the Chl-*a* values increased by 0.5 (0.7) mg m^{-3} within 8 to 16 days, but the coastal upwelling decreased. The temporal developments of the other four observed strong storms were similar to these both events. Between 2000 and 2008 many other Saharan storms were identified, but only these six events caused a biological response.

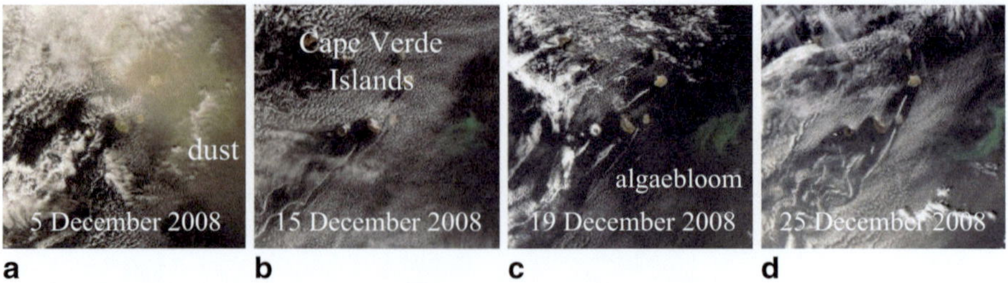

Fig. 6.5 The development of a large algal bloom between 15 and 25 December 2008 is shown in Fig. 6.5b–d using MODIS RGB images. A dust storm was observed before the bloom. See **a**

This result corresponded to the small correlation between AOD_{dust} and Chl-*a* derived from statistical analysis. The upwelling processes are most significantly responsible for the seasonal cycle of surface Chl-*a* concentration in the study area of NW Africa.

Not only the satellite derived Chl-*a* concentration can be used for study of biological response to dust deposition. Algae blooms can also be identified by satellite RGB images which combine the red, green and blue channels of ocean colour sensors. End of 2008 an algal bloom was observed by MODIS for about 25 days east of Cape Verde Islands. The spatial and temporal development is shown in Fig. 6.5b–d. The bloom was probably initialized by a dust storm on 5 December 2008 (see Fig. 6.5a). The algal bloom could be a coccolithophore bloom because such algae were identified in the study area during a cruise in February 2008 using different methods like HPLC (High Performance Liquid Chromatography) pigment analysis, SEM (Scanning Electronic Microscopy) and spectral particle absorption.

6.3.2 *Dust Impact on PAR*

The dust impact on PAR was investigated on the basis of Eq. 6.1. The detailed investigations were given in Ohde and Siegel (2012b). The Eq. 6.1 was used to determine the relative modifications of PAR ($\Delta_m Q_{PAR}/Q_{PAR}$) in the water column (see Fig. 6.6a) caused by the magnitude effect of atmospheric dust. The optical water properties and the meteorological data described in Sect. 6.2.2.2 were included in the calculations.

The clear sky case was characterized by the point ($N = 1, \Delta_m Q_{PAR}/Q_{PAR} = 0$). The normalization factors N covered the range from the minimum observed value $N^{min} = 0.81$ to the maximum value $N^{max} = 0.93$. Both values were determined from radiation measurements during a dust storm in February 2008. In the future they have to be adapted by the evaluation of other dust storms. The PAR was decreased between 5.2 and 19.3 % in the upper water column compared to the PAR of clear skies. However, the PAR can be reduced up to about 74 % in the case of cloudy skies (Ohde and Siegel 2012b).

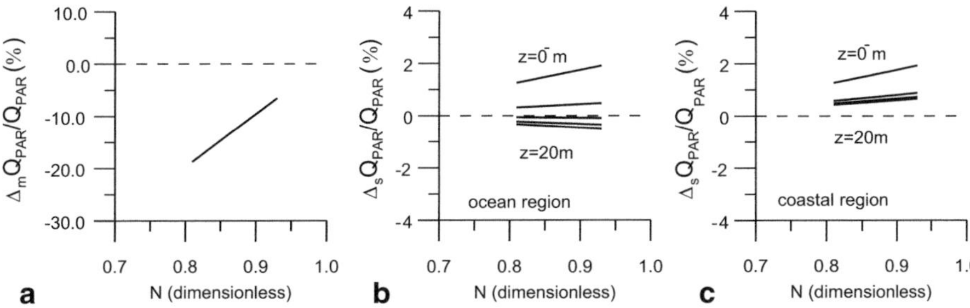

Fig. 6.6 The relative magnitude effect of dust on PAR in the ocean region are shown in Fig. 6.6a ($z = 5$ m). The relative modifications of PAR caused by the spectral effect are shown in dependence of the water depths (0^-, 5, 10, 15, 20 m) and water types in Figs. 6.1c and 6.6b (according to Ohde and Siegel 2012b).

Equation 6.2 was the basis for the calculation of the relative modifications of PAR ($\Delta_s Q_{PAR}/Q_{PAR}$) in the water column (see Fig. 6.6b, c) caused by the spectral effect of atmospheric dust. The spectral effects in ocean and coastal regions were distinguished by the optical water properties described in Sect. 6.2.2.2. The PAR including the spectral effect was about 1.3 % ($N^{min} = 0.81$) and 1.9 % ($N^{max} = 0.93$) higher than the PAR without this effect for water depths near the surface and for both considered water types. On the other hand, the spectral effect of clouds can reduce the PAR between 4 and 6 % for nearly overcast skies (Bartlett et al. 1998; Ohde and Siegel 2012b).

The magnitude effect of dust can be much higher than the spectral effect (cf. Fig. 6.6a with Fig. 6.6b, c). The influences of the spectral effect decreased with increasing water depth. The PAR was higher compared to the clear sky PAR in water depths near the water surface in ocean and coastal regions. In contrast, the PAR was lower in ocean regions in water depths higher than about 7 m. That means a compensation of the impacts of spectral effects was found in water depths of about 7 m in ocean waters. This was not observed in coastal areas because of different light attenuation.

6.3.3 Dust Impact on Satellite Derived Ocean Surface Wind Speed

The impact of atmospheric dust on the wind speed product derived from QuikSCAT microwave satellite data was analyzed by a Saharan dust storm in May 2007 (see Fig. 6.7a, details in Ohde 2010). A quasi-true color image of MODIS of 9 May 2007 showed the yellow-brownish dust aerosols in the atmosphere above the ocean areas (Fig. 6.7b, Ohde 2010). The 3d means of w_{METAR}, $w_{QuikSCAT}$ and AOD_{dust} illustrated the temporal development of the dust storm.

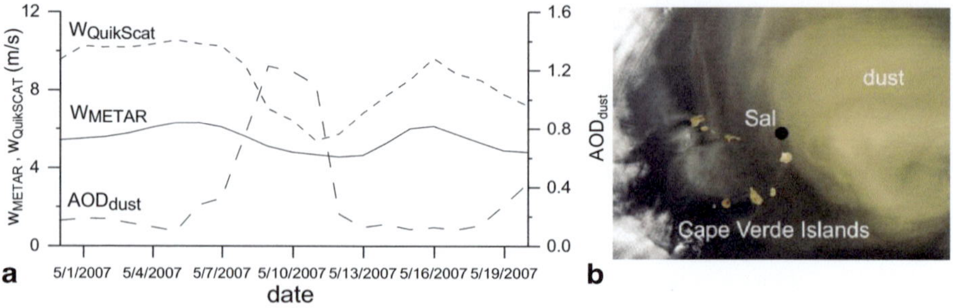

Fig. 6.7 The dust component of aerosol optical depth (AOD_{dust} in Fig. 6.7a) demonstrates the temporal development of the dust storm (according to Ohde 2010) which was also identified in the RGB-MODIS image of 9 May 2007 (Fig. 6.7b). The impact of dust aerosols on satellite derived wind speeds is clearly seen (Fig. 6.7a) in the QuikSCAT data ($w_{QuikSCAT}$) compared to the ground truth wind speed (w_{METAR})

The maximum of the storm was reached between 9 and 10 May 2007 with AOD_{dust} values between 1.2 and 1.3. The observed dust storm was a strong storm according to the classification by Ohde (2010).

The impact of dust on wind speeds was clearly seen (cf. Fig. 6.7a). The ground truth wind speeds (w_{METAR}) were decreased up to 1.5 m/s during the storm, but the satellite derived wind speeds ($w_{QuikSCAT}$) were stronger decreased up to 5 m/s.

A considerable difference between w_{METAR} and $w_{QuikSCAT}$ was observed even in the absence of dust storms (cf. Fig. 6.7a). The mean bias between them was 2.89 m/s for the time period 2001 to 2008 (Ohde 2010). Reasons for the deviations were different sampling methods in relation to position and time, radiometer noise and errors in the atmospheric model. The METAR wind speeds were measured onshore and the QuikSCAT wind speeds were determined offshore. Probably the roughness of the land surface caused the smaller METAR wind speeds. Further details were given in Ohde (2010) where even the mean biases of different microwave sensors were determined.

The mean relative errors (see Fig. 6.8, introduced in Sect. 6.2.3) depended on the wind speed, the microwave frequency and the dust component of aerosol optical depth. The relative error decreased with increased wind speed. The relative errors were higher at higher microwave frequencies of the satellite sensors. The relative errors increased with increasing dust aerosol optical depths.

The mean ground truth wind speed and the mean dust component of aerosol optical depth in the study area 2 (Fig. 6.1) between the years 2001 and 2008 were 4.9 m/s and 0.31, respectively. Both values were used to determine the mean relative error for the wind speed products of microwave sensors. An error of smaller than about 10 % followed from Fig. 6.8. It means, the wind speeds derived by QuikSCAT and TMI were averaged determined smaller than 10 % to small in the study area (Ohde 2010).

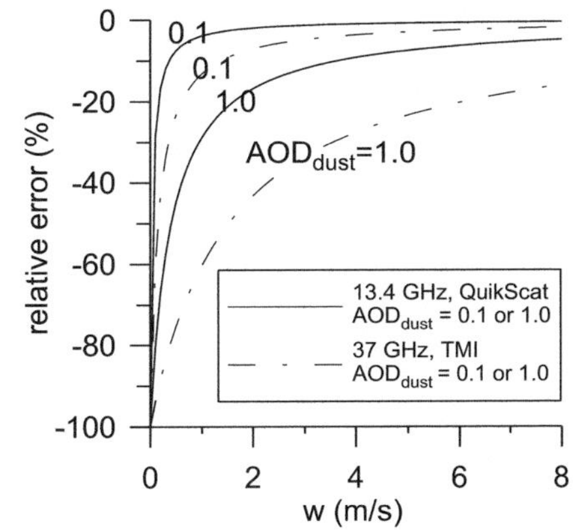

Fig. 6.8 Plots of relative errors in dependence of the wind speed, the microwave frequency and the aerosol optical depth of the dust component are shown (according to Ohde 2010)

6.4 Conclusions

A wide range of impacts of Saharan dust on marine environment and satellite measurements exists. Only few aspects were considered in this chapter. The biological response of the marine environment to the nutrient supply in the surface water layer resulting from upwelling and dust deposition was investigated on the basis of remotely sensed proxies. Statistical analyses have shown that alongshore wind stress and induced upwelling were most significantly responsible for surface Chl-*a* variability mainly in winter and spring with delays of up to 16 days. Only a small part of the surface Chl- *a* variability could be ascribed to dust deposition. Only some of the identified strong dust storms between 2000 and 2008 related to enhanced phytoplankton growth. Time lags between dust deposition and Chl-*a* enhancement up to 16 days were determined.

The magnitude and spectral effects of atmospheric dust on PAR in the water column were quantified in relation to the clear sky case. These effects could be important because light dependent biological processes like phytoplankton growth and phytoplankton composition may be influenced. The magnitude effect of dust reduced the PAR in relation to the clear sky case. The spectral impacts on PAR in ocean regions were generally different in comparison to coastal areas. An interesting result was the compensation of the impacts of spectral effects on PAR at definite water depths in ocean regions.

The impact of atmospheric dust on the wind speed products of microwave sensors was investigated and quantified. The deviations of wind speeds to ground truth wind speeds increased with higher microwave frequency. The impact of dust was especially high for low wind speeds. The strength of dust storms expressed in terms of dust component of aerosol optical depth influenced the microwave signal. The impact was higher for strong dust storms.

Acknowledgments The authors thank the Leibniz Institute for Baltic Sea Research and the German Federal Ministry of Education and Research (BMBF) within the framework of the SOPRAN (Surface Ocean Processes in the Anthropocene) project (FKZ 03F0473E). This research was based on satellite data (MODIS, SeaWiFS) provided by NASA's Giovanni, an online data visualization and analysis tool. QuikScat, TMI and AMSR-E data were produced by Remote Sensing Systems. NCEP Reanalysis data were provided by NOAA (National Oceanic and Atmospheric Administration). METAR – data were contributed by NCAR/EOL under sponsorship of the National Science Foundation. The MSG products were kindly provided by EUMETSAT. The MODIS images were taken from MODIS Rapid Response Project. We also thank two anonymous reviewers for their helpful comments on the original manuscript.

References

Altshuler EE (1983) The effects of a low-altitude nuclear burst on millimeter wave propagation. Report, RADC-TR-83-286:1–19

Ansari AJ, Evans BG (1982) Microwave propagation in sand and dust storms. IEE Proc 129:315–322

Baker AR, Kelly SD, Biswas KF et al (2003) Atmospheric deposition of nutrients to the Atlantic Ocean. Geophys Res Lett 30(24):2296. doi:10.1029/2003GL018518

Bartlett JS, Ciotti AM, Davis RF et al (1998) The spectral effects of clouds on solar irradiance. J Geophys Res 103:31017–31031

Comparetto G (1993) The impact of dust and foliage on signal attenuation in the millimeter wave regime. J Space Commun 11:13–20

Croot PL, Streu P, Baker AR (2004) Short residence time for iron in surface seawater impacted by atmospheric dry deposition from Saharan dust events. Geophys Res Lett 31:L23S08. doi:10.1029/2004GL020153

Duarte CM, Dachs J, Llabrés M et al (2006) Aerosol inputs enhance new production in the subtropical northeast Atlantic. J Geophys Res 111:G04006

Frederick JE, Erlick C (1997) The attenuation of sunlight by high-latitude clouds: spectral dependence and its physical mechanisms. J Atmos Sci 54:2813–2819

Gao Y, Kaufman YJ, Tanre' D et al (2001) Seasonal distributions of aeolian iron fluxes to the global ocean. Geophys Res Lett 28:29–32

Gregg WW, Carder KL (1990) A simple spectral solar irradiance model for cloudless maritime atmospheres. Limnol Oceanogr 35:1657–1675

Jerlov NG (1976) Marine optics. Elsevier, Amsterdam

Kaufman YJ, Koren I, Remer LA et al (2005) Dust transport and deposition observed from the Terra-Moderate Resolution Imaging Spectroradiometer (MODIS) spacecraft over the Atlantic Ocean. J Geophys Res 110:D10S12

Kopelevich OV (2012) Application of data on seawater light scattering for the study of marine particles: a selective review focusing on Russian literature. Geo Mar Lett 32:183–193

Lee ZP, Ahn HW, Mobley C et al (2010) Removal of surface-reflected light for the measurement of remote-sensing reflectance from an above-surface platform, Optic. Express 18:26313–26324

Meskhidze N, Chameides WL, Nenes A (2005) Dust and pollution: a recipe for enhanced ocean fertizilation. J Geophys Res 110:D03301

Mobley CD (1999) Estimation of the remote-sensing reflectance from above-surface measurements. Appl Optics 38(36):7442–7455

Ohde T (2010) Impact of Saharan dust on ocean surface wind speed derived by microwave satellite sensors. J Infrared Millim Terahertz W 31(10):1225–1244. doi:10.1007/s10762-010-9695-z

Ohde T, Siegel H (2010) Biological response to coastal upwelling and dust deposition in the area off Northwest Africa. Cont Shelf Res 30(9):1108–1119. doi:10.1016/j.csr.2010.02.016

Ohde T, Siegel H (2012a) Impacts of Saharan dust and clouds on photosynthetically available radiation in the area off Northwest Africa. Tellus B64:17160. doi:10.3402/tellusb.v64i0.17160

Ohde T, Siegel H (2012b) Impacts of Saharan dust on downward irradiance and photosynthetically available radiation in the water column in the area off Northwest Africa. Adv Oceanogr Limnol 3(2):99–131. doi:10.1080/19475721.2012.716793

Ohde T, Siegel H (2013) Spectral effects of Saharan dust on photosynthetically available radiation in comparison to the influence of clouds. J. atmos. sol.-terr. phys. 102: 269–280, http://dx.doi.org/10.1016/j.jastp.2013.06.004

Otto S, Bierwirth E, Weinzierl B et al (2009) Solar radiative effects of a Saharan dust plume observed during SAMUM assuming spheroidal model particles. Tellus B61:270–296. doi:10.1111/j.1600-0889.2008.00389.x

Pradhan Y, Lavender SJ, Hardman-Mountford NJ et al (2006) Seasonal and inter-annual variability of chlorophyll-a concentration in the Mauritanian upwelling: Observation of an anomalous event during 1998–1999. Deep-Sea Res II 53:1548–1559

Prandtl L (1904) Über Flüssigkeitsbewegung bei sehr kleiner Reibung, Verhandlg. III Int Math Kong Teubner, Heidelberg

Prospero JM, Carlson TN (1972) Vertical and areal distribution of Saharan dust over the Western equatorial North Atlantic Ocean. J Geophys Res 77:5255–5265

Schepanski K, Tegen I, Macke A (2009) Saharan dust transport and deposition towards the tropical northern Atlantic. Atmos Chem Phys 9:1173–1189

Stramski D, Woniak SB, Flatau PJ (2004) Optical properties of Asian mineral dust suspended in seawater. Limnol Oceanogr 49:749–755

Van Camp L, Nykjær L, Mittelstaedt E, Schlittenhardt P (1991) Upwelling and boundary circulation off northwest Africa as depicted by infrared and visible satellite observations. Prog Oceanogr 26:357–402

Yelland M, Taylor PK (1996) Wind stress measurements from the open ocean. J Phys Oceanogr 26(4):541–558

Young RW, Carder KL, Betzer PR et al (1991) Atmospheric iron inputs and primary production: phytoplankton responses in the North Pacific. Global Biogeochem Cycles 5(2):119–134

Chapter 7
Water Constituents Assessment at the Sassandra River Mouth (Côte d'Ivoire): An Outline Based on Remote Sensing Reflectances

Eric V. Djagoua, Jeanne M. Kouadio, Cassandre Z. Dro, Mireille Bella, Jean B. Kassi, Brice A. Mobio, Kouadio Affian, Simon Belanger and Pierre Larouche

Abstract Optical remote sensing of surface waters has been used to monitor the impact of the Sassandra River—one of the four main rivers of Côte d'Ivoire—along the coast of the Gulf of Guinea. A 5-year (2003–2007) series of Sea-viewing Wide Field-of-view Sensor (SeaWiFS) images was chosen to identify (spectrally) the major constituents of coastal waters under the influence of river discharges. The percent contributions of yellow substance, i.e. coloured dissolved organic matter (*CDOM*), chlorophyll-like pigments of phytoplankton (*CHL*) and suspended detritus (*DET*) to the total reflectance (approximated by the sum of reflectances at 443, 555 and 670 nm) were calculated. Ternary plots of these percentages provided some general indications about the relative proportions of *CDOM*, *CHL* and *DET*, according to the impact attributed to each water constituent. The analysis showed that water constituents follow trends linked to the "marine" seasons of the area. The *CHL* and *DET* contributions seem to correlate with the timing of cooler (upwelling) periods, while *CDOM* seems to correlate better with that of warmer periods. The relationship with the Sassandra river runoff is not as evident, even though *CDOM* increased contributions appear to mirror the weakening of upwelling and the discharge of local rivers. This broad ecosystem outline should be extended to the entire Côte

E. V. Djagoua (✉) · J. M. Kouadio · C. Z. Dro · M. Bella · J. B. Kassi · B. A. Mobio · K. Affian
Centre Universitaire de Recherche et d'application en Télédétection (CURAT),
Université de Cocody, Abidjan, Côte d'Ivoire
e-mail: vdjagoua@yahoo.fr

K. Affian
Laboratoire de Géologie Marine et de Sédimentologie,
U.F.R. des Sciences de la Terre et des Ressources Minières,
Université de Cocody, Abidjan, Côte d'Ivoire

S. Belanger
Laboratoire d'optique Aquatique et de Télédétection (AquaTel),
Université de Québec à Rimouski (UQAR), Rimouski (Québec), Canada

P. Larouche
Maurice Lamontagne Institute, Fisheries and Oceans Canada,
Mont Joli (Québec), Canada

V. Barale, M. Gade (eds.), *Remote Sensing of the African Seas*,
DOI 10.1007/978-94-017-8008-7_7,

d'Ivoire coastline, to better classify marine waters impacted by upwelling processes or anthropogenic pollution due to river runoff.

7.1 Introduction

Environmental management, in the sustainable development framework, is an absolute necessity for our society. Historically, narrow and fragile coastal zones have been strategically chosen by large resident populations to establish their economic activities. Today, the safeguard of coastal, semi-enclosed and enclosed water bodies represent one of the major challenges for environmental mangers, due to the combined impact of both anthropogenic and natural factors (Jaquet 1989).

In addition to human activities, river discharges constitute important sources of nutrients, organic matter and inorganic particles to the coastal ocean (Dagg et al. 2004; Geyer et al. 2004). These materials have great influence on the bio-geochemistry and bio-optical characteristics of coastal waters (Froidefond et al. 2002; Hu et al. 2004), as they can stimulate primary productivity, but also cause pollution and/or oxygen depletion, thereby hampering the very survival of halieutic species. As a consequence, careful monitoring of coastal zones by means of continuous and long-term observations is required, in order to ensure that their ecological balance is maintained and their natural resources preserved.

Given the optical impact of continental runoff onto marine waters, Remote Sensing (RS) from satellite sensors, operating in the visible part of the electromagnetic spectrum, is well suited for monitoring basic ecosystem dynamics of coastal zones. It is well known (see e.g. Robinson 2004) that the radiometric signatures of coastal waters depend mainly on three types of optically-active water constituents, namely phytoplankton and their derivative products; dissolved organic material, also known as yellow substance, in the RS jargon; and suspended sediments, as well as other particles of inorganic nature. All of these can occur in different proportions, highly variable in both space and time (see e.g. Lahet et al. 2001 and references therein). The assessment of such parameters by means of RS techniques allows deriving invaluable information for the sustainable management of coastal regions, in particular when *in situ* data are lacking.

In the following, optical RS of water constituents will be used to monitor the impact of the Sassandra River—one of the four major rivers of Côte d'Ivoire—onto the coastal waters of the Gulf of Guinea, in which it flows directly without passing through a lagoon system (see Fig. 7.1 for a geographical description of the area of interest in the African continental context). The Sassandra River has its origin in the high grounds of the north, where the Tiemba River joins the Férédougouba River, flowing from the Guinea highlands. It is joined by the Bagbé, Bafing, Nzo, Lobo, and Davo rivers and winds through shifting sandbars to form a narrow estuary (which is navigable for about 80 km inland from the port of Sassandra). Like many other rivers, the Sassandra River also serves as a receptacle for domestic sewage, agricultural runoff (often carrying a sizeable load of fertilizers and pesticides), industrial production waste and other uncollected materials (CEDA 1997). In addition,

Fig. 7.1 The Sassandra River mouth, in southwestern Côte d'Ivoire, outflowing in the Gulf of Guinea, within the African continental context

its waters are invaded in places by floating aquatic plants. All of this exposes the river basin to severe environmental stress, and can be the origin of uncontrolled and environmentally dangerous pollution in coastal waters.

The goal of this paper is to provide a broad ecosystem outline of the Sassandra River mouth and adjacent marine area, trying to identify (spectrally) the major constituents of coastal waters under the influence of river discharges, in the different seasonal periods of the Côte d'Ivoire littoral.

7.2 Methodology

The basic optical properties of the Sassandra River outflow, and of the Gulf of Guinea coastal waters that it impacts, have been explored by means of an image time series generated by the Sea-viewing Wide Field-of-view Sensor (SeaWiFS) over a period of 5 years (McClain et al. 2004). SeaWiFS (level 3) data, at a 4 km spatial resolution, were obtained from the US National Aeronautics and Space Administration (NASA) GIOVANNI data portal[1]. This on-line RS data access tool is a web-based application that allows accessing, visualizing, and analyzing vast amounts of Earth Observation

[1] The data portal **GES-DISC** (Goddard Earth Sciences Data and Information Services Center) **I**nteractive **O**nline **V**isualization and **A**nalysis **I**nfrastructure, also known as GIOVANNI, is available at the address: http://gdata1.sci.gsfc.nasa.gov/daac-bin/G3/gui.cgi?instance_id=ocean_month.

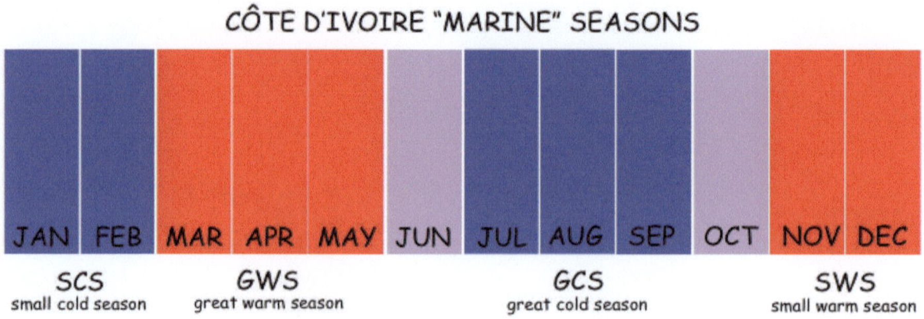

Fig. 7.2 Typical "marine" seasons of the Côte d'Ivoire coastal region

data and data products, originated by a number of remote sensors, over the past 3 decades. RS reflectances (R_{rs}), at 443, 555 and 670 nm, in the coastal area impacted by the Sassandra River were acquired over 5 full annual cycles, from January 2003 to December 2007.

As will be detailed later, the data were processed on both an annual and a monthly basis, using images from all the 5 years considered, in order overcome the cloud cover problem originated by the local climate. In fact, along the Côte d'Ivoire coastal zone, set in a very humid equatorial region, almost constant cloud cover renders optical RS techniques rather ineffective in collecting long, continuous time series of cloud-free images. In addition to annual and monthly mean conditions, our analysis focused on those typical of the local "marine" seasons, i.e.:

- a small cool season (SCS), from January to February;
- a great warm season (GWS), from March to May;
- a great cool season (GCS), from July to September;
- a small warm season (SWS) from November to December.

This subdivision (Fig. 7.2) follows the recurrence of upwelling events, in the SCS and primarily in the GCS (Morlière 1970), due to complementary effects of the Guinea Current (Ingham 1970), its interaction with coastal morphology (Marchal and Picaut 1977), the equatorial wave regime (Picaut 1983), and the prevailing southwesterly winds (Colin et al. 1993).

7.2.1 Why Apparent rather than Inherent Optical Properties?

As introduced earlier, RS of marine optical features is based on the reflective properties of optically active seawater constituents: phytoplankton, the concentration of which is generally estimated by chlorophyll-a; colored dissolved organic matter; and non algal particles, which in coastal waters is mostly represented by suspended sediments or detritus (IOCCG 2000). Radiative transfer theory provides the link between Inherent Optical Properties (IOPs), like the absorption and scattering coefficients of seawater constituents, and Apparent Optical Properties (AOPs), like the R_{rs} measured above the sea surface (Chang et al. 2003).

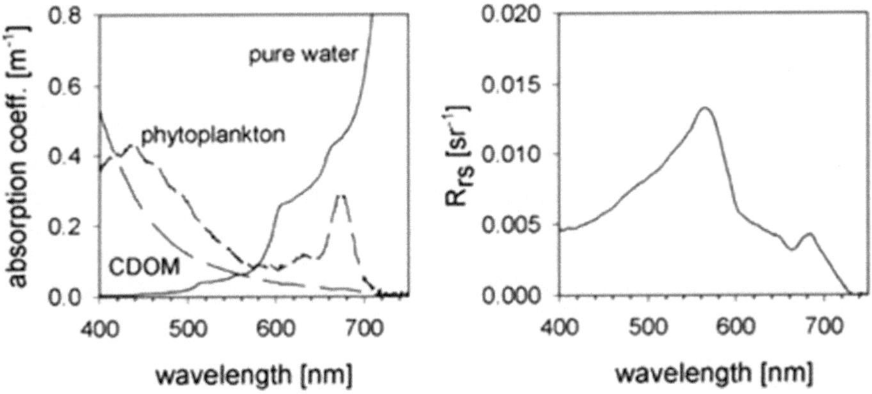

Fig. 7.3 *Left panel*: phytoplankton (*dashed line*) and *yellow* substance (em *dashed line*) absorption spectra for surface water samples at a coastal location, along with a pure water (*solid line*) absorption spectrum. *Right panel*: corresponding RS reflectance R_{rs} at the same location. Adapted from D'SA and Miller (2005)

D'Sa and Miller (2005) described the relationship of R_{rs} to scattering and absorption properties of seawater, which in turn are determined by the constituents above. Figure 7.3 (left panel) provides an example of the absorption by phytoplankton, yellow substance and pure water in the spectral range 400–750 for a coastal location (Pope and Fry 1997). The corresponding R_{rs} spectrum for the same location (Fig. 7.3, right panel) shows low values in the blue ($\sim$ 443 nm) that can be attributed to high absorption by phytoplankton and yellow substance, while high values in the green ($\sim$ 555 nm) may be attributed to scattering by suspended material. The secondary peak observed in the far red ($\sim$ 683 nm) is related to phytoplankton chlorophyll fluorescence.

These considerations indicate that the simple analysis of R_{rs} data, used in the following to typify AOPs, can be considered an efficient instrument for the detection of seawater constituents, without resorting to IOPs, in the coastal area impacted by the Sassandra River.

7.2.2 Why did we choose R_{rs} at 443, 555 and 670 nm?

The R_{rs} data chosen to describe surface colour variations account for suspended particles, both organic and inorganic (i.e. phytoplankton and sediments, or detritus in general, referred to in the following by *CHL* and *DET)*, and dissolved organics (i.e. yellow substance), also known as Colored Dissolved Organic Matter (*CDOM* in the following). As seen above, each of these constituents has its own typical spectral signature, which can be used to determine its concentration by means of satellite observations (Morel and Prieur 1977; Gordon and Morel 1983). Hence, the remote assessment of water constituents, by virtue of their reflectance, requires the choice of suitable RS wavelengths, capable of ensuring their discrimination.

The similarity of the Sassandra river mouth with other test cases, in which the spectral characteristics of water constituents have been highlighted by comparable studies (see e.g. Miller et al. 2005), suggested to select 443, 555 and 670 nm as the wavelengths of choice for our water constituents classification. The basic reasoning underlying this choice is that the chlorophyll-like pigments of phytoplankton reflect less at the shorter wavelengths, i.e. 443 nm (and 490 nm), and more at 555 nm. Moreover, at 443 nm (and 490 nm), the shape of the reflection curve is not constant, as it depends on the *CHL* concentration (Gower et al. 1988). Absorption by detritus is strong at shorter wavelengths (i.e. 412 nm) and weak at longer wavelengths (i.e. 670 nm; Froidefond and Doxaran 2004) or, in other words, the reflectance associated to *DET* is low at the shorter wavelengths and high at the longer wavelengths.

An increase in *CDOM* concentration leads to a lower reflectance at the shorter wavelengths, from UV to blue-green (Bricaud et al. 1981). *CDOM* absorbs visible light, especially below 500 nm, and its absorbance increases exponentially with decreasing wavelength. Nonetheless, Strömbeck and Pierson (2001) report that absorbance of red light can be significant at high *CDOM* concentrations. So, as suggested by the literature above, the wavelengths 443, 555 and 670 nm appear to be those that best allow the discrimination of the constituents of a given water body, and they were selected for the present study.

7.2.3 How did we Compute the Percent Contribution to the Total Reflectance of each major Constituent from R_{rs} (443, 555, 670 nm)?

The RS reflectances at each of the selected wavelengths (i.e. 443, 555 and 670 nm) were averaged pixel by pixel (i.e. points of constant latitude and longitude) over the 5 years considered (i.e. for the whole 2003–2007 period, as well as for each month in the same multi-annual period), according to the equation:

$$R_{rs\,MEAN}(\lambda, i, j) = \frac{1}{n} \sum_{t=1}^{n} R_{rs}(\lambda, i, j, t) \tag{7.1}$$

where:

- $R_{rs\,MEAN}(\lambda, i, j)$ is the mean (annual or monthly) RS reflectance, at the wavelength λ and at the pixel location *i, j;*
- $R_{rs}(\lambda, i, j, t)$ are the RS reflectance values, at the wavelength λ and at the pixel location $\boldsymbol{i}, \boldsymbol{j}$, which are being summed over the time $t = 1, 2, 3 \ldots n$ (i.e. all images in the 2003–2007 data set, or all images for each climatological month in the 2003–2007 data set);
- n is the number of valid pixels used in the sum.

(see also Fig. 7.4, for a graphical description of these variables).

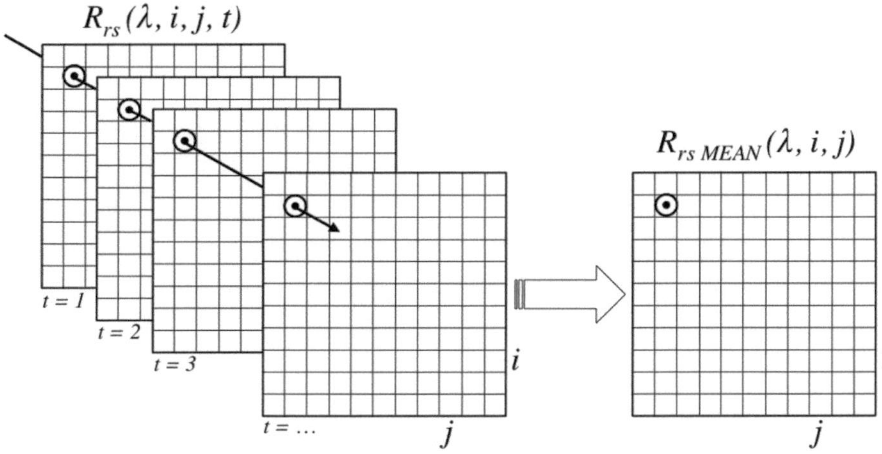

Fig. 7.4 Schematics of the approach used to compute the (2003–2007) annual/monthly mean image, where $R_{rs\ MEAN}$ (λ, i, j) is the image for the wavelength λ, generated by averaging pixel-by-pixel the single (daily) images R_{rs} (λ, i, j, t), for all time steps $t = 1, 2 \ldots n$ (i.e. for all input images)

Based on the above considerations about wavelength selection, the percent contribution of each water constituent to the total reflectance observed in coastal waters (approximated by the sum of reflectances at 443, 555 and 670 nm) was calculated according to the following equations:

$$\%[CDOM]_{ij} = \{R_{rs\ MEAN}(443, i, j) / \sum\nolimits_{\lambda=443,555,670} R_{rs\ MEAN}(\lambda, i, j)\} \times 100 \quad (7.2)$$

$$\%[CHL]_{ij} = \{R_{rs\ MEAN}(555, i, j) / \sum\nolimits_{\lambda=443,555,670} R_{rs\ MEAN}(\lambda, i, j)\} \times 100 \quad (7.3)$$

$$\%[DET]_{ij} = \{R_{rs\ MEAN}(670, i, j) / \sum\nolimits_{\lambda=443,555,670} R_{rs\ MEAN}(\lambda, i, j)\} \times 100 \quad (7.4)$$

where:

- $\%[CDOM]_{ij}$, $\%[CHL]_{ij}$ *and* $\%[DET]_{ij}$ represent the percent contribution due to dissolved organic matter, phytoplankton and detritus, respectively, at either the annual or the monthly scale;
- $R_{rs\ MEAN}(443, i, j)$, $R_{rs\ MEAN}(555, i, j)$ *and* $R_{rs\ MEAN}(670, i, j)$ are respectively the means of the reflectance at 443, 555 and 670 nm, for the pixel (i, j), at the annual/monthly scale, as defined in (1);
- $\sum\nolimits_{\lambda=443,555,670} R_{rs\ MEAN}(\lambda, i, j)$ is the sum of all mean reflectances at 443, 555 and 670 nm, still for the pixel (i, j) and at the annual/monthly scale.

The same approach was used to derive also seasonal statistics, subdividing the 5-year data set according to the local prevailing "marine" seasons, and then averaging pixel by pixel the reflectances at each of the selected wavelengths (i.e. 443, 555 and 670 nm), as in 7.1, for each season, over the 5 years considered. The percent

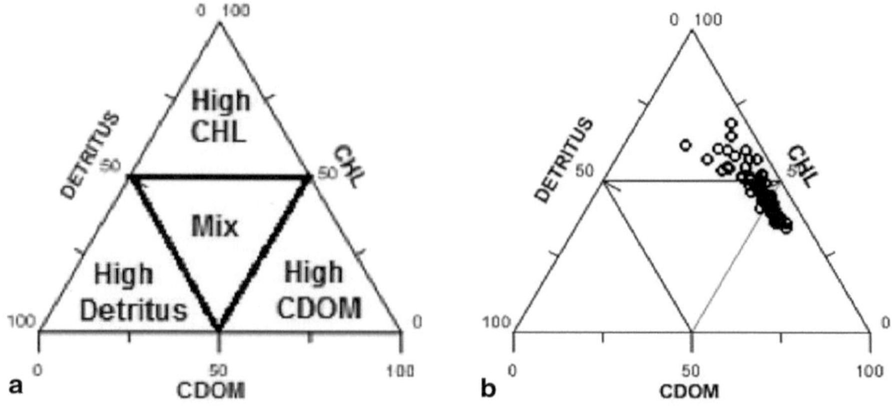

Fig. 7.5 **a** General interpretation of the relative proportions of *CDOM*, *CHL* and *DET*, using a ternary plot, according to the percent contribution to the total reflectance attributed to each water constituent. **b** Percent contributions obtained for the 2003–2007 annual case

Table 7.1 Mean percent contributions of *CDOM*, *CHL* and *DET* to the total reflectance off the Sassandra River mouth at the annual (2003–2007) scale and during the main "marine" seasons (SCS, GWS, GCS, SWS)

Seasonal Period	Months	CDOM	CHL	DET
Annual 2003–2007	January to December	46.42	45.75	7.83
SCS	January–February	10.52	74.77	14.71
GWS	March–May	60.84	37.27	1.89
1st transition	June	47.91	43.89	8.20
GCS	July–September	28.09	58.12	13.78
2nd transition	October	47.60	43.70	8.70
SWS	November–December	59.21	37.22	3.57

contribution of each component to the total reflectance was computed for each season, using again 7.2, 7.3 and 7.4. The mean values of the *CDOM*, *CHL* and *DET* percent contributions to the total reflectance, computed as an average over the whole area of interest, for the annual case and for all seasonal periods, are given in Table 7.1.

7.3 Seasonal Variability of Seawater Constituents

The percent contributions to the total reflectance attributed to *CDOM*, *CHL* and *DET*, as derived from the multi-annual SeaWiFS data, allowed assessing the relative proportions of water constituents at the Sassandra River mouth, according to the general scheme in Fig. 7.5 (following Arnone et al. with modifications, in IOCCG 2006). Figure 7.5a shows the kind of ternary plot – a barycentric plot of three variables, the proportions of which sum to a constant – used to graphically depict the ratios of the *CDOM*, *CHL* and *DET* percent contributions as positions in an equilateral triangle. Figure 7.5b shows the ratios distribution obtained for the 2003–2007 annual cases, and suggests that *CDOM* and *CHL* have similar higher contributions, in comparison to *DET*. However, the contribution of *CDOM* would appear to be much higher, if not for the SCS/GCS upwelling seasons, which increase substantially that of CHL.

As shown in Table 7.1, the percent contributions of the relative proportions of the main water constituents does vary quite a bit in each of the "marine" seasons described above. The relative proportion of *CDOM*, *CHL* and *DET* in each of the seasons and transition periods considered is shown in the ternary plots of Fig. 7.6, panels a to f. In general, we observed three different sets of conditions: periods of *CHL* prevalence (as in the SCS and GCS); periods of *CDOM* prevalence (as in the GWS and SWS); and periods in which we find the presence of all three components (as in two transition periods, between the other seasons).

The first set of conditions concerns the "cold" seasons, which take place from January to February (SCS) and from July to September (GCS). These periods display a high percent contribution of *CHL*, in both cases greater than 50 %. However, the SCS percentage (Fig. 7.6a) varies between 60 % and 90 %, while the GCS percentage (Fig. 7.6d) varies between 50 and 70 % only (see also Table 7.1). Both the cold seasons are characterized by upwelling episodes, which increase nutrients availability and phytoplankton productivity (Demarcq and Aman 2002; Djagoua 2003). The higher percent contribution of *CHL* in the SCS might suggest that this period can be more productive than the GCS. As this is not the case (Bakun 1978; Binet 1997), it would seem that the estimated *CHL* contribution doesn't always represent real conditions along the Cote d'Ivoire coastal zone. Possibly, the high *CDOM* contribution in the GSC, with respect to the SCS—one third (28,09 %) *vs* one tenth (10.52 %) of the total, in Table 7.1, while the *DET* contribution remains rather constant, at ~ 14 %—might indicate that the *CHL* contribution was underestimated in this particular season.

The second set of conditions is linked to a high percent contribution of *CDOM* and concerns the "warm" seasons, which take place from March to May (GWS) and from November to December (SWS). During these periods (Fig. 7.6d, f), the *CDOM* contribution is nearly the same (~ 60 %; see also Table 7.1). But that of *CHL* (~ 37 %) is not negligible either. Indeed, during both the SWS and especially the GWS, the Cote d'Ivoire coastal waters are characterized by nutrients too low to sustain a high productivity. However, in spite of the generally poor phytoplankton field, one or more episodes of higher productivity may occur during the GWS, when the thermocline rises to the surface (Sevrin-Reyssac 1993).

The third set of conditions is that of the two transition periods (June and October), immediately before and after the GCS, which present similar percent contributions of both *CDOM* and *CHL* (Fig. 7.6c, e). The relatively high *CHL* values may be related to (early or late) upwelling events of the GCS, and/or high river discharges occurring in these periods (especially in October). Data about the monthly mean discharge from the Sassandra River, for the period from 1986 to 1999, are shown in Fig. 7.7, lower panel. The seasonal variations of river runoff can be compared with the relative proportions of RS-derived water constituents plotted in Fig. 7.6, upper panel, at the same monthly scale.

Le Loeuff and Marchal (1993) ascribe the prevalence of one or the other water constituent, during the Cote d'Ivoire "marine" seasons, to either the shift between

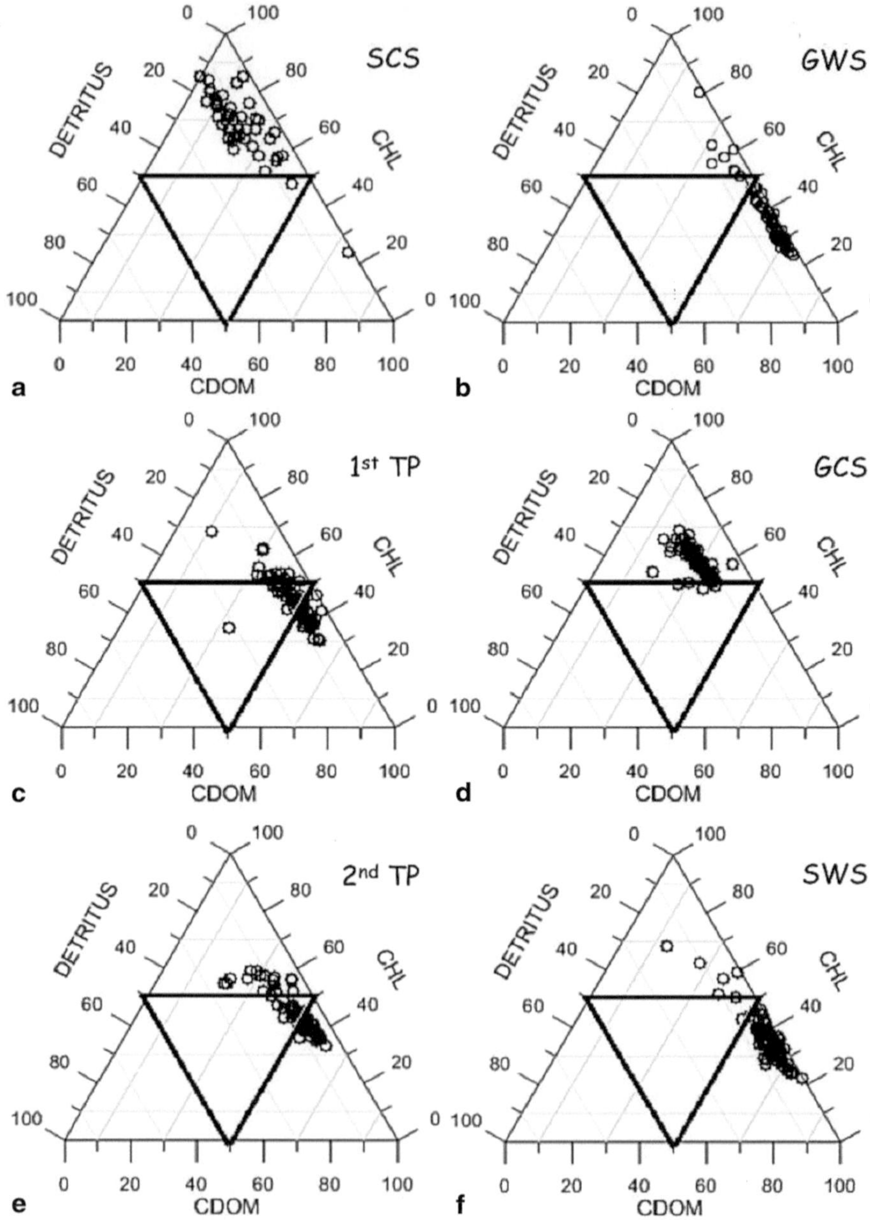

Fig. 7.6 Ternary plots of the percent reflectance attributed to *CDOM*, *CHL* and *DET* for each pixel to identify the dominant components during the different "marine" seasons of the Sassandra river mouth, i.e. **a** SCS. **b** GWS. **c** 1st transition period. **d** GCS. **e** 2nd transition period. **f** SWS

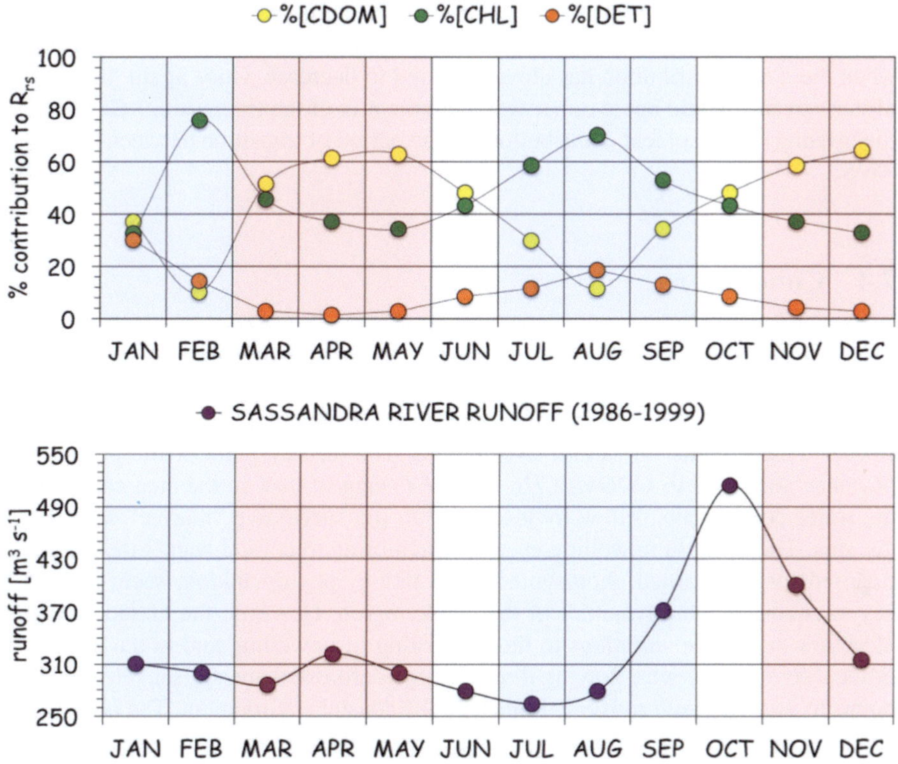

Fig. 7.7 Seasonal variability of the Sassandra River runoff, monthly means, 1986–1999 (*lower panel*), and of the percent contribution to the total reflectance attributed to *CDOM*, *CHL* and *DET*, monthly means, 2003–2007 (*upper panel*). The colored shading represents the "marine" seasons of the Côte d'Ivoire coastal region: SCS and GCS in *blue*, GWS and SWS in *red*

higher and lower nutrient availability in the cold and warm periods, or to the impact of river runoff. In the RS-derived data set considered here, while the increase of both the *CHL* and *DET* contribution seems to correlate rather well with the timing of the cooler (upwelling) periods (i.e. SCS, GCS), the *CDOM* trend appears to correlate better with that of the warmer periods (i.e. GWS, SWS). The relationship with the Sassandra river runoff is not as evident (see Fig. 7.7), and deserves more attention.

The regime of all Côte d'Ivoire major rivers, with most catchment areas located in the tropical zone, is under the influence of a "savannah climate", generally characterized by a pronounced dry season. The low-water period of the Sassandra River extends from December to April, but is particularly severe in February and March. This (relative) minimum of fluvial runoff, early in the year, would agree with the low *CDOM* contribution observed in the coastal zone during the SCS.

Similarly, the following spring (relative) maximum would go along with the increase of the *CDOM* contribution during the GWS. During the high-water period, from late summer to early fall, the Sassandra river runoff increases substantially

(Herbland and Le Loeuff 1993), reaching a peak outflow in October. This period of high river discharge corresponds also to the end of the main upwelling period, when the *CHL* contribution has already started to decrease. Once again, the timing of these events would agree rather well with the mix of the (increasing) *CDOM* and (decreasing) *CHL* percent contributions in the period of transition between GCS and SWS.

7.4 Conclusions

The analysis of the R_{rs} spectral components, in the annual, seasonal and monthly composite images derived from a 2003–2007 time series of SeaWiFS data, has shown the relative proportions of the main water constituents present in the Côte d'Ivoire coastal waters at the Sassandra river mouth. The ternary plots of the percent reflectance attributed to *CDOM*, *CHL* and *DET* suggest that, in the area considered, the water constituents follow trends linked to the successive "marine" seasons of the area. The onset of upwelling events, which bring to coastal waters the nutrients required for a sustained photosynthetic activity of phytoplankton, seem to be the key to environmental dynamics of the whole region. However, the hinted coupling of yellow substance variations to the weakening of upwelling and to the discharge of local rivers (otherwise flowing through large deforested areas), is an element that points to a strong anthropogenic impact on the coastal environment. The indications emerging from the present study, albeit approximate and very preliminary, provide an example of the important role to be played by RS in managing the area adjacent to the Sassandra river mouth (where *in situ* data are sparse). In the future, such a role could be extended to the entire coastline of the Côte d'Ivoire, in order to better classify marine waters impacted by natural processes such as coastal upwelling or by anthropogenic pollution due to river runoff.

Acknowledgments The SeaWiFS data used in this paper were obtained through the GIOVANNI online data system, developed and maintained by the NASA Goddard Earth Sciences Data and Information Services Center (GES DISC). The authors wish to express their gratitude to Global Ocean Observing System (GOOS) Africa, Interim Guinea Current Commission (IGCC) Guinea Current Large Marine Ecosystem (GCLME) project, and United Nations Educational, Scientific and Cultural Organization (UNESCO) for the support granted to the present work. Special thanks are due to the Joint Research Centre (JRC), European Commission (EC) for the opportunity to publish in this volume devoted to the African Seas.

References

Bakun A (1978) Guinea current upwelling. Nature 271:147–150

Binet D (1997) Climate and pelagic fisheries in the Canary and Guinea currents 1964–1993: the role of trade winds and the southern oscillation. Oceanologica Acta 20:177–190

Bricaud A, Morel A, Prieur L (1981) Absorption by dissolved organic matter of the sea (yellow substance) in the UV and visible domains. Limnol Oceanogr 26:43–53

CEDA (1997) Rapport sur le profil environnemental de la zone côtière. Ministère du Logement, du Cadre de Vie et de l'Environnement de Côte d'Ivoire, p 59

Chang GC, Dickey TD, Jiang S, Manov DV, Spada FW (2003) Optical methods for interdisciplinary research in the coastal ocean, Recent research developments in optics. In: Kawasaki M, Ashgriz N, Anthony R (eds) Research Signpost, India, 3, pp 249–270

Colin C, Gallardo Y, Chuchla R, Cissoko S (1993) Environnements climatique et océanographique sur le plateau continental de la Côte d'Ivoire. In: LeLoeuf P, Marshal E, Kothias JBA (eds) Environnement et Ressources Aquatiques de la Côte d'Ivoire, I—Le milieu marin. ORSTOM, Paris, pp 75–110

Dagg M, Benner R, Lohrenz S, Lawrence D (2004) Transformation of dissolved and particulate materials on continental shelves influenced by large rivers: plume processes. Continental Shelf Res 24(7–8):833–858

Demarcq H, Aman A (2002) A multi-data approach for assessing the spatio-temporal variability of the Ivoirian-Ghanaian coastal upwelling: understanding pelagic fish stock dynamics. In: McGlade JM, Cury P, Koranteng KA, Hardmann-Mountford NJ (eds) The Gulf of Guinea large marine ecosystem. Elsevier Science, Amsterdam, pp 83–92

Djagoua EV (2003) Contribution de l'imagerie satellitaire visible et infra à rouge et l'étude de la variabilité spatio-temporelle des phénomènes physiques de surface du littoral marin ivoirien et implication dans la variabilité du phytoplancton et des prises de *Sardinella aurita*. Thèse de Doctorat Unique, N° d'ordre 383, Université de Cocody-Abidjan, p 117

D'Sa EJ, RL Miller (2005) Bio-optical properties of coastal waters. In: Miller RL, Del Castillo C, McKee B (eds) Remote sensing of coastal aquatic environments. Springer, New York, pp 129–155

Froidefond D, Doxaran D (2004) Télédétection optique appliquée à l'étude des eaux côtières. Télédétection 4(2):579–597

Froidefond JM, Gardel L, Guiral D, Parra M, Ternon JF (2002) Spectral remote sensing reflectances of coastal waters in French Guiana under the Amazon influence. Remote Sens Environ 80:225–232

Geyer WR, Hill PS, Kineke GC (2004) The transport, transformation and dispersal of sediment by buoyant coastal flows. Continental Shelf Res 24(7–8):927–949

Gordon H, Morel A (1983) Remote assessment of ocean color for interpretation of satellite visible imagery: a review. Lecture notes on coastal and estuarine studies, vol 4. Springer, New York, p 114

Gower JFR, Borstadt GA, Gray LH, Edel HR (1988) The fluorescence line imager: high-resolution imaging spectroscopy over water and land. In: Guyenne TD, Hunt JJ (eds) Spectral signatures of objects in remote sensing. Proceedings of the conference held 18–22 January 1988 in Aussois (Modane), France. European Space Agency, ESA SP-287, pp 273–287

Herbland A, Le Loeuff P (1993) Les sels nutritifs au large de la Côte d'Ivoire. In: LeLoeuf P, Marshal E, Kothias JBA (eds) Environnement et Ressources Aquatiques de la Côte d'Ivoire, I—Le milieu marin. ORSTOM, Paris, pp 123–148

Hu C, Montgomery ET, Schmitt RW, Muller-Karger FE (2004) The dispersal of the Amazon and Orinoco River water in the tropical Atlantic and Caribbean Sea: observation from space and S-PALACE floats. Deep-Sea Res II 51(10–11):1151–1171

Ingham MC (1970) Coastal upwelling in the northwestern gulf of Guinea. Bulletin Marine Sci 20:1–34

IOCCG (2000) Remote sensing of ocean colour in coastal, and other optically-complex, waters. In: Sathyendranath S (ed) Reports of the International Ocean-Colour Coordinating Group, No 3, IOCCG, Dartmouth, p 140

IOCCG (2006) Remote sensing of inherent optical properties: fundamentals, tests of algorithms and applications. In Lee ZP (ed) Reports of the international ocean-colour coordinating group, No 5, IOCCG, Dartmouth, p 126

Jaquet JM (1989) Limnologie et télédétection: situation actuelle et développements futures. Revue des sciences de l'eau/. J Water Sci 2(4):457–481

Lahet F, Forget P, Ouillon S (2001) Application of a colour classification method to quantify the constituents of coastal waters from in situ reflectances sampled at satellite sensor wavebands. Int J Remote Sens 22(5):909–914

Le Loeuff P, Marchal E (1993) Géographie littorale. In: LeLoeuf P, Marshal E, Kothias JBA (eds) Environnement et ressources aquatiques en Côte D'Ivoire: I—Le milieu marin. ORSTOM, Paris, pp 15–22

Marchal E, Picaut J (1977) Répartition et abondance évaluées par écho-intégration des poissons du plateau ivoiro-ghanéen en relation avec les upwellings locaux. J de Recherche Oceanographique 2:39–57

McClain CR, Feldman GC, Hooker SB (2004) An overview of the SeaWiFS project and strategies for producing a climate research quality global ocean bio-optical time series. Deep-Sea Res II 51:5–42

Miller RL, Del Castillo CE, McKee BA (eds) (2005) Remote sensing of coastal aquatic environments: technologies, techniques and applications. Remote Sensing and Digital Image Processing, vol 7. Springer, Dordrecht, p 345

Morlière A (1970) Les saisons marines devant Abidjan. Doc Scient Centre Rech Océanogr Abidjan 1(2):1–15

Morel A, Prieur L (1977) Analysis of variations in ocean color. Limnol Oceanogr 22:709–722

Picaut J (1983) Propagation of the seasonal upwelling in the eastern equatorial Atlantic. J Phys Oceanogr 13:18–37

Pope RM, Fry ES (1997) Absorption spectrum ~380–700 nm of pure water. II. Integrating cavity measurements. Applied Optics 36(33):8710–8723

Robinson IS (2004) Measuring the oceans from space: the principles and methods of satellite oceanography. Springer & Praxis, Heidelberg, p 669

Sevrin-Reyssac J (1993) Phytoplancton et production primaire dans les eaux marines ivoiriennes. In: LeLoeuf P, Marshal E, Kothias JBA (eds) Environnement et Ressources Aquatiques de la Côte d'Ivoire, I—Le milieu marin. ORSTOM, Paris, pp 151–166

Strömbeck N, Pierson DC (2001) The effects of variability in the inherent optical properties on estimations of chlorophyll a by remote sensing in Swedish reservoirs. Sci Total Environ 268:123–137

Chapter 8
Oil Pollution in Coastal Waters of Nigeria

Andrey G. Kostianoy, Olga Yu. Lavrova and Dmytro M. Solovyov

Abstract The petroleum industry in Nigeria has become the largest industry and the main Gross Domestic Product (GDP) generator in Western Africa. Nigeria's proven oil reserves make Nigeria the most affluent country in Africa and the tenth most petroleum-rich one in the world. The main oil fields are concentrated in and around the Niger River Delta, which is seriously damaged by oil leakages. It has been estimated that between 9 and 13 million barrels have been spilled since oil drilling started in 1958. Between 1976 and 2001 the Government documented 6817 spills, but analysts suspect that the real number may be ten times higher. Little is known about oil spills in the coastal zone of Nigeria, but offshore spills contaminate coastal environments and cause a decline in local fishing production. We present a general view on oil pollution in the coastal waters of Nigeria, based on the analysis of historical information and recent Advanced Synthetic Aperture Radar (ASAR) data, with a focus on the Bonga oil spill case study, which occurred on 20 December 2011.

8.1 Introduction

Since the late 1950s, when the British discovered oil in the Niger Delta, the petroleum industry in Nigeria has become the largest industry and the main Gross Domestic Product (GDP) generator in Western Africa (see Fig. 8.1). In 2000, oil and gas exports in Nigeria accounted for 95 % of foreign exchange earnings, 65 % of Government budgetary revenue, and 40 % of its GDP (Wikipedia 2012a). Nigeria's proven oil reserves, estimated by the US Energy Information Administration, vary between 16 and 22 billion barrels, but other sources raise this value to 35 billion barrels. These reserves make Nigeria the most richest country in Africa and the tenth most

A. G. Kostianoy (✉)
P. P. Shirshov Institute of Oceanology, Russian Academy of Sciences, Moscow, Russia
e-mail: kostianoy@gmail.com
Also at: I. Kant Baltic Federal University, Kaliningrad, Russia

O. Y. Lavrova
Space Research Institute, Russian Academy of Sciences, Moscow, Russia

D. M. Solovyov
Marine Hydrophysical Institute, National Academy of Sciences, Sevastopol, Ukraine

V. Barale, M. Gade (eds.), *Remote Sensing of the African Seas*,
DOI 10.1007/978-94-017-8008-7_8,

Fig. 8.1 Oil and gas industry in the Niger Delta area and coastal zone of Nigeria (Africa Confidential 2008)

petroleum-rich one in the world. Nigeria has a total of 159 oil fields and 1481 wells in operation (Wikipedia 2012a). The main oil fields are concentrated in and around the Niger River Delta, but off-shore rigs are also encountered in the coastal zone. The environment of the Niger River Delta is seriously damaged by oil leakages. The Government documented 6,817 spills between 1976 and 2001– practically one per day for 25 years, but analysts suspect that the real number may be ten times higher. It has been estimated that between 9 and 13 million barrels have been spilled since oil drilling started in 1958 (Wikipedia 2012a, b). The main causes include corrosion of pipelines and tankers (accounts for 50 % of all spills), sabotage (28 %), and oil production operations (21 %), non-functional production equipment (1 %) (Wikipedia 2012b). A massive parade of major tankers can be observed daily in Cawthorne Channel, which are filled by the Bonny Light at export terminal at Bonny Island. Little is known about the oil spills in the coastal zone, but offshore spills, which are usually much greater in scale, contaminate coastal environments and cause a decline in local fishing production.

Fig. 8.2 Map of Nigeria. (In http://nigeriamasterweb.com/blog/index.php/2012/01/26/nigeria-it-s-a-matter-of-upbringing)

On 20 December 2011 the Bonga offshore oil spill resulted from the routine transfer of crude from the Floating Production, Storage And Offloading (FPSO) vessel Bonga to a waiting oil tanker. An export line linking the FPSO and the tanker was identified as the likely source, and has been closed and de-pressurized. But the next day Shell estimated the spill volume as of 40,000 barrels. It was likely the largest oil spill which has occurred offshore Nigeria since 1998, when about 40,000 barrels of oil leaked from a Mobil field, producing an oil slick extending for more than 180 km to Lagos. The next days the Bonga spill stretched in the direction of the coast. The spill was clearly detected in the image acquired on 21 December 2011 by the Advanced Synthetic Aperture Radar (ASAR), onboard the Envisat orbital platform of the European Space Agency (ESA), as well as quite well visible in an optical image of the Moderate Resolution Imaging Spectroradiometer (MODIS), onboard the Terra satellite (MODIS Terra) of the US National Aeronautics and Space Administration (NASA). It was 80 km long, 15 km wide and had a total surface of 923 km^2. In the following we present a general view on oil pollution in the country and coastal waters of Nigeria, based on the analysis of historical and recent ASAR data, with a focus on the Bonga oil spill case study.

8.2 Niger Delta and Offshore: Environmental Issues

The Niger Delta covers about 70,000 km^2 including wetlands, which is about 7.5 % of Nigeria territory (Figs. 8.2 and 8.3). It can be divided in four zones: coastal barrier islands, mangrove swamp forests, freshwater swamps, and lowland rainforests

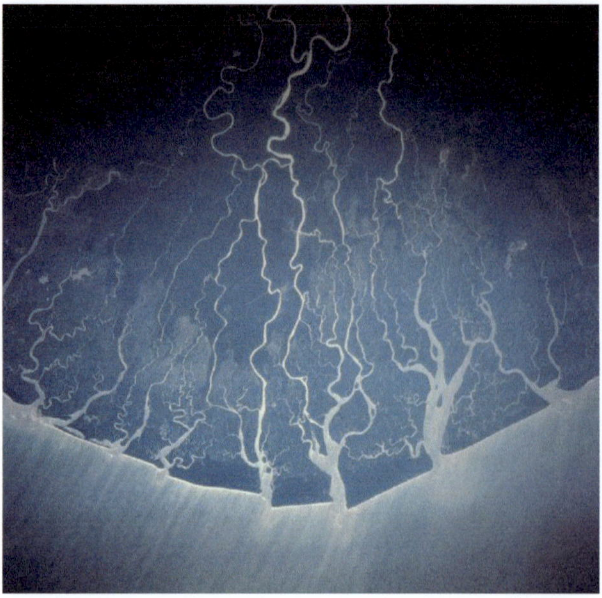

Fig. 8.3 NASA Space Shuttle Overflight photo. (Image: STS61C-42-72.jpg) of the Niger Delta, 19 November 2005

(Wikipedia 2012b). Its ecosystem (flora, fauna, freshwater fish) has one of the highest biodiversities on the planet. This is a densely populated region sometimes called the Oil Rivers because it was once a major producer of palm oil. Environmental issues of the Niger Delta and offshore are mainly related to the petroleum industry in Nigeria.

Nigeria has become the biggest producer of petroleum in Africa since 1956, when the British Royal Dutch Shell discovered crude oil at Oloibiri, a village in the Niger Delta, and started commercial production in 1958 (Nwilo and Badejo 2001, 2007; Egberongbe et al. 2006). In 2006 it was 606 oil fields in the Niger Delta, of which 360 were on-shore and 246 offshore, with a total crude oil production averaging around 2.5 million barrels (400,000 m^3) per day (Nwilo and Badejo 2001, 2007; Egberongbe et al. 2006; Wikipedia 2012a). The same oil production rate was estimated for 2010. Together oil and natural gas extraction comprise 95 % of Nigeria's foreign exchange revenues. Multinational corporations (Royal Dutch Shell, Chevron, Exxon-Mobil, Agip, Total, Texaco) play a key role in the Nigerian oil and gas industry with necessary technological and financial resources. Joint ventures between foreign multinational corporations and the Nigerian Federal Government account for approximately 95 % of all crude oil production (Wikipedia 2012a). Nigeria has six export terminals including Forcados and Bonny (operated by Shell), Escravos and Pennington (by ChevronTexaco), Qua Iboe (by Exxon Mobil), Brass (Agip), and Bonny Island LNG (by Nigeria LNG).

Natural gas extracted from oil wells in the Niger Delta is immediately burned, or flared, into the air at a rate of approximately 70 mln m^3 per year. This is equivalent to 41 % of African natural gas consumption, and forms the largest single source of greenhouse gas emissions on the planet. Gas flaring in Nigeria releases large amounts of methane and carbon dioxide, which have very high global warming effect. Gas flares release a number of poisonous chemicals such as nitrogen dioxides, sulphur dioxide, volatile organic compounds like benzene, toluene, xylene and hydrogen sulfide, as well as carcinogens like benzapyrene and dioxin (Nwilo and Badejo 2001, 2007; Wikipedia 2012a). It is well known that gas flaring is a major contributor to air pollution and acid rains.

The Nigerian officials (Department of Petroleum Resources in Nigeria) estimated 1.89 mln barrels of petroleum were spilled into the Niger Delta between 1976 and 1996 out of a total of 2.4 mln barrels spilled in 4,835 incidents. The UNDP report shows a total of 6,817 oil spills between 1976 and 2001, which account for a loss of 3 million barrels of oil, of which more than 70 % was not recovered. The Nigerian National Petroleum Corporation reports that an average 300 individual large oil spills of a total 2,300 m^3 annually poison the environment. World Bank estimates this quantity as ten times larger. In 2010 Baird reported that between 9 and 13 million barrels have been spilled in the Niger Delta since 1958. Maximal estimates of the total amount of petroleum spilled between 1960 and 1997 reach 100 million barrels, or 16,000,000 m^3 (Wikipedia 2012b).

There are different and even unusual (for example, sabotage) causes for oil pollution in Nigeria. They include: corrosion of pipelines and tankers (50 % of all spills), sabotage (28 %), oil production operations (21 %), and inadequate or non-functional production equipment (1 %) (Nwilo and Badejo 2001, 2007; Wikipedia 2012b). Corrosion of pipelines is a main reason, which is explained by an extensive network of pipelines and small flowlines connecting the small size oil fields and oil flowstations. In onshore areas most pipelines and flowlines are laid above ground for 30–50 years. With a life span of about fifteen years, corrosion allows many opportunities for leaks. Sabotage is performed primarily through what is known as "bunkering", whereby the saboteur attempts to tap the pipeline. Theft of oil through oil siphoning has become a major issue also. In the process of oil extraction sometimes the pipeline is damaged or destroyed with leakages for several days (Nwilo and Badejo 2001, 2007; Wikipedia 2012b).

Oil spills have a major impact on the environment and ecosystem of the Niger Delta and coastal zone. Available records for the period 1976 to 1996 indicate that approximately 6, 25, and 69 % respectively, of total oil spilled in the Niger Delta area, were in land, swamp and offshore environments (Nwilo and Badejo 2001, 2007; Egberongbe et al. 2006). Oil kills plants and animals in the estuarine zone, settles on beaches and kills organisms that live there, settles on the ocean floor and kills benthic organisms such as crabs. Oil poisons algae, disrupts major food chains, decreases the yield of edible crustaceans, affects birds, contaminates commercially valuable fish in coastal zone (Nwilo and Badejo 2001, 2007). Oil contamination affects the population of nearly 250 species of fish, of which 20 are endemic. The mangrove and rain forests, especially vulnerable to oil, have been significantly destroyed. In the populated areas agriculture, aquaculture and fishery are developing under the pressure of hydrocarbons contamination of soils, groundwater, freshwater and seawater.

Water quality, breathing problems and skin lesions are the main health issues for people living in the affected areas (Wikipedia 2012b). In April 1997, samples taken from water used for drinking and washing by local villagers were analyzed in USA. A sample from Luawii, in Ogoni, where there had been no oil production for four years, had 18 ppm of hydrocarbons in the water, 360 times the level allowed in drinking water in the European Union (E.U.). A sample from Ukpeleide, Ikwerre, contained 34 ppm, 680 times the E.U. standard (Nwilo and Badejo 2007).

8.3 Major Oil Spill Incidents in Nigeria

Oil spillages in Nigeria are categorized into four groups: minor, medium, major and disaster. The spill is classified as minor when the oil discharge is less than 25 barrels in inland waters or less than 250 barrels on land, offshore or coastal waters and does not pose a threat to the public health or welfare. Medium spill must be 250 barrels or less in the inland water or 250–2,500 barrels on land, offshore and coastal water while for the major spill, the discharge to the inland waters is in excess of 250 barrels on land, offshore or coastal waters. The disaster refers to any uncontrolled well blowout, pipeline rupture or storage tank failure which poses an imminent threat to the public health or welfare (Nwilo and Badejo 2001, 2007; Egberongbe et al. 2006).

The largest individual oil spills in Nigeria include (Nwilo and Badejo 2001, 2007; Egberongbe et al. 2006; Evoh 2009; Wikipedia 2012b):

- The Escravos spill in 1978, which involved about 300,000 barrels.
- In 1978 a failure of the Forcados Terminal tank (Royal Dutch Shell) produced a spill of about 580,000 barrels (92,000 m^3).
- One of the most disastrous spill in Nigeria occurred on January 17, 1980 when about 200,000 barrels spilled into the Atlantic Ocean from a Texaco facility and destroyed 340 ha of mangroves.
- The blowout of a Texaco Funiwa-5 offshore station in 1980 which dumped an estimated 400,000 barrels (64,000 m^3) of crude oil into the Gulf of Guinea.
- The 1982 Abudu pipeline oil spillage of 18,818 barrels.
- On 12 January 1998 an oil pipeline ruptured at Mobil-Idoho platform on the inner shelf of the Atlantic coast of southeastern Nigeria releasing over 40,000 barrels (approximately 6,000 tones) of Qua Iboe light crude oil into the marine environment over the next three days. Oil fire incident claimed about a thousand lives and produced the spillage extending for more than 180 km to Lagos. More than 700 km of the Nigerian coastline including estuaries were impacted.
- Finally, on 20 December 2011 the Shell's Bonga offshore oil spill of 40,000 barrels resulted from the routine transfer of crude oil from the Bonga FPSO to a waiting oil tanker, mentioned earlier.

Oil spill incidents of smaller size occur almost daily in Nigeria. We did not find a statistics for marine oil spills separately, but there is an official statistics for all oil spill incidents in the country provided by the Department of Petroleum Resources

Table 8.1 Oil spill incidents in Nigeria (1976–1998). (Source: Department of Petroleum Resources of Nigeria; Nwilo and Badejo 2001)

Year	Number of spill incidents	Quantity spilled (barrels)
1976	128	26,157.00
1977	104	32,879.25
1978	154	489,294.75
1979	157	694,117.13
1980	241	600,511.02
1981	238	42,722.50
1982	257	42,841.00
1983	173	48,351.30
1984	151	40,209.00
1985	187	11,876.60
1986	155	12,905.00
1987	129	31,866.00
1988	208	9,172.00
1989	195	7,628.161
1990	160	14,940.816
1991	201	106,827.98
1992	367	51,131.91
1993	428	9,752.22
1994	515	30,282.67
1995	417	63,677.17
1996	430	46,353.12
1997	339	59,272.30
1998	390	98345.00
Total	5724	2,571,113.90

of Nigeria (Table 8.1) for 1976–1998 (Nwilo and Badejo 2001). Taking in mind that for almost the same time period (1976–1996) approximately 69 % of total oil has spilled into offshore environment (Nwilo and Badejo 2001, 2007; Egberongbe et al. 2006), it is easy to understand that the pollution of the coastal ocean is enormous in numbers and quantity (Table 8.1).

An interesting statistics for marine oil spills observed in the Gulf of Guinea, including the coastal zone of Nigeria, can be found in (Terleeva et al. 2012). Based on the analysis of ASAR Envisat data (Medium Resolution Image Product) for 2003–2009 it was shown that the largest number of oil spills was observed along the coast of Nigeria (148), then along the coasts of Cameroon and Equatorial Guinea (98), Ghana (49), Cote d'Ivoire (31), and Togo/Benin (9). Oil spill area varied from 0.5 to 80 km^2. This recent statistics, based on irregular satellite information, confirms that coastal waters of Nigeria are highly polluted.

8.4 The Bonga Oil Spill Case Study

On 20 December 2011 the Bonga offshore oil spill resulted from the routine transfer of crude from the Bonga FPSO vessel (Fig. 8.4) to a waiting oil tanker. An export line linking the FPSO unit and the tanker was identified as the likely source, and has been closed and de-pressurized. But the next day Shell estimated the spill volume as of

Fig. 8.4 The Bonga floating production, storage and offloading vessel. (From http://gcaptain.com/gulf-guinea-filthy-barrels-leak/)

40,000 barrels. It was likely the largest oil spill which has occurred offshore Nigeria since 1998, when about 40,000 barrels of oil leaked from a Mobil field, producing an oil slick extending for more than 180 km to Lagos. The next days the Bonga spill stretched in the direction of the coast. The spill was clearly detected on the Envisat ASAR image acquired on 21 December 2011, 09:30 UTC (Fig. 8.5), as well as quite well visible on the MODIS Terra optical image (Fig. 8.6).

Driven by the southwesterly wind, the spill had stretched in the direction to the coast. At the moment of image acquisition, the spill was 80 km long, 15 km wide. It had a total area of 923 km^2 and a perimeter of 175 km (see Fig. 8.5). Its minimal distance from the coastline was 52 km. It is noteworthy that beside this gigantic oil spill in the region of Bonga platform, a number of other cases of oil pollution of the sea surface near the coastline can be identified in the same ASAR image of 21 December 2011 (Fig. 8.5). The total area of 11 large spills, excluding the Bonga one, was over 100 km^2 (five of them are shown in red circles in Fig. 8.5). Moreover, three river plumes from the Niger River arms polluted with oil products were also detected (shown in red rectangles in Fig. 8.5). Their area amounted to 78 km^2.

One can see that the analysis of just one ASAR image demonstrates the high degree of oil pollution typical for coastal waters of Nigeria. It is interesting to note that the accident at the Bonga FPSO was regarded by Nigerian officials and Shell almost as a catastrophe, possibly the largest oil spill which has occurred offshore Nigeria since 1998. The same day we found a series of routine oil spills of a total area of 178 km^2, which was not observed and announced by anybody. If such a large total polluted area was observed during one day in one of the European seas, it would certainly be regarded as an exceptional case (Kostianoy 2008; Kostianoy and Lavrova 2013).

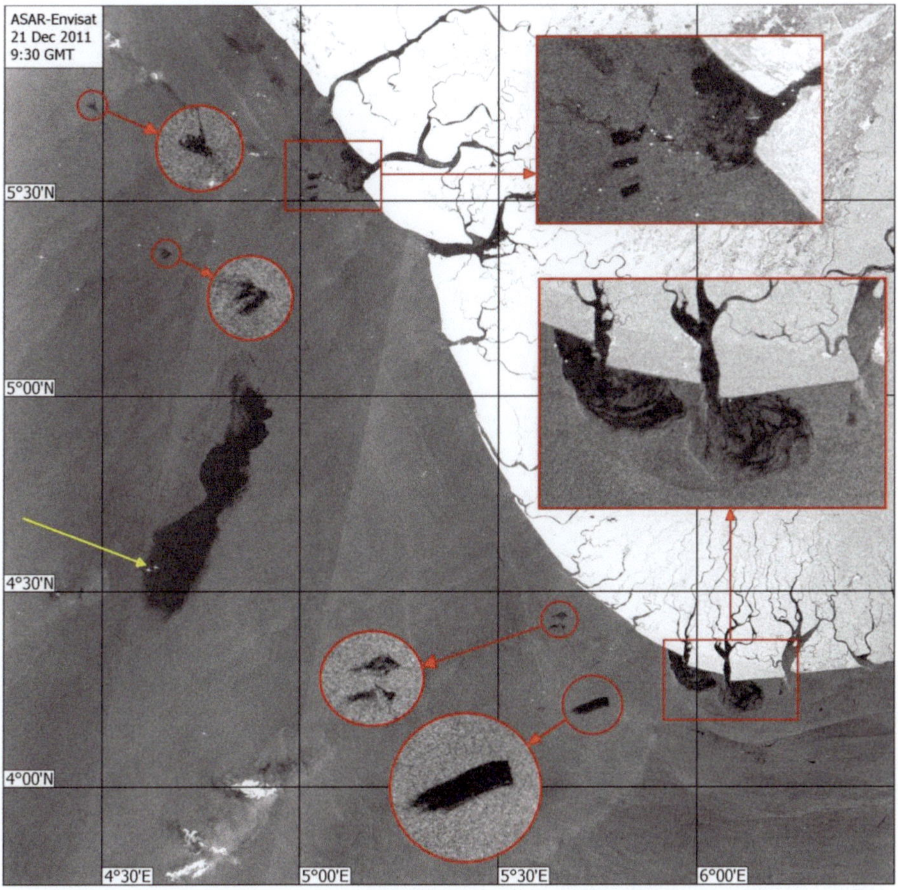

Fig. 8.5 Envisat ASAR image of the coastal zone of Nigeria acquired on 21 December 2011, 09:30 UTC (©ESA 2011). Oil spills are visible as *black patches*. The zooms on the specific sea areas with oil spills are shown in *red circles* and *rectangles*. Ships in the sea are visible as bright white dots. *Yellow arrow* shows the position of the Bonga floating production, storage and offloading vessel, and a very large Bonga oil spill which is drifting northeastward

As shown in several investigations oil spills can under certain conditions also be seen on optical imagery (Lavrova and Kostianoy 2010; Lavrova et al. 2011). For example, such a SAR/optical combined approach was very effective during our monitoring of oil pollution in the Gulf of Mexico in April-July 2010 after an explosion at the BP Deepwater Horizon oil platform which created a long-lasting leakage of oil from the sea floor (Lavrova and Kostianoy 2010; Lavrova et al. 2011). The MODIS Terra image of the same area as in Fig. 8.5 acquired on 21 December 2011, 10:20 UTC, i.e. 50 min after the Envisat pass over the area, showed the Bonga spill due to different optical characteristics of the sea surface in the spill and in the surrounding waters (Fig. 8.6).

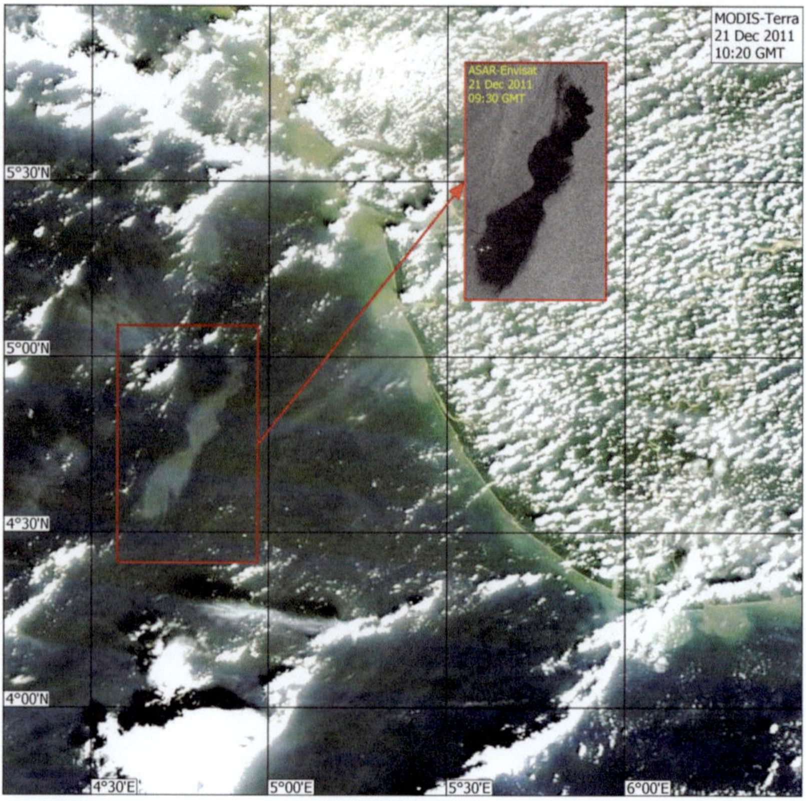

Fig. 8.6 MODIS-Terra image of the same area as in Fig. 8.5 acquired on 21 December 2011, 10:20 UTC. *Red rectangles* show a location of the Bonga oil spill and its view on Envisat ASAR image acquired 50 min before the overflight of the Terra satellite

As concerns other offshore oil spills, they are too small to be identified on the optical image (even with known position on SAR image) due to low optical contrasts, worse spatial resolution and cloudiness, which is a main obstacle for use of optical imagery in this region. As concerns river plumes, which bring a large amount of oil via several arms of the Niger River (see Fig. 8.5), optical imagery is useless also, because of large concentration of suspended matter inside the plumes (light brown colors along the coasts of Nigeria in Fig. 8.6), which has the same color as oil at the sea surface on MODIS images. Thus, as for ocean or coastal waters of Nigeria, SAR imagery is much more effective instrument to detect oil (see Figs. 8.5 and 8.6).

8.5 Main Regions of Oil Pollution

Analysis of a set of ASAR images of the coastal zone of Nigeria acquired in 2011–2012 has revealed three main regions of permanent oil pollution: the region of port of Lagos and regions close to the southern and western parts of the Niger River Delta.

All these areas correspond to main oil production regions on the land and export oil terminals in the coastal zone of Nigeria (see Figs. 8.1 and 8.2). Port of Lagos is one of the largest ports in Africa with increasing volumes of crude oil export. Delta Port, Rivers Port and Onne Port are located in the oil and gas producing Niger River Delta region. Delta Port Complex in Delta State includes the ports of Warri, Burutu, Sapele, and oil terminals at Escravos, Forcados and Pennington (see Fig. 8.1). The Rivers Port Complex in coastal Rivers State has Port Harcourt (see Fig. 8.1), Okrika Refined Petroleum Oil Jetty, Haastrup/Eagle Bulk Cement Jetty, Kidney Island Jetty, Ibeto Jetty, Macobar Jetty and Bitumen Jetty. Rivers State is also a principal petroleum-producing region in the country. Rivers State also has Onne Port Complex on Ogu Creek near the Bonny River, 20 km from Port Harcourt. It has two major facilities, the Federal Ocean Terminal and the Federal Lighter Terminal. The port has been designated as an Oil and Gas Free Zone by the Government. It serves as the main port for oil and gas operations with Western and Central Africa (Nwilo and Badejo 2001, 2007; Egberongbe et al. 2006; Wikipedia 2012b).

In all the SAR images of the Lagos area, many patches of anthropogenic pollution are observed (Figs. 8.7 and 8.8). These are probably due to discharges of polluted water from numerous ships anchored near the port. An example is presented in Fig. 8.7 featuring an ASAR Envisat image on 5 April 2012 of the coastal zone of Benin and Nigeria with a zoomed focus on the marine areas in the vicinity of Cotonou and Lagos ports.

As usual, on SAR images ships are displayed as white bright dots. In front of Cotonou, which is the economic capital and main port of Benin, we see a dozen of large ships in the sea with a little oil pollution. The dark area to the right of the outlet of the canal (see Fig. 8.7), which divides the city in two parts and links Lake Nokoué and the Atlantic Ocean, is explained by biogenic films. These are found in the canal plume and dampen gravity-capillary waves.

In front of Port of Lagos we see several dozens of ships with over 30 ship discharges of different size (Fig. 8.7). Total polluted area is estimated to 7 km^2. Affected by the local coastal current, they stretch out along the coastline. This area is much more polluted than Cotonou Port. Westward of this outlined area we observe another 20 km long oil slick. Other coastal and open ocean areas on this satellite frame are free of oil pollution (Fig. 8.7). A similar picture is observed almost on every Envisat ASAR image of the Lagos area (Fig. 8.8). For example, in Fig. 8.8 we again observe several dozen ships in front of the port and a dozen oil spills of different size around.

In the coastal region off the western Niger Delta, oil pollution has many different causes. These include operational activities at sea oil platforms, bunker works at oil terminals, ship discharges, and outflows from the river arms. All these types of pollution are evident, for example, in Envisat ASAR images of 21 December 2011 (Fig. 8.5) and 28 March 2012 (Fig. 8.8). Beside anthropogenic pollution, the region of the western Niger Delta has many locations where natural oil seeps in the sea. Their ASAR patterns on the sea surface are typical of similar petroleum seeps in other oil rich marine areas such as the Gulf of Mexico, Persian Gulf, Caspian Sea, etc. (Lavrova and Kostianoy 2010; Lavrova et al. 2011).

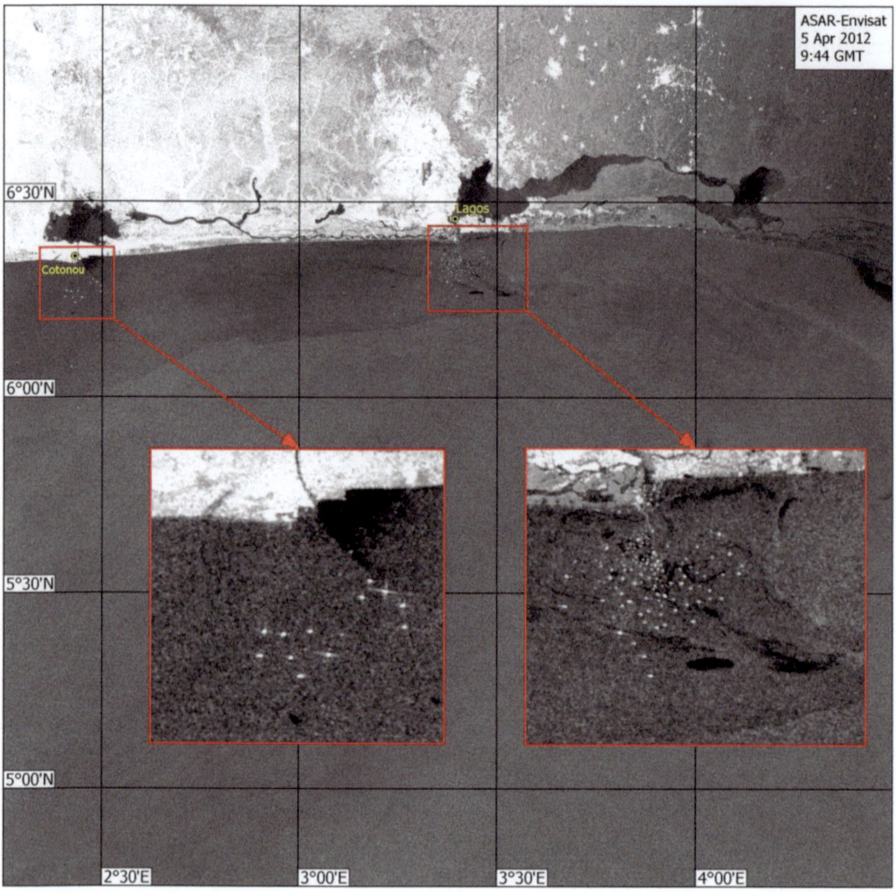

Fig. 8.7 Envisat ASAR image of the coastal zone between Cotonou (Benin) and Lagos (Nigeria) acquired on 5 April 2012, 09:44 UTC (©ESA 2012). *Red rectangles* show a zoom in the coastal zone of Cotonou and Lagos ports

A petroleum seep is a natural phenomenon. This is a place where natural liquid or gaseous hydrocarbons escape to the surface of the land or to water from the bottom of the sea or a lake. Seeps generally occur above either terrestrial or offshore petroleum accumulation structures. For example, in the Gulf of Mexico there are more than 600 natural oil seeps from the bottom with estimated volume of leakage up to 140,000 tones per year (Committe on Oil in the Sea 2003). There are four regions offshore North America with known seeps: the Gulf of Mexico, southern California, the Gulf of Alaska, and waters of Canada (at Scott Inlet and Buchan Gulf). Recent estimates of the annual amounts of oil seepage occurring offshore from North America shows 160,000 tones, while the global oil seepage rate is now estimated to be between 200,000 and 2,000,000 tones (Committee on Oil in the Sea 2003). The Envisat

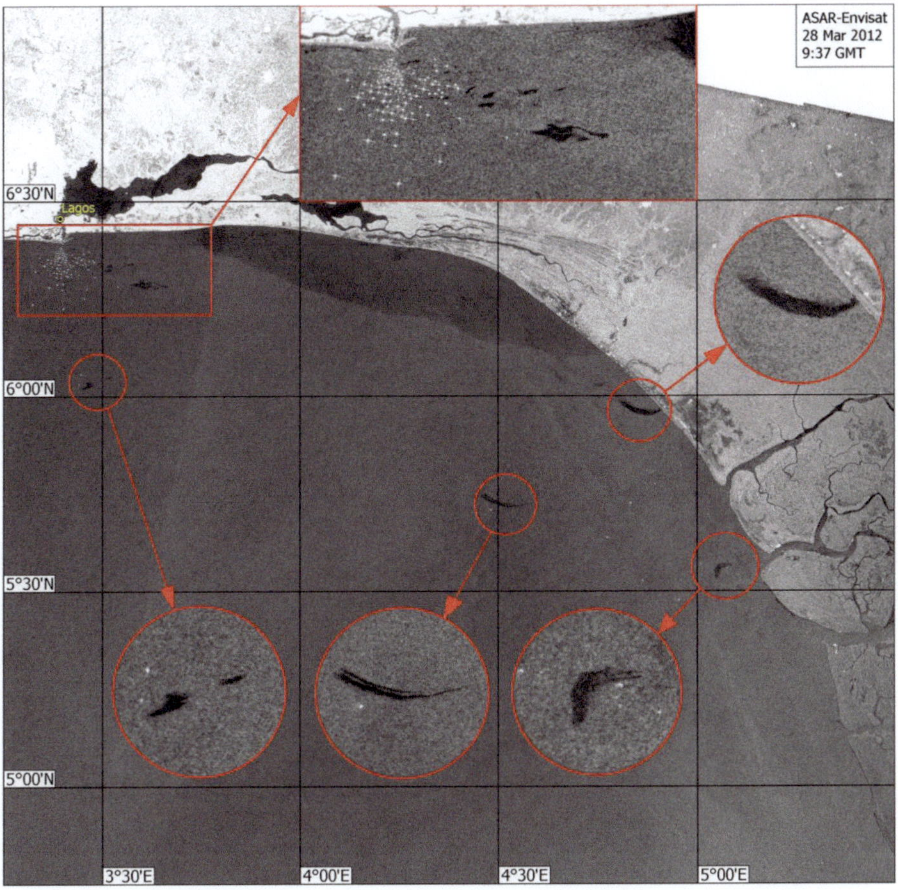

Fig. 8.8 ASAR Envisat image of the coastal zone of Nigeria acquired on 28 March 2012, 09:37 UTC (©ESA 2012). *Red circles* show a zoom in the observed oil spills

ASAR image of 28 March 2012 with two distinct curved oil slicks presents a good example of natural seeps in the coastal zone of Nigeria (Fig. 8.8).

Oil pollution in the coastal region off the southern Niger Delta, which is densely populated with oil rigs, is associated primarily with the outflows of polluted waters from the river arms to the Atlantic Ocean (Figs. 8.5 and 8.9). This type of pollution is detected in all available ASAR images of the region. For instance, in the image of 3 April 2012 (Fig. 8.9), two river plumes with high concentration of oil in water can be easily detected in an ASAR image. The size of each polluted plume is around of 40 km^2. Almost identical situation is observed in the ASAR Envisat image of 21 December 2011 (Fig. 8.5).

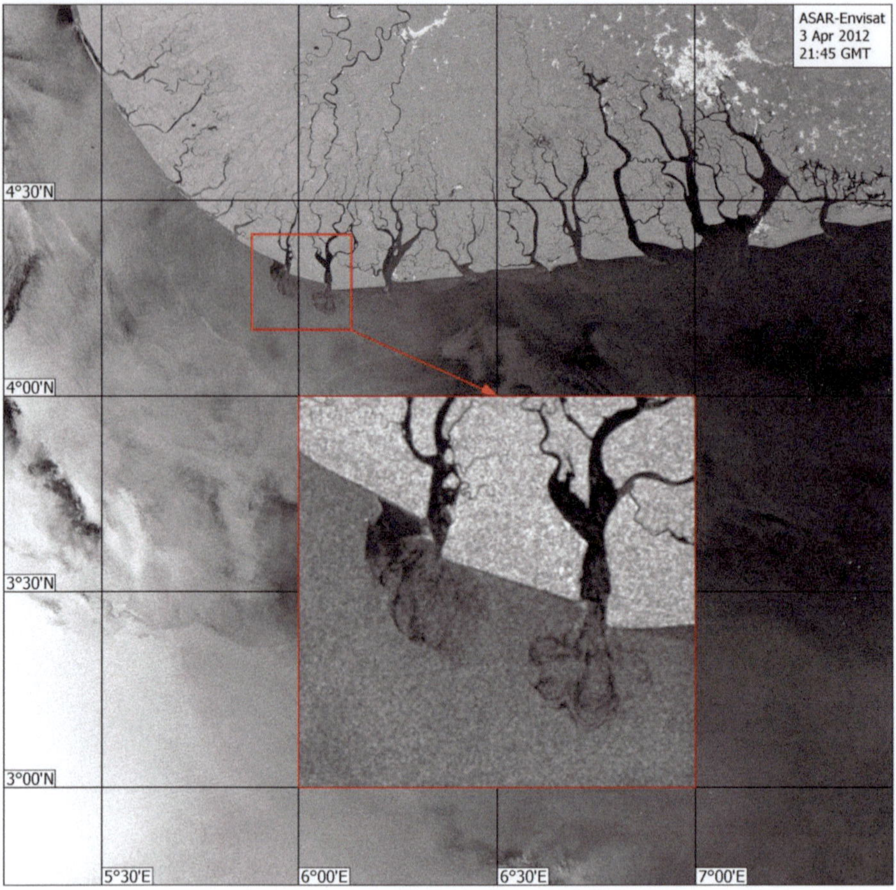

Fig. 8.9 ASAR Envisat image of the coastal zone of Nigeria acquired on 3 April 2012, 21:45 UTC (©ESA 2012). *Red rectangle* shows a zoom in the observed area

8.6 Conclusions

The Ministry of Environment in Nigeria is legally empowered with the responsibility of protecting and sustaining the Nigerian environment. The Environment Impact Assessment Decree (1992) controls activities that have environmental impact. In 2002, the Nigerian Government has issued the Environmental Guidelines and Standards for the Oil Industry and has ordered the oil companies operating in the country to comply with it. The authority that is responsible for the oil spill clean up is the National Oil Spill Detection and Response Agency (NOSDRA) that has been created in 2006. The Ministry of Environment is responsible for oil spill planning and response with NOSDRA, NMISA (Nigerian Maritime and Safety Agency), the Urban and Regional Planning, the Department of Petroleum Resources (DPR) and the Coastal

State Ministries of Environment. A National Oil Spill Contingency Plan has been set up by the National Oil Spill Response Advisory Committee. Each operating oil company is required to have a contingency plan for the prevention, control and removal of spilled oil from its own facilities. In addition, Clean Nigeria Associates (CNA) founded in 1981, a consortium of eleven oil companies operating in Nigeria, can be called upon by members when faced with spills beyond the capability of own resources. Despite the level of the Government and oil industries activities, oil spill management in Nigeria is relatively undeveloped (Nwilo and Badejo 2001, 2007; Egberongbe et al. 2006).

In 2000 the Federal Government of Nigeria established the Niger Delta Development Commission (NDDC) which aims to suppress the impact of petroleum pollution on the Niger Delta environment. Governmental and nongovernmental organizations have also utilized some technologies to identify the source, volume and trajectories of oil spills (Nwilo and Badejo 2001, 2007; Egberongbe et al. 2006; Wikipedia 2012b). Unfortunately there is little information about these technologies, thus it is difficult to compare them with satellite, aerial and marine technologies applied, for example, by western countries in the European seas (Kostianoy 2008; Kostianoy and Solovyov 2010, 2011; Lavrova and Kostianoy 2010; Kostianoy and Lavrova 2013). Satellite monitoring systems based on the operational reception and analysis of SAR data is a key instrument for the detection of oil spills in the mostly cloudy conditions over the region (Lavrova et al. 2006, 2008). Unfortunately, Nigeria, although it has several satellites with optical sensors, has no SAR instrument onboard. This seriously restricts their ability to monitor oil spills in the coastal ocean.

The National Space Research and Development Agency (NASRDA) is the Nigerian Space Agency (established in 1998), which on 27 September 2003 has launched NigeriaSat-1 with a lifespan of five years. NigeriaSat-1 is a satellite in the standard Disaster Monitoring Constellation (DMC) design. The DMC consists of a number of remote sensing satellites constructed by Surrey Satellite Technology Ltd (SSTL) and operated for the Algerian, Nigerian, Turkish, British and Chinese governments by DMC International Imaging. The DMC provides emergency Earth imaging for disaster relief under the International Charter for Space and Major Disasters, which the DMC formally joined in November 2005. NigeriaSat-1 carries an optical imaging payload (similar to Landsat TM + bands 2, 3 and 4) developed by SSTL to provide 32 m ground resolution with a wide swath width of over 640 km. The data are used within Nigeria to monitor pollution, land use and other medium-scale phenomena. In August 2011 Nigeria launched two satellites: NigeriaSat X with a 2.5-metre high spatial resolution sensor and NigeriaSat-2 to replace NigeriaSat-1.

Creation of the Nigerian national or international oil spill response centre would help in managing oil spills along the coastlines and offshore. Bearing in mind the number of oil spills and their total volume in the country, the centre should use regular operational satellite radar information from several satellites, operational weather and oil spill drift forecasts, based on the operational atmosphere and ocean models. The Swedish Seatrack Web model, used in the Baltic Sea is a good example of such an oil spill drift model, which can be adapted to local conditions (Ambjörn et al. 2013; Kostianoy et al. 2013a). An operational, multisensor and multiplatform approach to constructing an integrated daily satellite monitoring system (Kostianoy et al. 2005, 2006, 2013b; Lavrova et al. 2011) for Nigeria is the only way to build a modern monitoring infrastructure capable of solving one of the most serious environmental problems in Nigeria—oil pollution of the marine environment (Evoh 2009).

Acknowledgments The authors are grateful to European Space Agency for the Envisat ASAR data obtained in the framework of projects C1P6342.

References

Africa Confidential (2008) Wasteful wars, foreign friends. 49(17) http://www.africa-confidential.com/article-preview/id/2744/Wasteful_wars%2c_foreign_friends. Accessed 10 June 2012

Ambjörn C, Liungman O, Mattsson J, Håkansson B (2013) Seatrack Web: the HELCOM tool for oil spill prediction and identification of illegal polluters. In: Kostianoy AG, Lavrova O Yu (eds) Oil pollution in the Baltic Sea. Springer, Berlin

Committee on Oil in the Sea (2003) Oil in the Sea III: inputs, fates, and effects. The National Academies Press, Washington, DC, p 280

Egberongbe FOA, Nwilo PC, Badejo OT (2006) Oil spill disaster monitoring along Nigerian coastline. Proceedings, 5th FIG Regional Conference: promoting land administration and good governance, Accra, Ghana, March 8–11, 2006

Evoh CJ (2009) Green crimes, petro-violence and the tragedy of oil: the case of the Niger-Delta in Nigeria. In-Spire J Law, Polit Soc 4(1):40–60

Kostianoy AG (2008) Satellite monitoring of oil pollution in the European Coastal Seas. OCEANIS 34(½):111–126

Kostianoy AG, Solovyov DM (2010) Operational satellite monitoring systems for marine oil and gas industry. Proceedings of 2010 Taiwan Water Industry Conference, Tainan, Taiwan, 28–29 October 2010, B173–B185

Kostianoy A, Solovyov D (2011) Operational satellite monitoring systems for marine oil and gas industry. Proceedings of 34th International Symposium on Remote Sensing of Environment, Sydney, Australia, 10–15 April 2011

Kostianoy AG, Lavrova OYu (eds) (2013) Oil pollution in the Baltic Sea. Springer, Berlin

Kostianoy AG, Lebedev SA, Soloviev DM, Pichuzhkina OE (2005) Satellite monitoring of the Southeastern Baltic sea. Annual Report 2004. Lukoil-Kaliningradmorneft, Kaliningrad, p 36

Kostianoy AG, Ambjorn C, Solovyov DM (2013a) Seatrack web: a numerical tool for environmental risk assessment in the Baltic Sea. In: Kostianoy AG, Lavrova OYu (eds) Oil pollution in the Baltic Sea. Springer, Berlin

Kostianoy AG, Lavrova OYu, Mityagina MI, Solovyov DM, Lebedev SA (2013b) Satellite monitoring of oil pollution in the Southeastern Baltic sea. In: Kostianoy AG, Lavrova Oyu (eds) Oil pollution in the Baltic Sea. Springer, Berlin

Kostianoy AG, Litovchenko KTs, Lavrova OYu, Mityagina MI, Bocharova TYu, Lebedev SA, Stanichny SV, Soloviev DM, Sirota AM, Pichuzhkina OE (2006) Operational satellite monitoring of oil spill pollution in the southeastern Baltic Sea: 18 months experience. Environ Res, Eng Manage 4(38):70–77

Lavrova OYu, Kostianoy AG (2010) A catastrophic oil spill in the Gulf of Mexico in April-May 2010. Issledovanie Zemli iz Kosmosa (Russian Journal for Remote Sensing), 6:67–72 (in Russian). Izvestiya, Atmospheric and Oceanic Physics, Pleiades Publishing Ltd. 2011, 47(9):1114–1118 (in English)

Lavrova O, Bocharova T, Kostianoy A (2006) Satellite radar imagery of the coastal zone: slicks and oil spills. In: Marcal A (ed) Global developments in environmental earth observation from space. Millpress, Rotterdam, The Netherlands, pp 763–772

Lavrova O, Mityagina M, Bocharova T, Kostianoy A, Krovotyntsev V (2008) Multisensor approach to operational oil pollution monitoring in coastal zones. In: 2008 IEEE International Geoscience & Remote Sensing Symposium. Proceedings of IGARSS 2008, 6–11 July 2008, Boston, Massachusetts, U.S.A, ISBN 978-1-4244-2808-3, IEEE Catalog Number CFP08IGA-CDR, Library of Congress 2008906761, III 1386–1389

Lavrova OYu, Kostianoy AG, Lebedev SA, Mityagina MI, Ginzburg AI, Sheremet NA (2011) Complex satellite monitoring of the Russian seas. Moscow, IKI RAN, p 470 (in Russian)

Nwilo PC, Badejo OT (2001) Impacts of oil spills along the Nigerian coast. http://www.aehsmag.com/issues/2001/october/impacts.htm. Accessed 10 June 2012

Nwilo PC, Badejo OT (2007) Impacts and management of oil spill pollution along the Nigerian coastal areas. International Federation of Surveyors. http://www.fig.net/pub/figpub/pub36/chapters/chapter_8.pdf. Accessed 10 June 2012

Terleeva NV, Kamagate SA, Ivanov AYu (2012) Oil pollution in the Gulf of Guinea based on satellite radar data. Earth Space 13:51–56 (in Russian)

Wikipedia (2012a) Petroleum industry in Nigeria. http://en.wikipedia.org/wiki/Petroleum_industry_in_Nigeria. Accessed 10 June 2012

Wikipedia (2012b) Environmental issues in the Niger Delta. http://en.wikipedia.org/wiki/Environmental_issues_in_the_Niger_Delta. Accessed 10 June 2012

Chapter 9
The Upwelling Area Off Namibia, the Northern Part of the Benguela Current System

Herbert Siegel, Thomas Ohde and Monika Gerth

Abstract In the area off Namibia satellite remote sensing data of the visible and infrared spectral range were used to investigate the upwelling processes and the biological response. Satellite derived Sea Surface Temperature was applied to study upwelling processes in relation to the driving trade winds. The investigations were focussed on the intensity and horizontal extent, the temporal and spatial variability including inter-annual and climate scales. Ocean colour satellite data allow investigation of the response of surface water to the nutrient input into the euphotic zone by upwelling processes. Observations in the area revealed, that the phytoplankton development starts with species absorbing light in the visible spectral range (diatoms and dinoflagellates) and may end in persistent shallow surface filaments with light scattering algae blooms. These blooms of coccolithophores identified by different *in situ* methods changed the water colour by strong particle scattering to milky turquoise discolorations. Further milky turquoise discolorations were identified as sulphur plumes. Low wind periods support the degradation of organic matter and the development of hydrogen sulphide in the bottom layer. After the onset of the trade winds and the offshore transport of surface water hydrogen sulphide enriched waters reaches with the near-bottom counter current the coast. Upwelling transports the water to the surface where the hydrogen sulphide will be oxidised to elemental sulphur. An algorithm was developed to identify and separate the sulphur plumes from algae blooms and other features on the basis of highly spectrally resolved satellite data from the MERIS sensor. The algorithm identified only coastal plumes as sulphur and that offshore plumes are formed by coccolithophores. The sulphur season is the boreal spring starting in February and reaching the maximum in April.

9.1 Introduction

The Namibian coastal waters are hydrographically highly variable. Upwelling processes, plankton blooms and sulphur events are characteristics of the region which have been intensively investigated by satellite remote sensing methods. This chapter

H. Siegel (✉) · T. Ohde · M. Gerth
Leibniz Institut für Ostseeforschung Warnemünde, Rostock, Germany
e-mail: herbert.siegel@io-warnemuende.de

V. Barale, M. Gade (eds.), *Remote Sensing of the African Seas*,
DOI 10.1007/978-94-017-8008-7_9,

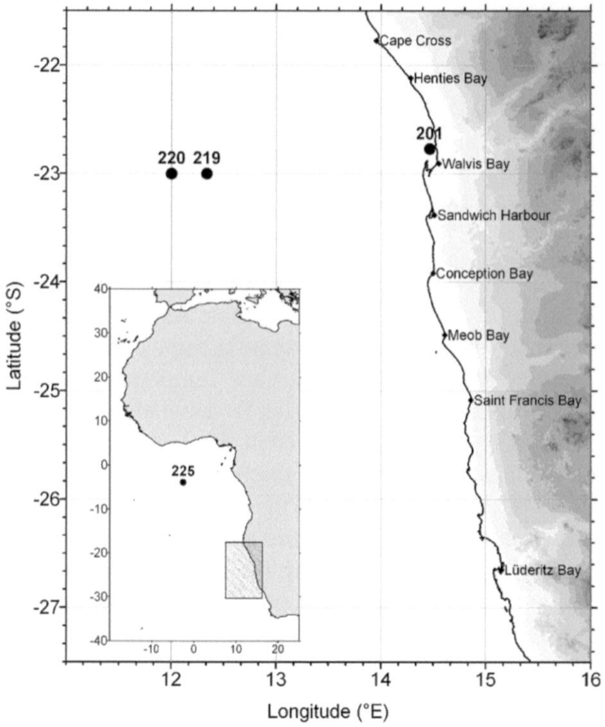

Fig. 9.1 Map of Namibian upwelling area including selected stations

provides examples of applications of satellite data in the infrared and visible spectral range in the region off Namibia (Fig. 9.1). Satellite data in the infrared spectral range provide information about the Sea Surface Temperature (SST) distribution to investigate the upwelling processes. Sensors in the visible domain measure the water colour modified by the regional specific composition of optically active water constituents and their inherent optical properties.

9.2 The Upwelling Area Off Namibia

The Namibian upwelling area is the northern part of the Benguela current system which forms one of major eastern boundary current systems of the world ocean (Fig. 9.1). The system was intensively investigated by dynamical modelling, *in situ* measurements as well as by remote sensing methods (Nelson and Hutchings 1983; Shannon 1985; Fennel 1999; Barlow et al. 2001; Carr 2002; Barlow et al. 2003; Campillo-Campbell and Gordoa 2004; Heymans et al. 2004; Bartholomae and van der Plas 2007; Colberg and Reason 2007; Mohrholz et al. 2008; Pardo et al. 2011). The extent is limited in the north by the southward directed warm Angola current forming the Angola-Benguela front. In the south it extends to the warm Agulhas current coming from the Indian Ocean, passing South Africa and entering the Atlantic Ocean. The coastal Benguela current propagates northward along the African coast and passes the upwelling area off Namibia. Prevailing south-east trade winds are blowing the whole year with spatial and temporal variations. The winds induce Ekman offshore transport of surface water, a compensation current in the near bottom layer resulting in upwelling of cold nutrient-rich waters along the coast. Due to the physical and biological characteristics the entire area can be divided into two regions. The northern region which mainly include the Namibian waters is characterised by seasonal warming during late summer and early autumn (Boyd et al. 1987; Shannon et al. 1987) and intense upwelling throughout the year (Boyd and Agenbag 1985; Campillo-Campbell and Gordoa 2004). The upwelling is stronger during the cooler months. The seasonal effect caused by the intrusion of warm saline water from the equatorial region reduces the upwelling to the Lüderitz area and increases the permanent SST latitudinal gradient (Gordoa et al. 2000; Campillo-Campbell and Gordoa 2004). The upwelling filaments may extend up to 625 km offshore (Lutjeharms and Meeuwis 1987; Lutjeharms and Stockton 1987) and over 1000 km off Lüderitz (Lutjeharms et al. 1991). The region around Lüderitz is the most active upwelling cell (Boyd and Agenbag 1985; Shannon 1985; Agenbag and Shannon 1988).

The Namibian waters are very productive because of the continuous nutrient input into the euphotic zone by upwelling processes. The phytoplankton development is highly variable in space and time and occurs in a succession starting with plankton groups absorbing visible light. Algae groups which are dominantly scattering light may develop in persistent shallow surface filaments (Siegel et al. 2004). The absorbing phytoplankton is dominated by diatoms and dinoflagellates and the scattering phytoplankton bloom consist of coccolithophores (Siegel et al. 2007). After the blooms dead organisms sink down and are decomposed by bacteria under consumption of dissolved oxygen. Denitrification and sulphate reduction lead to hydrogen sulphide in the lower water layers particularly during calm conditions. Onset of trade winds cause Ekman offshore transport again and the compensation current transports the hydrogen sulphide enriched waters onshore. The upwelling processes along the coast carry this water into the surface layer where hydrogen sulphide can be oxidised finally to elemental sulphur changing the water colour milky turquoise.

An example of Advanced Very High Resolution Radiometer (AVHRR) SST and Sea-viewing Wide Field-of-view Sensor (SeaWiFS) chlorophyll-a shows the highly variable spatial distributions (Fig. 9.2). The core of the upwelling area is located in front of Lüderitz with minimum temperature lower than 10 °C. The cold upwelled water is distributed offshore in filaments with a meridional extent of up to 400 km. The temperature of the open ocean waters outside the influence of upwelling is about 25 °C. The chlorophyll maps show the highest concentrations north of the main upwelling area due to transport of nutrient enriched water by the northward directed Benguela current and the response time of phytoplankton.

The large extent of the upwelling area, the high spatial and temporal variability connected with phytoplankton blooms favor remote sensing methods to investigate these processes synoptically.

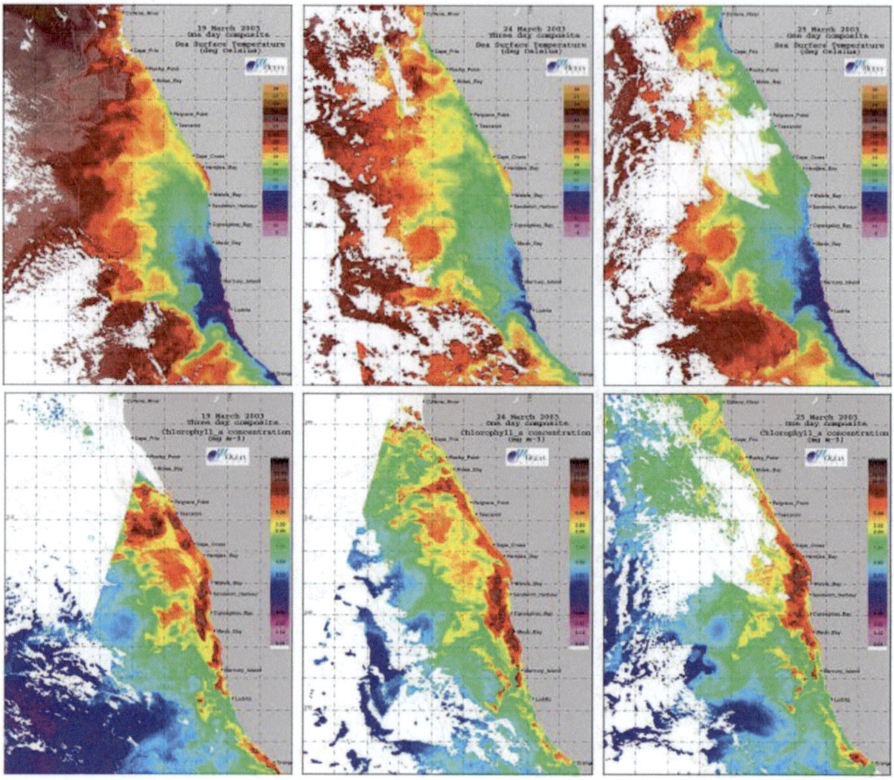

Fig. 9.2 AVHRR SST maps (*upper panel*) and SeaWiFS Chlorophyll-a maps (*lower panel*) of the upwelling area off Namibia coast on 19, 24 and 25 March 2003. Images: OCEAN Space (Cape Town, South Africa)

9.3 Remote Sensing Methods

The investigation of the highly variable temperature distribution along the Namibian coast is based on data of the infrared channels of the AVHRR, operated on the weather satellites of the US National Oceanic and Atmospheric Administration (NOAA). The sensor has a spatial resolution of 1 km. The systems and data are described in detail in (Siegel and Gerth 2008).

Ocean colour satellite data and derived products were used to investigate the distribution of absorbing and scattering phytoplankton and sulphur plumes as the biological response to the upwelling. The applied sensors are Coastal Zone Color Scanner (CZCS), SeaWiFS, Moderate-resolution Imaging Spectroradiometer (MODIS) and Medium Resolution Imaging Spectrometer (MERIS). CZCS was operated onboard of the satellite Nimbus 7 and SeaWiFS onboard OrbView-2. The sensors had a spatial resolution of about 1 km and a repeating rate of 1 day. MODIS is operated by the US National Aeronautics and Space Administration (NASA) on board the Terra (EOS AM) and Aqua (EOS PM) satellites. Both instruments and their level-2 products have a spatial resolution of 1 km and cover the entire Earth every 1–3 days. MERIS was operated until April 2012 by the European Space Agency (ESA) on board the Environmental Satellite (Envisat) covering the earth in 3 days. The Full Resolution (FR) and the Reduced Resolution (RR) products had spatial resolutions of 300 m and 1200 m, respectively. True colour images derived from MODIS and MERIS high resolution data (250 m and 300 m spatial resolution) are implemented for detailed studies of filaments.

Broad band data of higher spatial resolution of both the Landsat 7 Enhanced Thematic Mapper (ETM+) and SPOT were implemented to investigate the structure of coastal sulphur plumes. The visible channels of ETM+ and SPOT have a spatial resolution of 30 m and 20 m, respectively, and the repetition rates of 16 and 23 days allowed only case studies. Wind and sea level data were derived from Topex/Poseidon, the TRMM Microwave Imager (TMI), where TRMM stands for the Tropical Rainfall Measuring Mission satellite, the AMSR-E (Advanced Microwave Scanning Radiometer—Earth Observing System on MODIS Aqua, or the SeaWinds Scatterometer on QuikBird (later dubbed QuikSCAT).

9.4 Bio-Optical Characterisation of the Area

The described features resulting from oceanographic processes cause high variability in bio-optical properties. Chlorophyll-a concentrations between 0.5 and 18 mg m^{-3} were presented for the area south of Walvis Bay for austral winter and spring (Barlow et al. 2001; Barlow et al. 2006). *In situ* measurements of optically active water constituents were performed during two cruises in 2003 and 2004 (Ohde et al. 2007; Siegel et al. 2007). Chlorophyll-a, Suspended Particulate Matter (SPM) and Coloured Dissolved Organic Matter (CDOM) were characterised by strong gradients from the open sea to the coast. Chlorophyll-a varied between 0.3 and 42 mg m^{-3} partly also with a high portion of phaeopigment, an indicator for degradation of chlorophyll-a. Similar patterns were measured for SPM ranging from 1.1 to 24 g m^{-3} and for the absorption by CDOM measured at 440 nm from 0.06 to 0.5 m^{-1}. These variations in the concentration of optically active water constituents strongly influence the transparency and water colour. Secchi disc depth as measure of water transparency varied between 2.8 m and 32.5 m. The measured spectral reflectances quantify the differences in water colour (Fig. 9.3).

The reflectance of clear open ocean water is characterised by a reflectance maximum in the blue spectral range. In blue green waters the maximum shifted to 480–500 nm. Coastal stations were characterized by very turbid brownish water with reflectance maxima between 550 and 580 nm. This was old productive water with high amount of chlorophyll-a, but also with high concentration of degradation products such as detritus and CDOM. The absorption of phytoplankton decreases the reflectance in the blue and red spectral range because of their absorption maxima around 440 nm and 674 nm. The absorption in the short wavelength range is also

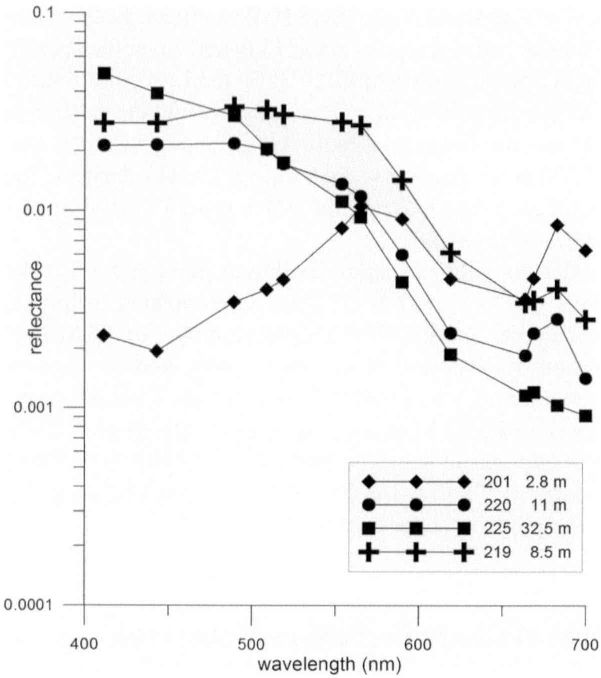

Fig. 9.3 Spectral reflectances measured in March 2003 in the area off Namibia representing the main water masses, clear ocean water (Stat. 225) offshore implement, the upwelling area (Stat. 220) and productive near coastal water (Stat. 201), characterized by Secchi disc depth of 32.5, 11, and 2.8 m, respectively. Station 219 was characterized by a Coccolithophore bloom

influenced by the absorption of yellow substance and detritus. Their exponential increase to shorter wavelength led to the slope in the reflectance of coastal waters and shifted the reflectance maximum to longer wavelengths. Usually high concentrations of CDOM due to a terrigenous origin are transported via rivers to the ocean. Along the Namibian coast the coastal runoff is negligible and the CDOM is produced by degradation of organic matter. The particle scattering by phytoplankton increases the reflectance in the entire visible spectral range. This is only visible in the long wavelength range because of the strong absorption in the blue shortwave part (Stat. 201, 220, 225). The station with the highest absorption (lowest reflectance) in the short wavelength range exhibits the highest scattering (reflectance) in the long spectral range. In contrast, in bright milky turquoise waters enhanced scattering by molecular sulphur or by coccolithophores increases the reflectances in the entire spectral range (Fig. 9.3) because of the very low phytoplankton absorption in that waters (Stat. 219).

9.5 Applications

The summary of applications in the region off Namibia is mainly focussed on SST and water colour and derived products. The synoptic coverage of the entire region and the daily repeating rate supported the capture of processes of fast spatial and temporal variability. Hardman-Mountford et al. (2003) gave a detailed overview about the first applications of satellite data in the Benguela region in the mid 1980s up to the end of the 1990s. The authors used NOAA-AVHRR, CZCS and SeaWiFS data to investigate the extent and seasonality of upwelling, to describe the distribution of chlorophyll-a, and to develop algorithms for the retrieval of chlorophyll-a and suspended sediments from the ocean colour satellite data (Shannon 1985; Lutjeharms and Meeuwis 1987; Lutjeharms and Stockton 1987; Shannon et al. 1987; Shelton and Hutchings 1990; Bailey 1991). Several publications are related to environmental conditions and fish recruitment (Shelton and Hutchings 1990; Cole and McGlade 1998; Cole 1999). More recent publications combined time series of NOAA-SST and SeaWiFS- derived chlorophyll-a to investigate physical and biological variability in the Namibian upwelling area including the Angola-Benguela Front (Campillo-Campbell and Gordoa 2004; Veitch et al. 2006; Bartholomae and van der Plas 2007). Silió-Calzada et al. (2008) improved a model approach to estimate primary productivity with MERIS and AATSR data with application to the Benguela upwelling region. Demarcq (2009) investigated trends in primary production, sea surface temperature and wind in upwelling systems for the period 1998–2007 including Benguela. Shutler et al. (2010) developed an approach to identify coccolithophore blooms in other regions. Validation is based on SeaWiFS data in the period 2003–2004. A model development (Williamson et al. 2011) contributes to the evaluation of the vertical distribution of chlorophyll-a. Pardo et al. (2011) investigated the long-term development of SST in eastern boundary upwelling systems.

9.5.1 Upwelling and Biological Response

Monthly mean maps of NOAA-SST were used to derive the Intense Benguela Upwelling (IBU) index during the period 1982–1999 (Hagen et al. 2001). The IBU was calculated for the region 9–34°S and 8–20°E. The index was defined as the area between the 13 °C isotherm and the coast. The IBU indicated a decreasing trend and a quasi-cycle of about 27 months. Seasonality was discussed for the entire cold water area and for the mean offshore and alongshore extent. The main upwelling season occurred during the austral winter (July until September) with a maximum extent of about $30 \times 10^3 km^2$ in August. The most intense upwelling filaments develop around 26 and 29°S with a mean offshore extension of 210 and 130 km. During weak upwelling years the area decreased by a factor of two and expanded by a factor of 1.5 during strong years. Sea level changes derived from measurements at four coastal stations (1982–1987) showed the lowest sea levels in the Lüderitz cell near 26°S. Decreasing cold water areas are accompanied by increasing sea levels and vice versa. Mean seasonal cycles in the total cold water area has a time lag of about 1 month behind those in the sea level along the entire south-west African coast.

9.5.2 Coccolithophore Blooms

The occurrence and bloom of coccolithophores in front of Namibia were intensively investigated by Siegel et al. (2007). Coccolithophores are the most productive calcifying organism on earth and play an important role in the marine carbon cycle effecting the carbon and carbonate content. A summary about the research on the coccolithophore and particularly on the species *Emiliania huxleyi* is presented in the monograph edited by (Thierstein and Young 2004). After the nutrient input into the euphotic zone by upwelling the phytoplankton development starts with diatoms and dinoflagellates. These algae are characterized by conspicuous absorption in the visible spectral range and can be studied by satellite derived chlorophyll-a concentrations. Under certain conditions the succession may develop to coccolithophores which discolour the surface water milky turquoise due to their scattering properties. The enhanced scattering allows to apply quasi true colour images to follow the spatial and temporal development. Coccolithophore blooms are known from different regions of the world ocean (Weeks et al. 2004b; Tyrrell et al. 2008; Holligan et al. 2010; Painter et al. 2010; Garcia et al. 2011). Milky turquoise discolouration were found off the coast of the Republic of South Africa which were identified as coccolithophore blooms (see also Mitchell-Innes and Winter 1987). These blooms have been mainly associated with sub-polar regions as in the global maps by Brown and Yoder (1994) based on SeaWiFS data have shown.

The optimal growth conditions are high incident solar radiation, stable shallow top-layer, silicate depletion, specific nutrient conditions, and reduced grazing by zooplankton (Rost and Riebesell 2004; Tyrrell and Merico 2004). Nanninga and Tyrrell (1996) and Garcia et al. (2011) determined a mixed layer depth of less than 30 m, that the coccolithophores remain under high solar radiation (Iglesias-Rodriguez et al. 2002). *Emiliania huxleyi* is a fast growing phytoplankton which develops in eutrophic environments, when the fastest growing diatoms are excluded by silicate depletion (Humborg et al. 1997). *Emiliania huxleyi* can grow if phosphate is exhausted because it is able to synthesize enzyme alkaline phosphatase and may use dissolved organic phosphorus (Riegman et al. 2000). Start conditions are high N:P ratios but during the bloom the ratio is highly variable (Fanning 1992; Townsend et al. 1994; Tyrrell and Taylor 1996).

A milky turquoise offshore plume was identified as a coccolithophore bloom for the first time in the region off Namibia in March/April 2003. The plume was followed by satellite imagery for about 10 days (Siegel et al. 2007). An example of the time series is presented in Fig. 9.4 (left panel) and the SeaWiFS derived calcite flag (right panel). Near real-time satellite data access enabled the research crew to perform interdisciplinary investigation in the patch. Different methods such as HPLC pigment analysis, Energy Dispersive X-Ray analysis, Scanning Electronic Microscopy, and measurements of inherent and apparent optical properties were used to validate the coccolithophore bloom of *Emiliania huxleyi*.

The water was characterized by high reflectances, the occurrence of the marker pigment 19'-Hexanoyloxyfucoxanthin and a double maximum of the spectral

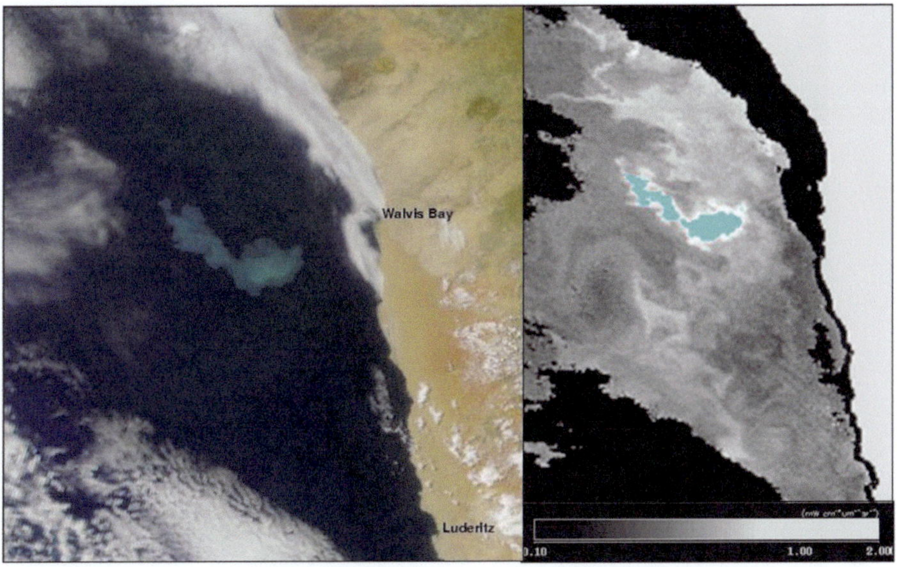

Fig. 9.4 Quasi-True colour image showing coccolithophore bloom on 31 March 2003 and the corresponding SeaWiFS derived calcite flag (According to Siegel et al. 2007)

absorption in the blue spectral range. *Emiliania huxleyi* caused the high calcite content. The bloom occurred during high solar radiation and shallow surface layer of about 10 m with low chlorophyll-a and high turbidity during relatively long low wind period. The nutrient conditions consisted of low silicate content, high N:P ratio with low phosphate concentration. The measurements at a shore normal transect confirmed that the bloom was the result of a succession of phytoplankton in a stable water body with low vertical mixing. In the transition zone between diatom enriched waters to coccolithophore dominated waters destroyed phytoplankton was observed above hydrogen sulphide enriched bottom layers. Weak transport of hydrogen sulphide may partly force the succession.

Ocean colour satellite data of SeaWiFS, MODIS and MERIS were used to investigate the blooms. During the intense bloom in 2003 the SeaWiFS coccolithophore flags identified the bloom very well. MERIS- derived spectral reflectances were in good agreement with *in situ* measurements. They were applied to study the occurrence of coccolithophores in the period from January to June 2004. The observed blooms were located at a distance of 75–120 km and 110–300 km from the coast. The blooms occurred in two different background waters. Offshore the spectral shape of the reflectances is dominated by the blue ocean water increased by scattering of the coccolithophores. Directly outside the active upwelling area the reflectances are dominated by the spectral shape of the absorbing waters.

The marker pigment for coccolithophores, 19′-Hexanoyl-oxyfucoxanthin, which was determined during two research cruises, showed that the pigment occurred frequently, but the blooms were observed only sporadically.

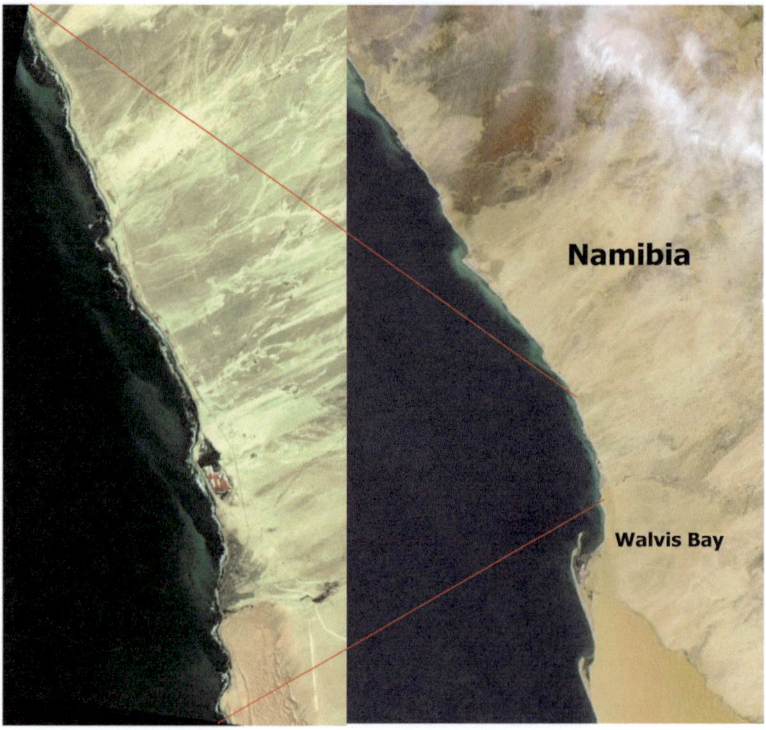

Fig. 9.5 SPOT (*left*) and MODIS (*right*) images, collected on 9 April 2004, showing sulphur plumes with filaments along the coast around and north of Walvis Bay, with filaments in the high resolution SPOT image

9.5.3 Occurrence of Sulphur Plumes

Sulphur plumes are striking features in Namibian waters (Ohde et al. 2007). Scenes of SPOT and MODIS from 9 April 2004 show sulphur plumes along the coast around and north of Walvis Bay with filaments in the high resolution SPOT image (Fig. 9.5). Hydrogen sulphide can be produced at the shelf in the sediments and in near bottom waters (Copenhagen 1953; Bailey 1991; Emeis et al. 2004). The high productivity in the upwelling area is dominated by diatoms and dinoflagellates and sporadic blooms of coccolithophores or red tides (Pitcher et al. 1995; Cockroft et al. 1999; Weeks et al. 2004b; Siegel et al. 2007; Pitcher et al. 2010). After the blooming the dead organic material sinks down, accumulate at the seafloor and form muddy sediment. The organic material is decomposed by bacteria under consumption of dissolved oxygen starting already in the water column. Further degradation is realised by nitrate and sulphate reducing bacteria (Widdel 1988; Jørgensen 1990), sulphate is reduced to hydrogen sulphide. These processes occur particularly during low wind periods. The production of hydrogen sulphide can be intensified by waters of low oxygen content from north of the Angola-Benguela front transported below the halocline into the northern Benguela system (Monteiro et al. 2006).

The SE trade winds generate Ekman offshore transport and the near bottom compensation current transported hydrogen sulphide-rich waters towards the coast. The upwelling processes transport this water to the surface (e.g. Bailey 1991; Weeks et al. 2002; Brüchert et al. 2004; Emeis et al. 2004) where the hydrogen sulphide can be oxidised finally to elemental sulphur. This oxidation depends on water temperature, pH-value, dissolved oxygen content, salinity, the presence of chemical catalysts and the biological system (Train 1979; Dohnalek and FitzPatrick 1983; Hill 1984; Millero et al. 1987). The oxidation of hydrogen sulphide to elemental sulphur increases the light scattering and makes the water milky (Weeks et al. 2004a). Besides the depletion of oxygen hydrogen sulphide has a strong influence on the marine ecosystem due to the toxic effect on organisms (e.g. Evans 1967). This can decrease the amount of benthos, fish stocks and other marine organisms or can cause mass mortalities (Grindley and Sapeika 1969; Horstman 1981; Matthews and Pitcher 1996; Cockroft et al. 1999; Cockcroft 2001). Lobsters are leaving for example the ocean and can be found in large numbers on the beach. The hydrogen sulphide gas is also reaching the atmosphere causing "rodden egg" like smell and leading to erosion effects on metallic objects. Potential regions for the occurrence of sulphur outbreaks are oxygen minimum zones related to the eastern boundary currents in the ocean with strong upwelling and high bio-productivity. The Namibian upwelling area is a prominent example where the sulphur plumes were verified in surface water.

Because of the consequences for the marine ecosystem the identification and registration of colloidal sulphur events is important. The sporadic occurrence of the events makes the study by land observations or ship cruises difficult. Satellite remote sensing is the only method to investigate the occurrence of sulphur plumes systematically. Validation was performed by contacting colleagues from Swakopmund during such an event. The contact was realised from both side, when we found indications in satellite data and when the colleagues recognised the "rodden egg" like smell. An example is presented in a quasi-true colour image (Red-Green-Blue, or RGB) of a MERIS scene on 10 April 2004 with a spatial resolution of 300 m (Fig. 9.6). The plume is clearly visible in Saint Francis Bay. In early studies (Weeks et al. 2004a) quasi-true colour images derived from SeaWiFS and combinations of spectral channels were applied to identify sulphur plumes. Thus, sulphur plumes could not be distinguished from other phenomena changing the water colour such as coccolithophore blooms. Therefore, discrepancies occurred in the interpretation of satellite images (Weeks et al. 2004a; Siegel et al. 2007).

A method to identify colloidal sulphur in the area off Namibia was developed on the basis of detailed analyses of MERIS derived water-leaving reflectances (Ohde et al. 2007). Sulphur patches in the area off Namibia have specific spectral reflectance signatures. They significantly differ from the spectral characteristics of the typical water types in this area and such as different algae blooms, river discharge, and sediment resuspension. The algorithm derived from MERIS data of the first half of the year 2004 eliminates all other events changing the water colour. Four conditions and thresholds were implemented in the classification algorithm and have to be executed consecutively to flag the pixels of sulphur spectra:

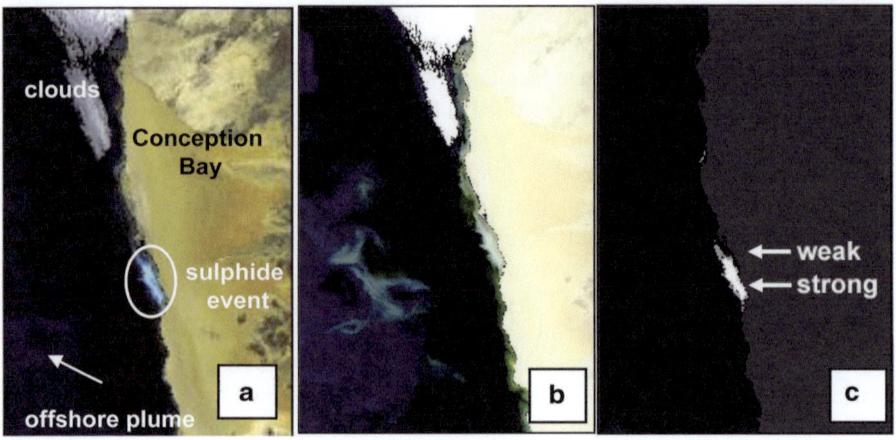

Fig. 9.6 Quasi-True colour image of a sulphur plume (**a**), false colour image showing onshore and offshore plume (**b**) and the identified sulphur (**c**) (according to Ohde et al. 2007)

- The 1st condition eliminates all water pixels of low productive blue ocean water and of high productive water with the typical chlorophyll-a reflectance minimum at 442.5 nm.
- The 2nd condition is a threshold value for a reflectance ratio which includes the reflectance minimum due to the chlorophyll-a absorption maximum at 664.6 nm. This very effective condition eliminates most pixels which do not correspond to sulphur spectra.
- The 3rd condition is a threshold for the reflectance at 559.6 nm and eliminates the transition areas between sulphur plumes and unaffected areas and high reflective algae blooms. It does not consider water pixels of high-suspended matter concentration from rivers or resuspensions.
- The 4th condition considers different reflectance slopes of sediment-loaded water and the sulphur plumes to eliminate sediment influence in the image.

The algorithm was applied to all available scenes from January 2004 to August 2005 to study the occurrence of sulphur plumes, the temporal and spatial development, the extension and the strength of such events as well as inter-annual difference (Fig. 9.7). Only near-shore plumes were identified as sulphur with a lifetime of 1–6 days and a zonal extent of up to 21 km. Up to 12 % of the incident radiation is reflected. The patches propagated in north-westerly direction with a current velocity of up to 14 cms^{-1}. The strongest events started in both years end of January and reached their highest intensities in April. The sulphur season was more intensive in 2005 than in 2004 with a maximum daily extension of up to 1600 and 800 km^2, respectively. The strongest events appeared near-shore in the Meob- and Saint Francis Bay between February 2004 and May 2004. The classification algorithm identified only coastal patches and excluded offshore plumes which differed from corresponding results of other investigations.

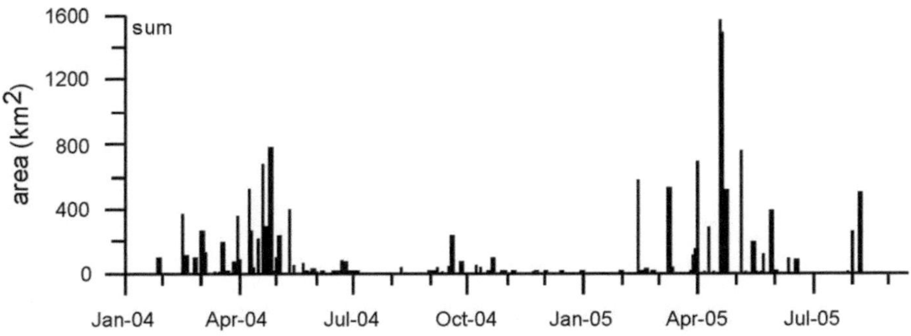

Fig. 9.7 MERIS derived sulphur events in the period Jan 2004 until July 2005. (According to Ohde et al. 2007)

9.6 Summary

Combined remote sensing methods are powerful tools to investigate the ecosystem off Namibia from the highly variable forcing and oceanographic conditions to the ecosystem response. Microwave data allow tracking the forcing parameters, SST data show the characteristics of upwelling processes and ocean colour derived chlorophyll-a deliver information about the phytoplankton development. True colour images and special derived products support to study special ecosystem responses enhancing the light scattering in the surface water. Persistent shallow filaments at the border between upwelling and open ocean waters may support coccolithophore blooms. After calm conditions, the developed H_2S will be transported to the coast and into the surface layer, where H_2S will be oxidized to elemental sulphur. In–situ measurements of inherent and apparent optical properties in combination with satellite imagery verify the different observed features and can explain their origin.

Acknowledgments The MODIS data were received from NASA MODIS Rapid Response Project at the NASA Goddard Space Flight Center; the SeaWiFS data from the SeaWiFS project of NASA; and the MERIS data from ESA.

References

Agenbag JJ, Shannon LV (1988) A suggested physical explanation for the existence of a biological boundary at 2430S in the Benguela system. S Afr J Mar Sci 6(1):119–132

Bailey GW (1991) Organic carbon flux and development of oxygen deficiency on the modern Benguela continential shelf south of 22{degrees}S: spatial and temporal variability. In: Tyson R, Pearson TH (eds) Modern and Ancient Continental Shelf Anoxia, vol 58. vol 1. Geological Society, Special Publications, London, pp 171–183. doi:10.1144/gsl.sp.1991.058.01.12

Barlow R, Louw D, Balarin M, Alheit J (2006) Pigment signatures of phytoplankton composition in the northern Benguela ecosystem during spring. Afr J Mar Sci 28(3–4):479–491

Barlow R, Sessions H, Silulwane N, Engel H, Hooker SB, Aiken J, Fishwick J, Vicente V, Morel A, Chami M, Ras J, Bernard S, Pfaff M, Brown JW, Fawcett A (2003) BENCAL Cruise Report. SeaWiFS Postlaunch Tech Report Series, vol 27. NASA Goddard Space Flight Center, Greenbelt, Maryland

Barlow RG, Aiken J, Sessions HE, Lavender S, Mantel J (2001) Phytoplankton pigment, absorption and ocean colour characteristics in the southern Benguela ecosystem. S Afr J Sci 97(5–6):230–238

Bartholomae CH, van der Plas AK (2007) Towards the development of environmental indices for the Namibian shelf, with particular reference to fisheries management. Afr J Mar Sci 29(1):25–35

Boyd AJ, Agenbag JJ (eds) (1985) Seasonal trends in the longshore distribution of surface temperatures off South-western Africa 18–34°S, and their relation to subsurface conditions and currents in the area 21–24°S. International Symposium on the Most Important Upwelling areas off Western Africa (Cape Blanco and Benguela), vol I. Instituto de Investigaciones Pesqueras, Barcelona

Boyd AJ, Salat J, Maso M (1987) The seasonal intrusion of relatively saline water on the shelf off northern and central Namibia. S Afr J Mar Sci 5(1):107–120

Brown CW, Yoder JA (1994) Coccolithophorid blooms in the global ocean. J Geophys Res (C Oceans) 99 (C4):7467–7482

Brüchert V, Endler R, Vogt T, Emeis KC (2004) Dynamics of methane and hydrogen sulphide in Namibian shelf sediments. Geophysical Research Abstracts 6

Campillo-Campbell C, Gordoa A (2004) Physical and biological variability in the Namibian upwelling system: October 1997-October 2001. Deep Sea Research Part II: Topical Studies in. Oceanography 51(1–3):147–158

Carr ME (2002) Estimation of potential productivity in eastern boundary currents using remote sensing. Deep-Sea Res (II Top Stud Oceanogr) 49(1–3):58–80

Cockcroft AC (2001) Jasus lalandii 'walkouts' or mass strandings in South Africa during the 1990s: an overview. Mar Freshwat Res 52(8):1085–1094

Cockroft AC, Schoeman DS, Pitcher GC, Bailey GW, Van Zyl DL (1999) A mass stranding or "walkout" of West Coast rock lobster Jasus lalandii in Elands Bay, South Africa: causes, results and implicatio. In: Von Kaupel Klein JC, Schram FR (eds) The Biodiversity Crises and Crustacea. Crustacean Issues, vol 11. pp 673–688

Colberg F, Reason CJC (2007) A model investigation of internal variability in the Angola Benguela Frontal Zone. J Geophys Res 112 (C7):C07008

Cole J (1999) Environmental conditions, satellite imagery, and clupeoid recruitment in the northern Benguela upwelling system. Fish Oceanogr 8(1):25–38

Cole J, McGlade J (1998) Clupeoid population variability, the environment and satellite imagery in coastal upwelling systems. Rev Fish Biol Fish 8(4):445–471. doi:10.1023/a:1008861224731

Copenhagen WJ (1953) The periodic mortality of fish in the Walvis region. A phenomenon within the Benguela Current. Investigational Report, vol 14. Division of Sea Fisheries of South Africa, Cape Town, South Africa

Demarcq H (2009) Trends in primary production, sea surface temperature and wind in upwelling systems (1998–2007). Prog Oceanogr 83(1–4):376–385. doi:10.1016/j.pocean.2009.07.022

Dohnalek DA, FitzPatrick JA (1983) The chemistry of reduced sulphur species and their removal from groundwater supplies. J Am Water Works Assn 6:298

Emeis KC, Bruchert V, Currie B, Endler R, Ferdelman T, Kiessling A, Leipe T, Noli-Peard K, Struck U, Vogt T (2004) Shallow gas in shelf sediments of the Nambian coastal upwelling ecosystem. Cont Shelf Res 24(6):627–642

Evans CL (1967) The toxicity of hydrogen sulphide and other sulphides. Q J Exp Psychol 52:231–248

Fanning KA (1992) Nutrient provinces in the sea: concentration ratios, reaction rate ratios, and ideal covariation. J Geophys Res (C Oceans) 97(C4):5693–5712

Fennel W (1999) Theory of the Benguela Upwelling System. J Phys Oceanogr 29(2):177–190

Garcia CAE, Garcia VMT, Dogliotti AI, Ferreira A, Romero SI, Mannino A, Souza MS, Mata MM (2011) Environmental conditions and bio-optical signature of a coccolithophorid bloom in the Patagonian shelf. J Geophys Res 116(C3):C03025

Gordoa A, Masó M, Voges L (2000) Satellites and fisheries: the Namibian hake, a case study. In: Halpern D (ed) Satellites, Oceanography and Society. Elsevier Science B.V, Amsterdam, pp 193–205

Grindley JR, Sapeika N (1969) The cause of mussel poisoning in South Africa. S Afr Med J 43:275–279

Hagen E, Feistel R, Agenbag JJ, Ohde T (2001) Seasonal and interannual changes in Intense Benguela Upwelling (1982–1999). Oceanol Acta 24(6):557–568

Hardman-Mountford NJ, Richardson AJ, Agenbag JJ, Hagen E, Nykjaer L, Shillington FA, Villacastin C (2003) Ocean climate of the South East Atlantic observed from satellite data and wind models. Prog Oceanogr 59(2–3):181–221

Heymans JJ, Shannon LJ, Jarre A (2004) Changes in the northern Benguela ecosystem over three decades: 1970s, 1980s, and 1990s. Ecol Model 172(2–4):175–195

Hill JM (1984) The acute toxicity of hydrogen sulphide—a literature review and application to storm sewage discharges. Report ER, vol 675-M. Water Research Centre, Medmenham, United Kingdom

Holligan PM, Charalampopoulou A, Hutson R (2010) Seasonal distributions of the coccolithophore, Emiliania huxleyi, and of particulate inorganic carbon in surface waters of the Scotia Sea. J Mar Syst 82(4):195–205

Horstman DA (1981) Reported Red-Water Outbreaks and Their Effects on Fauna of the West and South coasts of South Africa, 1959–1980. Fish Bull S Afr 15(1):71–88

Humborg C, Ittekkot V, Cociasu A, Bodungen BV (1997) Effect of Danube River dam on Black Sea biogeochemistry and ecosystem structure. Nature 386(6623):385–388

Iglesias-Rodriguez MD, Brown CW, Doney SC, Kleypas J, Kolber D, Kolber Z, Hayes PK, Falkowski PG (2002) Representing key phytoplankton functional groups in ocean carbon cycle models: Coccolithophorids. Global Biogeochemical Cycles 16 (4):[np]

Jørgensen BB (1990) A thiosulfate shunt in the sulfur cycle of marine sediments. Science (Wash) 249(4965):152–154

Lutjeharms JRE, Meeuwis JM (1987) The extent and variability of South-east Atlantic upwelling. S Afr J Marine Sci/Suid-Afrikaanse Tydskrif vir Seewetenskap 5:51–61

Lutjeharms JRE, Shillington FA, Duncombe Rae CM (1991) Observations of extreme upwelling filaments in the Southeast Atlantic Ocean. Sci (Wash) 253(5021):774–776

Lutjeharms JRE, Stockton PL (1987) Kinematics of the upwelling front off southern Africa. S Afr J Marine Sci/Suid-Afrikaanse Tydskrif vir Seewetenskap 5:35–49

Matthews SG, Pitcher GC (1996) Worst recorded marine mortality on the South African coast. Harmful and Toxic Algal Blooms. UNESCO, Paris (France)

Millero FJ, Hubinger S, Fernandez M, Garnett S (1987) Oxidation of H2S in seawater as a function of temperature, pH and ionic strength. Environ Sci Technol 21:439–443

Mitchell-Innes BA, Winter A (1987) Coccolithophores: a major phytoplankton component in mature upwelled waters off the Cape Peninsula, South Africa in March, 1983. Marine biology. Heidelberg 95(1):25–30

Mohrholz V, Bartholomae CH, van der Plas AK, Lass HU (2008) The seasonal variability of the northern Benguela undercurrent and its relation to the oxygen budget on the shelf. Cont Shelf Res 28(3):424–441

Monteiro PMS, Van der Plas A, Mohrholz V, Mabille E, Pascall A, Joubert W (2006) Variability of natural hypoxia and methane in a coastal upwelling system: oceanic physics or shelf biology? Geophys Res Lett 33 (16):[np]

Nanninga HJ, Tyrrell T (1996) Importance of light for the formation of algal blooms by Emiliania huxleyi. Marine ecology progress series Oldendorf 136(1–3):195–203

Nelson G, Hutchings L (1983) The Benguela upwelling area. Prog Oceanogr 12:333–356

Ohde T, Siegel H, Reißmann J, Gerth M (2007) Identification and investigation of sulphur plumes along the Namibian coast using the MERIS sensor. Cont Shelf Res 27(6):744–756

Painter SC, Poulton AJ, Allen JT, Pidcock R, Balch WM (2010) The COPAS'08 expedition to the Patagonian Shelf: physical and environmental conditions during the 2008 coccolithophore bloom. Cont Shelf Res 30(18):1907–1923

Pardo PC, Padín XA, Gilcoto M, Farina-Busto L, Pérez FF (2011) Evolution of upwelling systems coupled to the long-term variability in sea surface temperature and Ekman transport. Clim Res 48(2–3):231–246

Pitcher GC, Agenbag JJ, Calder DA, Horstman DA, Jury MR, Taunton-Clark J (1995) Red tides in relation to the meteorology of the southern Benguela upwelling system. In: Lassus P, Arzul G, Erard E, Gentien P, Marcaillou C (eds) Harmful marine algal blooms. Technique et Documentation, Lavoisier Intercept Ltd, Paris, pp 657–662

Pitcher GC, Figueiras FG, Hickey BM, Moita MT (2010) The physical oceanography of upwelling systems and the development of harmful algal blooms. Progress In Oceanography 85(1–2):5–32

Riegman R, Stolte W, Noordeloos AAM, Slezak D (2000) Nutrient uptake and alkaline phosphatase (EC 3:1:3:1) activity of Emiliania huxleyi (Prymnesiophyceae) during growth under N and P limitation in continuous cultures. J Phycol 36(1):87–96

Rost B, Riebesell U (2004) Coccolithophores and the biological pump: responses to environmental changes. In: Thierstein HR, YoungJR (eds) Coccolithophores—From Molecular Processes to Global Impact. Springer Verlag, New York, 99–125

Shannon LV (1985) The Benguela Ecosystem, Part I., Evolution of the Benguela, physical features and processes. Oceanogr Mar Biol Annu Rev 23:105–182

Shannon LV, Agenbag JJ, Buys MEL (1987) Large- and meso-scale features of the Angola-Benguela. S Afr J Mar Sci 5:11–34

Shelton PA, Hutchings L (1990) Ocean stability and anchovy spawning in the southern Benguela Current region. Fish Bull 88(2):323

Shutler JD, Grant MG, Miller PI, Rushton E, Anderson K (2010) Coccolithophore bloom detection in the north east Atlantic using SeaWiFS: algorithm description, application and sensitivity analysis. Remote Sens Environ 114(5):1008–1016

Siegel H, Gerth M (2008) Optical remote sensing applications in the enclosed Baltic Sea. In: V.Barale MG (ed) Remote Sensing of European Seas. Springer Science + Business Media B.V., pp 91–102

Siegel H, Gerth M, Ohde T (2004) Optical *in situ* measurements of a coccolithophoride bloom in the SE Atlantic Ocean off Namibia identified and followed by satellite ocean colour data. In: Ocean Optics Conference, Fremantle

Siegel H, Ohde T, Gerth M, Lavik G, Leipe T (2007) Identification of coccolithophore blooms in the SE Atlantic Ocean off Namibia by satellites and *in situ* methods. Cont Shelf Res 27(2):258–274

Silió-Calzada A, Bricaud A, Uitz J, Gentili B (2008) Estimation of new primary production in the Benguela upwelling area, using ENVISAT satellite data and a model dependent on the phytoplankton community size structure. J Geophys Res 113 (C11):C11023

Thierstein HR, Young JR (2004) Coccolithophores. From Molecular Processes to Global Impact, Springer, pp 565

Townsend DW, Keller MD, Holligan PM, Ackleson SG, Balch WM (1994) Blooms of the coccolithophore Emiliania huxleyi with respect to hydrography in the Gulf of Maine. Cont Shelf Res 14(9):979–1000

Train RE (1979) Quality Criteria for Water. Castle House Publ. Ltd, London

Tyrrell T, Merico A (2004) Emiliania huxleyi: bloom observations and the conditions that induce them. In: Thierstein HR, Young JR (eds) Coccolithophores. From molecular processes to global impact. Springer, New York, pp 75–97

Tyrrell T, Schneider B, Charalampopoulou A, Riebesell U (2008) Coccolithophores and calcite saturation state in the Baltic and Black Seas. Biogeosciences 5(2):485–494

Tyrrell T, Taylor AH (1996) A modelling study of Emiliania huxleyi in the NE atlantic. J Mar Syst 9(1–2):83–112

Veitch JA, Florenchie P, Shillington FA (2006) Seasonal and interannual fluctuations of the Angola-Benguela Frontal Zone (ABFZ) using 4.5 km resolution satellite imagery from 1982 to 1999. Int J Remote Sens 27(5–6):987–998. doi:10.1080/01431160500127914

Weeks SJ, Currie B, Bakun A (2002) Satellite imaging: Massive emissions of toxic gas in the Atlantic. Nature 415(6871):493–494

Weeks SJ, Currie B, Bakun A, Peard KR (2004a) Hydrogen sulphide eruptions in the Atlantic Ocean off southern Africa: implications of a new view based on SeaWiFS satellite imagery. Deep Sea Res (I Oceanogr Res Pap) 51(2):153–172

Weeks SJ, Pitcher GC, Bernard S (2004b) Satellite Monitoring of the Evolution of a Coccolithophorid Bloom in the Southern Benguela Upwelling System. Oceanography [Oceanography] 17(1):83–89

Widdel F (1988) Microbiology and ecology of sulfate-and sulfur-reducing bacteria. In: Zehnder AJB (ed) Biology of anaerobic microorganisms. Wiley, New York, pp 469–585

Williamson R, Field JG, Shillington FA, Jarre A, Potgieter A (2011) A Bayesian approach for estimating vertical chlorophyll profiles from satellite remote sensing: proof-of-concept. ICES J Mar Sci 68(4):792–799. doi:10.1093/icesjms/fsq169

Chapter 10
Ocean Colour Remote Sensing of Harmful Algal Blooms in the Benguela System

Stewart Bernard, Grant Pitcher, Hayley Evers-King, Lisl Robertson, Mark Matthews, Andy Rabagliati and Christelle Balt

Abstract The Benguela, as a highly productive upwelling system, suffers from the occurrence of a variety of harmful algal blooms, most of which are associated with elevated biomass; a feature common to the shelf environment of upwelling systems. Most harmful blooms have in the past been attributed to one or another dinoflagellate species, but more recently harmful impacts have also been ascribed to other groups of phytoplankton, including diatom and autotrophic ciliate species. Typical bloom assemblages, forcing mechanisms and harmful impacts are outlined, and bloom types most amenable to detection with ocean colour radiometry are identified. Inherent and apparent optical properties of these algal assemblage types are described, and a preliminary evaluation is made of the suitability of available ocean colour data and algorithms. The evolution of several bloom events is described using various algorithms applied to ocean colour data from the Medium Resolution Imaging Spectrometer (MERIS), and recommendations are made about optimal ocean colour usage for high biomass algal blooms in coastal zones.

S. Bernard (✉)
Council for Scientific and Industrial Research, Natural Resources and Ecosystems, Cape Town, South Africa
e-mail: sbernard@csir.co.za

S. Bernard · H. Evers-King · L. Robertson · A. Rabagliati
Department of Oceanography, University of Cape Town, Cape Town, South Africa

M. Matthews
Department of Oceanography, University of Cape Town, Cape Town, South Africa
Also at: University of Cape Town, Cape Town, South Africa

G. Pitcher
Fisheries Research and Development, Department of Agriculture, Forestry and Fisheries, Cape Town, South Africa

C. Balt
Graduate School of Oceanography, University of Rhode Island, Narragansett, RI, USA

V. Barale, M. Gade (eds.), *Remote Sensing of the African Seas*,
DOI 10.1007/978-94-017-8008-7_10,

10.1 Introduction

The upwelling systems of the eastern boundaries of the world's oceans support high biological productivity and a high incidence of Harmful Algal Blooms (HABs) (Kudela et al. 2005). The majority of species contributing to HABs in upwelling systems constitute regular components of the annual succession of phytoplankton, and their harmful impacts are associated with either their toxic properties or the high biomass such blooms can achieve (Trainer et al. 2010). Negative consequences include contaminated seafood, the mortality of fish and other animals, and habitat or ecosystem degradation, all to the detriment of coastal and fishing communities.

There is a strong seasonality to phytoplankton community structure in the southern Benguela, with larger-celled diatoms and dinoflagellates typically dominating inshore waters during the upwelling-driven summer months, and smaller-celled flagellates more dominant in winter (Barlow et al. 2005). Most HABs in the Benguela upwelling system have been attributed to one or another dinoflagellate species, but harmful impacts have also been ascribed to other groups of phytoplankton, including the raphidophytes, diatoms and the ciliate *Myrionecta rubra* (Trainer et al. 2010). Of the impacts associated with toxigenic phytoplankton, those dinoflagellate species responsible for the shellfish poisoning syndromes Paralytic Shellfish Poisoning (PSP) and Diarrhetic Shellfish Poisoning (DSP) pose the greatest risk to human health in the Benguela region (Pitcher and Calder 2000). As a producer of saxitoxins *Alexandrium catenella* is the primary cause of PSP, whereas several species of *Dinophysis*, through the production of okadaic acid and its derivatives, are the causative agents of DSP (Pitcher et al. 2011; Hubbart et al. 2012). Lesser risks to human health in the Benguela are presented by the production of yessotoxins by the dinoflagellate *Protoceratiun reticulatum* (Krock et al. 2008), and by the production of domoic acid by diatoms of the genus *Pseudo-nitzschia* (Fawcett et al. 2007; Hubbart et al. 2012). The risk to seafood safety presented by these species can result in significant losses in harvestable resources and represents a significant threat to the aquaculture industry within the region.

In addition to their accumulation in shellfish, some dinoflagellate toxins, such as those produced by species of *Karlodinium* and *Karenia* can lead directly to mortalities of marine life. In the northern Benguela *Karlodinium veneficum* produces a suite of toxic compounds characterised by haemolytic, ichthyotoxic and cytotoxic properties, and has long been associated with massive fish kills (Copenhagen 1953), whereas in the southern Benguela *Karenia cristata*, is not only responsible for faunal mortalities but may impact human health through eye, nose, throat and skin irritations (Botes et al. 2003). Other fish-killing phytoplankton in the Benguela include the raphidophyte *Heterosigma akashiwo*, although the underlying toxicological mechanisms remain uncertain (Pitcher and Calder 2000).

Events of anoxia, and in some cases associated increases in sulphide concentration, represent one of the more conspicuous and well documented impacts of HABs in the Benguela ecosystem (van de Lingen et al. 2006). Anoxia follows the decay of high biomass dinoflagellate blooms, referred to as red tides. Often dominated by species of *Ceratium* and *Prorocentrum,* these blooms develop during late summer and autumn as seasonal stratification strengthens. Wind relaxation and consequent downwelling force the inshore accumulation of these blooms. Here bloom decay, driven by the exhaustion of nutrients, creates a very high oxygen demand though the entire water column (Pitcher and Probyn 2011). The resulting onset of hypoxia and anoxia may cause major perturbations to the diversity, structure and functioning of the coastal environment as witnessed on occasions by mass mortalities of tons of rock lobster and other marine life (Cockcroft et al. 2000). Other ecosystem impacts may be more subtle and difficult to quantify, and may include altered food web interactions and habitats. The poor food source provided by blooms of the pelagophyte *Aureococcus anophagefferens,* owing either to their small size or to a toxic entity, has impacted food chains in this way, with dramatic reductions in the growth rates of filter feeding bivalves (Probyn et al. 2001). Further impacts associated with high biomass blooms, often dominated by a single species, include unsightly discolouration of the water, which may negatively influence the important economic sectors of tourism and recreation.

Globally, more HABs are recorded now than in the past placing a greater need for resource managers and public health officials to be provided with better tools to monitor and possibly forecast imminent HAB events. The association of HABs in upwelling systems with elevated biomass allows spaceborne surveillance of ocean colour to contribute to such an operational capability through the provision of rapid and spatially broad information relating to the development and progression of HABs (Bernard et al. 2006). Ocean colour may therefore contribute fundamental information for the establishment of early warning systems allowing better assessment of the incidence of blooms, and better planning and management options in different coastal regions.

10.2 Aims of Ocean Colour Application to Harmful Algal Blooms

Ocean colour radiometry offers routine synoptic data pertaining to the phytoplankton biomass and assemblage type in the upper optical depths, and thus offers systematic observations from the event to decadal time scales. On the event scale ocean colour derived observations in upwelling systems offer enhanced, even unique, capabilities in several domains: freely-available, near real time synoptic data for operational bloom monitoring; the detection of precursive and formative bloom conditions; an enhanced capacity to determine whether a bloom is of a potentially harmful nature; an accessible facility to monitor bloom growth, movement and decay; and a greater understanding of the bio-physical dynamics underlying bloom formation (Bernard et al. 2006, Pitcher et al. 2008a). Over longer time scales, such observations offer insight into upwelling system functionality, inter- and intra-seasonal variability and perhaps even environmental regime shifts (Kahru et al. 2009, Pitcher and Weeks 2006, Weeks et al. 2006). However it must be realised that ocean colour data, dependent on the gross bio-optical characteristics of surface water constituents, is by nature imprecise by comparison with the species-level *in situ* data typically used to determine HAB assemblage characteristics and potentially harmful impacts.

Ocean colour application to HABs is therefore most valuable when contextualised from an ecosystem-specific perspective. Of particular importance is the need to contextualise observations from an ecosystem understanding of phytoplankton succession; to utilise satellite derived information through time and space to identify and understand the competitive success of a particular phytoplankton species or functional type in a given ecological niche. Conceptual succession schemes (Margalef 1978, Smayda 2002) provide a useful framework for ocean colour application in upwelling systems in ecological space: a key aspect is the distinction between diatom and dinoflagellate dominance, primarily controlled by turbulence and nutrient availability (Margalef 1978). Upwelling systems are physically driven, but with a somewhat stochastic determination of phytoplankton functional type dominance within a given ecological niche (Kudela et al. 2010). Beyond the obvious need to detect the magnitude, extent and transport of blooms, one of the key aims of ocean colour application should thus be to elucidate assemblage structure within an environmentally determined ecological niche. HAB application of ocean colour radiometry therefore necessarily introduces the need for a species or at least phytoplankton functional type identifier from ocean colour: a bloom cannot be identified as potentially harmful without some form of phytoplankton assemblage information, even if indirectly derived or highly probabilistic in nature.

Such identifiers can be as simple as an ecosystem-contextualised threshold for phytoplankton abundance, typically determined using chlorophyll *a* concentration as a biomass proxy: such an approach has been successfully used in the Gulf of Florida for the operational detection of blooms of the toxic dinoflagellate *Karenia brevis* using commonly available ocean colour products (Tomlinson et al. 2009). The southern Benguela system is highly productive; diatom blooms of up to 20 mg m^{-3} chlorophyll *a* are not uncommon in inshore waters (Barlow 1982, Brown and Hutchings 1987), with diatom-dominated biomass occasionally reaching values of > 50 mg m^{-3} (Mitchell Innes and Walker 1991; Fawcett et al. 2007). However, typically the majority of reported blooms displaying enhanced biomass concentrations of > 30–50 mg m^{-3} are dinoflagellate dominated (Fawcett et al. 2007; Pitcher et al. 2008a, b; Pitcher and Probyn 2011; Pitcher and Weeks 2006), and thus more likely to result in a harmful impact. Such enhanced biomass is associated with cellular motility and the ability to regulate position in the water column, resulting in enhanced near-surface aggregation of flagellated cells (Franks 1992; Smayda 1997). The biomass range threshold of 30–50 mg m^{-3} chlorophyll *a* can be used as a probabilistic first-order identifier of bloom type: it is likely that a bloom with biomass above this range is dinoflagellate dominated. However, the accurate determination of chlorophyll a concentrations > 20 mg m^{-3} is well outside the scope of commonly available algorithms for the Medium Resolution Imaging Spectrometer (MERIS), both for Case 1 waters (Morel et al. 2007), and Case 2 waters (Schiller and Doerffer 1999, 2005), and the demonstration of new empirical and semi-analytical algorithms designed for such high biomass application is a primary aim of this study.

The provision of some form of direct assemblage description from ocean colour measurements offers significant advantages over descriptors derived indirectly from abundance estimates; namely an ability to distinguish a greater variety of assemblage types, with greater confidence and in a wider variety of bloom conditions. There are several algorithm types that could be used to achieve this (IOCCG 2006): inherent optical property based reflectance inversion algorithms will be focused on here as amongst the least constrained and therefore most likely to provide output pertinent to a variety of bloom types (*ibid.*). Such algorithms are necessarily dependent upon an ecosystem-specific characterisation and parameterisation of the inherent optical properties (IOPs) of the phytoplankton assemblage, and the effects of these upon the water-leaving radiance/reflectance. Use of such algorithms requires some consideration of the major cellular and assemblage-specific parameters causal to bio-optical variability; and quantitative consideration of the ability to detect the effects of such phenomena in the water-leaving radiance signal across a range of phytoplankton biomass—a complex issue, and one as yet unresolved (*e.g.* Alvain et al. 2012; Dupouy et al. 2011). Nevertheless key assemblage characteristics potentially identifiable through determination of phytoplankton IOPs from the water-leaving radiance spectral signal can be considered as:

1. Average assemblage size. Size, as determined through mean assemblage metrics such as the effective diameter (Bernard et al. 2006), is arguably one of the most important gross determinants of the optical properties of the algal assemblage (Morel and Bricaud 1981; Roy et al. 2011; Zhou et al. 2012). In addition, size-based ocean colour algorithms have shown effective application in both the coastal (Ciotti and Bricaud 2006) and open ocean (Kostadinov et al. 2010). Detection of changing assemblage size is of considerable potential value in HAB applications: bloom onset is often characterised by changes in the gross size of the algal assemblage, and arguably blooms dominated by large dinoflagellate cells are most commonly associated with harmful impact in the southern Benguela (Pitcher and Calder 2000).
2. Gross changes in intra-cellular accessory pigments. Whilst the effects of pigment changes on cellular absorption are significant and well understood (Johnsen et al. 1994; Lohrenz et al. 2003), there is considerable debate as to how detectable such effects are on the water-leaving reflectance across different biomass and other optical regimes (Dierssen et al. 2006; Jackson et al. 2011; Alvain et al. 2012). One important determinant in detecting absorption-related effects, in a reflectance signal, is phytoplankton biomass. There is little doubt that, while a detailed sensitivity study still must be conducted, such effects are much more visible, and perhaps only detectable, at considerable biomass concentration (Dierrsen et al. 2006). In the southern Benguela, it is highly unlikely that it is possible to distinguish even high biomass blooms dominated by diatoms or dinoflagellates based only on pigment-related absorption effects, given that fucoxanthin and peridinin (as the primary respective accessory pigments) are spectrally similar (Bidigare et al. 1990). However, the photosynthetic ciliate *Myrionecta rubra*, a large-celled, bloom-forming species displaying considerable phycoerythrin absorption (Kyewalyanga 2002), is potentially detectable based on absorption effects on the water leaving reflectance, and such an approach is investigated during this study.

3. Ultrastructure. The presence of intracellular organelles of appreciably different refractive index to the surrounding cell, *e.g.* gas vacuoles in cyanobacterial species, or calcium carbonate platelets associated with coccolithophores, can have considerable impact on cellular scattering properties (Volten et al. 1998; Whitmire et al. 2010) and thus ocean colour (Bracher et al. 2009; McKinna et al. 2011). Blooms dominated by vacuolate species are not of ecological consequence in the southern Benguela and will not be considered further here. Whilst coccolithophore blooms occur in the Benguela, and can be discriminated using ocean colour radiometry (Henderiks et al. 2012; Weeks et al. 2004), such blooms are not associated with harmful impacts and are not considered further here.
4. Fluorescence line height and quantum yield. Both MERIS and the Moderate Resolution Imaging Spectometer (MODIS) have a set of bands, centred at the chlorophyll *a* fluorescence emission peak of ± 683 nm, to measure the sun induced fluorescence of the phytoplankton assemblage *e.g.* Gower et al. 1999. This signal has a complex dependency on phytoplankton biomass, taxonomy and physiological status (Behrenfeld et al. 2009; MacIntyre et al. 2010). There is evidence indicating that diatoms may exhibit elevated fluorescence quantum yields due to photo-machinery able to rapidly respond to short term variability in photoacclimation in a high-mixing environment (Lavaud et al. 2002; Nymark et al. 2009). There is thus potential to use fluorescence quantum yield to discriminate between diatom- and dinoflagellate-dominated blooms, of considerable value to HAB application in the Benguela, and such an approach is explored further below.

An example of the large differences in the spectral reflectance, with associated and causal differences in the measured phytoplankton absorption (as per Bernard et al. 2006) and modelled backscattering coefficients (as per Bernard et al. 2009)—as the principle IOPs determining ocean colour—can be seen in Fig. 10.1. Data from a diverse range of bloom types, dominated by a range of assemblages, are shown: a mono-specific bloom (Chl $a = 12.8$ mg m^{-3}) of the very small-celled pelagophyte *Aureococcus anophagefferens* (Probyn et al. 2010); a very high biomass bloom (Chl $a = 309.0$ mg m^{-3}) of the PSP-toxic, large-celled, chain-forming dinoflagellate *Alexandrium catenella* (Bernard et al. 2009); a high biomass bloom (Chl $a = 172.5$ mg m^{-3}) of the relatively small celled dinoflagellate *Prorocentrum triestinum (ibid.)*; a moderate biomass bloom (Chl $a = 18.0$ mg m^{-3}) of the autotrophic, large celled, phycoerythrin containing ciliate *M. rubra (*unpublished*)*; and a moderate biomass bloom (Chl $a = 20.7$ mg m^{-3}) dominated by a variety of diatom species.

The *Aureococcus* bloom (with cell counts of $\pm 1.5 \times 10^9$ cells l^{-1}) is notable for very high reflectance values (red spectra, Fig. 10.1a), caused by the enhanced backscattering resulting from the ± 2 μm effective diameter cell size (Fig. 10.1c), which also results in the notably high Chl-specific phytoplankton absorption values (Fig. 10.1d) (Quirantes and Bernard 2006). This can be contrasted with the *Alexandrium* bloom, where the large effective diameter of ± 30 μm results in Chl *a* values > 20 times larger than the *Aureococcus* bloom, despite cell counts ($\pm 9.8 \times 10^6$ cells l^{-1}) $\pm$ two orders of magnitude smaller. The *Alexandrium* bloom

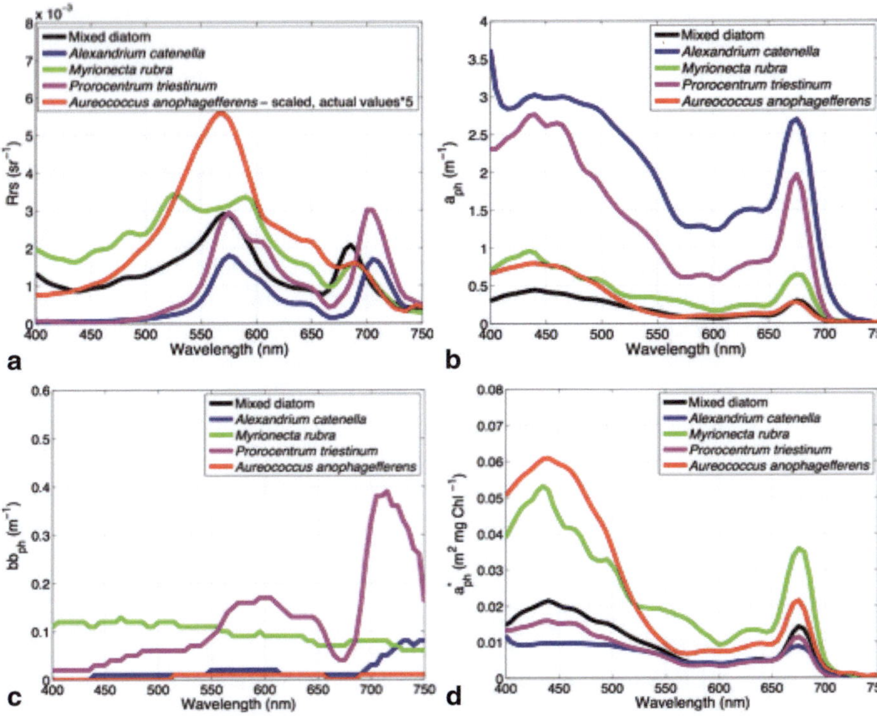

Fig. 10.1 Optical properties of five diverse blooms occurring between 1998 and 2005 from various locations in the southern Benguela: remotely-sensed reflectance R_{rs} (**a**), phytoplankton absorption coefficients a_{ph} (**b**), phytoplankton backscattering coefficients b_{bph} (**c**), and Chl-specific phytoplankton absorption coefficients $a*_{ph}$ (**d**)

is notable for attributes associated with large cells at high biomass: the high magnitude, highly-packaged (*i.e.* spectrally relatively flat) phytoplankton absorption values (blue spectra, Fig. 10.1d); the relatively low backscattering (Fig. 10.1c) for such elevated biomass, and the relatively low reflectance values dominated by peaks at ± 570 nm and 709 nm (Fig. 10.1a).

Again, there is interesting contrast with the *Prorocentrum* bloom, of effective diameter ± 13 μm and cell counts of $\pm 1.2 \times 10^8$ cells l^{-1}. The smaller cell size relative to the *Alexandrium* assemblage results in reflectance of similar shape (pink spectra, Fig. 10.1a) but enhanced magnitude, an obvious result of the much greater phytoplankton backscattering (Fig. 10.1c) associated with the smaller cell size (Bernard et al. 2009). The *Myrionecta* bloom displays reflectance features associated with considerable phycoerythrin absorption (the bifurcated peaks at ± 530 and 600 nm, green spectra, Fig. 10.1a), and the effects of slightly lower biomass relative to the two dinoflagellate assemblages: greater reflectance at blue wavelengths, and a longer wavelength peak < 700 nm where Chl *a* fluorescence is still observable (Fig. 10.1a).

Such data demonstrate the potential utility of using *in situ* bio-optical and ocean colour-analogous data to distinguish between bloom types. However, there is still a considerable challenge to utilising such data from space-borne ocean colour sensors. The retinue of methodological challenges to ocean colour application in the coastal environment are compounded by HAB application: atmospheric correction in turbid waters and an aerosol-laden atmosphere; distinction of bio-optical constituents in optically complex waters; and adjacency effects from land. The phytoplankton dominated waters of the Southern Benguela can in the large part be conventionally classified as Case 1 (Morel and Prieur 1977), excluding the typically short-lived case of freshly upwelled waters where there are enhanced concentrations of small inorganic particulate through the water column (Kudela et al. 2006). However, the considerable phytoplankton biomass observed under bloom conditions in the Benguela provide the same functional obstacles to atmospheric correction algorithms as sediment-influenced Case 2 waters: enhanced turbidity and resulting non-zero signal in the near-infrared, thereby compromising atmospheric correction algorithms assuming null water-leaving radiance signal in NIR bands (Gordon and Wang 1994; Wang 2006). Past use in the Benguela has shown that the Bright Pixel Atmospheric Correction (BPAC) algorithm (Moore and Lavender 2011), with an ability to adjust for non-zero NIR reflectance, performs relatively well on data from the MERIS in high biomass Benguela waters. In addition, the typical high biomass reflectance signals of interest here contain a strong signal, if not a distinct peak, in the 700–730 nm spectral range (Matthews et al. 2012; Pitcher et al. 2008a): the only recent ocean colour mission carrying an appropriately located spectral band, at 709 nm, is the MERIS (and forthcoming OLCI on the Sentinel 3 series). Reduced resolution ($\pm$ 1 km ground resolution) data from the MERIS will therefore be exclusively focused on here; in addition to optimal band placement and routine availability of suitably derived water-leaving reflectance data, the sensor has excellent sensitivity with a large signal:noise ratio, and suitably high revisit time with global coverage every 2 days. This study seeks to demonstrate the utility of products derived from these ocean colour data to identify bloom types based on cell size and intracellular pigment characteristics, and show ocean colour application to bloom observation at the event scale.

10.3 Methods: Bio-Optical Data and Algorithms

St Helena Bay in the Southern Benguela is focussed on due to its high biomass levels and the frequent occurrence of HABs within this region. A retentive circulation, linked with the upwelling dynamics of the bay, plays an important role in the transport, concentration and dissipation of blooms (Pitcher and Calder 2000). As such, St Helena Bay has been the site of numerous *in situ* data collection campaigns during the last decade. In addition to the satellite data and algorithm output presented here, *in situ* radiometry from a Satlantic Hyperspectral Tethered Surface Radiometer Buoy (H-TSRB) and *in situ*, fluorometric chlorophyll values are provided for comparison. Absorption measurements were made from filtered samples using a Shimadzu UV-2501 spectrophotometer. Identification of species and estimates of effective diameter from particle size distributions were made through microscopy and use of a Beckman Coulter Counter respectively. Sampling was generally conducted from Lamberts Bay at stations around − 32.08 N and 18.26 E, with additional sampling where bloom patches were evident. Detailed laboratory and processing methodology can be found in Bernard et al. 2009.

Two algorithms are used in the study. The Maximum Peak Height (MPH) algorithm (Matthews et al. 2012) is an empirical top-of-atmosphere algorithm designed for MERIS in high biomass waters, returning only Chl *a* concentrations. The Equivalent Algal Population (EAP) algorithm (variant of Bernard 2005) is a semi-analytical reflectance algorithm, based upon inversion of IOPs from modelled algal populations, which gives a more comprehensive suite of returns.

The EAP algorithm can be summarised as follows. Inputs are MERIS atmospherically corrected multi-spectral normalised water leaving reflectance, on a pixel-by-pixel basis. The model uses five solvable unknowns: chlorophyll a concentration (Chl a, mg m^{-3}), algal effective diameter (D_{eff}, μm), the relative concentration of two representative algal groups (diatoms/dinoflagellates and nanoflagellates/chlorophytes), combined gelbstoff and detrital absorption (*e.g.* $a_{gd}(400)$, m^{-1}), and small particle backscattering (*e.g.* $b_{bs}(550)$, m^{-1}). Gelbstoff/detrital absorption and small particle backscattering employ constant spectral shapes and variable magnitude (Bernard 2005).

The EAP algorithm uses an unconstrained non-linear minimisation (Nelder and Mead 1965), with constant initial values except for chlorophyll, for which an initial estimate is provided via an empirical, band-ratio switching algorithm on a per pixel basis. This algorithm extends the maximum band ratio approach used in the standard MERIS algal 1 product, by adding an additional ratio between 665 and 709 nm bands. The EAP algorithm and all subroutines are coded in Matlab R14. As the underlying model specifically does not account for sun-induced natural fluorescence, the convergence weighting for the Nelder-Mead solution is set to negligible values between 665 and 715 nm—the spectral region affected by natural algal fluorescence. The algorithm thus does not seek to match spectral reflectance values at these fluorescence wavelengths, and in effect offers a means of discriminating fluorescence effects, as have earlier models of a similar nature (Roesler and Boss 2003). This allows the derivation of algal fluorescence quantum yield as an additional algorithm product, utilising integrated fluorescence (calculated from fitting a Gaussian distribution to modelled R_{rs} subtracted from measured R_{rs} at fluorescence wavebands), algal absorption as returned by the algorithm and incident scalar irradiance calculated independently (Gregg and Carder 1990).

10.4 Case Studies: Application to Bloom Types

10.4.1 Transitions Between Dominant Species Detected Through Changes in Effective Diameter Using the EAP Algorithm

The EAP algorithm approach allows for determination of a wide range of chlorophyll concentrations and an "effective diameter" relating to the particle size distribution of the algal assemblage. In the case of mono-specific blooms this will equate approximately to the average cell size of the dominant species. Imagery from 30 March and 5 April 2005 processed with the EAP algorithm allows for detection of the elevated biomass levels within the greater St Helena Bay region through this period (*in situ* values between 172.5 and 39.6 mg m^{-3}). In addition, a significant change in effective diameter of the assemblage is detected (Fig. 10.2e, f). Concurrent *in situ* measurements (microscopy and coulter counter derived size distributions) suggest that this change in size represents a change in dominance of this bloom from the small-celled dinoflagellate *Prorocentrum triestinum* to the large-celled dinoflagellate *Ceratium furca*.

Comparisons between *in situ* (Satlantic H-TSRB) and satellite (MERIS) derived radiometry (Fig. 10.2a, b) show that the satellite captures both the magnitude and shape of the spectra in each case. The significant signal in the NIR bands seen here triggers the use of the bright pixel atmospheric correction (BPAC) over large portions of the images, particularly in areas associated with high chlorophyll concentrations.

10.4.2 Use of Fluorescence and an Assessment of Pigment Detection Using Hyperspectral and Multispectral Radiometry Using the EAP Algorithm and Spectral Flags

Whilst detection of cell size provides a useful ecosystem parameter and potential indicator of taxonomic shifts, the assessment of HAB impact often requires species identification, particularly where toxic effects are being considered. As previously mentioned, a shift from diatom to dinoflagellate dominance forms a key part of many conceptual frameworks for phytoplankton succession in upwelling systems (Margalef 1978). An ability to detect this shift using satellite data may help provide early prediction for the onset of HABs, as well as for comparison to oceanographic data to validate these succession schemes.

Though differentiating between diatoms and dinoflagellates is useful, an ability to identify particular species would be optimal in cases where toxicity is suspected. The existence of accessory pigments with specific absorption characteristics provides a way that individual species can be identified from radiometric data. However the extent to which these spectral characteristics can be exploited is dependent on both the level of biomass and the specifications of instruments used, particularly in the case of satellite derived radiometry. Figure 10.3a shows match-up *in situ* and MERIS reflectance associated with a relatively low biomass (Chl a = 3.5 mg m^{-3}) bloom in the St Helena Bay area containing *Myrionecta rubra*. Figure 10.3b displays a variety of *in situ* R_{rs} spectra from a 3-week period during the same bloom event; spectra chosen to show naturally occurring admixtures of diatoms and *M. rubra* at a range of biomass. Peaks related to phycorethyrin absorption can be seen in the data from the *in situ* Satlantic H-TSRB between 500 and 620 nm. However, from satellite it is difficult to fully resolve these peaks due to a paucity of bands in this region of the spectra (Fig. 10.3a), leading to difficulties during the inversion procedure. In lower

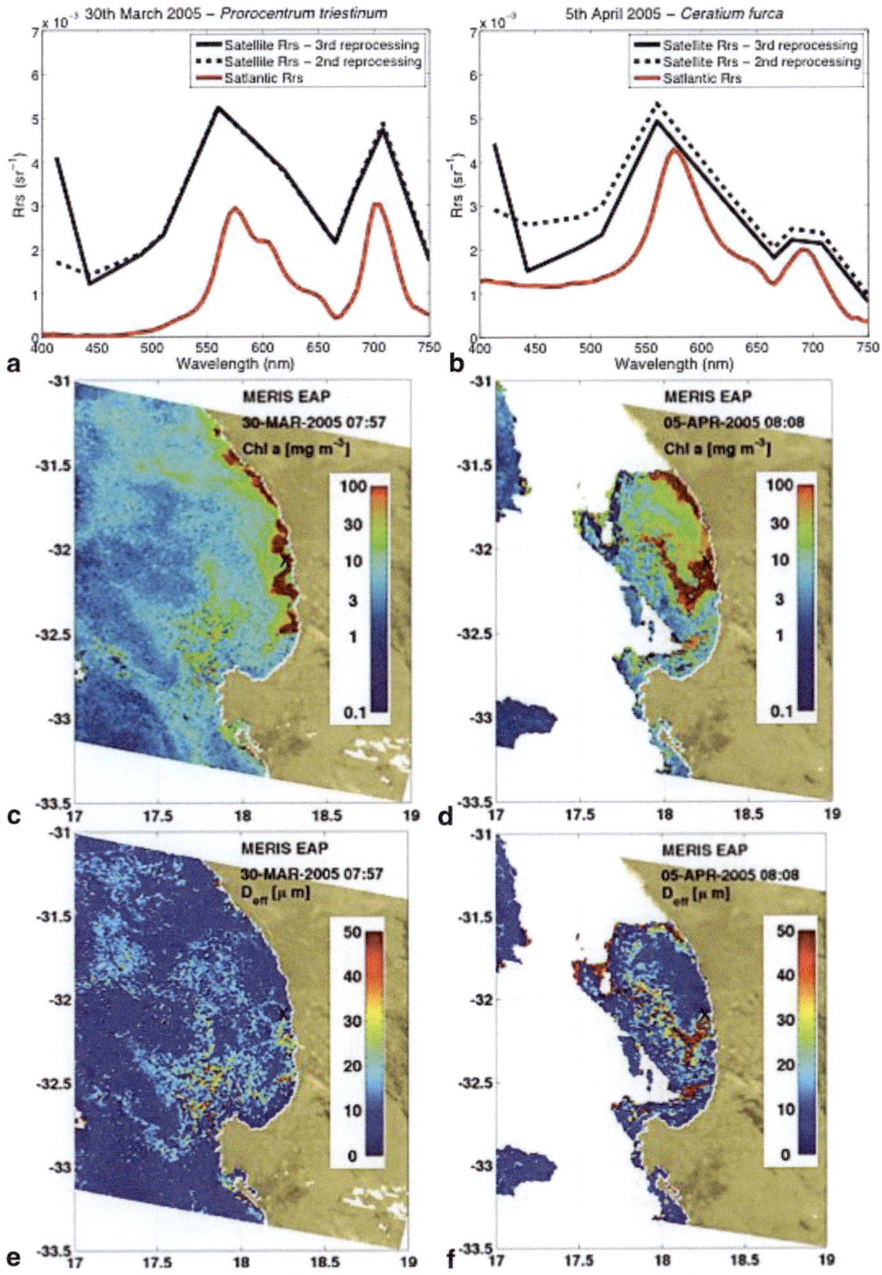

Fig. 10.2 MERIS RR radiometric match-ups and algorithm outputs from the EAP semi-analytical algorithm, demonstrating the effectiveness of this approach for detecting high biomass (**c** and **d**) and distinguishing dominance by small (*Prorocentrum triestinum*) and large (*Ceratium furca*) dinoflagellates (**e** and **f** respectively) using the effective diameter product. Radiometric match ups show consistent Bright Pixel Atmospheric Correction performance across 2nd and 3rd MERIS reprocessing, though some overestimation is seen (**a** and **b**)

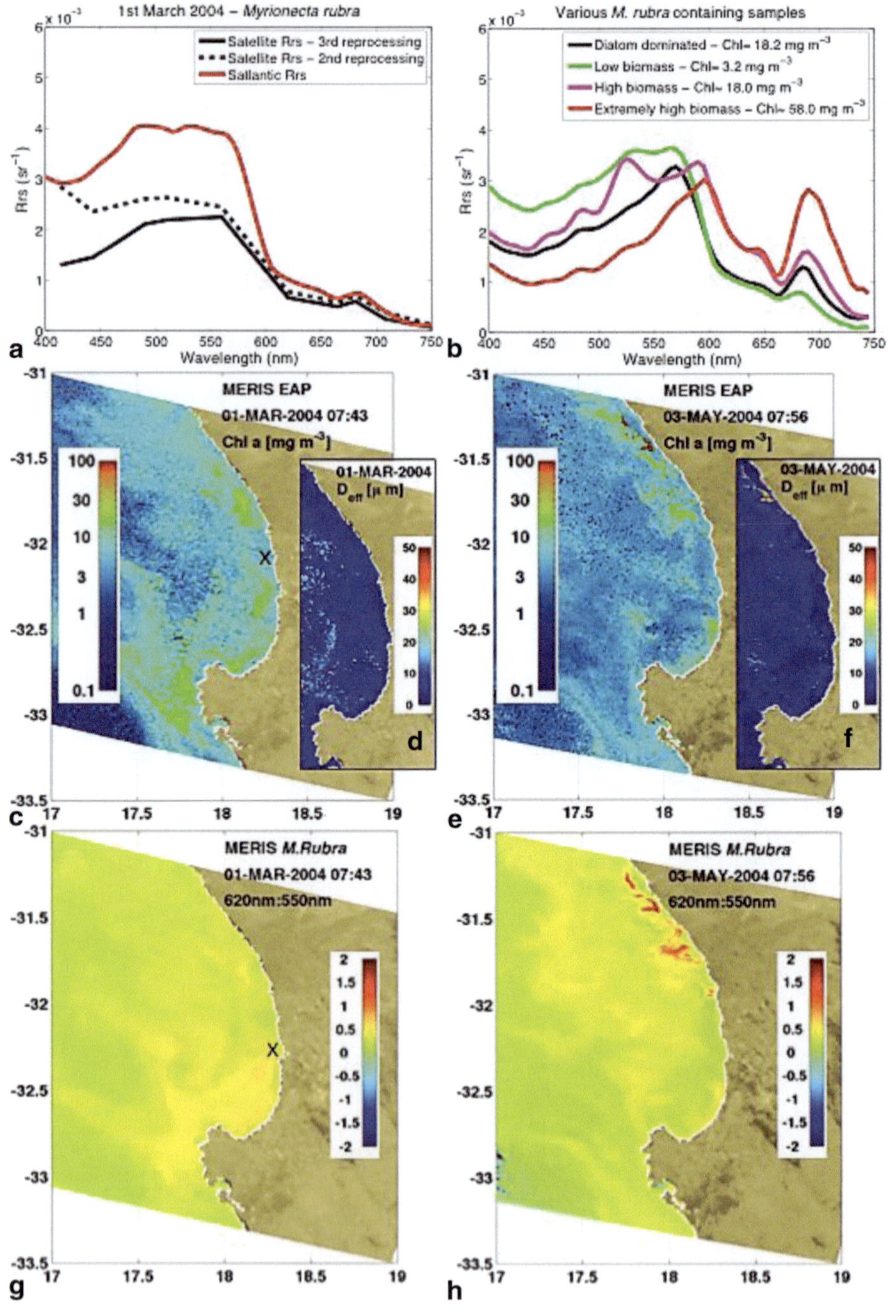

Fig. 10.3 MERIS RR radiometric match-ups (**a**) for 1st March, and a variety of Rrs spectra from a 3 week period from the same bloom event (**b**). Algorithm outputs from the EAP semi-analytical algorithm: Chl a concentration (**c**) and effective diameter (**d**) from 1st March 2004; Chl a concentration (**e**) and effective diameter (**f**) from 3rd May 2004; and 620:550 nm R_{rs} ratio from 1st March (**g**) and 3rd May (**h**)

biomass cases (Fig. 10.3c), the generation of the effective diameter fails to retrieve a reasonable estimate (around 30 μm for mono-specific blooms) (Fig. 10.3d). In high biomass cases however, such as for the bloom shown in Fig. 10.3e, where more reasonable estimate of effective diameter can be retrieved (Fig. 10.3f), a ratio between bands 620 and 550 nm can be used to further identify the presence of *Myrionecta rubra* (Fig. 10.3h) from a background of mixed diatom blooms. These examples show the complexities of detecting accessory pigment-based optical variability with multi-spectral ocean colour data, in particular the very strong effects of biomass on the useable reflectance signal.

10.4.3 Tracking the Development of a Ceratium Balechii Bloom and Consequent Anoxia Using the MPH Algorithm

In the autumn of 2009 the development and ultimate decay of an exceptional bloom of the dinoflagellate *Ceratium balechii* lead to an anoxia-induced mass mortality of macrofauna in St Helena Bay (Pitcher and Probyn 2011). Initial build-up of the bloom was evident in mid-February from chlorophyll concentrations derived by application of the MPH algorithm to MERIS data (Fig. 10.4). The bloom was first observed to the north of St Helena Bay in the nearshore region of the southern Namaqua shelf and was on occasions shown to extend in a narrow band over a distance of 100 km (Fig. 10.4; 14 March 2009). Under diminished upwelling activity during late summer and early autumn inshore counter currents resulted in the southward progression of the bloom and its entrainment into the Bay (Fig. 10.4; 11 April 2009). In early May the bloom was shown to have accumulated in the shallow, southern reaches of the Bay (Fig. 10.4; 1 May 2009). Degradation of the bloom occurred in these shallow waters where subthermocline nutrients, necessary for bloom maintenance, are inaccessible. The exceptional organic loading of the system as afforded by the bloom resulted in anoxia through the entire water column. Large fish and lobster mortalities were consequently observed off the Berg River estuary and adjacent beaches on the 5 May 2009.

Termination of the boom was associated with persistent downwelling conditions resulting in advection of the bloom from St Helena Bay shortly after the anoxic event (Fig. 10.4; 4 May 2009). Surveillance of these blooms as depicted in Fig. 10.4 enables continual assessment by fisheries and coastal managers of the risk posed by these phenomena to marine resources and the coastal environment.

The case study demonstrates the ability of relatively simple algorithms, such as the MPH, to effectively track very high biomass blooms. It should further be noted that several of the images ($\pm$ 40 %) suffered from high sun-glint, which would prohibit employing EAP-type algorithms that use the full spectrum of water leaving reflectance data after aerosol and Rayleigh atmospheric correction. The MPH is less susceptible to glint effects, as it uses only a small set of green to NIR wavebands and does not employ an aerosol correction (Matthews et al. 2012): this allows increased frequency of image utility at the event scale, a significant advantage for operational application.

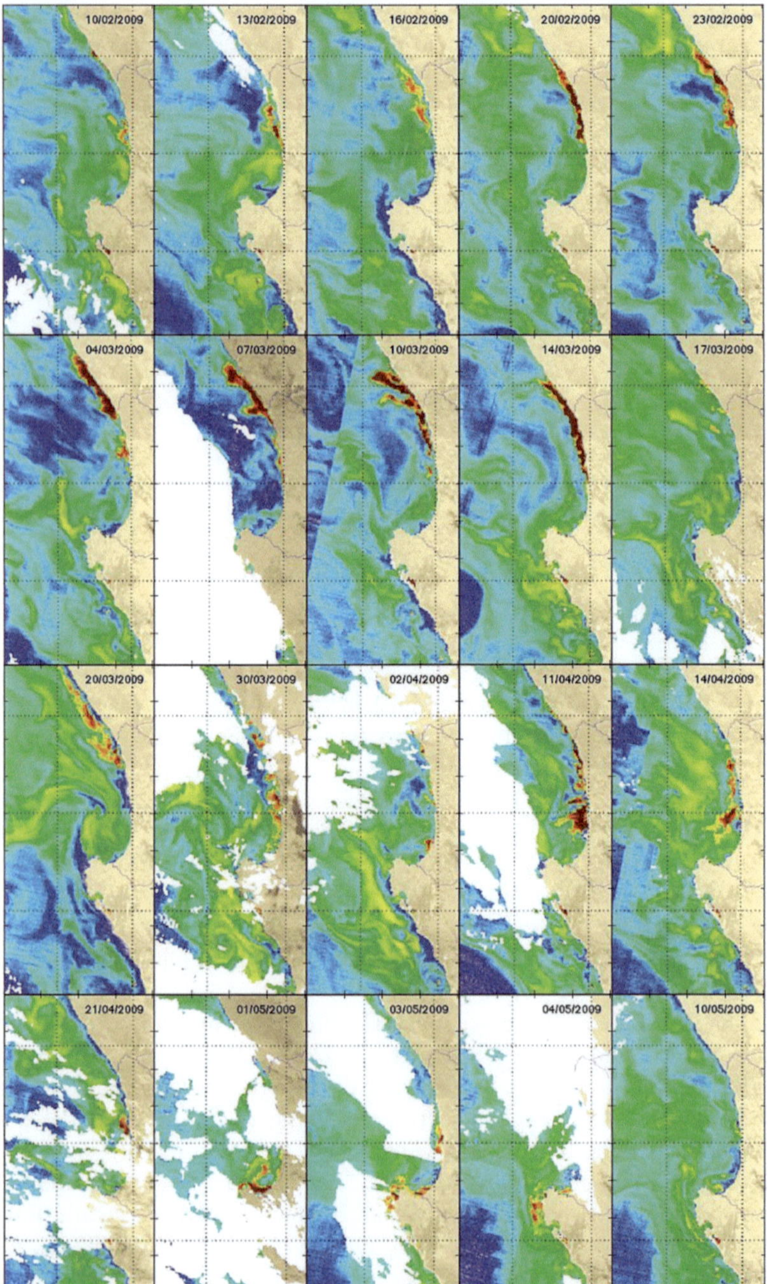

Fig. 10.4 A montage of MPH derived Chl *a* output showing the development and transport of a bloom of *C. balechii* into St Helena Bay. Consequent decay of the bloom in the nearshore environment led to conditions of anoxia and the mortality of marine life. (Pitcher and Probyn 2011)

10.5 Summary and Outlook

Ocean colour products show great potential for enhancing our ability to both detect and quantify HABs, in particular for the very high biomass blooms typically encountered in the Bengeula. It should be stressed that many toxic algal species can have harmful effects at low biomass (*e.g.* McGillicuddy et al. 2005); the algorithms presented here, and arguably perhaps even ocean colour data in general, may not be effective for such bloom types. Combined with an understanding of ecosystem dynamics, satellite derived products can be used to detect elevated biomass and indicate potential for occurrence of toxic species at appropriate temporal and spatial scales. This work highlights the importance of band selection on satellite-mounted sensors, particularly when wishing to exploit pigment-related effects and fluorescence-based products. It is apparent that accurate atmospheric correction in turbid waters must be a first priority to fully utilise these spectrally dependent algorithms. Alternatively, top of atmosphere approaches such as the MPH algorithm used here can circumvent this requirement. The frameworks provided in this study should be adaptable for use in other systems where HABs occur. For systems where species have broadly similar inherent optical properties (IOPs) to those in the Benguela, a semi-analytical inversion approach such as that used in the EAP algorithm, demonstrated with the case study in Sect. 4.1, should be able to provide improved chlorophyll estimates and an indication of the effective diameter of the assemblage. Where mono-specific blooms are prevalent, this parameter is most useful in detecting changes in dominance. This method may prove less successful where other constituents dominate the optical signal, *e.g.* in severely sediment loaded waters. In cases where species are present which have obscure spectral signatures, resulting from accessory pigments or ultrastructure, new IOPs would need defining, however empirical, spectral ratio techniques such as those used here to detect *Myrionecta rubra* in Sect. 4.2, could also be applied, given sufficient biomass.

References

Alvain S, Loisel H, Dessailly D (2012) Theoretical analysis of ocean color radiances anomalies and implications for phytoplankton groups detection in case 1 waters. Opt Express 20(2):1070–1083

Barlow R, Sessions H, Balarin M, Weeks S, Whittle C, Hutchings L (2005) Seasonal variation in phytoplankton in the southern Benguela: pigment indices and ocean colour. Afr J Mar Sci 27:275–287

Barlow RG (1982) Phytoplankton ecology in the southern Benguela Current. I. Biochemical composition. J Exp Mar Biol Ecol 63(3):209–227

Behrenfeld MJ, Westberry TK, Boss ES, O'Malley RT, Siegel DA, Wiggert JD, Franz BA, McClain CR, Feldman GC, Doney SC, Moore JK, Dall'Olmo G, Milligan AJ, Lima I, Mahowald N (2009) Satellite-detected Fluorescence reveals global physiology of Ocean Phytoplankton. Biogeosciences 6:779–794

Bernard S (2005). The bio-optical detection of harmful algal blooms. PhD Thesis. University of Cape Town

Bernard S, Kudela R, Franks P, Fennel W, Kemp A, Fawcett A, Pitcher G (2006) The requirements for forecasting harmful algal blooms in the Benguela. In: Shannon V, Hempel G, Malanotte-Rizzoli P, Moloney C, Woods J (eds) Benguela: predicting a large marine ecosystem. Elsevier 14:273–294

Bernard S, Probyn TA, Quirantes A (2009) Simulating the optical properties of phytoplankton cells using a two-layered spherical geometry. Biogeosci Discuss 6(1):1497–1563

Bidigare RR, Ondrusek ME, Morrow JH, Kiefer DA (1990) In vivo absorption properties of algal pigments. In: Spinrad RW (ed) Ocean optics X. Proc Soc Photo-Opt Instrum Eng 1302:90–302

Botes L, Sym SD, Pitcher GC (2003) *Karenia cristata* sp. Nov. and *Karenia bicuneiformis* sp. Nov. (Gymnodiniales, Dinophyceae): two new *Karenia* species from the South African coast. Phycologia 42:563–571

Bracher A, Vountas M, Dinter T, Burrows JP, Röttgers R, Peeken I (2009) Quantitative observation of cyanobacteria and diatoms from space using PhytoDOAS on SCIAMACHY data. Biogeosciences 6:751–764. doi:10.5194/bg-6–751-2009.

Brown PC, Hutchings L (1987) The development and decline of phytoplankton blooms in the southern Benguela upwelling system. 1. Drogue movements, hydrography and bloom development. In: Payne AL, Gulland A, Brink KH (eds) The Benguela and comparable ecosystems. S Afr J Marine Sci 5:357–391

Ciotti AM, Bricaud A (2006) Retrievals of a size parameter for phytoplankton and spectral light absorption by Colored Detrital Matter from water-leaving radiances at SeaWiFS channels in a continental shelf region off Brazil. Limnol Oceanogr Method 4:237–253

Cockcroft AC, Schoeman DS, Pitcher GC, Bailey GW, van Zyl DC (2000) A mass stranding, or "walkout" of west coast rock lobster Jasus lalandii in Elands Bay, South Africa: causes, results and implications. In: Von Kaupel Klein JC, Schram FR (eds) The biodiversity crises and crustacea. Crustacean Issues 11:362–688

Copenhagen WJ (1953) The periodic mortality of fish in the Walvis region: a phenomenon within the Benguela Current. Division of Fisheries Investigational Report 14, p 35

Dierssen H, Kudela RM, Ryan JP, Zimerman RC (2006) Red and black tides: Quantitative analysis of water-leaving radiance and perceived color for phytoplankton, colored dissolved organic matter, and suspended sediments. Limnol Oceanogr 51(6):2646–2659

Dupouy C, Benielli-Gary D, Neveux J, Dandonneau Y, Westberry TK (2011) An algorithm for detecting *Trichodesmium* surface blooms in the South Western Tropical Pacific. Biogeosciences 8:3631–3647. doi:10.5194/bg-8–3631-2011

Fawcett A, Pitcher GC, Bernard S, Cembella AD, Kudela RM (2007) Contrasting wind patterns and toxigenic phytoplankton in the southern Benguela upwelling system. Mar Ecol Prog Ser 348:19–31

Franks P (1992) Sink or swim: accumulation of biomass at fronts. Mar Ecol Prog Ser 82:1–12

Gordon HR, Wang M (1994) Retrieval of water-leaving radiance and aerosol optical thickness over the oceans with SeaWiFS: a preliminary algorithm. Appl Optics 33(3):443–452

Gower JFR, Doerffer R, Borstad GA (1999) Interpretation of the 685 nm peak in water-leaving radiance spectra in terms of fuorescence, absorption and scattering, and its observation by MERIS. Int J Remote Sens 20(9):1771–1786

Gregg WW, Carder KL (1990) A simple spectral solar Irradiance model for cloudless maritime atmospheres. Limnol Oceanogr 35(8):1657–1675

Henderiks J, Winter A, Elbrächter M, Feistel R, van der Plas A, Nausch G, Barlow R (2012) Environmental controls on *Emiliania huxleyi* morphotypes in the Benguela coastal upwelling system (SE Atlantic). Mar Ecol Prog Ser 448:51–66

Hubbart B, Pitcher GC, Krock B, Cembella AD (2012) Toxigenic phytoplankton and concomitant toxicity in the mussel *Choromytilus meridionalis* off the west coast of South Africa. Harmful Algae 20:30–41

IOCCG (2006) Remote sensing of inherent optical properties: Fundamentals, tests of algorithms, and applications. IOCCG, Dartmouth, pp 1–126

Jackson T, Bouman HA, Sathyendranath S, Devred E (2011) Regional-scale changes in diatom distribution in the Humboldt upwelling system as revealed by remote sensing: implications for fisheries. ICES J Mar Sci 68(4):729–738

Johnsen G, Samset O, Granskog L, Sakshaug E (1994) In-vivo absorption characteristics in 10 classes of bloom-forming phytoplankton: taxonomic characteristics and responses to photoadaptation by means of discriminant and HPLC analysis. Mar Ecol Prog Ser 105(1/2):149–157

Kahru M, Kudela R, Manzano-Sarabia M, Mitchell BG (2009) Trends in primary production in the California Current detected with satellite data. J Geophys Res 114(CO2004):7

Kostadinov TS, Siegel DA, Maritorena S (2010) Global variability of phytoplankton functional types from space: assessment via the particle size distribution. Biogeosciences 7:3239–3257

Krock B, Alpermann T, Tillmann U, Pitcher GC, Cembella AD (2008) Yessotoxin profiles of the marine dinoflagellates Protoceratium reticulatum and Gonyaulax spinifera. In: Moestrup O et al (eds) Proceedings of the 12th International Conference on Harmful Algae. International Society for the Study of Harmful Algae and Intergovernmental Oceanographic Commission of UNESCO, Copenhagen

Kudela R, Pitcher G, Probyn T, Figueiras F, Moita T, Trainer V (2005) Harmful algal blooms in coastal upwelling systems. Oceanography 18:84–197

Kudela RM, Garfield N, Bruland KW (2006) Bio-optical signatures and biogeochemistry from intense upwelling and relaxation in coastal California. Deep-Sea Res Part II 53(25–26):2999–3022

Kudela RM, Seeyave S, Cochlan W (2010) The role of nutrients in regulation and promotion of harmful algal blooms in upwelling systems. Prog Oceanogr 85:122–135

Kyewalyanga M, Sathyendranath S, Platt T (2002) Effect of *Mesodinium rubrum* (= *Myrionecta rubra*) on the action and absorption spectra of phytoplankton in a coastal marine inlet. J Plankton Res 24(7):687–702

Lavaud J, Rousseau B, van Gorkom HJ, Etienne AL (2002) Influence of the diadinoxanthin pool size on photoprotection in the marine planktonic diatom *Phaeodactylum tricornutum*. Plant Physiol 129:1398–1406

Lohrenz SE, Weidemann AD, Tuel M (2003) Phytoplankton spectral absorption as influenced by community size structure and pigment composition. J Plankton Res 25(1):33–61

MacIntyre HL, Lawrenz E, Richardson TL (2010) Taxonomic discrimination of phytoplankton by spectral fluorescence. In: Suggett DJ et al (eds) Chlorophyll a fluorescence in aquatic sciences: methods and applications. Developments in Applied Phycology 4, Springer, p 129

Margalef R (1978) Life-forms of phytoplankton as survival alternatives in an unstable environment. Oceanol Acta 1:193–509

Matthews MW, Bernard S, Robertson L (2012) An algorithm for detecting trophic status (chlorophyll-a), cyanobacterial-dominance, surface scums and floating vegetation in inland and coastal waters. Remote Sens Environ 124:637–652

McGillicuddy DJ Jr, Anderson DM, Solow AR, Townsend DW (2005). Interannual variability of *Alexandrium fundyense* abundance and shellfish toxicity in the Gulf of Maine. Deep Sea Res Part II 52(19–21):2843–2855

McKinna LIW, Furnas MJ, Ridd PV (2011) A simple, binary classification algorithm for the detection of Trichodesmium spp. within the Great Barrier Reef using MODIS imagery. Limnol Oceanogr Methods 9:50–66

Mitchell-Innes BA, Walker DR (1991) Short-term variability during an Anchor Station study in the southern Benguela upwelling system. Phytoplankton production and biomass in relation to species changes. Prog Oceanogr 28(1–2):65–89

Moore G, Lavender S (2011) Algorithm identification: Case II. S Bright pixel atmospheric correction MERIS ATBD Plymouth Marine Laboratory, Issue 5.0

Morel A, Bricaud A (1981) Theoretical results concerning light absorption in a discrete medium, and application to specific absorption of phytoplankton. Deep-Sea Res 28A(11):1375–1393

Morel A, Prieur L (1977) Analysis of variations in ocean color. Limnol Oceanogr 22(4):709–722

Morel A, Claustre H, Antoine D, Gentili B (2007) Natural variability of bio-optical properties in Case 1 waters: attenuation and reflectance within the visible and near-UV spectral domains, as observed in South Pacific and Mediterranean waters. Biogeosciences 4:913–925

Nelder JA, Mead R (1965) A Simplex Method for Function Minimization. Comp J 7:308–313

Nymark M, Valle KC, Brembu T, Hancke K, Winge P et al (2009) An integrated analysis of molecular acclimation to high light in the marine diatom *Phaeodactylum tricornutum*. PLoS ONE 4(11):e7743

Pitcher GC, Calder D (2000) Harmful Algal Blooms of the southern Benguela current: a review and appraisal of monitoring from 1989–1997. S Afr J Mar Sci 22:255–271

Pitcher GC, Probyn TA (2011) Anoxia in southern Benguela during the autumn of 2009 and its linkage to a bloom of the dinoflagellate *Ceratium balechii*. Harmful Algae 11:23–32

Pitcher GC, Weeks SJ (2006) The variability and potential for prediction of Harmful Algal Blooms in the southern Benguela ecosystem. In: Shannon, V, Hempel G, Malanotte-Rizzoli P, Moloney C, Woods J (eds) Benguela: Predicting a large marine ecosystem. Elsevier 14:125–146

Pitcher GC, Bernard S, Fawcett A (2008a) Real-time coastal observing systems for ecosystem dynamics and harmful algal blooms: the needs and expectations of users. In: Babin M, Roesler CS, Cullen JJ (eds) Real-time coastal observing systems for marine ecosystem dynamics and harmful algal blooms. UNESCO, Paris, France

Pitcher GC, Bernard S, Ntuli J (2008b) Contrasting bays and red tides in the southern Benguela upwelling system. Oceanography 21(3):82–91

Pitcher GC, Krock B, Cembella AD (2011) Accumulation of diarrhetic shellfish poisoning toxins in the oyster Crassostrea gigas and the mussel Choromytilus meridionalis in the southern Benguela ecosystem. Afr J Mar Sci 33:273–281

Probyn TA, Pitcher GC, Pienaar RN, Nuzzi R (2001) Brown tides and mariculture in Saldanha Bay, South Africa. Mar Pollut Bull 42(5):405–408

Probyn TA, Bernard S, Pitcher GC, Pienaar RN (2010) Ecophysiological studies on *Aureococcus anophagefferens* blooms in Saldanha Bay, South Africa. Harmful Algae 9:123–133

Quirantes A, Bernard S (2006) Light-scattering methods for modelling algal particles as a collection of coated and/or nonspherical scatterers. J Quant Spectrosc Radiat Transfer 100(1–3):315–324

Roesler CS, Boss E (2003) Spectral beam attenuation coefficient retrieved from ocean color inversion. Geophys Res Lett 30(9):1–4

Roy S, Sathyendranath S, Platt T (2011) Retrieval of phytoplankton size from bio-optical measurements: theory and applications. Journal of the Royal Society Interface 8(58):650–660

Schiller H, Doerffer R (1999) Neural network for emulation of an inverse model operational derivation of Case II water properties from MERIS data. Int J Remote Sens 20(9):1735–1746

Schiller H, Doerffer R (2005) Improved determination of coastal water constituent concentrations from MERIS data. IEEE Trans Geosci Remote Sens 43(7):1585–1591

Smayda TJ (1997) What is a bloom? A commentary. Limnol Oceanogr 42:1132–1136

Smayda TJ (2002) Turbulence, watermass stratification and harmful algal blooms: an alternative view and frontal zones as "pelagic seed banks". Harmful Algae 1:95–112

Tomlinson MC, Wynne TT, Stumpf RP (2009) An evaluation of remote sensing techniques for enhanced detection of the toxic dinoflagellate, *Karenia brevis*. Remote Sens Environ 113(3):598–609

Trainer VL, Pitcher GC, Reguera B, Smayda TJ (2010) The distribution and impacts of harmful algal bloom species in eastern boundary upwelling systems. Prog Oceanogr 85:33–52

Van derLCD, Freón P, Hutchings L, Roy C, Bailey GW, Bartholomae C, Cockcroft AC, Field JG, Peard KR, Van derPAK (2006) Forecasting shelf processes of relevance to living marine resources in the BCLME. In: Shannon, V, Hempel G, Malanotte-Rizzoli P, Moloney C, Woods J (eds) Benguela: Predicting a large marine ecosystem. Elsevier 14:309–347

Volten AH, Haan JFD, Hovenier JW, Schreurs R, Vassen W, Dekker AG, Hoogenboom J et al (1998) Laboratory measurements of angular distributions of light scattered by Phytoplankton and Silt. Limnol Oceanogr 43(6):1180–1197

Wang M (2006) Aerosol polarization effects on atmospheric correction and aerosol retrievals in ocean color remote sensing. Appl Optics 45(35):8951–8963

Weeks SJ, Pitcher GC, Bernard S (2004) Satellite monitoring of the evolution of a coccolithophorid bloom in the southern Benguela ecosystem. Oceanography 17:83–89

Weeks SJ, Barlow R, Roy C, Shillington FA (2006) Remotely sensed variability of temperature and chlorophyll in the southern Benguela: upwelling frequency and phytoplankton response. Afr J Mar Sci 28:493–509

Whitmire AL, Pegau WS, Karp-boss L, Boss E, Cowles TJ (2010) Spectral backscattering properties of marine phytoplankton cultures. Opt Express 18(14):1680–1690

Zhou W, Wang G, Sun Z, Cao W, Xu Z, Hu S, Zhao J (2012) Variations in the optical scattering properties of phytoplankton cultures. Opt Express 20:11189–11206

Chapter 11
Satellite Observations of Oceanic Eddies Around Africa

Werner Alpers, Dominique Dagorne and Peter Brandt

Abstract Oceanic eddies having scales from several hundred meters to several hundred kilometers are ubiquitous phenomena in the World's ocean. This became evident only after they could be observed from satellites and space shuttles. Here we present several images taken in different spectral bands which show signatures of eddies of different spatial scales in sea areas around Africa. In particular, we present a series of satellite images showing the propagation of a small-scale cyclonic (cold) eddy generated at Cap-Vert at the coast of Senegal into the open ocean. We show that this small-scale eddy transported nutrients from the Senegal upwelling region westward into the oligotrophic North Atlantic thus giving rise to enhanced chlorophyll-a concentration there. Since eddies are also areas of high fish population, knowledge of their position and properties is of great importance for fishery.

11.1 Introduction

Oceanic eddies of horizontal scales ranging from several hundred meters to several hundred kilometers are ubiquitous phenomena in the World's ocean (McWilliams 1985). This came as a great surprise to many oceanographers when the first images taken over the ocean from spacecraft became available. (see e.g., Cresswell and Legeckis 1986; Stevenson 1998; Munk et al. 2000; Siegel et al. 2001). Boundaries of strong currents, like the Gulf Stream and the Agulhas Current, and upwelling regions, like the Canary upwelling area, are preferred regions of eddy generation.

W. Alpers (✉)
Institut för Meereskunde, Universität Hamburg,
Hamburg, Germany
e-mail: werner.alpers@zmaw.de

D. Dagorne
US Imago, Institut de Recherche pour le Développement,
Plouzané, France

P. Brandt
GEOMAR Helmholtz-Zentrum für Ozeanforschung Kiel,
Kiel, Germany

V. Barale, M. Gade (eds.), *Remote Sensing of the African Seas*,
DOI 10.1007/978-94-017-8008-7_11,

Eddies strongly affect the dynamics of the ocean and can transfer heat, salt, trace gases, nutrients, and CHL across frontal zones (see e.g., Olson 1991; Morrow et al. 2003). In particular, they also can transport nutrients from upwelling regions into oligotrophic ocean regions causing enhanced CHL concentration there (Falkowski et al. 1991; Levy et al. 2001; Lin et al. 2010). Although already more than 30 years ago, Gower et al. (1980) have pointed out that phytoplankton patchiness is linked to mesoscale eddies, it has been realized only recently that eddies play a key role in the variability of the CHL distribution in the World's ocean (Williams 2011). Chelton et al. (2011) state that most of the variability of CHL distribution results from redistribution of CHL caused by advection with the mesoscale flow field and not from changes in the local phytoplankton growth.

Often oceanic eddies are divided into mesoscale and submesoscale eddies with horizontal scales above and below the first baroclinic Rossby radius of deformation, respectively. The Rossby radius is larger than 100 km in the inner tropics, around 50 km in the upwelling region off West Africa and becomes smaller towards north reaching about 10 km in the subpolar gyre (Chelton et al. 1998). Mesoscale eddies having spatial scales above 100 km have been studied intensively in the last two decades by using radar altimeter (RA) data and by assimilating them in global ocean circulation models. They are measured by altimeters via sea surface height anomalies from which current velocities are retrieved by using the geostrophic approximation (see e.g., Fu and Holt 2008; Chelton et al. 2011). Present-day global ocean circulation models have horizontal resolutions of the order of 10 km and thus are suited for modeling mesoscale oceanic eddies, but not submesoscale ones (Maltrud and McClean 2005; Le Galloudec et al. 2008; Capet et al. 2008).

Mesoscale eddies can be detected from satellites also by infrared and optical sensors, like the Moderate Resolution Imaging Spectroradiometer (MODIS) onboard the Terra and Aqua satellites, and by Synthetic Aperture Radars (SARs). However, these sensors are not well suited to study the dynamics of mesoscale eddies systematically. Optical/infrared sensors provide useful data only when there are no or only few clouds in the imaged scene. On the other hand, SAR data are not dependent on cloud coverage and the time of the day, but their application is limited by the fact that a given ocean area is imaged very infrequently by a SAR on a single satellite and that it is often difficult to identify unambiguously features visible on SAR images as radar signatures of oceanic eddies.

Eddies with horizontal scales below 100 km, which we shall call in the following small-scale eddies, are not detectable by present-day space-borne RAs, but they can be detected from satellites by high-resolution optical/infrared sensors and by SARs. These small-scale eddies have been observed from space first on sunglint images (Soules 1970; Scully-Power 1986) and later also on synthetic aperture radar (SAR) images (Fu and Holt 1983; Johannessen et al. 1993, 1996; Munk et al. 2000; DiGiacomo and Holt 2001; Ivanov and Ginzburg 2002; Yamaguchi and Kawamura 2009; Karimova 2012). Unlike mesoscale eddies, small-scale eddies cannot be modeled by using traditional quasi-geostrophic theory (Thomas et al. 2008). Simulations carried out with models with horizontal resolution of the order of 1 km show that small-scale eddies are generated by instabilities of geostrophic surface flows and that they are confined to the upper layer of the ocean (Eldevik and Dysthe 2002). Furthermore, they are a source of kinetic energy for the larger scale flow. Like mesoscale eddies, also small-scale eddies can be generated by several processes, such as (1) interaction of large-scale currents with bottom topography, islands or headlands, (2) barotropic or baroclinic instability of currents and fronts, or (3) by atmospheric forcing (vorticity input from wind stress). Recently Golitsyn (2012) suggested that the majority of the small-scale eddies are generated by convection in the water caused by surface cooling.

Another detailed study of small-scale eddies was recently carried out by Karimova (2012). She analyzed over 2000 SAR images acquired between 2009 and 2010 by the European Remote Satellites ERS-1 and ERS-2 and the European Envisat satellite over the Baltic Sea, the Black Sea, and the Caspian Sea. She detected on them more than 14,000 radar signatures of vortical structures with diameters between 1 and 75 km. About 99 % of them had diameters in the range 1–20 km and 98 % had a cyclonic rotation.

In this paper we present several satellite images showing sea surface signatures of meso- and small-scale eddies around Africa. Prominent areas where such eddies are encountered are the upwelling regions off West Africa and the Agulhas Current at the east coast of Africa which frequently sheds eddies.

This paper is organized as follows: In Sect. 11.2 we present three photographs taken by a handheld camera from Space Shuttle over areas of the Mediterranean close to the Egyptian coast, On these photographs, eddies are imaged in the sunglint. They become visible on these photographs via surface films which trace the eddy's surface current field and modify the reflected light. In Sects. 11.3 and 11.4 we present SAR and optical and infrared images, respectively, showing signatures of eddies in the waters surrounding Africa. In Sect. 11.5 we present results from a case study showing the propagation of a small-scale eddy from its birth place in the upwelling region near Cap-Vert, Senegal, into the open North Atlantic. It transported nutrients from the Senegal upwelling region westward into the oligotrophic North Atlantic, where it gave rise to enhanced CHL concentration. Finally, in Sect. 11.6 we discuss and summarize the results.

11.2 Sunglint Images

Sunglint refers to the reflection of incoming solar radiation from the sea surface observed by optical (visible/near-infrared) sensors under specular or near-specular conditions, where the sunlight's angle of incidence equals the angle of reflection to the observer. On a flat, undisturbed sea surface the sun's reflection would simply appear as a mirror image of its disk in the sky. However, the surface of the ocean is almost always roughened due the action of the wind such that the "mirror" is broken into many small facets tilted at a variety of different angles and directions. This sea surface roughening makes it possible to observe specular reflected sunlight by suitably tilted facets even when its viewing angle is not directed at the specular

Fig. 11.1 Photograph taken from the space shuttle Challenger at 12:02:55 UTC on 7 October 1984 (STS 41-G mission) by a handheld camera showing sea surface signatures of a cyclonic eddy in the Mediterranean Sea adjacent to the Egyptian coast. The center point of the photograph is at 31.5°N, 27.0°E. The eddy with a diameter of approximately 35 km lies in inner sunglint region, where the lines of reduced surface roughness (caused by surface films) are imaged *bright*

point on the water surface. The size and shape of the sunglint pattern depends on the probability distribution of the slopes of the facets caused by the sea surface roughness, the direction of incident sunlight and the viewing direction of the sensor (Cox and Munk 1954; Jackson and Alpers 2010). Satellite optical sensors with spatial resolutions of meters to hundreds of meters do not see the small individual glittering facets, but rather averages over all facets that are contained within a resolution cell. Thus differences in the sea surface roughness in and around the sunglint region manifest themselves as variations in the image brightness.

On the photographs shown in Figs. 11.1, 11.2 and 11.3, the change of the sea surface roughness is caused by surface films, which are entrained in the eddy's

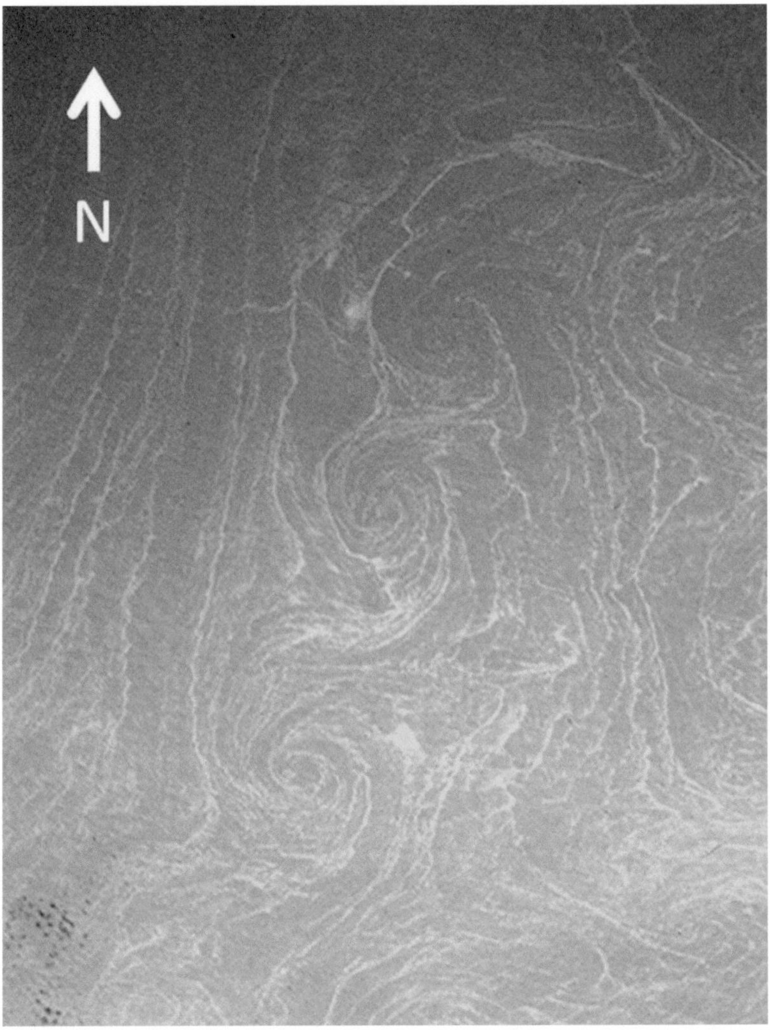

Fig. 11.2 Photograph taken from the space shuttle Challenger at 12:02:43 UTC on 7 October 1984 (STS 41-G mission) by a handheld camera showing sea surface signatures of three interconnected cyclonic eddy in the Mediterranean Sea near the Egyptian coast. The center point of the photograph is at 32.0°N, 26.0°E. The eddies lie in inner sunglint region, where the lines of reduced surface roughness (caused by surface films) are imaged *bright*

surface current field. In the sunglint region, areas with film coverage appear brighter on the photo than areas with no film coverage, because the surface film reduces the sea surface roughness such that more light can be specular reflected to the camera. But eddies are also visible outside the sunglint region. Here the areas with film coverage appear darker on the photo than areas with no film coverage. The dividing line between the inner and outer sunglint regions, where brightness reversal occurs,

Fig. 11.3 Photograph taken from the space shuttle Challenger at 11:45:57 UTC on 8 October 1984 (STS 41-G mission) showing sea surface signatures of a cyclonic eddy in the Mediterranean Sea adjacent to the Egyptian coast. The center point of the photograph is at 31.5°N, 27.5°E. Visible on the right Cape Alan el Rum. The eddy with a diameter of approximately 40 km lies in the outer sun glint region, where lines of reduced sea surface roughness (caused by surface films) are imaged *dark*

is determined by the critical angle which depends on the viewing geometry, i.e. on the relative elevation and azimuth angles between the sun and the sensor, and also on the sea surface roughness (Jackson and Alpers 2010). The photographs depicted in Figs. 11.1, 11.2 and 11.3 were taken by the astronaut-oceanographer Paul Scully-Power with a Hasselblad camera with a focal length of 250 mm during the Space Shuttle Challenger mission 41-G, which lasted from 5 to 13 October 1984 (Scully-Power 1986). During the times when the photographs shown here, always calm wind conditions prevailed with wind speeds below 2 ms^{-1}. Figure 11.1 shows a photograph of the Mediterranean Sea near the Egyptian coast, which were taken at 12:02:55 UTC on 7 October 1984 (orbit: 33, roll: 35, frame: 95).

The center point of the photograph (determined to the nearest half degree) is at 31.5°N, 27.0°E. The Space Shuttle was located at the time of the data take at 32.8°N, 29.1°E, where the elevation angle of the sun was 40°. The photograph shows in the upper section sea surface signatures of a cyclonic eddy in the sunglint region. The bright lines on the photograph are smooth lines on the sea surface covered by surface films. Based on the scale of land features, we estimate the diameter of the eddy to be 35 km.

The photograph depicted in Fig. 11.2 was taken 12 s earlier at 12:02: 43 UTC (orbit: 33, roll: 35, frame: 94). It also shows a photograph of the Mediterranean Sea near the Egyptian coast, approximately half a degree further north and one degree further east than the previous photograph. The center point of the photograph is at 32.0°N, 28.0°E. The Space Shuttle was located at the time of the data take at 33.5°N, 28.5°E, where the elevation angle of the sun was 40°. The photograph shows three interconnected cyclonic eddies in the sunglint region. This photograph is also shown in the paper by Munk et al. (2000) and has since then been reproduced in many papers, e.g., in the paper by Eldevik and Dysthe (2002). Munk et al. (2000) have estimated the width of the imaged area to be 30 km, from which follows that the diameter of the largest eddy is about 8 km. They attribute the eddy generation to horizontal current shear induced by wind shear. Although calm wind conditions prevailed during the time of the data take, strong winds with speeds up to 10 ms^{-1} were blowing the week before from a north-northwesterly direction over the Mediterranean Sea. An argument in favor of this generation mechanism is that the centers of the spirals are aligned approximately in wind direction which should have been also direction of the shear line.

However, in a recent paper Golitsyn (2012) argues that the majority of the small-scale eddies visible on sunglint and SAR images is not generated by horizontal shear instability, but by convection in the water caused by surface cooling. Since most of the observed small-scale eddies are cyclonic, this points to the fact that eddy generation is affected by the Earth's rotation. The sinking of cold water leads to convergence of surface waters, i.e. to a concentration of angular momentum associated with the Earth's rotation. Golitsyn (2012) estimated that the time it takes for generating a small-scale eddy is of the order of hours, and not of 5–15 days as in the case of eddy generation by shear instability (Munk et al. 2000). According to Golitsyn, most of the cyclonic eddies in the ocean can be considered as hydrodynamic analogs to tropical and polar hurricanes in the atmosphere. This would explain the dominance of cyclonic small-scale eddies in the World's ocean. On the other hand, small-scale eddies generated by velocity shear should be rare phenomena, since they require special wind conditions, like constant winds blowing over the water for a sufficient long time, which is seldom the case.

Figure 11.3 shows another photograph of the Mediterranean Sea near the Egyptian coast with the Cape Alan el Rum to the right, which was taken at 1145 UTC on 8 October 1984 (orbit: 49, roll: 38, frame: 65). The center point of the photograph is at 31.5°N, 27.5°E. The space shuttle was located at the time of the data take at 32.2°N, 28.3°E, where the elevation angle of the sun was 43°. The photograph shows a cyclonic eddy with a diameter of approximately 40 km in the outer sunglint region, where the dark lines are smooth lines caused by damping of the short-scale sea surface waves by surface films. Thus, on this photograph, the eddy pattern is traced out by dark lines caused by the accumulation of surface films in convergent surface current regions. Surface films damp the short-scale sea surface waves and thus modify the reflection of sunlight from the sea surface. Sea surface films are either of natural or anthropogenic origin.

Natural surface films consists of mineral oil released from oil seeps at the sea bottom or of surface active material secreted by biota in the water column. In the latter case, they are called biogenic surface films, which form monomolecular surface films at the sea surface. They can damp the short surface waves as strongly as mineral oil films. Anthropogenic or man-made surface films usually consist of oily substances released from ships (bunker or sludge oil from the ship's engine room, crude oil from washing the tanks of tankers), from oil platforms or from land-based industrial plants. Since it is well known that the biological productivity is very low in the Mediterranean waters west of the Nile estuary (Siokou-Frangou et al. 2010), we suspect that in this case not biogenic surface films served as tracer for the eddy's surface current field, but anthropogenic oily substances released from ships. Since we are not aware of any oil seeps in this area, we suspect that the surface films originated from ships. Since the imaged area is located close to the shipping lane leading to the Suez Canal, the probability to encounter anthropogenic surface films in this area is quite high.

11.3 Synthetic Aperture Radar Images

Mesoscale and submesoscale eddies become detectable on SAR images by modulation of the short-scale sea surface waves, which causes a modulation of the backscattered radar power or the Normalized Radar Cross Section (NRCS). Such modulation can be induced (1) by surface current variations associated with the eddy, (2) by cold upwelled water in cyclonic eddies or (3) by surface films.

In the first case, the interaction of short surface with a variable current enters into the SAR imaging mechanism for which theories have been developed (Alpers and Hennings 1984; Kudryavtsev et al. 2005). Since the variations of the surface currents associated with eddies are small, also the corresponding radar signatures are small. This makes it usually difficult to identify sea surface signatures of eddies on SAR images.

In the second case, the stability of the air-sea interface enters into the SAR imaging mechanism. If the Sea Surface Temperature (SST) over an eddy is cooler than over the ambient waters, then the stability of marine boundary layer changes from (usually) neutrally stable (over the ambient water) to unstable (over the cold eddy) which causes a reduction of the friction velocity (or wind stress) and thus a reduction of the capability of the wind to generate short surface waves (see e.g., Large and Pond 1981; Koslov et al. 2011). As a result, the NRCS is reduced over the cold eddies, typically by 2–3 dB.

In the third case the damping of short surface waves by biogenic or anthropogenic surface films enters into the SAR imaging mechanism (Valenzuela 1978; Huehnerfuss et al. 1996; Espedal et al. 1998; Alpers and Espedal 2004). This requires the presence of surface films, which is not always the case. The presence of surface films

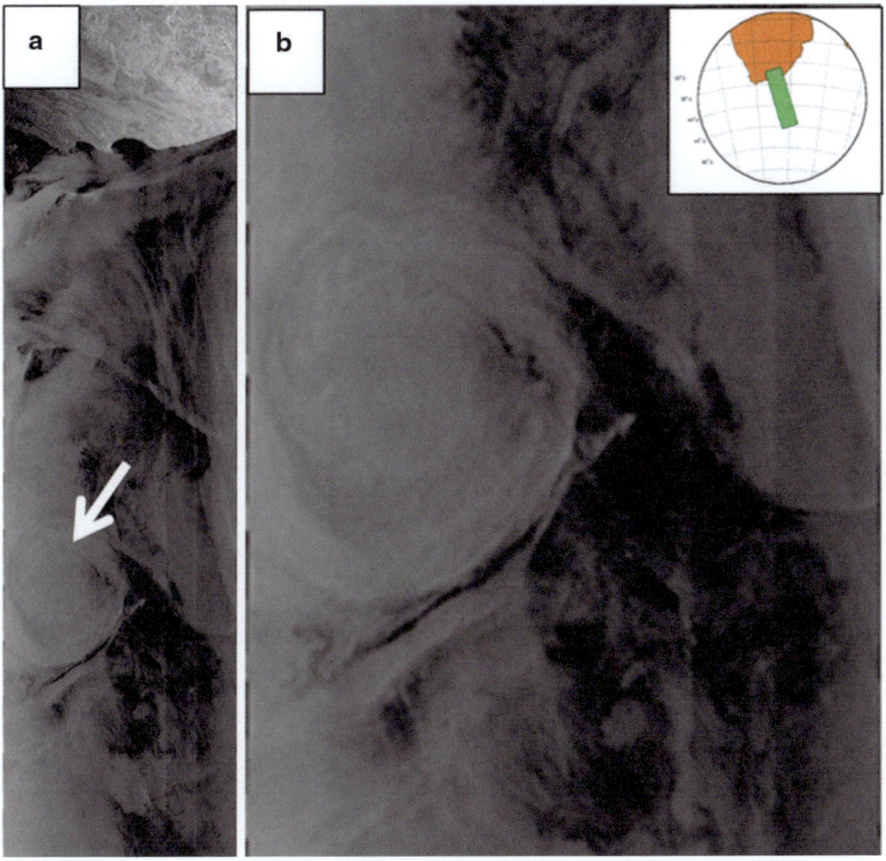

Fig. 11.4 **a** SAR image acquired by the ASAR onboard the Envisat satellite in the Wide Swath Mode (WSM) at 0113 UTC on 6 December 2011 showing in the lower left section the radar signature of an anticyclonic eddy southeast of South Africa. The imaged area is 400 km × 1450 km. **b** Zoom on the eddy whose diameter is approximately 210 km. The inset shows the position of the entire imaged scene

causes a reduction of the NRCS, and thus areas covered with surface films manifest themselves as dark patches on SAR images. When the sea surface is partially covered with surface films, then the surface films accumulate in convergent surface current regions. An eddy is then visualized on SAR images through a characteristic surface pattern consisting of dark (i.e., low radar backscatter), narrow, curvilinear, concentric bands that spiral inward into the pattern (see e.g., Johannessen et al. 1993; Yamagushi and Kawamura 2009). This imaging mechanism yields very strong radar signatures of small-scale eddies, which are similar to sea surface signatures of small-scale eddies seen in sunglint images (Sect. 2).

In Figs. 11.4 and 11.5 two SAR images are depicted which show radar signatures of a large-scale (mesoscale) anticyclonic (Fig. 11.4) and a large-scale (mesoscale) cyclonic (Fig. 11.5) eddy in the Indian Ocean off the east coast of South Africa. Note

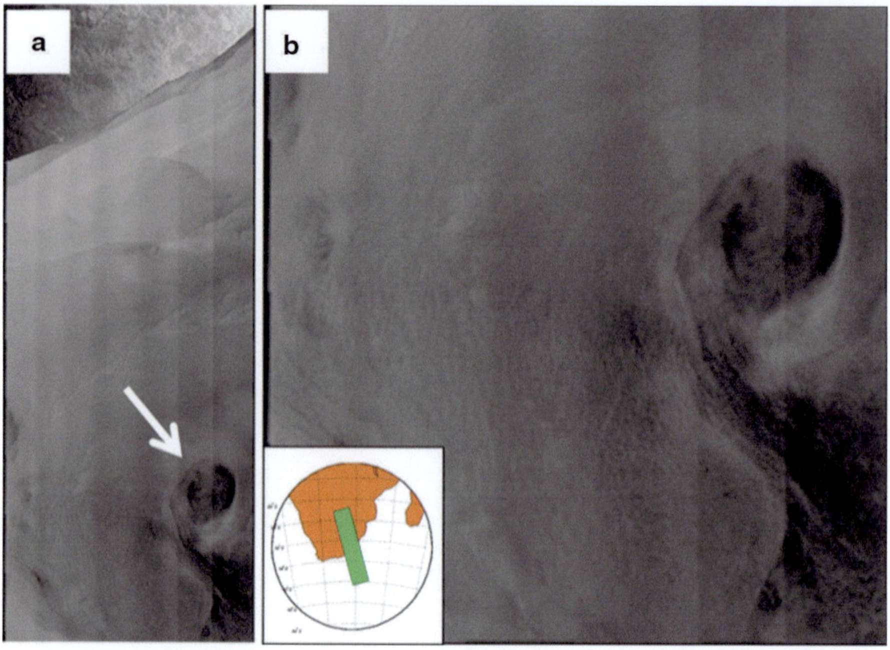

Fig. 11.5 **a** Section of an SAR image acquired by the ASAR onboard the Envisat satellite in the Wide Swath Mode (WSM) at 0113 UTC on 6 December 2011 in the Wide Swath Mode (WSM) at 2041 on 27 October 2006 during a ascending satellite pass showing in the lower right-hand section the radar signature of a cyclonic eddy southeast of South Africa. The imaged area is 400 km × 1000 km. **b** Zoom on the eddy whose diameter is approximately 110 km. The inset shows the position of the entire imaged scene

that in the southern hemisphere cyclonic eddies rotate clockwise and anti-cyclonic eddies counterclockwise. Cyclonic eddies give rise to upwelling of cold water and anticyclonic eddies to downwelling of warm water.

Both images were acquired during ascending satellite passes, i.e. when the satellite was heading north. The resolution of the Wide Swath Mode (WSM) ASAR images is 150 m and the swath width is in this mode 400 km. In Sect. 11.5 we shall present another Envisat ASAR image which shows the radar signature of a small-scale eddy. The image depicted in Fig. 11.4a was acquired at 0113 UTC on 6 December 2011 shows in the lower left-hand section the radar signature of an anticyclonic eddy. It is located southeast of South Africa (see inset) and has a diameter of approximately 210 km. A zoom on the eddy is shown in Fig. 11.4b. Note that in this case the radar signature of the large-scale eddy is very weak due to the weak hydrodynamic modulation by horizontal current shear. On the other hand, the ASAR image depicted in Fig. 11.5, which was acquired at 2041 UTC on 27 October 2006, shows in the lower right-hand section the radar signature of a cyclonic eddy. The radar signature of this cyclonic eddy is much stronger than that of the eddy depicted in Fig. 11.4. The reason is that in this case surface films and the change of the stability of the air-sea interface have contributed to the imaging of the eddy. Since this is a cyclonic eddy, it is associated with upwelling of cold water. Cold eddies are prone to carry with them much biota which are sources of surface active material, which ascends to the sea surface and forms there biogenic surface films.

11.4 Infrared and Optical Images

Oceanic eddies can also be detected from space by infrared and optical sensors provided that there is no or only little cloud coverage. Infrared sensors measure SST and optical sensors the ocean color from which the CHL distribution is derived. Figure 11.6 shows a reflection map derived from data acquired by the MODIS sensor onboard the Terra satellite in the 555 nm band at 0910 UTC on 26 December 2011. It shows an anti-cyclonic mesoscale eddy (rotating anticlockwise in the southern hemisphere) with a diameter of approximately 150 km located 800 km south of South Africa. The eddy becomes visible in light blue due to high scattering by phytoplankton (probably coccolithophorids). Observations and computer simulations have shown that such eddies are frequently encountered in this ocean region and that they spin off from the Agulhas current.

Figure 11.7 shows a map of the CHL distribution in the waters west of the coasts of Mauretania, Senegal and Guinea. It was derived from data acquired by the MODIS sensor onboard the Aqua satellite at 1440 UTC on 16 November 2006. Visible is a large-scale anti-cyclonic eddy (which rotates clockwise in the northern hemisphere) by means of a band of high CHL concentration (brownish colors). It is located west of the coast of Senegal/Guinea and has a diameter of approximately 150 km. This band of high CHL concentration forming the rim of the eddy is connected with the area of high CHL concentration at the coast of Guinea. This suggests that the anticyclonic eddy has drawn its high CHL concentrated in the spiral arm from the upwelling area at the coast of Guinea. Note that the CHL concentration in the center of the eddy is very low. Such anti-cyclonic eddies with high CHL concentration at the rim of anticyclonic eddies have also been observed in other parts of the World's ocean, among them in the Oyahio Current region in the western North Pacific. Kusakabe et al. (2002) observed that boundaries of anticyclonic eddies in the Oyashio Current region contain coastal Oyashio waters with high biological activity, like in our case (Fig. 11.7), where the rim of the anticyclonic eddy contains waters with high biological activity.

Figures 11.8, 11.9 and 11.10 show SST and CHL maps based on MODIS data on which signatures of small-scale cyclonic eddies are visible. All of them are located off the coast of Senegal and were captured between November and December when upwelling starts at the Senegalese coast. In Sect. 11.5 we shall show more images of this type. The two SST maps depicted in Fig. 11.8 have been derived from MODIS data acquired at 1440 UTC on 7 November 2006 and at 1440 UTC on 16 November 2006, respectively. They show signatures of a small-scale eddy that was generated at Cap-Vert (Fig. 11.8a), presumably by flow separation (see Sect. 11.5), which then propagated northwestward into the Atlantic. Also visible are on Fig. 11.8a, west of this eddy, signatures of several very small-scale eddies. They are not visible anymore on Fig. 11.8b, presumably because they have dissipated.

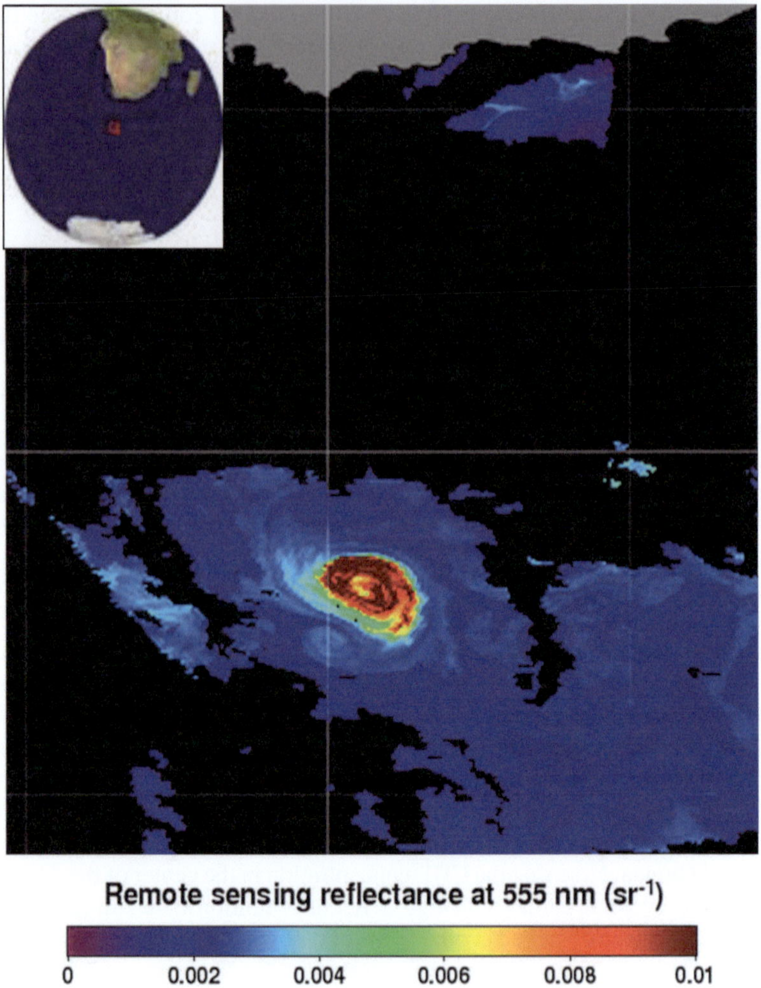

Fig. 11.6 Reflectance map at 555 nm obtained from MODISTerra data acquired at 0910 UTC on 26 December 2011 showing an anticyclonic mesoscale eddy in the southern hemisphere 800 km south of Africa with diameter of approximately 150 km. The eddy becomes visible in red by enhanced phytoplankton concentration. The *gray* area at the *top* is southern part of South Africa. The inset shows the position of the eddy. (Source: http://oceancolor.gsfc.nasa.gov, 4 km resolution data)

Figure 11.9 shows an SST map (Fig. 11.9a) and a CHL map (Fig. 11.9b) derived from MODIS data acquired at 1425 UTC on 25 December 2002. It shows the SST and CHL signatures of a small-scale eddy (marked by an arrow) attached to a filament of cold water and enhanced CHL concentration, which stretches from the upwelling region at the coast out into the North Atlantic. Figure 11.10 shows two SST maps retrieved from MODIS data acquired at 1440 UTC on 14 January 2011 (Fig. 11.10a) and at 1430 UTC on 16 January 2011 (Fig. 11.10b), respectively. Like in the SST map depicted in Fig. 11.9a, both maps show SST signatures of a small scale-eddy (marked

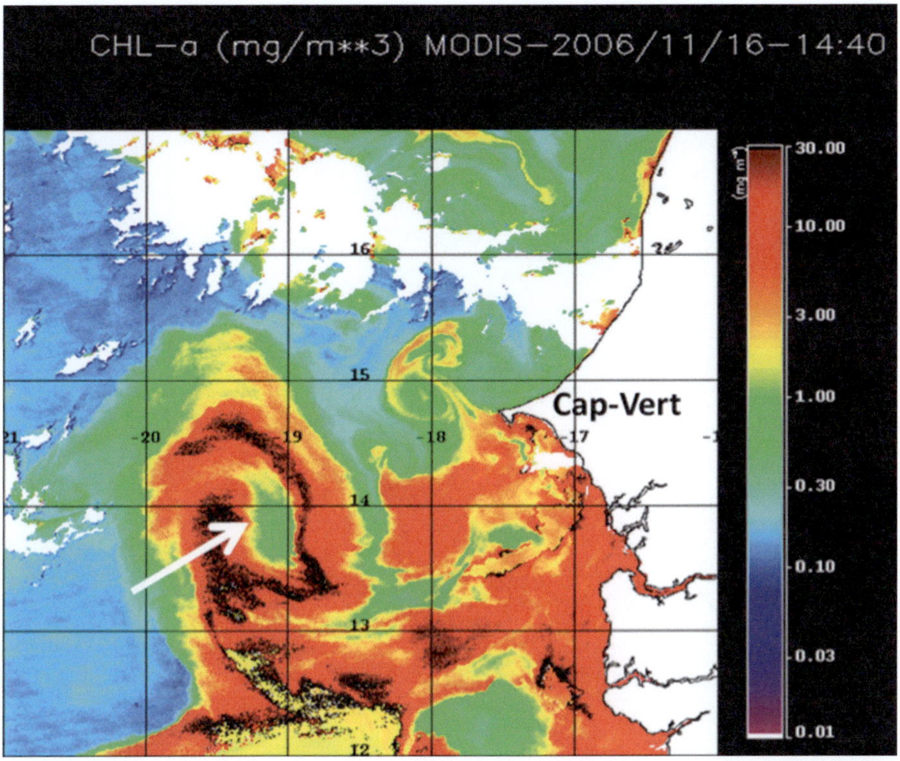

Fig. 11.7 MODIS Terra image acquired at 1440 UTC on 16 November 2006 showing an anticyclonic mesoscale eddy off the coast of Senegal/Guinea (marked by an *arrow*) with diameter of approximately 150 km. The eddy becomes visible in brownish colors by enhanced phytoplankton concentration. Note that the highest concentration is encountered at the rim of the eddy

by black arrows) attached to a filament of cold water stretching from the upwelling region out into the Atlantic. The two SST maps also show also another phenomenon: the birth of an anticylonic eddy north of Cap-Vert. Because of its location north of Cap-Vert, this headland cannot be involved in its generation process. These SST maps suggest that this small-scale eddy developed out of a bulge in the front of cold upwelled water at the coast.

11.5 A Case Study

In this section we present results from a case study of a small-scale eddy which originated in the upwelling area off the coast of Senegal in West Africa and then propagated westward into the North Atlantic (Alpers et al. 2013). The coast of West Africa between 12°N and 25°N is a well-known upwelling area, located in

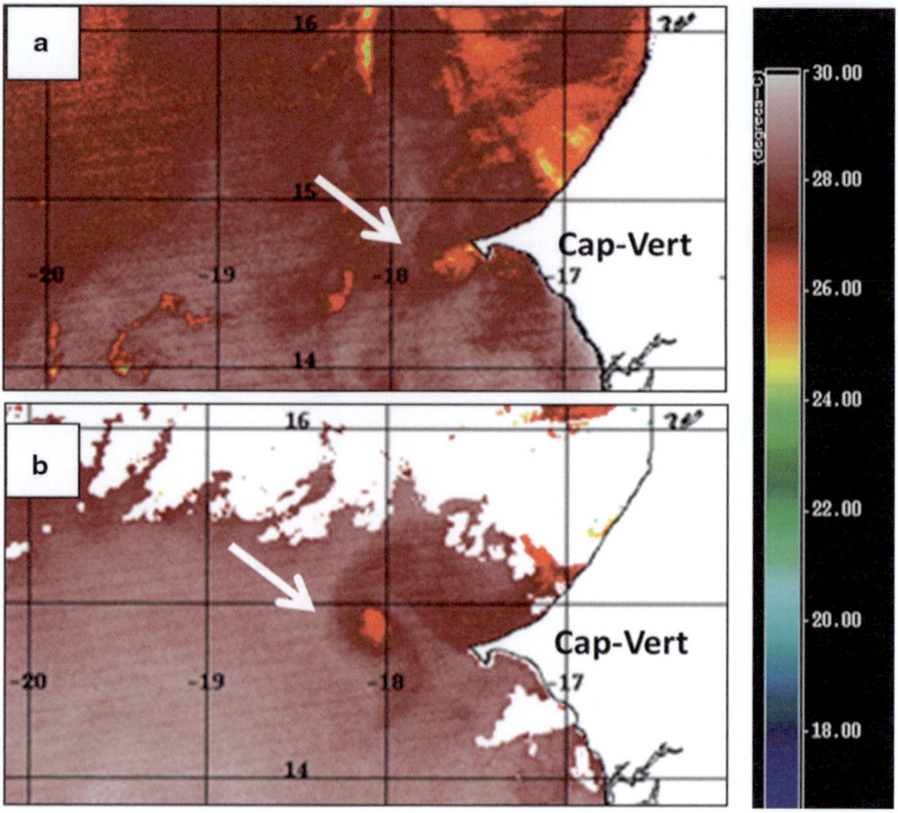

Fig. 11.8 **a** SST map in degrees Celsius retrieved from MODIS Terra data acquired at 1440 UTC on 7 November 2006. It shows SST signature of a small-scale eddy at Cap-Vert, Senegal (marked by an *arrow*) and of several very small-scale eddies further west. **b** Same as (a), but at 1440 UTC on 16 November 2006. During the 9 days between the two data takes the small-scale eddy has moved northwestward and the very small-scale eddies have disappeared. The *white* areas are land or clouds where no SST can be retrieved

the large marine system influenced by the Canary current and driven by the Trade winds. Between 20°N and 25°N, upwelling is a permanent phenomenon, but between 11°N and 20°N it occurs only in winter and spring (Demarcq 1998). Since the eddy investigated in this section originated from the coastal waters near Cap-Vert, we suspect that the headland of Cap-Vert played a key role in its generation. It is well known from other parts of the World's ocean that headlands are birthplaces of eddies (Pattiaratchi et al. 1987; Signell and Geyer 1991; Davies et al. 1995). Also Di Giacomo and Holt (2001) noted that most of the small-scale eddies in the Southern California Bight are observed in close proximity of islands and headlands, which suggests that they are topographically generated. Since eddy generation at Cap-Vert does not occur on a regular basis, it must have been caused by a sudden change in environmental conditions. One would expect that the most likely cause was a wind

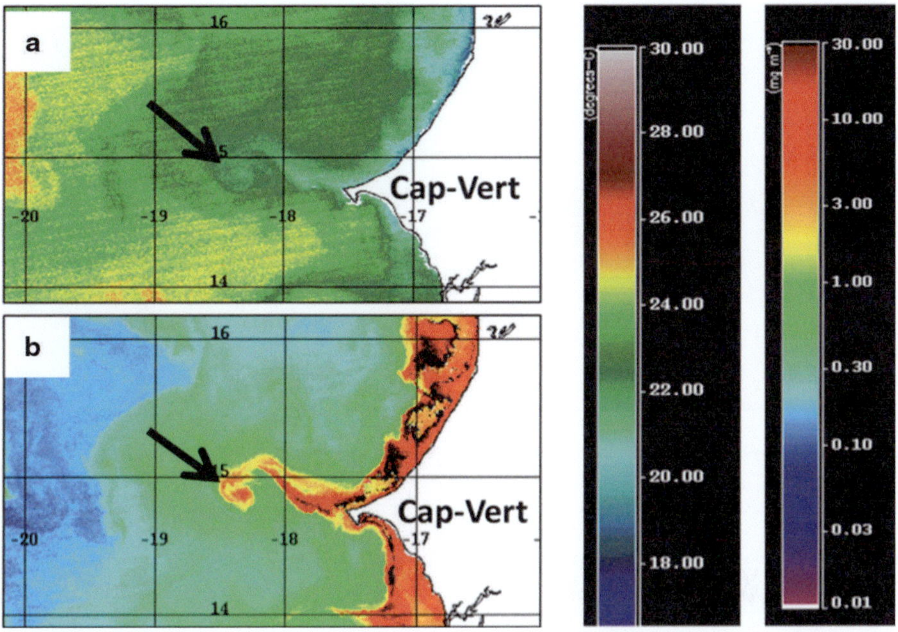

Fig. 11.9 **a** SST map in degrees Celsius retrieved from MODIS data acquired at 1425 UTC on 25 December 2002. **b** CHL distribution in mg/m^{-3} retrieved from MODIS data acquired at the same time. It shows the SST and CHL signature of a small-scale eddy (marked by *arrows*) attached to a filament of cold water and enhanced CHL concentration protruding jet-like from Cap-Vert

burst directed southward along the coast which caused an increase of the surface flow. This was indeed the case as shown in the next subsection.

11.5.1 Wind History

The wind field time history leading to the generation of the eddy is depicted in Fig. 11.11. The maps show near-sea surface wind fields provided by the National Climatic Data Center (NCDC) of the National Oceanic and Atmospheric Administration (NOAA), which generates them from a blend of satellite data, including data from the Advanced Scatterometer (ASCAT) onboard the European MetOp satellite. The four wind maps, show the time evolution of the wind field which caused the generation of the eddy. Figure 11.11a shows the wind field on 21 October 2011 before the onset of the wind burst and Fig. 11.11b–d the wind fields on 22, 23 and 24 October 2011, respectively. On 23 October (Fig. 11.11a) the maximum wind speed was 11 ms^{-1}, and on 24 October, when the wind burst had reached Cap-Vert, the wind speed had dropped to 9 ms^{-1} (Fig. 11.11d).

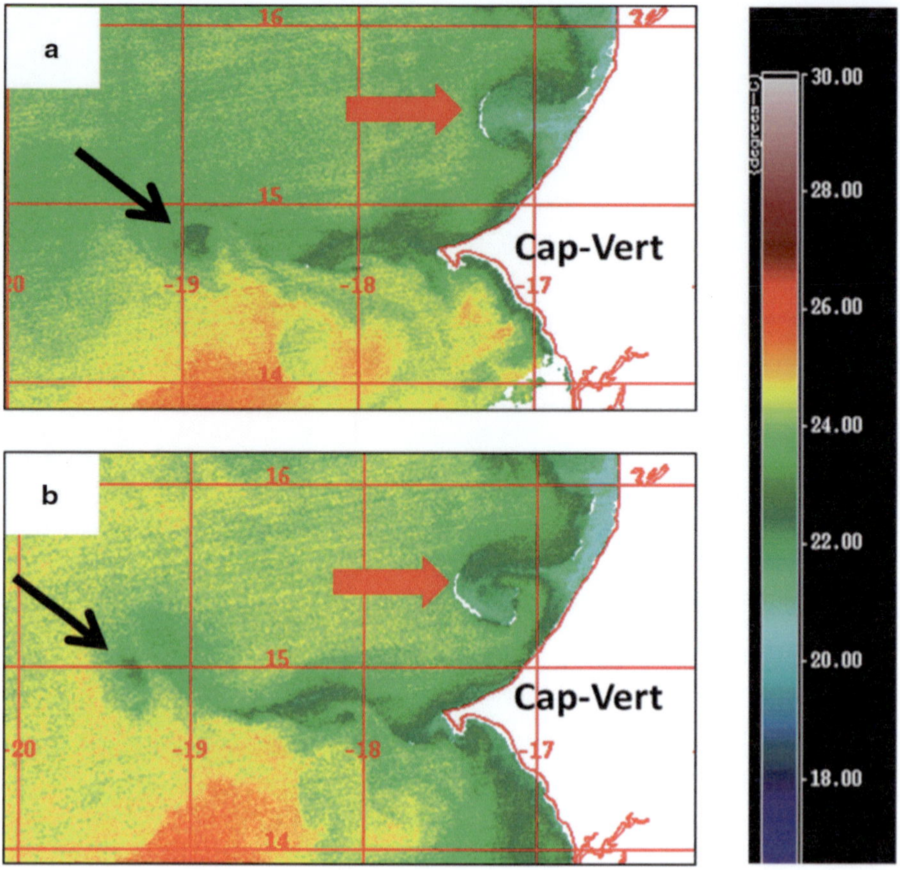

Fig. 11.10 Same as Fig. 11.8 (SST maps), but (a) at 1440 UTC on 14 January 2011 and (b) at 1430 UTC on 16 January 2011. Both maps show SST signatures of a small scale-eddy (marked by *black arrows*) attached to a filament of cold water stretching out into the Atlantic. It shows the development of another small-scale cyclonic eddy north of Cap-Vert (marked by a *red arrow*) the color coding denotes the wind speed in m/s

11.5.2 Ocean Circulation

In order to investigate the ocean circulation forced by the wind burst, we have used model calculations performed with the "Mercator Global Operational System PSY2V4R2" (Lellouche et al. 2013). This model assimilates sea level anomalies, SST, and temperature/salinity profiles. It is forced by wind and surface heat and freshwater fluxes provided by the European Centre for Medium-Range Weather Forecasts (ECMWF). It has a horizontal resolution of 1/12° (9 km at the equator and 3 km at 70 °N) and thus is well suited to study the generation and propagation of mesoscale eddies of scales above 50 km, but it cannot simulate small-scale eddy dynamics.

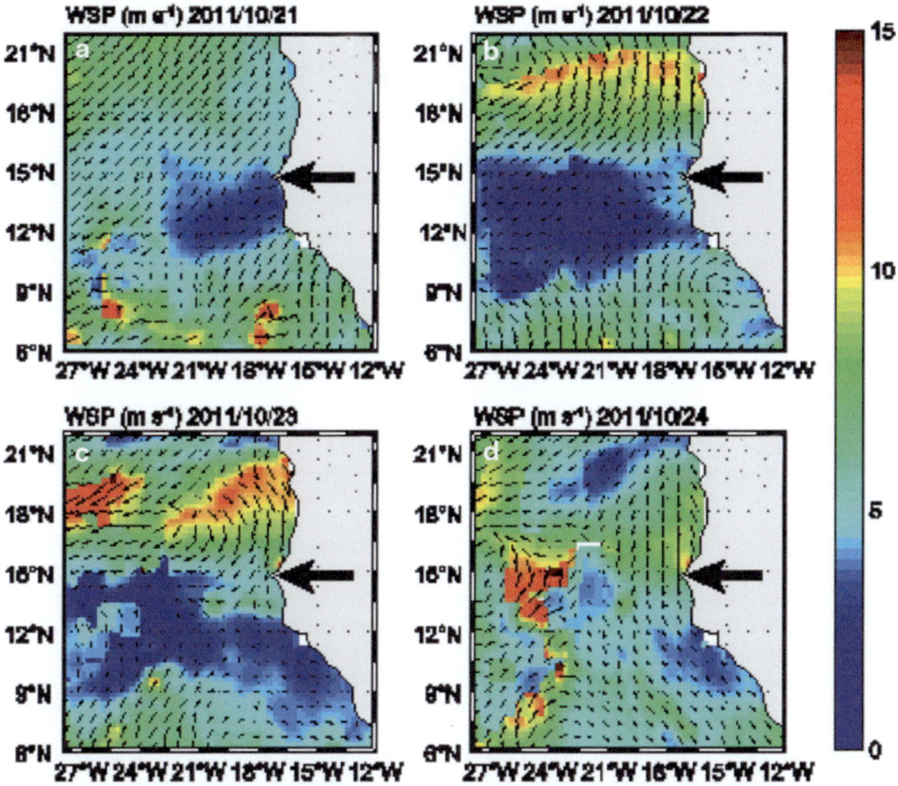

Fig. 11.11 Near sea surface wind fields provided by the National Climatic Data Center of NOAA generated from a blend of satellite data for 00 UTC on (**a**) 21 October, (**b**) 22 October, (**c**) 23 October, and (**d**) 24 October 2011. **a** shows a wind field typical for this time of the year and **b**, **c**, and **d** show the time evolution of the wind burst advancing southward along the coast of West Africa toward Cap-Vert (marked by an *arrow*)

However, here we use the Mercator model only to shed light on the onset of a southward wind-induced coastal current which we suspect to be responsible for the generation of the small-scale eddy at the headland of Cap-Vert. In Fig. 11.12a–c the surface current fields (represented by arrows) superimposed on the SST field (represented by colors) simulated by the Mercator model for 00 UTC on 23, 27, and 31 October 2011, respectively, are depicted. On 23 October (Fig. 11.12a), the flow at the approaches to Cap-Vert is dominated by northward flow associated with high SST values within the region. Four days later (Fig. 11.12b), a strong southward directed surface current, forced by the wind burst, dominated the coastal region north of Cap-Vert. Accordingly, upwelling occurred which is clearly visible on the SST map depicted in Fig. 11.12b as a coastal band of reduced SST. Note that maximum surface velocities are centered next to Cap-Vert, where also the SST is reduced. The map depicted in Fig. 11.12c shows that cold water has intruded the area southwest

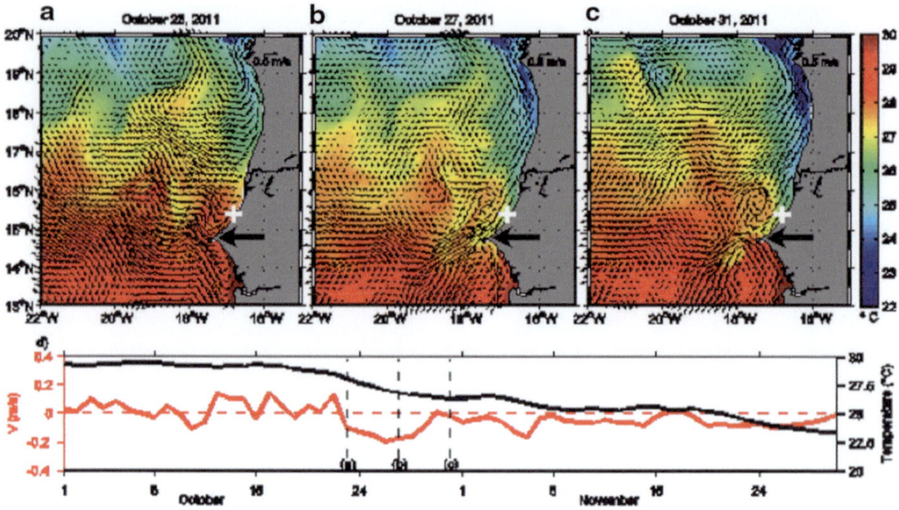

Fig. 11.12 Upper panels **a**, **b**, and **c**: simulated sea surface current field (*black arrows*) superimposed on the simulated SST field (*colors*) calculated by the Mercatormodel for 00 UTC on (**a**) 23 October, (**b**) 27 October, and (**c**) 31 October 2011, respectively. The maps show the surface current fields before and after the onset of the enhanced surface flow caused by the wind burst. Panel **c** shows that cold water has intruded the area southwest of Cap-Vert (marked by a *thick black arrow*) and that south of Cap-Vertwarmwater is flowing northward along the coast. This cyclonic circulation pattern suggests that flow separation has taken place. Lower panel: **d** time series of the sea surface temperature (*black curve*) and the along-shore surface current (*red curve*) at the grid point 15.39°N, 16.83°W, which is the grid point next to the coast slightly north of Cap-Vert (marked by a *white plus sign* in panels **a**, **b**, and **c**), simulated by the Mercator model. It shows the development of a strong southwestward near coastal current starting on 23 October 2011. The three dashed vertical lines mark the times at which the simulations shown in panels a, b. and c were performed

of Cap-Vert and that south of Cap-Vert warm water is flowing northward along the coast. This flow forms, together with the southward flow further west, a cyclonic flow pattern. Such a recirculation of water behind an obstacle is typical for eddy generation by flow separation.

The onset of the along-shore flow on 23 October can also be seen on the plot depicted in Fig. 11.12d, which shows how the simulated SST (black curve) and along-shore surface current (red curve) varied with time. The plot applies for the grid point 15.39°N, 16.83°W (marked by a white cross), which is the grid point next to the coast, slightly north of Cap-Vert (see Fig. 11.12). The three dashed vertical lines inserted in Fig. 11.12d mark the times at which the simulations shown in Fig. 11.12a–c were performed. The plot shows the onset of a strong near coastal current on 23 October (marked by the first dashed line from the left) and a drop in SST due to upwelling and southward advection of cold water masses. We suspect that flow separation at Cap-Vert, as visible in the simulated flow field depicted in Fig. 11.12b, represents the initial stage of the highly nonlinear eddy generation process. However, the model is not capable to reproduce the cyclogenesis in detail and cannot describe the generation of highly nonlinear eddies. But the model is capable of describing

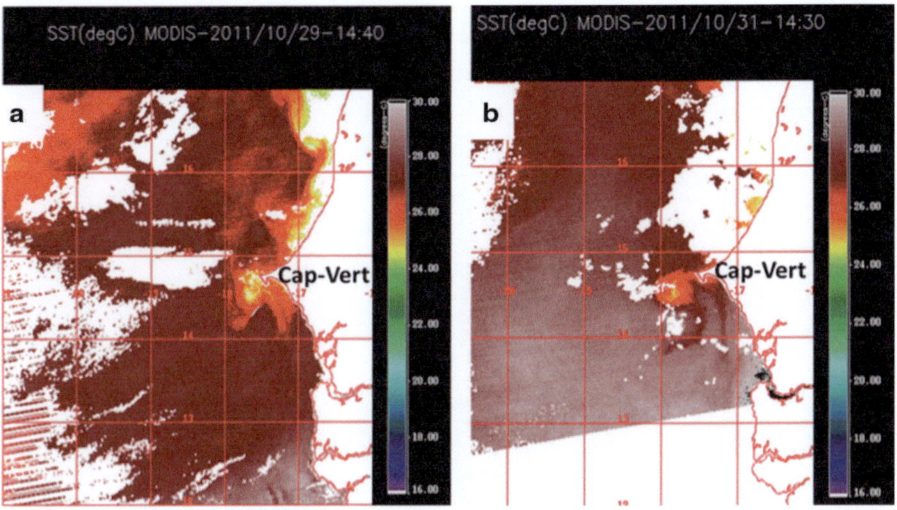

Fig. 11.13 (**a**) SST maps retrieved from MODIS data acquired (**a**) at 1440 UTC on 29 October and (**b**) at 1430 UTC on 31 October 2011. The *white* areas are land (on the *right*) or clouds where no SST could be retrieved

the upwelling induced by the northerly wind burst and the environment in which the small-scale eddy was generated.

11.5.3 Sea Surface Temperature and Chlorophyll-a Maps

Figures 11.13 and 11.14 show the time evolution of the small-scale eddy between 29 October and 7 November 2011 in the form of SST and CHL maps retrieved from Aqua MODIS data. The SST maps show sea surface signatures of the eddy, while the CHL maps show the CHL distribution in the upper layer of the ocean. Figure 11.13 shows two SST maps retrieved from MODIS data acquired (a) at 1440 UTC on 29 October 2011 and (b) at 1430 UTC on 31 October 2011 when the eddy was located close to Cap-Vert. The white areas are land (on the right) or areas with cloud coverage where no SST could be retrieved.

The maps depicted in Fig. 11.14 show that the cyclonic eddy drifted first westward into the open Atlantic. The SST maps depicted in the lower left panel of Fig. 11.14 show spiral arms emanating from the core of the eddy which are warmer than the core of the eddy, but still about 1 –1.5 °C cooler than the surrounding waters. When comparing the SST and CHL maps valid for the same times in Fig. 11.14, we see that the centers of the patches with strongly reduced SST (about 2 °C colder than the surrounding waters) and with strongly enhanced CHL concentration (about 3 mg m^{-3} higher) coincide.

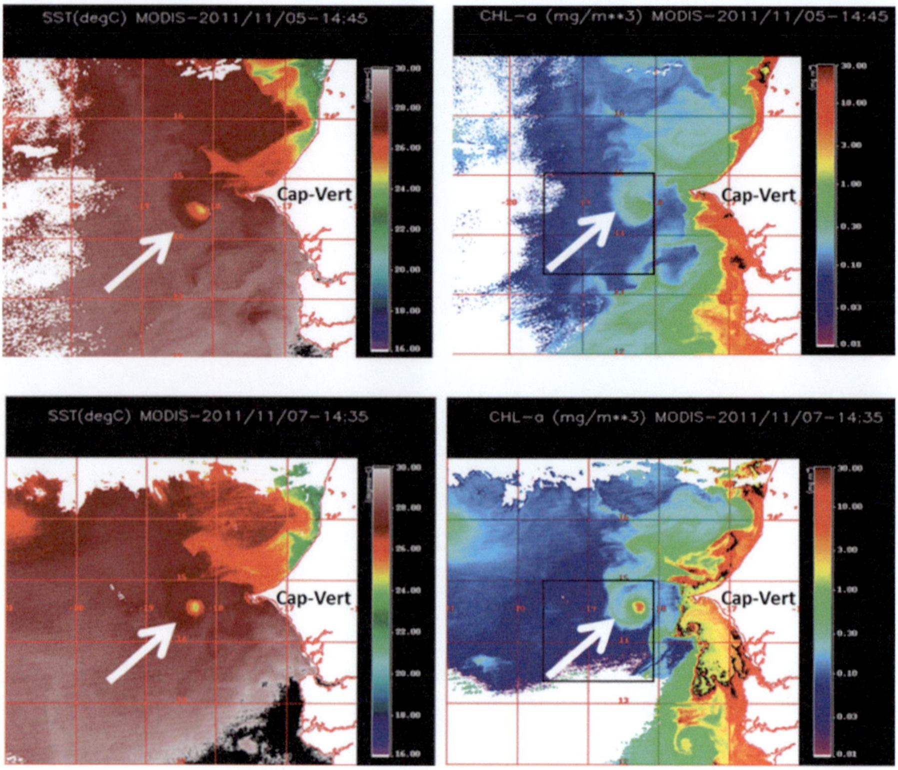

Fig. 11.14 Maps of SST in degrees Celsius and the CHL concentration in mg m^{-3} retrieved from MODIS data. Upper maps: at 1445 UTC on 5 November 2011. Lower maps: at 1435 UTC on 7 November 2011. The arrows point to the SST and CHL signatures of the small-scale eddy

While the diameter of the core of the eddy in the SST maps is estimated to vary between 15 and 30 km, the diameter of the eddy in the CHL maps is more difficult to estimate. In these maps the patch of strongly enhanced CHL distribution is surrounded by a broad band with medium enhanced CHL (about 1 mg m^{-3}). The SST maps depicted in Figs. 11.13 and 11.14, which were derived from MODIS data acquired during daytime, show that the maximum reduction of the SST in the core of the eddy relative to the surrounding waters is about 2.5 °C. However, this does not necessarily correspond to the maximum temperature difference between the upwelled water in the eddy core and the surrounding water. Typically, the SST measured during night time over the eddy is lower than the SST measured during day time and hence more representative for the true temperature of the upwelled water. Wang and Tang (2010) have studied this phenomenon and argued that, during daytime, absorption of solar radiation is enhanced by the presence of phytoplankton, which leads to higher daytime SST over phytoplankton bloom areas. These authors have estimated that the difference between daytime and nighttime SST depends on the CHL concentration and is of the order of 1°C for a CHL concentration of

3 mg m^{-}3. Our data confirm the presence of this diurnal SST variability in the Cap-Vert upwelling region.

11.5.4 Synthetic Aperture Radar Image

Figure 11.15 shows a SAR image which was acquired in the Wide Swath Mode (WSM) by the Advanced SAR (ASAR) onboard the European Envisat satellite at 2317 UTC on 31 October over the waters close to Cap-Vert. Similarly as on the SAR image shown in Fig. 11.5, also here the eddy becomes visible by reduced image brightness caused by reduction of the radar backscattering due to the presence of biogenic surface films. Since it is a cyclonic eddy causing upwelling of cold nutrient-rich water, very likely it contains much biota which secretes surface active material such that biogenic surface films were formed.

11.6 Discussion and Summary

In this paper we have presented several images collected from Space Shuttle, the Terra and Acqua satellites, and the Envisat satellite over the seas surrounding Africa showing sea surface signatures of mesoscale and small-scale eddies. These images were acquired in the visible, infrared and microwave (5.3 GHz) bands. The instrument most used from space to study the dynamics of mesoscale eddies is the radar altimeter. It is a microwave instrument, which yields data (almost) independent of cloud coverage and independent of the time of the day. However, it cannot resolve eddies with dimensions much smaller than100 km. Since the study of meso-scale oceanic phenomena using altimeter data is the subject of another chapter of this book, we have refrained here from discussing in detail the use of altimeter data in investigating eddies. Here we just want to mention a paper by Chaigneau et al. (2008), in which mesoscale eddy activity in the Canary upwelling area (10–45 °N; 40–50 °W) is investigated using 15 years of satellite altimetry data from several satellites. Restricting the analysis to long-lived eddies having sea surface height anomalies larger than 2 cm and lifetimes larger than 35 days, they found that, on the average, around 60–100 eddies with diameters of 140–320 km are present on weekly maps and that 4–7 eddies are generated each week in the Canary upwelling area.

Small-scale eddies can only be observed from space only by high resolution infrared/optical sensors and SAR. Both types of instruments have their limitations. Infrared and optical sensors are limited by cloud coverage, and spaceborne SARs by their poor coverage and the weak radar signatures of eddies (exception: when sea surface is covered partially by surface films and the film is entrained in the surface current field of the eddy).

As stated before, mesoscale and submesoscale eddies play an important role in ocean dynamics. In order to get a better understanding of this dynamics, a systemic

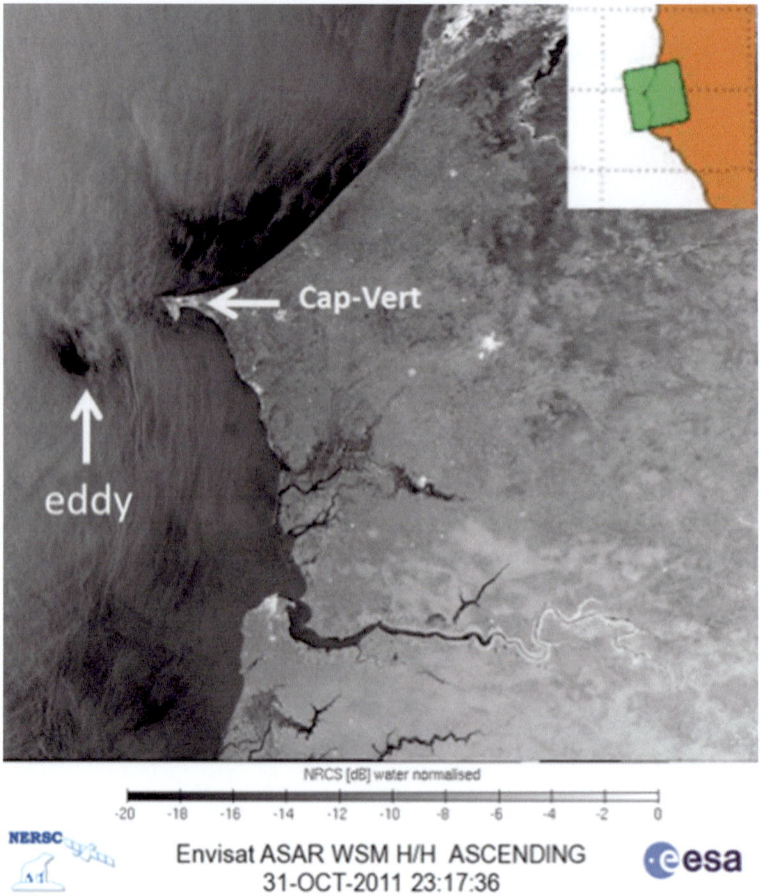

Fig. 11.15 SAR image acquired by the ASAR onboard the Envisat satellite in the Wide Swath Mode (WSM) at 2317 UTC on 31 October 2011 during an ascending satellite pass. The *solid white arrow* points to the radar signature of the small-scale eddy generated at Cap-Vert. The inset shows the position of the imaged scene of dimension 400 km × 400 km

monitoring of both types eddies is desirable. While mesoscale eddies can be measured on a global scale by altimeters flying presently on several satellites, small-scale eddies can, at present, not be monitored neither on a global nor on a regional scale. All present all studies of small-scale eddies are therefore necessarily case studies. Here we have presented some examples of satellite images showing sea surface signatures of eddies around Africa, which we found at random in archives of the space organizations. The hope is that with the future instrument Surface Water and Ocean Topography[1] (SWOT) also small-scale eddies can be measured on global scale independently of cloud coverage and time of the day (Fu and Ferrari 2008).

[1] See also http://swot.jpl.nasa.gov/

This instrument is scheduled to be launched in 2020. It is an altimeter having a wide swath which uses the SAR principle. Thus SWOT would allow in the future to monitor small-scale eddies on a global scale. This would greatly improve our understanding of the dynamics of the ocean and also would be of great benefit to fisheries.

Acknowledgments We thank NASA and ESA for providing the data and Sergey Stanichny for calling our attention to the NASA source of the MODIS reflectance map at 555 nm shown in Fig. 11.6. This study was supported by BMBF-Ib and AIRD grants obtained to build the Trilateral German–French–African Environmental research initiatives in Sub-Sahara Africa entitled AWA "Ecosystem Approach to the management of fisheries and the marine environment in West African waters".

References

Alpers W, Brandt P, Lazar A, Dagorne D, Sow B, Faye S, Hansen MW, Rubino A, Poulain PM, Brehmer P (2013) A small-scale oceanic eddy off the coast of West Africa studied by multi-sensor satellite and surface drifter data. Remote Sens Environ 129:132–143. doi: 10.1016/j.rse.2012.10.032

Alpers W, Espedal H (2004) Oils and Surfactants. In: Jackson CR, Apel JR (eds) Synthetic aperture radar marine user's manual, chapter 11. NOAA/NESDIS, Washington, DC, pp 263–276. http://www.sarusersmanual.com/ManualPDF/NOAASARManual_CH11_pg263-276.pdf

Alpers W, Hennings I (1984) A theory of the imaging mechanism of underwater bottom topography by real and synthetic aperture radar. J Geophys Res 89:10529–10546

Capet X, McWilliams JC, Molemaker MJ, Shchepetkin AF (2008) Mesoscale to submesoscale transition in the California Current System: I. Flow structure, eddy flux, and observational tests. J Phys Oceanogr 38:2256–2269

Chaigneau A, Eldin G, Dewitte B (2008) Eddy activity in the four major upwelling systems from satellite altimetry (1992–2007). Prog Oceanogr 83:117–123

Chelton DB, de Szoeke RA, Schlax MG, Naggar KE, Siwertz N (1998) Geographical variability of the first-baroclinic Rossby radius of deformation. J Phys Oceanogr 28:433–460

Chelton DB, Schlax MG, Samelson RM (2011) Global observations of nonlinear mesoscale eddies. Prog Oceanogr 91:167–216

Cox CC, Munk WH (1954) Measurement of the roughness of the sea surface from photography of the Sun's glitter. J Opt Soc Am 44(11):838–850

Cresswell GR, Legeckis R (1986) Eddies off southeastern Australia. Deep-Sea Res 33:1527–1562

Davies PA, Dakin JM, Falconer RA (1995) Eddy formation behind a coastal headland. J Coastal Res 11:154–167

Demarcq H (1998) Spatial and temporal dynamics of the upwelling off Senegal and Mauritania: local change and trend. In: Durand MH, Cury P, Mendelssohn R, Roy C, Bakun A, Pauly D (eds) Global versus local changes in upwelling systems: a report from the CEOS Workshop, Monterey, California, September 1994. ORSTOM Editions, Paris, pp 149–166. http://horizon.documentation.ird.fr/exl-doc/pleins_textes/pleins_textes_7/divers2/010015307.pdf

DiGiacomo PM, Holt B (2001) Satellite observations of small coastal ocean eddies in the Southern California Bight. J Geophys Res 106(C10):22521–22543

Eldevik T, Dysthe KB (2002) Spiral eddies. J Phys Oceanogr 32:851–869. doi: http://dx.doi.org/10.1175/1520-0485

Espedal H, Johannessen OM, Johannessen JA, Dano E, Lyzenga D, Knulst JC (1998) COAST-WATCH '95: a tandem ERS-1/2 SAR detection experiment of natural film on the ocean surface. J Geophys Res 103:24969–24982
Falkowski PG, Ziemann DZ, Kolbera Z, Bienfang PK (1991) Role of eddy of pumping in enhancing primary production in the ocean. Nature 352:55–58, doi:10.1038/352055a0
Fu LL, Ferrari R (2008) Observing oceanic submesoscale processes from space. Eos 89(48):488–489
Fu LL, Holt B (1983) Some examples of detection of oceanic mesoscale eddies by the Seasat synthetic aperture radar. J Geophys Res 88:1844–1852
Golitsyn GS (2012) On the nature of spiral eddies on the surface of seas and oceans. Izvestiya AN. Fizika Atmosfery i Okeana 48:391–395
Gower JFR, Denman KL, Holyer RL (1980) Phytoplankton patchiness indicates the fluctuations spectrum of mesoscale oceanic structure. Nature 288:157–159
Huehnerfuss H, Alpers W, Dannhauer H, Gade M, Lange PA, Neumann V, Wismann V (1996) Natural and man-made sea slicks in the North Sea, investigated by a helicopter-borne 5-frequency radar scatterometer. Int J Rem Sens 17:1567–1582
Ivanov AY, Ginzburg AI (2002) Oceanic eddies in synthetic aperture radar images. J Earth Syst Sci 111(3):281–295
Jackson CR, Alpers W (2010) The role of the critical angle in brightness reversals on sunglint images of the sea surface, J Geophys Res 115:C09019. doi:10.1029/2009JC006037
Johannessen JA, Roed LP, Wahl T (1993) Eddies detected in ERS-1 SAR images and simulated in reduced gravity model. Int J Rem Sens 14:2203–2213
Johannessen JA, Shuchman RA, Digranes G, Lyzenga D, Wackerman C, Johannessen OM, Vachon PW (1996) Coastal ocean fronts and eddies imaged with ERS-1 synthetic aperture radar. J Geophys Res 101:6651–6667
Karimova S (2012) Spiral eddies in the Baltic, Black and Caspian seas as seen by satellite radar data. Adv Space Res 50(8):1107–1124. http://dx.doi.org/10.1016/j.asr.2011.10.027
Kudryavtsev V, Akimov D, Johannessen JA, Chapron B (2005) On radar imaging of current features, part 1: model and comparison with observations. J Geophys Res 110:C07016
Kusakabe M, Andreev A, Lobanov V, Zhabin I, Kumamoto Y, Murata A (2002) Effects of the anticyclonic eddies on water masses, chemical parameters and chlorophyll distributions in the Oyashio current region. J Oceanogr 58:691–701
Le Galloudec O, Bourdalle-Badie R, Drillet Y, Derval YC, Bricaud C (2008) Simulation of mesoscale eddies in the Mercator global ocean high resolution model. Mercator Newsl 3
Lellouche JM et al. (2013) Evaluation of global monitoring and forecasting systems at Mercator Océan, Ocean Sci., 9, 57-81, doi:10.5194/os-9-57-2013
Levy M, Klein P, Treguier AM (2001) Impact of sub-mesoscale physics on production and subduction of phytoplankton in an oligotrophic regime. J Mar Res 59:535–565
Lin II, Lien CC, Wu CR, George TF, Wong GTF, Huang CW, Chiang TL (2010) Enhanced primary production in the oligotrophic South China Sea by eddy injection. Geophys Res Lett 37:L16602. doi:10.1029/2010GL043872
Maltrud ME, McClean JL (2005) An eddy resolving global 1/10° ocean simulation. Ocean Model 8(1–2):31–54
McWilliams JC (1985) Submesoscale, coherent vortices in the ocean. Rev Geophys 23:165–182
Morrow R, Fang F, Fieux M, Molcard R (2003) Anatomy of three warm-core Leeuwin current eddies. Deep-Sea Res Pt II 50:2229–2243
Munk W, Armi I, Fischer K, Zachariasen F (2000) Spirals on the sea. Proc R Soc Lon Ser-A 456:1217–1280
Olson DB (1991) Rings in the ocean. Annu Rev Earth Planet Sci 19:283–311
Pattiaratchi C, James A, Collins M (1987) Island wakes and headland eddies: a comparison between remotely sensed data and laboratory experiments. J Geophys Res 92:783–794

Scully-Power P (1986) Navy oceanographer shuttle observations, STS 41-G, mission report. Naval Underwater Systems Center Tech. Rep. NUSC TD 7611
Siegel A, Weiss JB, Toomre J, McWilliams JC, Berloff PS, Yavneh I (2001) Eddies and vortices in ocean basin dynamics. Geophys Res Lett 28:3183–3186
Signell RP, Geyer WR (1991) Transient eddy formation around headlands. J Geophys Res 96:2561–2575
Siokou-Frangou J et al (2010) Plankton in the open Mediterranean Sea: a review. Biogeosciences 7:1543–1586
Soules SD (1970) Sun glitter viewed from space. Deep-Sea Res 17:191–195
Stevenson RE (1998) Spiral eddies: the discovery that changed the face of the oceans. 21st Cent Sci Technol 11:58–71
Thomas, LN, Tandon A, Mahadevan A (2008) Submesoscale processes and dynamics. In Hecht M. and Hasumi H., editors, Ocean Modeling in an Eddying Regime, (AGU Monograph), American Geophysical Union, Washington DC, pages 17–38, 2008
Valenzuela GR (1978) Theories for the interaction of electromagnetic and ocean waves—a review. Bound Lay Meteorol 13:61–85
Wang S, Tang D (2010) Remote sensing of day/night sea surface temperature difference related to phytoplankton bloom. Int J Rem Sens 31:4569–4578
Williams RG (2011) Ocean eddies and plankton blooms. Nat Geosci 4:739–740
Yamaguchi S, Kawamura H (2009) SAR-imaged spiral eddies in Mutsu Bay and their dynamic and kinematic models. J Phys Oceanogr 65(4):525–539

Part III
Eastern Africa Near-Coastal Waters

Chapter 12
Observing the Agulhas Current With Sea Surface Temperature and Altimetry Data: Challenges and Perspectives

Marjolaine Krug, Paolo Cipollini and Francois Dufois

Abstract The Agulhas Current is a challenging region for satellite remote sensing observations. Strong evaporation rates above the current core and the Retroflection reduce the number of cloud-free observations from Infra-Red sensors, while microwave radiometers and altimeters measurements suffer from the proximity of the current to the coast in the northern region. Infra-Red observations of the Agulhas Current significantly improved with the launch of the Meteosat Second Generation satellite, but Infra-Red Sea Surface Temperature datasets still suffer from inadequate cloud masking algorithms, particularly in regions of strong temperature gradient. Despite both Sea Surface Height and Sea Surface Temperature observations being severely compromised in the northern Agulhas current, a synergetic use of merged altimetry and high frequency Infra-Red Sea Surface Temperature imagery provides a means to track deep-sea eddies, document their influence on the Agulhas Current and helps us improve our understanding of the Agulhas Current variability.

12.1 Introduction

The Agulhas forms the western boundary current of the South Indian subtropical gyre. It originates near the South-African/Mozambican border, at about 27°S and flows poleward to 40°S, transporting large volumes of sub-tropical water towards the southern ocean. The Agulhas Current can be divided into three regions (Fig. 12.1): the

M. Krug (✉)
Ecosystem Earth Observations, Council for Scientific and Industrial Research,
Cape Town, South Africa
e-mail: mkrug@csir.co.za
Also at: Nansen-Tutu Center for Marine Environmental Research, Oceanography Department,
University of Cape Town, South Africa

P. Cipollini
Marine Physics and Ocean Climate, National Oceanography Centre, Southampton, UK

F. Dufois
Oceanography Department, Mare Institute, University of Cape Town,
Cape Town, South Africa

CSIRO Marine and Atmospheric Research, Floreat, Australia

V. Barale, M. Gade (eds.), *Remote Sensing of the African Seas*,
DOI 10.1007/978-94-017-8008-7_12,

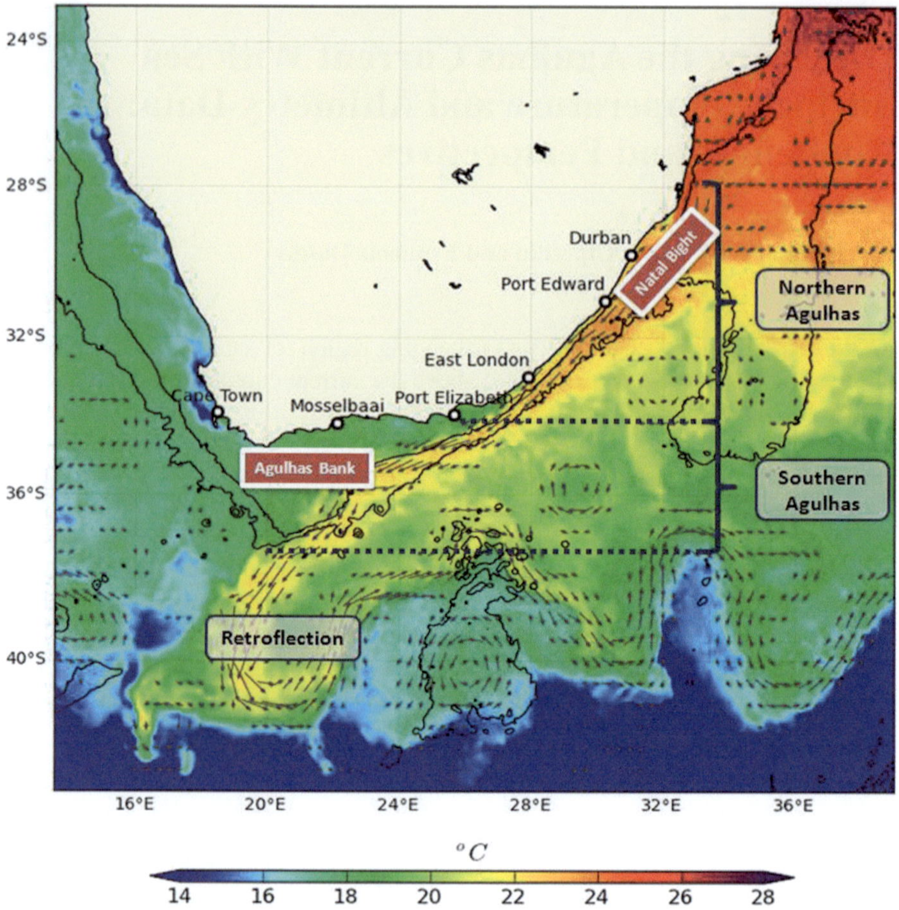

Fig. 12.1 The Agulhas current region. ODYSSEA SST with overlaid AVISO absolute geostrophic currents on 21 November 2010

northern Agulhas Current located between 26°S and 34°S where the current flows in close proximity to the coast, the southern Agulhas Current (east of the Agulhas Bank) where the Agulhas Current detaches from a wider continental slope, and the Agulhas Retroflection, between 36°S and 40°S, where the current undergoes a sudden change of direction (Lutjeharms 2006). Like other western boundary currents, the Agulhas Current is an intense and narrow flow characterised by strong velocity gradients and a central warm core, with isopycnal lines sloping steeply towards the coast (Goschen and Schumann 1990; Casal et al. 2009; Bryden et al. 2005). Thermal and sea level surface signatures associated with the Agulhas Current enable the mapping of the Agulhas Current's path from space, using observations of Sea Surface Height (SSH) from altimeters and Sea Surface Temperature (SST) from radiometers (see e.g. Fig. 12.1).

The Agulhas Current is the strongest western boundary current of the southern hemisphere and a major component of the global climate (Beal et al. 2011). On a regional scale, the Agulhas Current directly influences the oceanography of the continental shelves through a range of meso and sub-meso scale processes such as the shedding of rings, eddies or filaments, as well as the meandering or intrusion of the current onto the shelf (Lutjeharms 2006). Despite its importance at both the regional and global scales, the Agulhas Current remains poorly sampled. Only recently was an annual cycle observed in the Agulhas Current (Krug and Tournadre, 2012) and the long-term variability of the Agulhas Current remains largely unknown (Lutjeharms 2006).

Satellite remote sensing provides a powerful mean of observing the Agulhas Current over the time-scale required to resolve its seasonal and inter-annual variability. Over the last two decades, SST and SSH observations from space have been widely used by oceanographers to monitor change. Altimetry and Infra-Red (IR) measurements from space are mature techniques which provide extended observational records (with close to 20 and 30 years of measurements available for SSH and SST, respectively). Altimetry, through the use of the geostrophic approximation, has been used with great success to estimate the upper transport and magnitude of the world's major ocean currents, track eddies or even discover new currents (Siedler et al. 2006). SST observations through front-detection techniques have been used to estimate surface current speeds (Emery 2001) and monitor meso-scale activity in western boundary current regions (Rouault and Penven 2011).

The Agulhas Current system is a challenging environment for remote sensing observations as it is affected by a wide range of processes which originate near the coast or the open ocean and occur over synoptic to inter-annual time-scales. Over the Indian Ocean, large scale processes such as changes in the wind circulation, westward propagating Rossby waves or offshore eddies originating from the Mozambique or Madagascar regions constitute potential drivers of variability for the Agulhas Current. Variability intrinsic to the Agulhas Current also occurs on a wide range of scales from small filaments, shear-edge eddies or meanders to the shedding of large Agulhas Rings at the Retroflection (Lutjeharms 2006). Cloud formation over the Agulhas Current severely restricts observations from IR SST sensors. Altimeters or microwave SST sensors which can see through clouds, suffer from the proximity of the current to the coast over large regions of the Agulhas Current due to factors such as land contamination or atmospheric errors (Vignudelli et al. 2011).

In this paper we highlight some of the challenges of SST and altimetry observations in the Agulhas Current region and illustrate how the two measurements techniques can be combined to improve our understanding of the Agulhas Current dynamics. A presentation of the datasets used is given in Sect. 2. Section 3 discusses some of the limitations and advantages of IR and microwave observations in the Agulhas Current region while Sect. 4 provides an example of synergetic use of altimetry and SST to monitor the Agulhas Current. The conclusions to this Chapter are presented in Sect. 5 together with some perspectives for future observation of the Agulhas Current system from space.

12.2 Data

Three IR SST datasets are compared: the Pathfinder version 5.2 (hereafter referred to as Pathfinder v5.2), the Moderate Resolution Imaging Spectometer (MODIS) onboard the US National Aeronautics and Space Administration (NASA) satellite Terra and the Ocean and Sea Ice Satellite Application Facility[1] (OSI-SAF) experimental hourly SST products.

The Ocean Pathfinder SST project consists of a reprocessing of all Advanced Very High Resolution Radiometer (AVHRR) data on board the US National Atmospheric and Oceanic Administration (NOAA) satellites from 1981 to present with the same algorithm (Kilpatrick et al. 2001; Casey et al. 2010). The aim of this project is to produce consistent long term SST for research, modelling, and trend analysis (Casey et al. 2010). SST is derived using two channels in the thermal IR (10.8 and 11.4 μm). Monthly composites were computed from the daily daytime Pathfinder v5.2 at a 4 km spatial resolution. A quality flag of 4, considered as the lowest quality level for acceptable data (Kilpatrick et al. 2001), was imposed for our application.

SST data from MODIS Terra has been available since 2000. The MODIS SST dataset aims to provide high quality global measurements at a high spatial resolution. The MODIS SST is derived using two channels in either the thermal IR (11 and 12 um) or in the mid-IR region (3.8 and 4.1 um) (Minnett et al. 2002). Level-2 MODIS data were downloaded from the Ocean Color website[2] and processed at a 4 km resolution using the SeaWiFS Data Analysis System[3] (SeaDAS). The processing method is described in Dufois et al. (2012). Only the daytime passes were processed, allowing us to use the cloud flag (SST quality flags were not used). The daily data were then averaged to obtain monthly data. This product is hereafter referred to as Reprocessed MODIS.

The OSI-SAF hourly SST product is available on a 1/20° grid (5 km) and processed by the French Centre ERS d'Archivage et de Traitement (CERSAT). The OSI-SAF SST consists in the best hourly SST images sampled from the Spinning Enhanced Visible and Infrared Imager (SEVIRI), onboard the geo-stationary Meteosat Second Generation 2 (MSG-2) satellite. It is calculated using a classical multi-channel formulas applied to the 11 and 12 μm channels (Le Borgne et al. 2006). SEVIRI has been imaging the Earth since June 2004 at a 15 min sampling interval using 12 spectral channels (4 visible/near-IR and 8 IR channels) with spatial resolutions of 1 and 3 km. For comparisons purposes, our analysis of the OSI-SAF SST was restricted to daytime observations (from 7:00 to 19:00).

A weekly composite of SST data collected from the Advanced Microwave Scanning Radiometer—Earth Observing System (AMSR-E), on board the US NASA Aqua satellite, was used to illustrate some of the limitation of microwave radiometry in the Agulhas Current. Global maps of AMSR-E SST data are available on a 1/4°

---

[1] See www.osi-saf.org

[2] At http://oceancolor.gsfc.nasa.gov

[3] For details see http://seadas.gsfc.nasa.gov

uniform grid and can be downloaded from the Remote Sensing Systems website[4]. In this study, the AMSR-E data was flagged to only retain valid geophysical data (values above 250).

In this paper, the circulation in the Agulhas Current region is depicted using merged maps of the absolute geostrophic currents from the AVISO Maps of Absolute Dynamic Topography—Delayed Time (MADT-DT) product. The AVISO merged altimetry datasets are produced by Ssalto/Duacs and distributed by AVISO[5], with support from the French Centre National d'Etudes Spatiales (CNES). The MADT-DT product combines sea level anomaly signals from the Ocean Surface Topography Mission (OSTM)/Jason-2, Jason-1 and Envisat altimeters to the Mean Dynamic Topography (MDT) of Rio et al. (2011). The AVISO MADT-DT product is provided on a rectilinear grid with a spatial resolution of 1/3° and combines SSH observations collated by altimeters over a period of 1 week.

12.3 Some Challenges of SST and Altimetry Observations in the Agulhas Current Region

Over the Agulhas Current core, about 5 times as much water vapour is transferred to the atmosphere in comparisons to neighbouring waters (Rouault et al. 2000). The Agulhas Retroflection (Fig. 12.1) also constitutes one of the most significant region of heat flux loss globally (Lutjeharms 2006). Cloud contaminations can induce significant data loss in IR SST imagery as they obscure the ocean surface, absorb surface-leaving radiation, and re-emit this energy at lower temperatures. Most datasets of IR derived SST routinely available have been subjected to automatic cloud screening procedures to remove contaminated data. Clouds in IR radiometry are associated with colder and more reflective signals. Cloudy regions in IR images also display more spatial variance. Cloud detection algorithms usually involve masking pixels in the SST image which are significantly cooler than either neighboring pixels and/or the climatological value (Robinson 2004). In regions of strong thermal gradients, automatic cloud screening procedures often result in the loss of good geophysical data.

Monthly climatologies of the number of unclouded SST observations in the Agulhas Current region were derived from the Pathfinder v5.2, Reprocessed MODIS and OSI-SAF SST datasets to illustrate the importance of cloud contamination in the Agulhas Current region. Figure 12.2 shows the average number of valid SST observations for the summer and winter seasons and for all 3 daytime IR datasets. The Agulhas Current core and the Retroflection are poorly sampled by IR SST sensors due to high level of evaporation (Rouault et al. 2000). The northern Agulhas is one of the region most affected by cloud contamination. In the northern Agulhas, topographic steering confines the location of the current to the continental shelf break

---

[4] See http://www.remss.com/

[5] At http://www.aviso.oceanobs.com/duacs

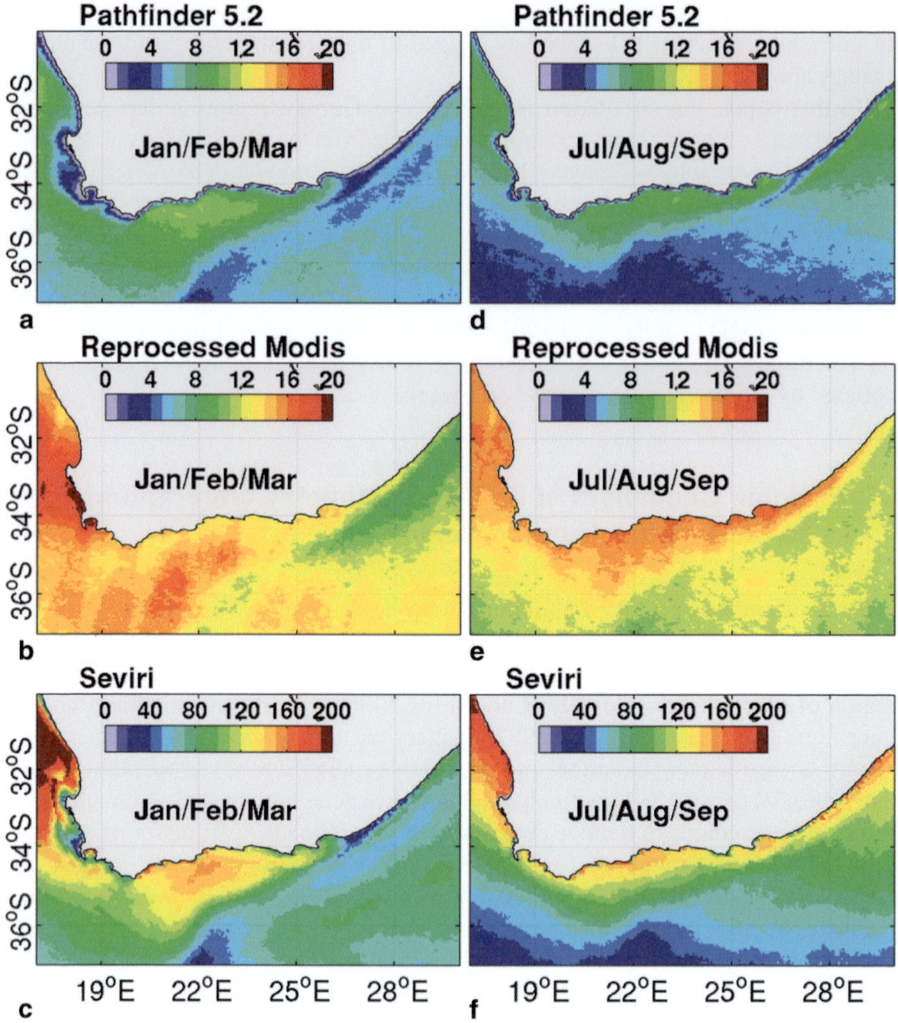

Fig. 12.2 Number of good observations per month derived using the Pathfinder v5.2, MODIS TERRA and OSI-SAF daytime SST. **a**, **b** and **c** show the average number of valid SST observations during the austral summer months of December, January and February. **d**, **e** and **f** show the average number of valid SST observations during the austral winter months of June, July and August. The MODIS and Pathfinder climatology were computed using 11 years of observations collected between 2000 and 2010, while the OSI-SAF climatology was computed over a 6 year period between 2005 and 2010

(Lutjeharms 2006) and strong evaporation constantly reduces the number of clear-sky observations from IR sensors. It is interesting to notice the strong differences in the sampling capacity of the Pathfinder v5.2 and Reprocessed MODIS SST datasets. Despite similar global daily coverage, the MODIS dataset provides a much larger number of valid observations over the Agulhas Current region. This can be explained by differences in both sensors measurement capacity and cloud masking algorithms. MODIS has a high spatial resolution and larger number of spectral bands compared to AVHRR and is therefore able to better detect clouds during daytime (Heidinger et al. 2002). In addition, the Pathfinder cloud mask algorithm relies on spatial uniformity test and background fields of reference SST (Hickox et al. 2000; Thomas et al. 2004). The Pathfinder cloud masking algorithm therefore tends to erroneously flag pixels in regions of strong thermal gradients (Dufois et al. 2012). Coastal regions in the southern Agulhas Current ecosystem often exhibit strong temperature gradients. Coastal regions west of Cape Agulhas (at the southernmost tip of Africa) are considered to form part of the Benguela upwelling ecosystem (Hutchings et al. 2009). West of Cape Agulhas, strong temperature fronts develop near the coast during the austral summer due to the predominance of upwelling-favorable southerly winds. The Agulhas Current also dynamically drives several coastal upwelling cells on the eastern Agulhas Bank (Lutjeharms 2006). The Pathfinder dataset is particularly bad at imaging the inshore front of the Agulhas Current and the coastal regions near Cape Town during the summer upwelling season. The OSI-SAF SST dataset seems to suffer from the same cloud masking limitations as the Pathfinder dataset. For the MODIS data presented here, no reference SST was used when masking clouds.

Figure 12.2 shows that the percentage of valid IR SST observations in the Agulhas Current region varies seasonally. North of the Retroflection, IR sensors provide a better coverage during the austral winter months (from June to August). Further south, at the Retroflection and within the Agulhas Return Current, the tendency is reversed with a greater number of valid SST observations available during the austral summer (December to February). Throughout the year, the high frequency OSI-SAF SST dataset provides 10 to 20 times more observations than the MODIS and Pathfinder datasets. Inadequate cloud masking algorithms can cause significant bias in the climatology derived from IR SST datasets, with important consequences in climatological or numerical modelling studies (Dufois et al. 2012). Figure 12.3 shows that during the austral summer months, the MODIS and Pathfinder SST climatology exhibits strong differences in the coastal regions of the Agulhas Current, near Port Elizabeth and around the Cape Peninsula.

Microwave radiometers which are able to "see" through clouds, provide an alternative to IR-based sensors in regions of strong air/sea interactions. The TRMM Microwave Imager (TMI), where TRMM stands for Tropical Rainfall Measuring Mission, launched in November 1997, was the first microwave radiometer to provide accurate measurements of SST in the Agulhas Current region. The TMI SST dataset was used successfully to monitor the Retroflection of the East Madagascar Current and track perturbations in the southern Agulhas Current region (Quartly and Srokosz 2002). Since 2002, the AMSR-E sensor (mentioned in Sect. 2) has provided additional microwave observations of SST in the Agulhas Current region. While the coverage from the TMI sensor is limited to regions north of 38°S, that of the AMSR-E dataset is global. The AMSR-E sensor is therefore well suited to the monitoring of the Agulhas Retroflection. Microwave SST observations suffer from 2 main limitations: a low spatial resolution of about 50 km and an inability to measure SST closer

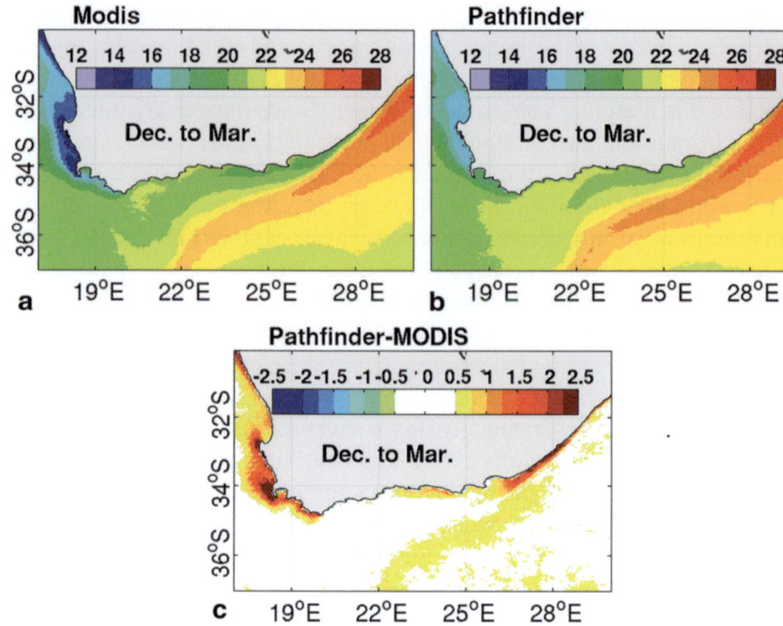

Fig. 12.3 Seasonal summer climatologies of SST derived from the Reprocessed MODIS **a** and Pathfinder v5.2 **b** datasets. **c** shows SST difference (°C) between Pathfinder v5.2 daytime and the reprocessed MODIS from 2000 to 2010 during Dec./Jan./Feb./Mar

than about 1.5 footprint from land (about 75 km) because of side-lobe contamination. A weekly composite of the SST observed with AMSR-E (Fig. 12.4) illustrates the inability of the microwave radiometer to sample near the coast. From 28 to 34°S, the microwave radiometer is unable to image the Agulhas Current due to side-lobe contamination extending as far offshore as the 3,000 m isobath. South of 35°S however, the AMSR-E sensor successfully images the Agulhas Current flowing along the widening continental shelf, the sudden Agulhas Current reversal at the Retroflection and the eastward flowing Agulhas Return Current. Both the AMSR-E and TMI sensors do not provide a complete coverage of the Agulhas Current region each day. Due to their spatial coverage characteristics and fairly low spatial resolution, the TMI and AMSR-E microwave sensors are best suited to imaging the southern Agulhas and Retroflection regions and to monitoring meso-scale features with spatial scales greater than 50 km.

Mapping the surface circulation with altimetry requires the merging of SSH observations from multiple altimeters. The footprint of an altimeter on the sea surface is limited by the length of the pulse (hence the wording 'pulse-limited' footprint) but also depends on the roughness of the surface due to wind waves. Typical values of the footprint diameter for operating altimeters range from ~2 km for very calm seas to ~10 km for a significant wave height of 10 m (Chelton et al. 1989). Measurements along the altimeter's tracks are normally averaged over 1 s of flight (1 Hz), implying

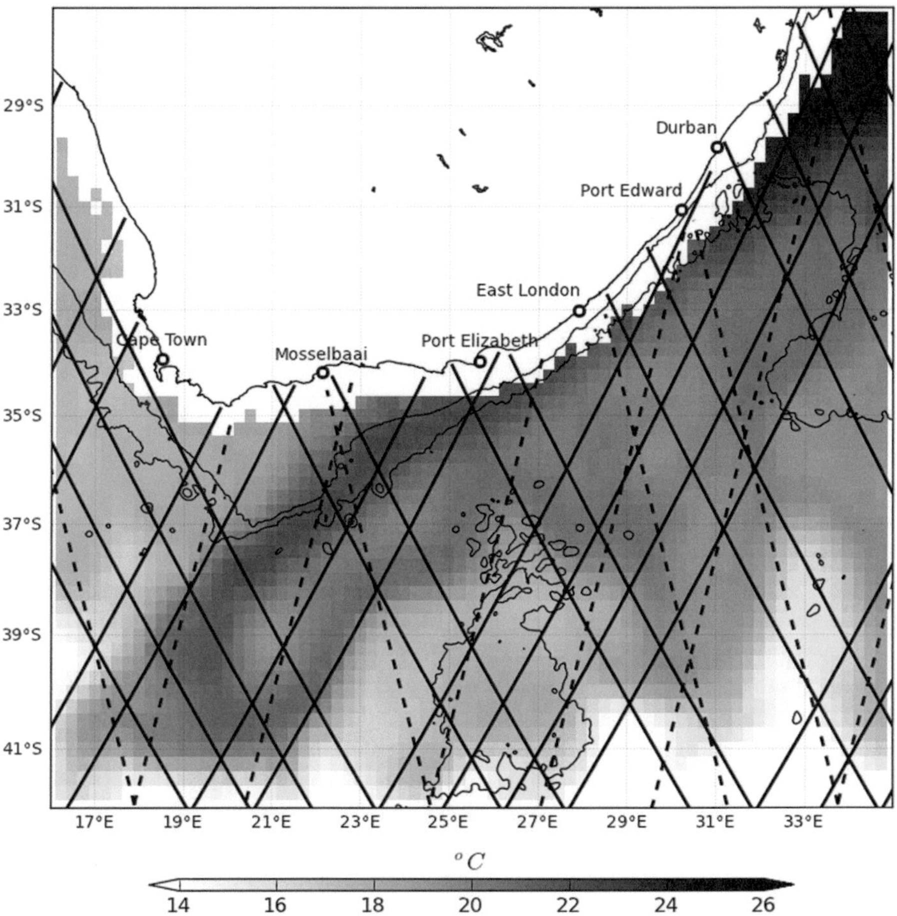

Fig. 12.4 Weekly composite of AMSRE-E microwave SST over Agulhas Current region over the period 18–24 November 2010. The *black lines* overlaid show altimetry tracks from all available altimeters (Jason-1, Jason-2 and Envisat) during the same week. The tracks of the Jason-1 and Jason-2 altimeters are shown as *solid black lines*. The *stippled black line* shows the track of the Envisat altimeter. The 1,000 and 3,000 m isobaths are plotted as *solid thin black contour lines*

that the spatial resolution along the altimeter track is about 7 km (the space travelled in 1 s by the projection of the satellite on the surface) plus the diameter of the footprint (the use of higher rate data, *i.e.* data with less along-track average such as 5 or 20 Hz, is sometimes attempted, especially in coastal regions). Altimeters have cross-track spacing varying between 30 and 300 km and repeat cycles ranging from 10 to 35 days (Vignudelli et al. 2011). The ability of merged altimetry products to capture the meso-scale circulation increases with the number of altimeters in space (Pascual et al. 2006). A two-altimeter configuration is typically not able to image sub-mesoscale features of the circulation (less than 30 km) or to resolve variability occurring within time-scales typically less than 10 days (LeTraon and Dibarboure

2002). Since the early 1990s, there have been a minimum of two altimeters in space. The AVISO data presented in this chapter merges SSH observations collected from the Jason-1, Jason-2 and Envisat altimeters. Ascending and descending tracks from these 3 altimeters over a period of 7 days have been plotted in Fig. 12.4 to illustrate the level of spatial and temporal smoothing required to produce maps of the oceanic surface circulation from multiple altimeters.

Like microwave radiometers, observations from radar altimeters suffer from contamination near the coast. Land contamination directly corrupts altimeter measurement, resulting in a loss of data near the coast. Inaccuracies in the wet tropospheric correction also restrict the use of altimetry near the coast. Some altimeters rely on an on-board microwave radiometer to correct for the presence of water vapour in the atmosphere. Since microwave radiometers typically have a footprint size of 30 km and can not accurately measure within 1.5 foot print size from the coast, accurate observations from altimeters can not be obtained within about 50 km of the coast (Vignudelli et al. 2011); models can be used to fill this gap but usually lack the shorter spatial scales of variability of water vapour.

Additional challenges faced by altimeters in the coastal regions are their inability to resolve high frequency signals from tidal or atmospheric forcing in the coastal regions. Finally, altimeters measure SSH variations in reference to a rough approximation of the Earth's surface, called the reference ellipsoid. To study ocean circulation from altimetry, it is necessary to refer SSH measurements to the geoid rather than to the reference ellipsoid. Recent measurements of the Earth gravity field from the highly successful Gravity Recovery and Climate Experiment (GRACE) and Gravity field and steady-state Ocean Circulation Explorer (GOCE) missions (and their combination) are able to resolve the geoid over length scales of the order of 100 km; Janjić et al. 2012), while global gravity models resolve the geoid with a resolution of a few 100 km (Rio and Hernandez 2004). Assuming the geoid is stationary, the time varying part of the ocean circulation can be reproduced by subtracting the mean SSH and working with height anomalies. However this procedure also removes the MDT which has a strong signature in western boundary current regions such as the Agulhas Current (Byrne and McClean 2008). The CNES-CLS09 MDT of Rio et al. (2011) which is integrated in the AVISO MADT-DT dataset presented here, is a hybrid MDT. It makes use of extended datasets of drifting buoy velocities (1993–2008) and dynamic heights (1993–2007) and adequately captures the time-averaged circulation of the Agulhas Current (Rouault et al. 2010).

The Agulhas Current region is a challenging region for space-based observations of SSH and SST. Cloud contamination and inadequate cloud masking procedures severely impact on the quality and density of IR SST observations over the Agulhas Current, particularly in the northern Agulhas region. Microwave radiometers provide a good alternative to IR sensors in the southern Agulhas and the Retroflection area but are not able to image most of the northern Agulhas Current. The proximity of the Agulhas Current to the coast from Durban to Port Elizabeth and its fairly invariant path (Gründlingh 1983) also limit the use of merged altimetry products in the northern Agulhas Current region. In Sect. 4, we demonstrate how despite their limitations, altimetry and SST observations can still be combined to improve our understanding of the northern Agulhas Current dynamics.

12.4 Combining SSH and SST Observations to Better Resolve the Circulation of the Northern Agulhas Current

Through its source regions, the Agulhas Current is connected to the variability of the Indian Ocean. Changes in the Indonesian throughflow as well as wind driven variability over the South Indian Ocean are transmitted to the source regions of the Agulhas Current through the South Equatorial Current (SEC) Ridderinkhof et al. 2010). Upon reaching the eastern shore of Madagascar at 15°E, the SEC branches into a northern and southern flow, feeding two main sources for the Agulhas Current: the Mozambique and the East Madagascar currents. In the Mozambique channel, the flow is dominated by the passage of large anticyclonic eddies propagating southwards towards the Agulhas Current (de Ruijter et al. 2002). These large anticyclonic eddies occur 5 to 6 times per year (van der Werf et al. 2010), they have spatial scales of 300 to 350 km and move downstream at speeds of 3 to 6 km/day. The contribution from the SEC to the Agulhas Current occurs in the form of cyclonic and anticyclonic eddies which tend to drift south-westward towards the northern Agulhas Current (Biastoch et al. 1999; de Ruijter et al. 2004). Analyses of altimeter data indicate that on average 4 anticyclonic eddies per year occur south of Madagascar (Schouten et al. 2003), with the eddy frequency increasing during negative phases of the Indian Ocean Dipole and El Niño cycles (de Ruijter et al. 2004). Previous observational and numerical modelling studies have shown that anticyclonic eddies reaching the eastern boundary of the Agulhas Current can trigger the formation of large offshore meanders in the Natal Bight region (Schouten et al. 2002; Tsugawa and Hasumi 2010). These large offshore meanders, named Natal Pulses due to their region of origin (Lutjeharms and Roberts 1988) are major drivers of variability in the Agulhas Current (Bryden et al. 2005; Lutjeharms 2006).

Altimetry and SST imagery can be combined to follow the trajectory of anticyclonic eddies in the southwest Indian Ocean and study their interaction as they reach the eastern edge of the northern Agulhas Current. In the next section, a case study of Natal Pulse's inception by an anticyclonic eddy is provided. The path and property of the anticyclonic eddy are extracted from the Chelton et al. (2011) global database of oceanic eddies. Daily composites of OSI-SAF SST are used to follow the evolution and fate of the Natal Pulse as it progresses downstream.

On 18 July 2007, an anticyclonic eddy is detected to the south east of Madagascar, at 48.5°E—27°S. The anticyclone then propagates westwards with a mean speed of about 8 km/day, until it reaches the eastern boundary of the northern Agulhas Current around 15 February 2008 (Fig. 12.5a). On15 February 2008 the anticyclonic eddy is centred at 34.5°E—28.8°S, it has diameter of about 230 km and maximum geostrophic rotational velocities of about 65 cm/s. Interactions between the eddy and the eastern edge of the Agulhas Current induce the development of a small perturbation at the inshore border of the Agulhas Current. This small perturbation in the Agulhas Current is captured in the daily composite map of the OSI-SAF SST (Fig. 12.5b) and is centred at 32.5°E—29.2°S. Due to its small spatial scale, this offshore perturbation is not successfully identified in the AVISO map of absolute geostrophic currents. By 10 March 2008, the small perturbation initiated on

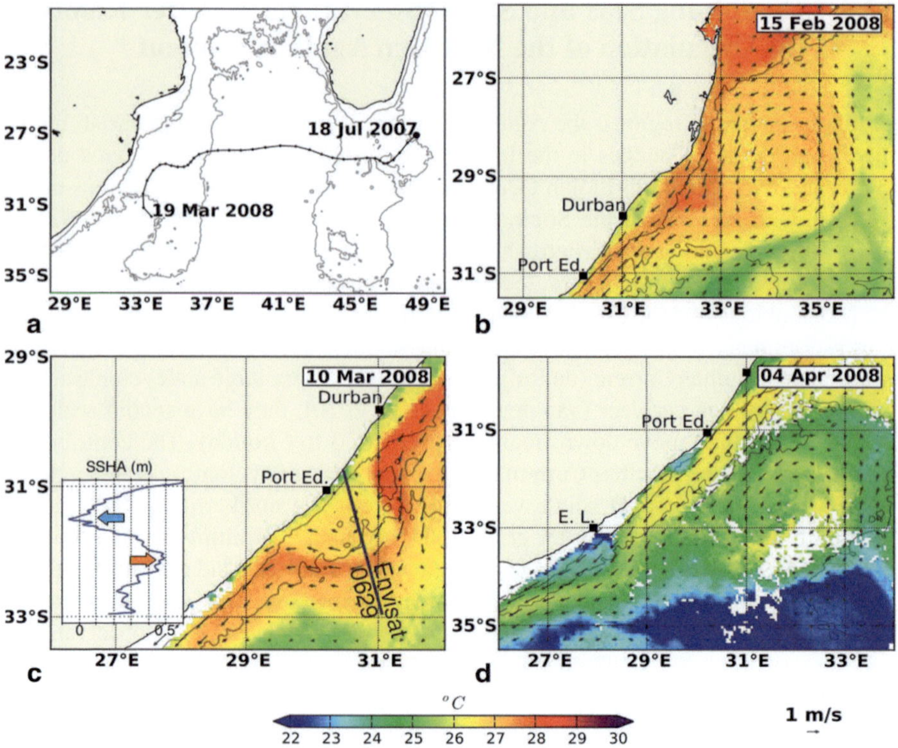

Fig. 12.5 Inception of a Natal Pulse from an anticyclonic eddy. **a** Path of the anticyclonic eddy extracted from the global eddy database of Chelton et al. (2011). The *large black dot* in the eddy trajectory marks the 1st location of eddy detection, with each subsequent dot showing the eddy's position at weekly intervals. Panels **b**, **c** and **d** are daily maps of OSI-SAF SST overlaid with the AVISO absolute geostrophic currents. The 1,000 and 3,000 m isobaths are plotted as *thin black lines*. *Black squares* show the cities of Durban, Port Edward (Port. Ed.) and East London (E. L.) on the South African coastline. The *thick black line* in **c** marks a section of Envisat pass 0629. The along-track SSHA measurement taken during cycle 66 overpass (4 March 2008) on that section are shown in the inset on the *left*, where the *blue arrow* marks the signature of the centre of Natal Pulse and the *orange arrow* marks its approximate offshore edge

15 February has developed into a well-defined cyclonic meander offshore Port Edward and is successfully identified in both the merged altimetry and SST imagery (Fig. 12.5c). The Natal Pulse observed on 10 March 2008 near Port Edward caused an offshore displacement of the Agulhas Current's front of about 100 km. This is also nicely captured by along-track altimetry, as seen in the inset of Fig. 12.5c which shows the SSH Anomaly (SSHA) profile along Envisat's ascending pass 0629 on 4 March 2008. The altimetric signature of the center of the meander cyclonic circulation is the low value marked by the blue arrow, while the 'bump' in SSHA indicated by the orange arrow marks the offshore edge of the meander. The increase of about

0.5 m in SSHA between 31.5°S and 32°S, *i.e.* over a distance of about 50 km, corresponds to a geostrophic current anomaly of about 1 m/s, directed SW. It is worth noting that between 31°S and 31.5°S, the along-track profile shows a distinct reversal in the direction of the current, with the flow switching from a SW direction away from the coast to a NE flowing current closer to the coast. The presence of a NE flow near the coast is not apparent in the interpolated map. On 10 March 2008, the anticyclonic eddy responsible for triggering the Natal Pulse is located at 33°E—31.3°S and has a diameter of about 200 km. Strong anticyclonic flow is observed within the anticyclonic eddy with maximum geostrophic rotational speed reaching about 80 cm/s. Between 18 and 20 March 2008 (not shown in figures), interactions between the trailing edge of the Natal Pulse and the topography induce an instability upstream of the Natal Pulse in the shape of a smaller secondary offshore meander. The development of such upstream instabilities during the southward progression of Natal Pulses has been described in detail in the study of Rouault and Penven (2011). SST imagery on 4 April 2008 (Fig. 12.5d) shows that the Natal Pulse has progressed south to 28°E—34°S with the secondary instability located just south of Port Edward (29.8°E—32°S). A decrease of about 50 % in the spatial extent of the Natal Pulse is observed between 15 February 2008 and 10 March 2008. Subsequent SST maps (not shown here) are not able to highlight the presence of the Natal Pulse and its upstream generated perturbation after 14 April 2008.

The case study presented here shows how altimetry can be used to track deep sea eddies and study their interaction with western boundary currents such as the Agulhas Current. Here we provide additional support to the study of Schouten et al. (2002), which linked Natal Pulses to the far eddy field. The inception of a Natal Pulse by an anticyclonic eddy follows a process similar to that described by Tsugawa and Hasumi (2010) in their modelling study. Closer to the shore the higher resolution and frequency acquisitions afforded by IR SST imagery proves useful when following the evolution of Natal Pulses. The initial growth of the Natal Pulse, its interaction with the coastal and shelf waters and its subsequent dissipation add weight to the hypothesis put forward by Rouault and Penven (2011), that part of the variability observed in the northern Agulhas Current is lost downstream.

12.5 Conclusions and Perspectives

Over the last 2 decades, satellite remote sensing observations have provided a cost-effective alternative to *in situ* measurements for many regions of the world's ocean. The sampling capabilities of the Agulhas Current from space have significantly improved in recent years due to an increase in satellite spatial and temporal coverage, the emergence of new remote sensing methods and the use of more robust algorithms for the derivation of ocean properties.

The launch of the MSG-2 geostationary satellite has provided a major boost for IR SST observations over the Agulhas Current region. High frequency acquisitions from the SEVIRI sensor onboard the MSG-2 satellite have markedly improved our ability to image the Agulhas Current from space. The SEVIRI SST imagery also provides an alternative to altimetry for tracking meso to sub-mesoscale features of the circulation, particularly in the northern Agulhas Current region where SSH observations from space are seriously compromised due to the proximity of the current to the coast. In addition, new generation of SST datasets have emerged under the GHRSST initiative. The currently available GHRSST dataset merge IR and microwave SST observations to improve the resolution and coverage of SST observations. Future research to validate and utilise these high resolution merged SST products could help improve our understanding of the Agulhas Current.

But cloud contamination and inadequate cloud masking procedures continue to hamper our ability to monitor the variability of the Agulhas Current from SST observations. The northern Agulhas Current and the coastal regions near Port Elizabeth remain poorly sampled by IR sensors and can not be resolved in Microwave SST imagery. Comparisons between the reprocessed MODIS and Pathfinder v5.2 datasets in Section 3 highlight some of the limitation of using global based cloud masking algorithm to assess long-term changes in the Agulhas Current. The better climatology and coverage of the reprocessed MODIS SST dataset, when compared to the Pathfinder v5.2 SST, clearly showed that regional-focused cloud masking procedures could significantly improve the quality of future SST products in the Agulhas Current region.

Altimetry remains very valuable for tracking deep-sea eddies and investigating their evolution at the seaward edge of the Agulhas Current. However in the northern Agulhas, the proximity of the current to the coast continues to challenge satellite-based observations of the Agulhas Current from altimeters. Projects such as the ESA-funded COASTALT and CNES-funded PISTACH, aiming at recovering useful altimetric measurements in the coastal zone, have demonstrated that is it possible to improve the quality of altimeter-based observations near the coast with anticipated benefits in the northern Agulhas and the coastal and shelf regions of the southern Agulhas. Surface current information derived from SARs could also contribute significantly to improving our knowledge of the northern Agulhas Current in the future. The principle of surface current measurements from SARs involves the extraction of a line of sight velocity from information contained in the frequency spectrum of the returned radar echoes. The line of sight velocity can then be projected onto a horizontal plane to provide a range-directed surface current velocity. Between July 2007 and April 2012, maps of range-directed surface current velocities in the Agulhas Current region were systematically recovered from the now defunct Envisat's ASAR, under the ESA funded SAR ocean wind-wave-current project (Collard et al. 2008; Johannessen et al. 2008). An assessment of the ASAR surface current velocities in the Agulhas Current region (Rouault et al. 2010) showed that the synoptic nature and relatively high resolution of ASAR acquisitions make the ASAR derived current velocities a good complement to altimetry for the study of sub-mesoscale processes in the Agulhas Current. SAR derived ocean surface currents will form part of the data stream routinely provided within the ESA driven Sentinel-1 satellite mission, which is due to be launched in 2013. With the planned deployments of SAR sensors in future Sentinel missions and the ongoing SAR observations available from Radarsat, the future of the SAR-derived ocean surface currents from space looks promising.

One of the issues of satellite observations in western boundary currents is the lack of match-up data to validate remote sensing algorithms. Short residence times in western boundary currents imply that few observations are available from platforms such as Argo floats which are typically used in match-up databases. In addition, the Agulhas Current is a region which has historically been poorly sampled. *In situ* observations across the Agulhas Current near East London (at 33°S) are currently being collected as part of the 3-year ACT program. The ACT measurement campaign (van Sebille et al. 2010) will provide a much needed validation database for future remote sensing studies in the Agulhas Current region.

References

Beal LM, de Ruijter WPM, Biastoch A, Zahn R (2011) On the role of the Agulhas system in ocean circulation and climate. Nature 472(7344):429–436. doi:dx.doi.org/10.1038/nature 09983

Biastoch A, Reason CJC, Lutjeharms JRE, Boebel O (1999) The importance of flow in the Mozambique channel to seasonality in the greater Agulhas current system. Geophys Res Lett 26(21):3321–3324. doi:dx.doi.org/10.1029/1999GL002349

Bryden HL, Beal LM, Duncan LM (2005) Structure and transport of the Agulhas current and its temporal variability. J Oceanogr 61:479–492

Byrne DA, McClean JL (2008) Sea level anomaly signals in the Agulhas Current region. Geophys Res Lett 35(13):L13601. doi:dx.doi.org/10.1029/2008GL034087

Casal TGD, Beal LM, Lumpkin R, Johns WE (2009) Structure and downstream evolution of the Agulhas current system during a quasi-synoptic survey in february—march 2003. J Geophys Res 114:C03001. doi:10.1029/2008JC004954

Casey KS, Brandon TB, Cornillon P, Evans R (2010) The past, present, and future of the AVHRR pathfinder SST program. In: Barale V, Gower J, Alberotanza (eds) Oceanography from space: revisited. Springer, pp 273–287. http://dx.doi.org/10.1007/978-90-481-8681-5_16

Chelton DB, Schlax MG, Samelson RM (2011) Global observations of nonlinear mesoscale eddies. Progrogr Oceanogr 91(2):167–216. doi:10.1016/j.pocean.2011.01.002

Chelton DB, Walsh EJ, MacArthur JL (1989) Pulse compression and sea level tracking in satellite altimetry. J Atmos Oceanic Technol 6:407–438

Collard F, Mouche A, Chapron B, Danilo C, Johannessen JA (2008) Routine high resolution observation of selected major surface currents from space. Proceedings of the Workshop SEASAR 2008, ESA SP-656

de Ruijter WPM, Ridderinkhof H, Lutjeharms JRE, Schouten MW, Veth C (2002) Observations of the flow in the Mozambique channel. Geophys Res Lett 29:1401–1403

de Ruijter WPM, van Aken HM, Beier EJ, Lutjeharms JRE, Matano RP, Schouten MW (2004) Eddies and dipoles around South Madagascar: formation, pathways and large-scale impact. Deep-Sea Res Part I 51(3):383–400. doi:10.1016/j.dsr.2003.10.011

Dufois F, Penven P, Whittle C, Veitch, J (2012) On the warm nearshore bias in Pathfinder monthly SST products over Eastern Boundary Upwelling Systems. Ocean Modelling, 47:13–118.

Emery WJ (2001) Estimating sea surface temperature from infrared satellite and in Situ temperature data. B Am Meteorol Soc 82(12):2773–2786

Goschen WS, Schumann EH (1990) Agulhas current variability and inshore structures off the Cape Province, South Africa. J Geophys Res 95:667–678

Gründlingh ML (1983) On the course of the Agulhas Current. S Afr Geogr J 65(1):49–57

Heidinger AK, Anne VR, Dean C (2002). Using MODIS to estimate cloud contamination of the AVHRR data record. J Atmos Ocean Tech 19(5):586–601. doi:10.1175/1520-0426(2002)019 <0586:UMTECC>2.0.CO;2

Hickox R, Belkin I, Cornillon P, Shan Z (2000) Climatology and seasonal variability of ocean fronts in the East China, Yellow and Bohai seas from satellite SST data. Geophys Res Lett 27(18):2945. doi:10.1029/1999GL011223

Hutchings L, van der Lingen CD, Shannon LJ, Crawford RJM, Verheye HMS, Bartholomae CH, van der Plas AK, Louw D, Kreiner A, Ostrowski M, Fidel Q, Barlow RG, Lamont T, Coetzee J, Shillington F, Veitch J, Currie JC, Monteiro PMS (2009) The Benguela current: an ecosystem of four components. Prog Oceanogr 83(1–4):15–32. doi:10.1016/j.pocean.2009.07.046

Janjić T, Schröter J, Savcenko R, Bosch W, Albertella A, Rummel R, Klatt O (2012) Impact of combining GRACE and GOCE gravity data on ocean circulation estimates. Ocean Sci 8:65–79. doi:10.5194/os-8-65-2012

Johannessen JA, Chapron B, Collard F, Kudryavtsev V, Mouche A, Akimov D, Dagestad KF (2008) Direct ocean surface velocity measurements from space: improved quantitative interpretation of Envisat ASAR observations. Geophys Res Lett 35:L22608. doi: 10.1029/2008GL035709

Kilpatrick KA, Podesta GP, Evans R (2001) Overview of the NOAA/NASA Advanced very high resolution radiometer pathfinder algorithm for sea surface temperature and associated matchup database. J Geophys Res 106(C5):9179–9197. doi:10.1029/1999JC000065

Le Borgne P, Legendre G, Marsouin A (2006) Operational SST retrieval from MSG/SEVIRI data. 2006 EUMETSAT Meteorological Satellite Conference.

Le Traon PY, Dibarboure G (2002) Velocity mapping capabilities of present and future altimeter missions: The role of high-frequency signals. J Atmos Ocean Tech 19(12):2077–2087. doi: abs/10.1175/1520–0426(2002) 019<2077:VMCOPA>2.0.CO;2

Lutjeharms JRE (2006) The Agulhas current. Springer, Berlin

Lutjeharms JRE, Roberts HR (1988) The natal pulse: an extreme transient on the Agulhas current. J Geophys Res 93:631–645

Minnett PJ, Evans RH, Kearns EJ, Brown OB (2002) Sea-surface temperature measured by the Moderate Resolution Imaging Spectrometer (MODIS). Proceedings IGARSS 1177–1179

Pascual A, Faugère Y, Larnicol G, Le Traon P-Y (2006) Improved description of the ocean mesoscale variability by combining four satellite altimeters. Geophys Res Lett 33(2):L02611. doi:10.1029/2005GL024633

Quartly GD, Srokosz MA (2002) SST observations of the Agulhas and East Madagascar retroflections by the TRMM microwave imager. J Phys Oceanogr 32(5):1585–1592

Ridderinkhof H, van der Werf PM, Ullgren JE, van Aken HM, van Leeuwen PJ, de Ruijter WPM (2010) Seasonal and interannual variability in the Mozambique channel from moored current observations. J Geophys Res 115:C06010. doi:10.1029/2009JC005619

Rio MH, Hernandez F (2004) A mean dynamic topography computed over the world ocean from altimetry, in situ measurements, and a geoid model. J Geophys Res 109:C12032

Rio MH, Guinehut S, Larnicol G (2011) New CNES-CLS09 global mean dynamic topography computed from the combination of GRACE data, altimetry, and in situ measurements. J Geophys Res 116(C7):C07018. doi:10.1029/2010JC006505

Robinson IS (2004). Measuring the oceans from space: The principles and methods of satellite oceanography. Springer-Praxis books in geophysical science. Springer-Verlag

Rouault M, Lee-Thorp AM, Lutjeharms JRE (2000) Observations of the atmospheric boundary layer above the Agulhas current during along-current winds. J Phys Oceanogr 30:70–85

Rouault MJ, Mouche A, Collard F, Johannessen JA, Chapron B (2010) Mapping the Agulhas current from space: an assessment of ASAR surface current velocities. J Geophys Res 115:C10026. doi:10.1029/2009JC006050

Rouault MJ, Penven P (2011) New perspectives on natal pulses from satellite observations. J Geophys Res 116:C07013. doi:10.1029/2010JC006866

Schouten MW, de Ruijter WP, van Leeuwen PJ (2002) Upstream control of Agulhas Ring shedding. J Geophys Res, 107(C8):3109

Schouten MW, de Ruijter WPM, van Leeuwen PJ, Ridderinkhof H (2003) Eddies and variability in the Mozambique channel. Deep-Sea Res II: Topical Studies in Oceanography 50(12–13): 1987–2003

Siedler G, Rouault M, Lutjeharms JRE (2006) Structure and origin of the subtropical South Indian Ocean countercurrent. Geophys Res Lett 33:L24609

Thomas SM, Heidinger AK, Pavolonis MJ (2004) Comparison of NOAA's operational AVHRR-derived cloud amount to other satellite-derived cloud climatologies. J Climate 17(24):4805–4822. doi:10.1175/JCLI-3242.1

Tsugawa M, Hasumi H (2010) Generation and growth mechanism of the natal pulse. J Phys Oceanogr 40(7):1597–1612. doi:dx.doi.org/10.1175/2010JPO4347.1

van der Werf PM, van Leeuwen PJ, Ridderinkhof H, de Ruijter WPM (2010) Comparison between observations and models of the Mozambique Channel transport: seasonal cycle and eddy frequencies. J Geophys Res 115:C02002. doi:10.1029/2009JC005633

van Sebille E, Beal LM, Biastoch A (2010) Sea surface slope as a proxy for Agulhas current strength. Geophys Res Lett 37(9) L09610 doi:10.1029/2010GL042847

Vignudelli S, Kostianoy AG, Cipollini P, Benveniste J, eds (2011) Coastal Altimetry. Springer. doi:10.1007/978-3-642-12796-0

Chapter 13
Use of SAR data to Monitor the Greater Agulhas Current

Johnny A. Johannessen, Bertrand Chapron, Fabrice Collard and Björn Backeberg

Abstract The strong and dynamic greater Agulhas Current is known to have significant influence on the local marine environment and ecosystem. It is also considered to play an important role in the global thermohaline circulation, notably due to the transport of heat and salt from the Indian Ocean into the South Atlantic Ocean. Regular monitoring of the dominant processes and variability within the greater Agulhas Current is thus highly needed. Recently Doppler-derived ocean surface velocities from the European Space Agency's (ESA) Envisat Advanced Synthetic Aperture Radar (ASAR) have demonstrated abilities to manifest the intensity of surface currents. In this paper the regular direct surface velocity estimates of the greater Agulhas Current from the ASAR since summer of 2007 are presented. These estimates are assessed by direct comparison to other independent estimation of the surface current in the greater Agulhas current obtained from surface drifter data, radar altimetry and an ocean circulation model. The results are promising and highlight that the Doppler-based ocean surface velocity retrieval offers a new innovative approach to monitor and advance the understanding of the dynamic processes of the greater Agulhas Current.

13.1 Introduction

The Agulhas Current is, together with the Gulf Stream and Kuroshio Current, among the strongest western boundary currents in the world's oceans (Lutjeharms 2006). It flows poleward along the eastern coast of southern Africa from about 27° to 40°S, where it retroflects and returns eastward back into the South Indian Ocean. In the retroflection region eddies are shed into the South Atlantic. The Agulhas Current core is steered along the continental shelf break, which in the north is very steep and close to the coast. In this region the Agulhas Current is rather stable, departing less than

J. A. Johannessen (✉)
Nansen Environmental and Remote Sensing Center, Bergen, Norway
e-mail: Johnny.Johannessen@nersc.no

B. Chapron · F. Collard
Ifremer, Plouzané, France

B. Backeberg
Nansen-Tutu Centre, University of Cape Town, Cape Town, South Africa

V. Barale, M. Gade (eds.), *Remote Sensing of the African Seas*,
DOI 10.1007/978-94-017-8008-7_13,

15 km from its mean path (Gründlingh 1983). Observations from a current meter mooring near 31°S have shown that the core of the current lies within 31 km off the coast almost 80 % of the time, displaying surface currents up to 200 cm s^{-1} (Bryden et al. 2005). Southward from 35°S the continental shelf widens forcing the current to separate from the coast and flow southwestward along the Agulhas Bank. In this region the current is becoming increasingly unstable exhibiting numerous meanders, plumes and eddies (Lutjeharms et al. 1989).

In recent years the importance of the greater Agulhas Current has become recognized (Lutjeharms and Bornman 2010). It plays a significant role in the Indo-Atlantic inter-ocean exchange and global thermohaline circulation (Lutjeharms 2006). Using in situ current meter measurements, Bryden et al. (2005) calculated the average poleward volume transport of the Agulhas Current (from 5 March–27 November 1995) to be about 70+/−22 Sv (1 Sv $= 10^6\ m^3\ s^{-1}$). Biastoch et al. (2009) associated the effect of Agulhas leakage dynamics of inter-basin heat and volume fluxes at decadal timescales to the decadal variability in the Atlantic overturning circulation. Recently Rouault et al. (2009) pointed out an augmentation of the transport since the 1980s in response to an increase in wind stress curl in the South Indian Ocean. Yet quantitative knowledge of the seasonal to inter-annual current and transport variability in the greater Agulhas Current regime is incomplete and not adequate to understand the spatial and temporal variability and eddy shedding (leakage) processes into the South Atlantic Ocean.

Modeling the greater Agulhas Current system is challenging, in particular in the very energetic and complex retroflection region, and validation of the model fields often lack in situ data with adequate spatial and temporal resolution. In this regard the regular monitoring of the surface geostrophic current from satellite altimetry has played a very important role both for validation and data assimilation, except near shore where the altimeter measurements are hampered by uncertainties in the correction terms. This can be compensated for by using the Advanced Synthetic Aperture Radar (ASAR)[1] range Doppler signals following the approach outlined by Chapron et al. (2005) and further applied by Johannessen et al. (2008) and Rouault et al. (2010).

In this paper we investigate and assess the surface velocity estimates of the greater Agulhas Current from the Doppler based method, in combination with surface drifter data, satellite altimetry, sea surface temperature measurements, and the simulated surface velocity field from the Hybrid Coordinate Ocean Model (HYCOM).

13.2 Retrieving Surface Velocities from ASAR

Since mid 2007 regular access to Envisat ASAR Wide Swath Mode (WSM) Doppler grid information has led to the creation of a comprehensive data set of Doppler anomalies including more than 1,000 synoptic wide coverage acquisitions with a high-resolution grid (12.5 km × 12.5 km) of the greater Agulhas Current. These data

[1] Onboard the Envisat orbital platform of the European Space Agency (ESA).

are associated with an overall range directed bulk velocity including the mean motion of the radar-detected wind scatters at the ocean surface and the desired ocean surface current.

Using Envisat ASAR observations, Chapron et al. (2005) demonstrated the capability to use the Doppler centroid information embedded in the radar signal to map surface velocity, including wind-generated waves and current, from SAR images. The difference between a predicted Doppler shift based on precise knowledge of the satellite orbit and attitude, and the Doppler centroid frequency estimate represents the geophysical Doppler shift experienced from the moving ocean surface. This geophysical Doppler shift in turn reflects the line-of-sight velocity of the scatterers, weighted by their contribution to the backscattered power (Romeiser and Thompson 2000; Romeiser et al. 2010). A detailed decription of the retrieval and subsequent error correction of the geophysical Doppler shift from the ASAR WSM product is presented in Hansen et al. (2011a). The accuracy of the geophysical Doppler shift is found to be about 5 Hz. This corresponds to a horizontal surface velocity of about 0.2 m/s at an incidence angle of 40°. As such, the accuracy is still an issue in single scenes, although temporal averaging has been shown to capture the mean circulation in e.g. the Agulhas region (Rouault et al. 2010) and in the Norwegian Sea (Hansen et al. 2011b).

The variation of the Doppler shift as a function of wind speed and radar configurations has been determined by collocating wind measurements at 10 m height from the European Centre for Medium-Range Weather Forecast (ECMWF), updated every 6 h with 0.5° spatial resolution, with C-band ASAR Doppler anomalies (Collard et al. 2008). In so doing a global match-up (collocated) database has been established between wind speed and direction from ECMWF with incidence angle and Doppler measurements from ASAR leading to 277211 collocations for VV polarizations (Mouche et al. 2012). Areas with strong currents have been excluded in order to avoid leakage of surface current signals into the wind speed-Doppler shift correlation. From this database an empirical function, named CDOP, that relates the Doppler shift at C-band to the wind field and the radar configuration was then determined as shown in Fig. 13.1 (Mouche et al. 2012). The spread indicated by the vertical bars can be explained by several factors such as the shift in time between ECMWF outputs and ASAR acquisition, errors in the wind from the model, other geophysical phenomena such as rain[2], errors in the Doppler estimates and the non-geophysical corrections applied to the Doppler centroid. Using the CDOP relationship the residual Doppler anomalies associated with the surface current can thus be determined with an r.m.s. error of 5 Hz (equivalent to 0.2 m/s) at a spatial resolution of 8 km (along-track) by 4 km (across-track) as demonstrated by Johannessen et al. (2008) and Hansen et al. (2011b).

The ascending Envisat tracks yield a more favorable imaging geometry with respect to the core flow direction of the greater Agulhas Current as shown in Fig. 13.2. Accordingly, the maximum mean radial Doppler velocity reaches about

[2] Rain impacts the roughness and the local wind speed at a scale not resolved by Numerical Weather Prediction (NWP) models.

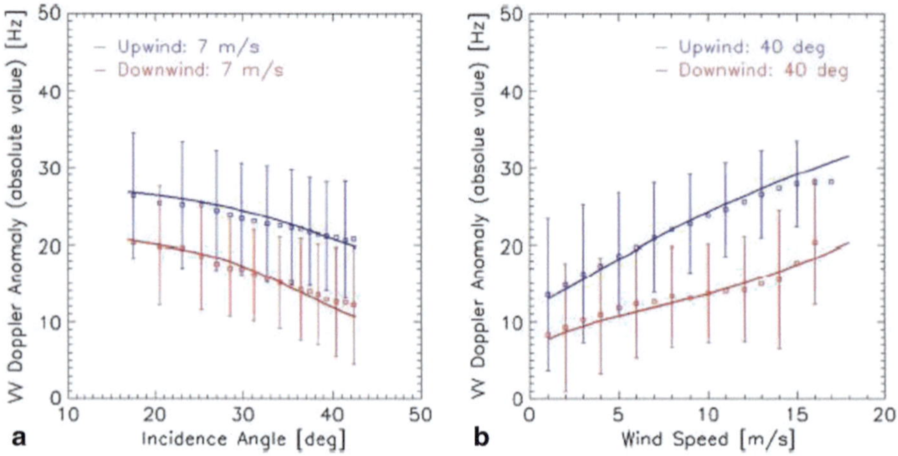

Fig. 13.1 **a** Doppler anomaly as a function of incidence angle for 7 m/s wind speed in upwind (*blue*) and downwind (*red*) configurations. **b** Doppler anomaly as a function of wind speed for upwind (*blue*) and downwind (*red*) at 40° incidence angle

1.5 m/s with a distinct mean current width reaching about 100 km east of 22°E. In comparison, the mean radial velocity obtained from the descending tracks (not shown) is only about 0.8 m/s with a mean width of 80 km. However, these velocity differences almost entirely vanish when the ascending and descending components are rotated into the distinct pathway of the Doppler velocity signal. Note that the CDOP corrected radial surface velocity obtained from individual ASAR acquisitions in the core of the Agulhas Current regularly exceeds 2 m/s (Johannessen et al. 2008).

The spatial structure of the mean radial Doppler velocities suggest that the separation from the coast occur slightly north of 33°S, growing with distances from the coast in the downstream direction. This is in agreement with the topographic steering of the Agulhas Current as reported by Gründlingh (1983). At about 36°S and 22.5°E the range Doppler based surface velocity reveals that the Agulhas Current undergoes an apparent shift in strength and flow direction. Downstream from this region the gradual weakening and discontinuity in the mean Doppler velocity pattern suggests that the Agulhas Current evolves into a less energetic flow towards the retroflection region centered at 38–39°S, 18°E. In this region the sign shifts of the range Doppler velocities are observed followed by the gradual appearance of moderately weaker and broader eastward velocities in the semi-permanent meandering of the Agulhas Return Current (Boebel et al. 2003). All in all the mean range Doppler velocity signature of the Agulhas Current is robust and displaying striking features. Although the Doppler velocity is not a direct surface current measurement, it inevitably suggests that the use of Doppler observations can help to derive new and innovative estimates of mesoscale dynamics provided the wind contribution is reliable removed.

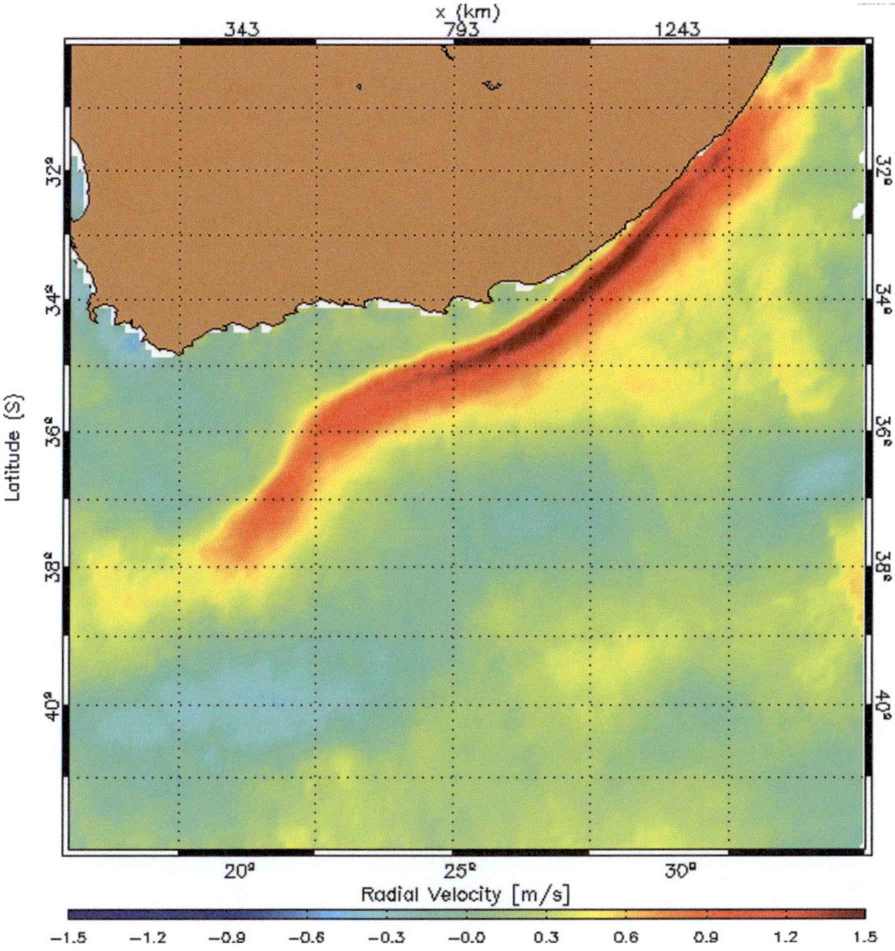

Fig. 13.2 Mean range Doppler velocities of the Agulhas Current derived from more than 600 ascending Envisat ASAR WSM acquisitions from mid-2007 to mid-2011. The color bar marks the speed in m/s with *warm* (*cold*) colors being directed towards west (east)

13.3 Intercomparison and Assessment

13.3.1 Surface drifter data

Using existing surface drifter data[3] the range Doppler based surface velocity estimates can be assessed as shown in Fig. 13.3. The corresponding trajectories of the drifters converge and closely follow the pathway of the core of the Agulhas Current

[3] Available at http://www.aoml.noaa.gov/phod/dac/gdp.html.

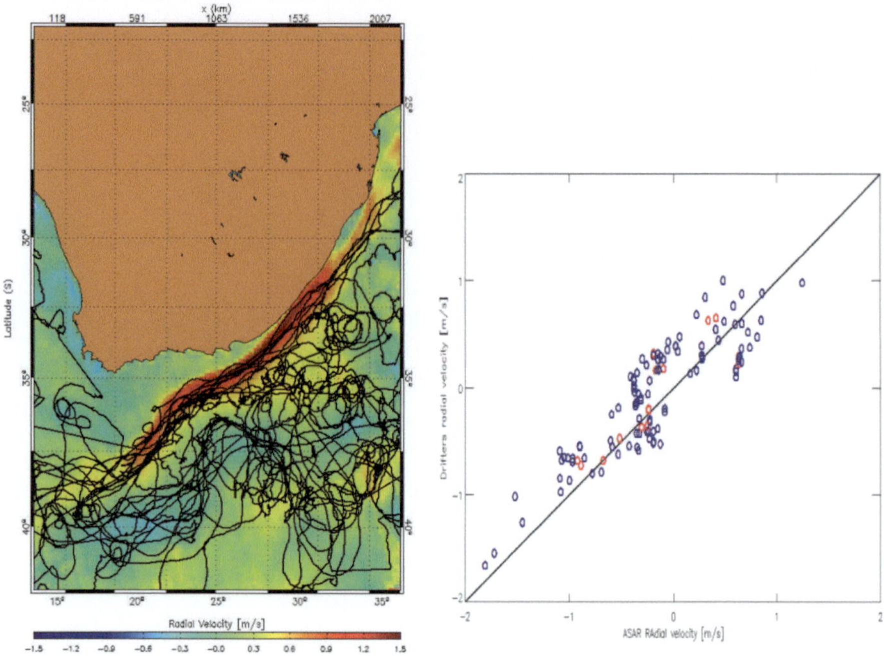

Fig. 13.3 (*left*) Surface drifter trajectories from the period 27 April 2003 to 30 August 2008 overlaid the mean range Doppler velocities shown in Fig. 13.2. (*right*) Scatter plot of Envisat ASAR range Doppler velocities and range component from 7 of the surface drifters from 17 May to 14 August 2008. *Red* (*blue*) symbols mark collocations within 3 (24) h separations

as depicted by the Doppler range velocities (Fig. 13.3, left), although only 7 of the drifters crossed the meridional section at 26°E in the core of the Agulhas Current bounded by 33–35°S. Note also the distinct curvature of the drifter trajectories that occur near 36°S and 22.5°E in agreement with the pattern of the radial Doppler velocities, although none of them were collocated in space and time with the SAR acquisitions in the core region.

On several occasions in the northeastern region the maximum speed of the drifters reached more than 1.7 m/s in very good agreement with the projected Doppler velocities as seen in the scatter diagram (Fig. 13.3, right). Consequently this Doppler shift technique offers the ability to infer valuable estimates of surface current in energetic ocean current regimes. There is also agreement in the location of the apparent merging of the drifters in the Agulhas Return Current and the range Doppler velocity.

13.3.2 Altimetry and Mean Dynamic Topography

The rather steep shelf break tends to stabilize the Agulhas Current and inhibits meander growth as it flows southwestward towards the retroflection area. Consequently

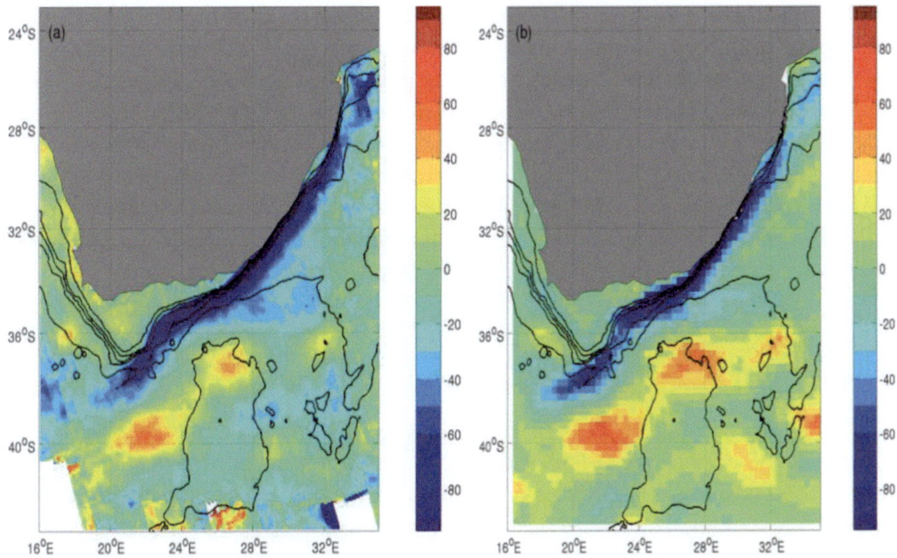

Fig. 13.4 **a** Time-average map of ASAR range velocities from 2007 to 2009. **b** Time-average map of surface geostrophic current derived from Rio09 at a spatial resolution of ¼° and rotated by 15° to conform with the range direction of the velocities observed from ASAR. Color bar indicates surface speed in cm/s. The 200, 500, 1,000, 2,000 and 4,000 m isobaths are plotted in *black*

the sea level anomalies (SLA) are minor and the corresponding estimates of the surface geostrophic current from the altimeter-derived SLAs are suppressed compared to the range Doppler velocities. This was reported by Johannessen et al. (2008) in which the maximum altimeter derived surface geostrophic current was only about 0.7 m/s compared to the Doppler velocity that reaches nearly 2 m/s. Comparing time integrated data, on the other hand, provides better results as illustrated in Fig. 13.4.

The near two decades of gridded absolute surface geostrophic velocity product (Fig. 13.4b) presently distributed by AVISO[4] combines the Sea Level Anomaly (SLA) measurements with the Mean Dynamic Topography (MDT) from Rio09[5] that incorporates an improved estimate of the geoid (using 4 ½ years of GRACE data), together with an updated dataset of drifting buoy velocities (1993–2009), Argo profiling floats and dynamic heights (Rio et al. 2011). The spatial resolutions of the velocity fields are 12.5 km × 12.5 km derived from the Doppler method (left) and about 25 km derived from Rio09. In the latter the maximum zonal speed of about 1.0 m/s is somewhat less than the ASAR derived speed of 1.2 m/s, while the separation from the coast occur in the same place and with the distinct bending of the main flow direction located at about 23°E. The signatures of the return currents are also

[4] See the Centre National d'Etudes Spatiales (CNES) website at www.aviso.oceanobs.com.

[5] Rio09 is the shorthand for Mean Dynamic Topography, after Rio et al. 2011. When surface geostrophic current is expressed as Rio09, it is implied that the mean dynamic topography has been inverted to surface geostrophic current assuming geostrophic balance.

quite comparable. However, in contrast to the comparison for the Agulhas Current the Rio09 surface geostrophic current is stronger than the ASAR based radial velocity for the Agulhas Return Current, reaching up to a zonal speed of 0.85 and 0.70 m/s respectively. This is probably resulting from more variability in the return current that increases the magnitude of the SLA and hence the surface geostrophic current. Finally the mean width of the Agulhas Current core is also fairly comparable being close to about 100 km. As the time integration of the two derived velocity fields are rather different (17 versus 2 years) the agreement suggest that the representation of the mean velocities of the greater Agulhas Current is rather robust.

13.3.3 Sea Surface Temperature Field

The surface thermal expression of the Agulhas Current as portrayed in cloud free infrared satellite images is characterized by a very sharp and rectilinear shore-ward boundary, extending southward from the separation at about 33°S with a Sea Surface Temperature (SST) change ranging from 4 to 6 °C (Lutjeharms 2006). Overlaying the mean SST field, obtained from the ODYSSEA global SST analysis products[6], on the mean range Doppler velocity field (Fig. 13.2) averaged for the same period reveals good agreement and consistency, as seen in Fig. 13.5.

The structure of SST maxima are clearly collocated and oriented with the mean range Doppler velocity maxima from the separation point to about 22°E. The gradual southward turning and spreading of the SST field southwestward from this region is also consistent with the pattern in the velocity map, as is the meandering SST contours representing the Agulhas Return Current eastward from the retroflection point to about 28°E. The combined effect of the strong surface current and SST front implies that significant modification of the surface stress (up to 25–30 %) will occur across this sharp and coupled front in surface current and temperature in agreement with Chelton et al. 2004.

13.3.4 Ocean Model Field

The ocean general circulation model used in this study is the Hybrid Coordinate Ocean Model (HYCOM). HYCOM combines the optimal features of isopycnic-coordinate and fixed-grid ocean circulation models in one framework (Bleck 2002). The name "hybrid" is derived from the models ability to efficiently change from isopycnal to Cartesian coordinates. The adaptive (hybrid) vertical grid conveniently resolves regions of vertical density gradients, such as the thermocline and surface

[6] Available at the Centre ERS d'Archivage et de Traitement (CERSAT), i.e. the French ERS Processing and Archiving Facility. See http://cersat.ifremer.fr/.

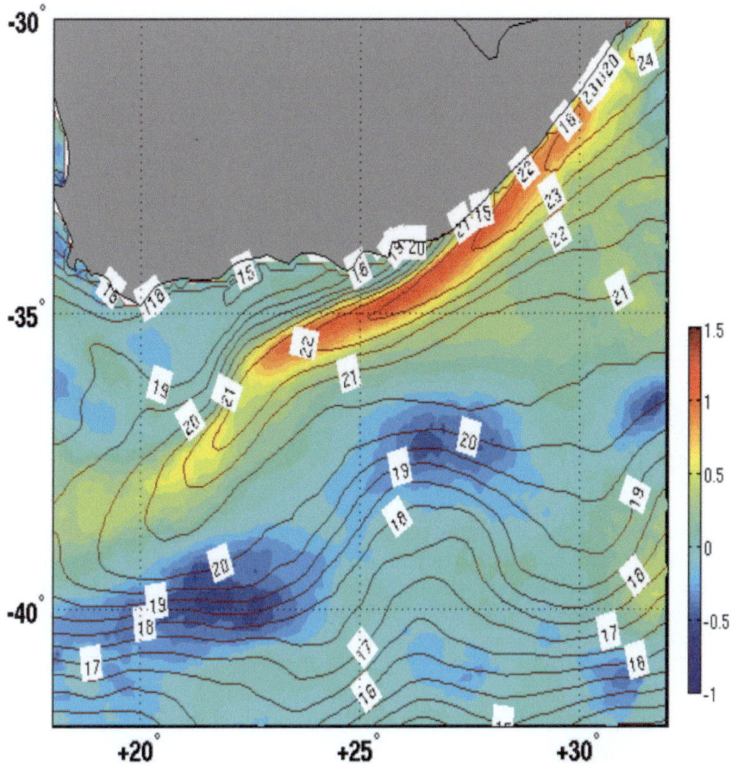

Fig. 13.5 Mean SST field in Celsius derived from the ODYSSEA global SST analysis products for the period mid-2007 to mid-2011 overlaid the mean range Doppler velocity field shown in Fig. 13.2. Color bar indicate surface speed in cm/s. Contour interval is every 0.5° from 14 to 24 °C. The color bar to the *right* represents the range Doppler velocity in m/s

fronts. An eddy resolving HYCOM simulation of the Agulhas region has been thoroughly validated previously using predominantly satellite altimetry measurements of the region (Backeberg et al. 2009).

The entire HYCOM derived velocity field was rotated by $-15°$ (from North) to conform with the range velocities derived from the ASAR measurements. From this comparison, it is evident that the 1/10° ($\sim$10 km) HYCOM (Fig. 13.6c) underestimates the surface velocities in the Agulhas Current core, with maximum poleward range velocities only reaching around 0.75–0.80 m/s, compared to the 1.20–1.30 m/s estimated from ASAR (left) and the $\sim$0.90 m/s derived from Rio09 (middle). Note that the spatial resolution varies from about 10 km for the ASAR velocity field, 25 km for the altimeter field and 10 km for the model simulation. The weaker mean HYCOM surface current which is in closer agreement with the mean altimeter derived surface geostrophic current, represents an integral of the current in the upper 5–10 m of the surface layer. The width of the Agulhas Current core compares well between the

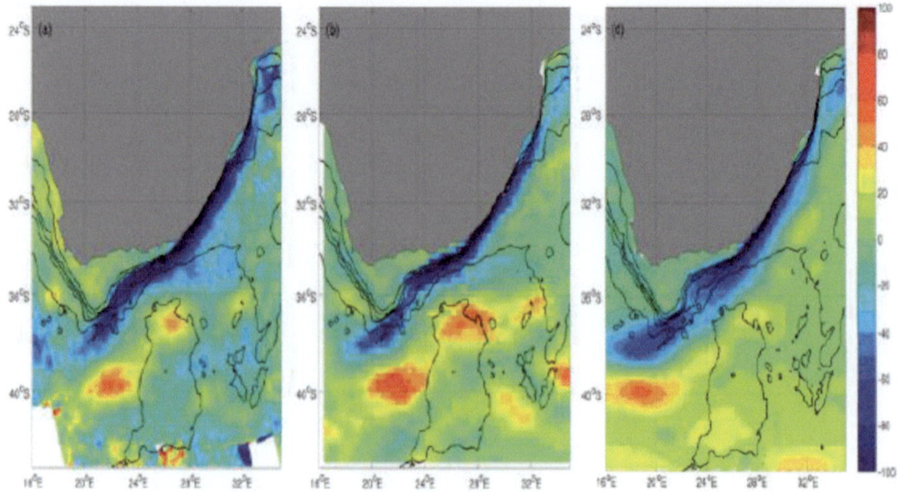

Fig. 13.6 Time-average maps of surface velocity from: **a** ASAR based range Doppler velocity from 2007–2009; **b** surface geostrophic current derived from Rio09 at a spatial resolution of 1/4°; **c** model surface velocities. Note that the velocities in the middle and right panels are rotated by 15° to conform with the ASAR range velocities. Color bar indicates surface speed in cm/s. The 200, 500, 1,000, 2,000 and 4,000 m isobaths are plotted in *black*

ASAR and altimeter derived fields (Fig. 13.6a, b) while the HYCOM has a tendency to simulate a too wide current.

In the northern part of the Agulhas Current the three fields are in fairly good agreement with each other, and following the separation of the continental shelf from the coast near 26°E and 34°S, the Agulhas Current core tends to follow the shelf break revealing a distinct bending pattern at about 22.5°E and 35°S as noticed in all the fields. Near the Agulhas retroflection HYCOM seems to have a too strong westward extension compared to the ASAR and altimeter based observations. Similarly the simulated maximum in the return current is unrealistically elongated and centered to far west compared to the ASAR and altimetry observations. Furthermore the observed maxima at the northern tip of the Agulhas Plateau are not visible in the model field. This study demonstrates the powerful ability to conduct model validation of the surface fields in the Agulhas Current using satellite observations. In view of the strong surface expression found in the satellite data it is also a promising region to carry out data assimilation.

13.4 Conclusions and Future Work

In this paper we have demonstrated that the ASAR mean range velocity estimates provide a new innovative approach to monitor and assess the strength and variability of the upper layer dynamics of the Agulhas Current, in particular, where it lies close to the coast, and the altimetry measurements are limited. Although a dedicated validation campaign is yet to be executed, the long-term persistence displayed in the Doppler velocity maps adds significant confidence in the retrieval method. When compared and assessed against other direct and indirect estimates of the strength of the greater Agulhas Current it is concluded that:

- trajectories and speed derived from surface drifter data provide an excellent source of data for validation of the range Doppler velocity field;
- the agreement between the zonal geostrophic current estimates derived from Rio09 MDTs and ASAR range velocities is very promising in view of both different sampling time and spatial resolution;
- there is a strong consistency between the structure and pathway of the range Doppler velocity field and the SST field which moreover implies that the region may be subject to abrupt anomalies in wind stress and momentum exchange;
- the strength and pathway of the surface velocities of the HYCOM simulation indicates that the model underestimates the mean velocity in the Agulhas Current, while the topographic steering effect of the Agulhas Current, especially in the southern region is not clearly depicted.

Routine provision of corrected range Doppler surface velocities, together with contemporary altimetry combined with new MDT field, surface drifter data and SST maps provide an extended and more comprehensive data set for studies and monitoring of the upper ocean dynamics and its variability in intense current regimes. Combined with in situ data the volume and heat transport estimations can be strengthen. It will also impose new invaluable possibilities for model validation and assimilation, not just how well the geostrophic current and transports are reproduced, but more importantly how their internal vorticity and divergence field maintain ageostrophic velocity and vertical motions. In agreement with Ljuteharms and Bornman (2010), this routine monitoring effort should stimulate scientists and also broaden the base of understanding of the marine environment and ecosystems to inform decision makers.

Acknowledgements This work was supported by the Research Council of Norway, project number 177441/V30 and by ESA through the study contract no. 18709/05/I-LG. ESA is also acknowledged for implementing systematic ASAR wide swath acquisitions over the greater Agulhas Current making this study possible. We are also grateful to the Global Drifter Program at their server at AOML in Miami http://www.aoml.noaa.gov/phod/dac/gdp.html and to CLS for their support to the range Doppler analyses. The modelling work presented in the paper was supported by the Mohn-Sverdrup Center for Global Ocean Studies and Operational Oceanography, through a private donation from Trond Mohn C/O Frank Mohn AS, Bergen, Norway and the National Research Foundation, South Africa, while the CPU time was covered by the Norwegian Super-computing project. Dr. Bjørn Backeberg is supported by the Nansen-Tutu Centre for Marine Environmental Research at University of Cape Town, Cape Town, South Africa.

References

Backeberg BC, Bertino L, Johannessen JA (2009) Evaluating two numerical advection schemes in HYCOM for eddy-resolving modelling of the Agulhas Current. Ocean Sci 5(2):173–190
Biastoch A, Böning CW, Schwarzkopf FU, Lutjeharms JRE (2009) Increase in Agulhas leakage due to poleward shift of southern Hemisphere westerlies. Nature 462:495–498
Bleck R (2002) An oceanic general circulation model framed in hybrid isopycnic-Cartesian coordinates. Ocean Model 37:55–88
Boebel O, Rossby T, Lutjeharms JRE, Zenk W, Barron C (2003) Path and variability of the Agulhas Return Current. Deep Sea Res 50:35–56
Bryden HL, Beal LM, Duncan LM (2005) Structure and transport of the Agulhas Current and its temporal variability. J Oceanogr 61:479–492
Chapron B, Collard F, Ardhum F (2005) Direct measurements of ocean surface velocity from space: interpretation and validation. J Geophys Res 110:C07008
Chelton DB, Schlax MG, Freilich MH, Milliff RF (2004) Satellite radar measurements reveal short-scale features in the wind stress field over the world ocean. Science 303:978–983
Collard F, Mouche A, Chapron B, Danilo C, Johannessen JA (2008) Routine high resolution observation of selected major surface currents from space. Proceedings SeaSAR 2008, p 8, ESA SP-656. ESA Communication Production Office, ESTEC, Noordwijk, Netherlands
Gründlingh ML (1983) On the course of the Agulhas Current. S Afr Geogr J 65(1):49–57
Hansen MW, Collard F, Dagestad KF, Johannessen JA, Fabry P, Chapron B (2011a) Retrieval of sea surface range velocities from Envisat ASAR Doppler centroid measurements. IEEE Trans Geosci Remote Sens 49(10):3582–3592
Hansen MW, Johannessen JA, Dagestad KF, Collard F, Chapron B (2011b) Monitoring the surface inflow of Atlantic water to the Norwegian sea using Envisat ASAR. J Geophys Res-Oceans 116:C12008. doi:10.1029/2011JC007375
Johannessen JA, Chapron B, Collard F, Kudryavtsev V, Mouche A, Akimov D, Dagestad KF (2008) Direct ocean surface velocity measurements from space: improved quantitative interpretation of Envisat ASAR observations. Geophys Res Lett 35:L22608
Lutjeharms JRE (2006) The Agulhas Current. Springer-Praxis Books
Lutjeharms JRE, Bornman TG (2010) The importance of the greater Agulhas Current is increasingly being recognised. S Afr J Sci 106(3/4):4 (#160). doi: 10.4102/sajs.v106i3/4.160
Lutjeharms JRE, Catzel R, Valentine HR (1989) Eddies and other boundary phenomena of the Agulhas Current. Cont Shelf Res 9:597–616
Mouche A, Dagestad KF, Collard F, Guitton G, Chapron B, Johannessen JA, Kerbaol V, Hansen MW (2012) On the use of Doppler shift for sea surface wind retrieval from SAR. IEEE Trans Geosci Remote Sens 50(7):2901–2909
Rio MH, Guinehut S, Larnicol G (2011) New CNES-CLS09 global mean dynamic topography computed from the combination of GRACE data, altimetry, and in situ measurements. J Geophys Res 116:C07018. doi: 10.1029/2010JC006505
Romeiser R, Johannessen JA, Chapron B, Collard F, Kudryavtsev V, Runge H, Suchandt S (2010) Direct surface current field imaging from space by along-track InSAR and conventional SAR. In: Barale V, Gower JFR, Alberotanza L (eds) Oceanography from space, revisited. Springer, Heidelberg, pp 73–91
Romeiser R, Thompson DR (2000) Numerical study on the along-track interferometric radar imaging mechanism of oceanic surface currents. IEEE Trans Geosci Remote Sens 38(I):446–458
Rouault M, Penven P, Pohl B (2009) Warming in the Agulhas Current system since the 1980's. Geophys Res Lett 36:L12602
Rouault MJ, Mouche A, Collard F, Johannessen JA, Chapron B (2010) Mapping the Agulhas current from space: an assessment of ASAR surface current velocities. J Geophys Res. Vol. 115, C10026 doi:10.1029/2009JC006050

Chapter 14
Internal Solitary Waves System in the Mozambique Channel

Jorge M. Magalhaes, José C. B. da Silva and Adrian L. New

Abstract Satellite images of the Mozambique Channel (MC), collected by Synthetic Aperture Radars (SARs) from the Envisat and European Remote Sensing (ERS) satellites, reveal sea surface signatures of oceanic Internal Solitary Waves (ISWs). The MC has been expected to be a major hotspot for the generation of Internal Tides (ITs) in the ocean. New results now show that the Sofala bank (or shelf, located near 20°S, 36°E) is indeed a strong and previously unknown source for the generation of ITs and ISWs. The hotspot region is located between 20 and 21°S, where the barotropic tidal forcing and tidal transports are highest in the area, due to the particular nature of the bathymetry there. Two distinct types of ISWs are observed to travel offshore, away from the shelf-break, and their possible generation mechanisms are discussed. Mode 2 ISWs propagating in MC are also observed in SAR imagery. It is discussed that they may be generated by a tidal beam impinging the thermocline from below, in a similar process to some of the Mode 1 ISWs generated off the Sofala bank. ISW refraction patterns, often observed in SAR imagery in MC are presented, and a particular example is shown where ISWs and eddy-like features are seen to interact. The influence of these rotating features in the refraction patterns is briefly discussed. Finally, it is shown that large-scale atmospheric gravity waves of solitary-like form are also frequently observed in the region, and thus care must be taken when discriminating oceanic and atmospheric ISW signatures in SAR images, if a correct interpretation is to be made.

J. M. Magalhaes (✉) · J. C. B. da Silva
Department of Geosciences, Environment, and Spatial Planning,
Interdisciplinary Centre of Marine and Environmental Research, CIMAR,
Porto, Portugal
e-mail: jmagalhaes@fc.ul.pt
Also at: Faculty of Sciences, University of Porto, Porto, Portugal

A. L. New
Marine Systems Modelling Group, National Oceanography Centre,
Southampton, Southampton, UK

V. Barale, M. Gade (eds.), *Remote Sensing of the African Seas*,
DOI 10.1007/978-94-017-8008-7_14,

14.1 Introduction

Internal Waves (IWs) are an important factor in several processes within geophysical fluid dynamics. They are ubiquitous features in the oceans and in the atmosphere, which can exist in a wide variety of time and space scales, and owe their existence to the vertical density stratification that characterizes both mediums.

In the ocean, tidal flow over irregular bottom topography can force vertical motions at the tidal frequency and generate IWs of tidal period, which can then propagate along the thermocline and are often referred to as Internal Tides (ITs). ITs can steepen and generate IWs of shorter period that are usually termed Internal Solitary Waves (ISWs). The term 'solitary' is used since they tend to occur in individual packets (usually trapped in the troughs of the IT), and have often been identified with the soliton solutions of nonlinear wave theory. In the atmosphere, these short-scale waves that are usually highly nonlinear, can also be observed to propagate in the lower troposphere, and are often accompanied by roll-cloud formations that resemble the packet-like structure of their oceanic counter-part. Their origin can be attributed to various generation mechanisms, often involving convection, wind-shear, gravity currents and different air mass collisions. In summary, any atmospheric process leading to the displacement of density surfaces can be a potential source of atmospheric IWs, providing that suitable propagation conditions are previously met.

During the last few decades remote sensing has played a key role, in providing the scientific community with an increasing number of IW observations (both of oceanic and atmospheric nature). Imaging sensors such as Synthetic Aperture Radars (SARs), as well as visible and near-infrared spectrometers, have proved to be particularly useful in providing new and valuable insights into these waves' physical characteristics. These new measuring capabilities boosted the number of observations, and led to several studies that indicated that the phenomenon is in fact much more frequent than previously thought (see e.g. Jackson 2007; Jackson et al. 2012; da Silva and Magalhaes 2009; Magalhaes et al. 2011). SARs in particular have proved very useful in revealing the two dimensional structure of IWs, as well as their spatial and temporal distributions. It is well known that both oceanic and atmospheric IWs are capable of producing roughness patterns on the ocean surface. Therefore, they can generate recognizable signatures in radar images. The oceanic roughness patterns are a result of either hydrodynamic modulation of Bragg waves by surface current gradients or wave damping due to surface films (Alpers 1985; da Silva et al. 1998). Similarly, the atmospheric IW signatures are a consequence of wind stress variability introduced by the disturbed wind velocity field associated with wave propagation (Alpers and Stilke 1996). SAR imagery has also been playing a key role in identifying and characterizing several IW hotspots, as well as in providing valuable insights into the generation mechanisms (see e.g. New and da Silva 2002; Vlasenko and Alpers 2005; Nash and Moum 2005; da Silva and Helfrich 2008; da Silva and Magalhaes 2009; da Silva et al. 2009, 2011; Magalhaes et al. 2011).

IWs can play an important role in several dynamic aspects and in a wide variety of geophysical processes. As already mentioned above, a more special case of these waves are the solitary waves of finite amplitude and permanent form, which owe their existence to a delicate balance between linear wave dispersion and nonlinear wave steepening. These are usually found in rank-order packets (a consequence of their speed being an increasing function of amplitude) and they have been observed to travel for considerable amounts of distance without significant changes in form. In turn, this is a clear indication that these waves can in fact be responsible for transferring considerable amounts of energy throughout large distances, which is one of the main reasons why they are so important in the dynamics of the coastal oceans and in the lower troposphere.

On the one hand, ISWs in the ocean have been proved to significantly contribute to ocean mixing, either during breaking processes in their final stages of propagation, or due to their highly turbulent character as they propagate along the pycnocline (Pinkel 2000; Moum et al. 2003; van Haren et al. 2005; Lamb and Farmer 2011). In fact, recent studies have shown that IWs contribute considerably to shelf dynamics in terms of vertical heat fluxes and ocean-shelf mass transports (Shroyer et al. 2010a, 2010b). Mixing due to IWs has also been shown to be important from a biological point of view. Several studies have shown that the dissipation of ISWs may constitute a primary mixing mechanism, and that it may be sufficient to induce strong upward fluxes of nutrients across the thermocline and produce bursts of primary production (see e.g. Sandstrom and Elliott 1984; Pingree et al. 1986). Da Silva et al. (2002) also suggested that enhanced primary production, could simply occur due to the increase and decrease in available light experienced by phytoplankton, during up and downward advection induced by IWs.

On the other hand, atmospheric IWs of solitary form—sometimes also termed Atmospheric Gravity Waves (AGWs)—have been shown to constitute a serious hazard to aircrafts operating at low altitudes and small speeds (see e.g. Christie and Muirhead 1983; Bedard et al. 1986). Some large-scale AGWs may also have a significant role in establishing low-level circulation and determining the vertical structure of wind, temperature and moisture. They can even modify and trigger convective motions and induce significant rain (Reeder et al. 1995). Their ability to transfer energy and momentum (together with the fact that they propagate for several hundreds of kilometres) makes them an important topic for research, especially since their influence can also extend into weather forecast and climate models.

Furthermore, ISWs can sometimes occur in regions where both mediums (ocean and atmosphere) can support simultaneous propagation. In these cases, care should be taken to avoid miss-interpretation. It is also important to recognize that atmospheric internal waves may not always be associated with cloud structures, since the ascending branches of the waves may not find the necessary conditions (high moisture in the atmosphere) for condensation to occur (see e.g. Crook 1986). In this case, there are no guaranteed methods to discriminate between oceanic and atmospheric ISW sea surface signatures in SAR images. However, recent research based on solitary wave and radar imaging theory has provided useful criteria (sometimes with complementary information on the ocean and atmosphere) to help in the distinguishing process (see Alpers and Huang 2011).

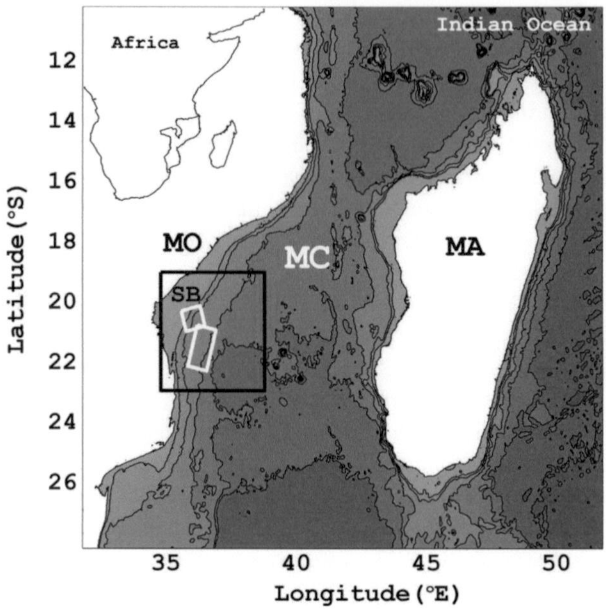

Fig. 14.1 Detailed bathymetry and location of the Mozambique Channel (labelled *MC*) in the southwest Indian Ocean. The labels *MO* and *MA* stand for Mozambique and Madagascar, respectively, and a reference location for the Sofala Bank is labelled *SB*. The inset in the *top-left* corner shows the locations of Mozambique and Madagascar in relation to the African continent. *White* areas represent land and *black* contours with *grey* areas mark isobaths for 200, 1,000, 2,000, 3,000 and 4,000 m. The *white rectangular* frames delineate the area surveyed in the SAR images shown in Figs. 14.2 and 14.3. The *black rectangular* frame is shown for reference and outlines the detailed area in Figs. 14.4 and 14.5

The Mozambique Channel (MC), sometimes also referred as the Strait of Madagascar, is a relative narrow passage centred at 20°S between the African continent (i.e. Mozambique) and the island of Madagascar (see Fig. 14.1). It extends approximately from the northeast to the southwest for more than 1,500 km. The width of the channel is approximately 400 km in the narrowest passage, while at 20°S it reaches a maximum width of about 1,000 km. The continental shelves on both sides are generally narrow, and are confined by supercritical continental slopes. However, near 20°S the shelf extends much further into the open ocean, for more than 100 km (in the Sofala Bank region, see Fig. 14.1 for location) and it is there that the tidal currents are the largest in MC.

This region is a privileged site for the observation of IWs both in the ocean and in the atmosphere. The existence of oceanic IWs is in agreement with Baines (1982), who estimated tidal energy fluxes from several shelf regions of the world's oceans, and showed that MC was among the 12 most energetic regions for IT generation. Here, the (semidiurnal) barotropic tidal ranges are over 7 m, and the barotropic currents typically exceed 0.50 ms^{-1} at spring-tides, with significant spring-neap variability. The shelf slope is also supercritical with respect to the M_2 barotropic tide (also known as the principal lunar semi-diurnal tide, with a period of 12.42 h), allowing the existence of critical regions for IT generation at the shelf break. Also noteworthy, is the fact that the flow in MC is influenced by large, southward propagating eddies, which feed their waters into the Agulhas Current system, and therefore constitute a major contribution to the overall flow within the channel.

14.2 Oceanic ISWs

14.2.1 Introduction

Internal tidal energy generated at the shelf-break may radiate away either horizontally in the form of interfacial ITs that propagate along the thermocline, or as internal tidal beams that propagate into the deep stratified ocean below. Therefore, large interfacial ITs may form in the thermocline directly above the shelf break and evolve (through nonlinear processes) to higher-frequency ISWs packets. Another possibility is that the IT energy, which propagates downward into the deep ocean, may give rise to a second generation mechanism known as "local generation" (New and Pingree 1992; Gerkema 2001; New and da Silva 2002; Akylas et al. 2007; Grisouard et al. 2011; Mercier et al. 2012). Here, a beam (or ray) of IT energy is generated at "critical" slope regions of the shelf-break where the bottom topographic slope matches the slope of the ray paths, and propagates at an angle (θ) to the horizontal into the deep ocean interior (see Eq. 14.1) where σ is the tidal frequency, f the Coriolis parameter and N the buoyancy frequency). These rays reflect from the seafloor (Pingree and New 1989, 1991), and interact with the thermocline from below, causing large IT oscillations there, and "locally" generating ISWs (far from the shelf-break where the beam is initially generated). We will show clear evidence (based on remote sensing SAR images and modelling results) that both generation mechanisms are at work in MC, which means that there are waves being formed directly at the shelf break, and further into the open ocean through local generation.

$$\tan(\theta) = \pm\left(\frac{\sigma^2 - f^2}{N^2 - \sigma^2}\right)^{1/2} \tag{14.1}$$

This hypothesis was originally proposed to explain some ISW observations in the northern Bay of Biscay, where it was studied for the first time, and where SAR imagery helped to clearly distinguish a bimodal distribution of ISWs. This distribution results from the direct generation mechanism near the shelf-break and from the local generation processes taking place some 150 km further into the ocean (New and da Silva 2002). Nonetheless, the widespread occurrence of the local generation mechanism in other regions of the world's oceans has been recently put forward with the contribution of in situ and satellite data. In fact, several other studies have emerged since, which have also identified locally generated waves. Namely, the southern Bay

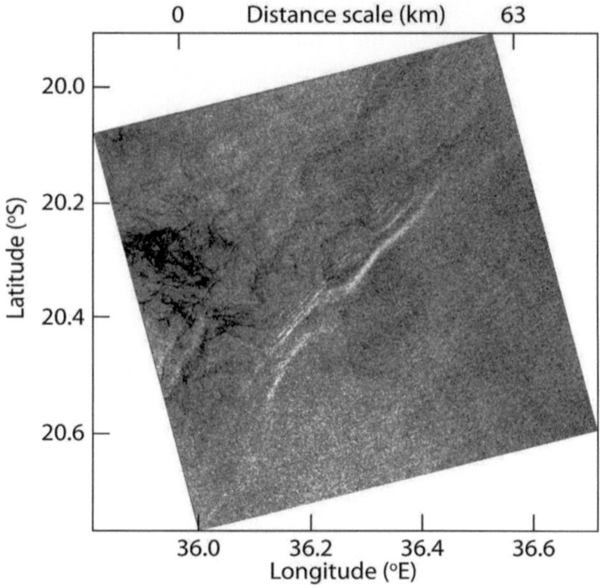

Fig. 14.2 Subset of an Envisat-ASAR image (WS mode) showing a case of an ISW packet generated directly over the shelf. The image is dated 4 December 2009 and was acquired at 20 h 07 m UTC. It is centred approximately at 36.26°E, 20.33°S, with an area of about 77 × 77 km^2 to the east of Mozambique, which is outlined in Fig. 14.1 by the *white square line*

of Biscay (Azevedo et al. 2006), the Estremadura Promontory (da Silva et al. 2007), and more recently the MC (da Silva et al. 2009), have also been found as major hotspot regions for the local generation mechanism.

14.2.2 SAR Evidence of ISWs

Figure 14.2 presents a subset of an Envisat ASAR image (in Wide Swath mode, WS) from 4 December 2009 (centred at 36.26°E, 20.33°S; see Fig. 14.1 for location), and corresponds to an area of approximately 77 × 77 km^2. Several ISWs can be identified as belonging to a packet resulting from direct generation over the shelf. Their sea surface signatures, which can be identified in the centre of the image, are consistent with mode 1 waves of depression.

At least 3 individual solitons are already formed in the northeast end of the packet, but their coherence is lost towards the other end. This together with the short wavelengths is an indication that these waves have been recently formed. Another example of ISWs propagating farther off the Sofala Bank is given in Fig. 14.3. This image shows a subset of an Envisat ASAR image, in Wide Swath (WS) mode, dated 7 April 2003 (centred at 36.6°E, 21.3°S; see Fig. 14.1 for location), which corresponds to an area of approximately 76 × 156 km^2.

It is a typical example of the local generation mechanism with waves being generated locally and farther off-shelf than the ones shown in the previous example. Three well-developed waves can be clearly distinguished propagating towards the southeast with larger crestlengths compared with those in Fig. 14.2 (the first of which

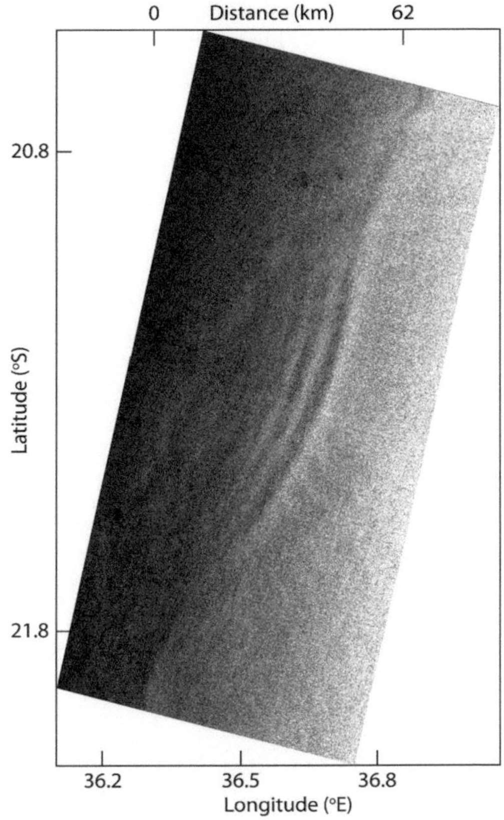

Fig. 14.3 Subset of an Envisat-ASAR image (WS mode) showing a typical case of locally generated ISWs. The image is dated 7 April 2003 and was acquired at 07 h 09 m UTC. It is centred approximately at 36.6°E, 21.3°S, with an area of about $76 \times 156\,\text{km}^2$ to the east of Mozambique, which is outlined in Fig. 14.1 by the *white rectangle*

with more than 100 km). It can also be seen that these individual waves are organized in a rank-ordered fashion, and that there are brighter bands preceding darker ones in the direction of propagation (a typical feature of mode 1 waves of depression).

Several other examples of ISW signatures have been found in similar images acquired over the MC. A composite map of these examples has been assembled (to be presented in the next Section) where it will be clearly seen that (similarly to the Bay of Biscay) there are two distinct concentrations of ISW signatures. These local maxima are consistent with simultaneous generation near the shelf-edge and farther offshore, owing to the local generation mechanism.

14.2.3 Tidal Forcing

We now examine the generation issue more closely in order to investigate the most likely locations for the generation of these IWs. To do so the methodology that is usually presented in the literature consists of two main steps. On the one hand, the

Baines (1982) barotropic tidal forcing is often used to search and identify hotspot regions of ISWs within a given study area. It has proved to be a valuable indicator in several other independent studies, particularly to study the possibility of the local generation mechanism. For instance, Azevedo et al. (2006) and da Silva et al. (2007) have used this method to investigate the local generation mechanism in the southern Bay of Biscay (near Cape Finisterre), and in the southern slopes of the Estremadura Promontory (off the west Iberian shelf). In addition, there have been other authors that also used this technique over the last decade, as a valuable indicator of where large ITs may be generated (e.g. Colosi et al. 2001; Merrifield and Holloway 2002; Niwa and Hibiya 2004). On the other hand, ray-tracing techniques are used to see if the locations where ISWs first appear are consistent with the local generation mechanism (i.e. if the first ISW packets appear a few kilometers ahead of the expected impact of an IT beam with the pycnocline). Note that, ITs in a continuously stratified ocean can be described by beams or rays that follow characteristic pathways (along which their energy can propagate—see Eq. 14.1). In other words, this means that when the bottom slopes match the local value of tan(θ) (called a critical region) the generation of ITs is more pronounced, particularly if the barotropic currents are strong. In such cases, the direction of the forcing barotropic flow is then coincident with the motion plane for free internal waves, resulting in resonant conditions and enhanced generation of the ITs and ISWs.

The barotropic forcing term (following Baines 1982) for ITs, resulting from the interaction of bottom topography and tidal flow, can be defined as

$$F = -z\, N^2(z) \left[\int Q_x dt \frac{1}{h^2} \frac{\partial h}{\partial x} + \int Q_y dt \frac{1}{h^2} \frac{\partial h}{\partial y} \right] \tag{14.2}$$

where z is the vertical coordinate (positive upwards), Q is the barotropic mass flux vector $Q = (uh, vh)$ with u and v being the zonal and meridional components of the barotropic velocity, and h is the ocean depth. This means that F can be analytically integrated provided that Q is previously known, and for that purpose the components of the barotropic velocity vector were taken from the 1/8° resolution OTIS model (Oregon state university Tidal Inversion Software, developed by Egbert and Erofeeva 2002). The model included the M_2 and S_2 tidal constituents (periods of 12.42 and 12.00 h, respectively) of the barotropic tide since these are the most important in this study region (see da Silva et al. 2009). The bathymetry data is part of the one minute global bathymetry from Smith and Sandwell (1997), and N (assumed to be spatially constant) is the same used in da Silva et al. (2009) for the October stratification (see their Fig. 2a).

Figure 14.4 shows the tidal ellipses for a complete semi-diurnal tidal cycle corresponding to the image in Fig. 14.2 (4 December 2009). These current ellipses, that are needed to calculate Q, were derived using least-squares fits to the data in each grid point. Note that the original outputs do not form perfectly closed ellipses due to the slowly varying nature of the tide.

It can be seen that the barotropic tidal currents achieve their biggest values over the continental shelf (in the Sofala Bank region), after crossing the 200 m depth contour.

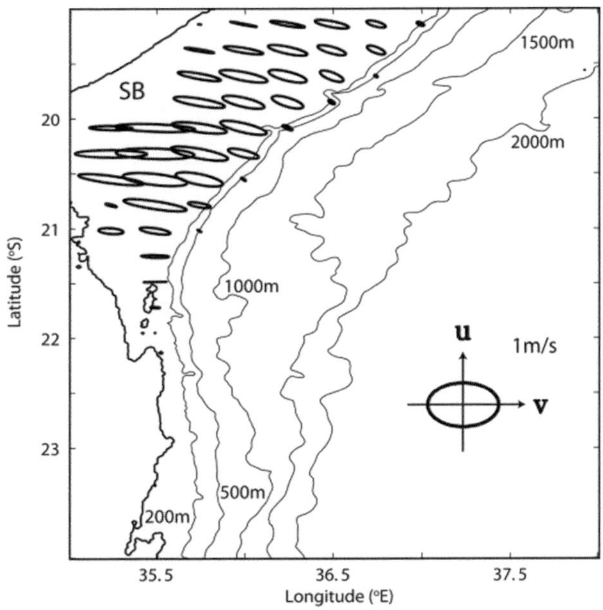

Fig. 14.4 Detailed bathymetry of the area outlined by the *black rectangle* in Fig. 14.1 together with ellipse tidal currents for the date of the image in Fig. 14.2 (4 December 2009). A reference ellipse is shown on the *right-hand side* (with both axes representing a speed of 1 ms^{-1}). For clarity ellipses with axes below 0.05 ms^{-1} are not represented. Depth contours are also shown in meters (starting at 200, 500, 1,000, 1,500 and 2,000 m). A reference location for the Sofala Bank is labelled *SB*

It can also be seen that these are approximately aligned with the bathymetry gradient. This indicates that there will be a strong tidal forcing owing to strong tidal currents passing directly over steep slopes. The magnitudes of these currents (see reference ellipse on the right) can reach relatively large values, some close to 1 ms^{-1} farther inside the inner-shelf. However, there is a rapid decrease in the tidal currents towards the open ocean areas that starts immediately after crossing the shelf-break (with an average depth of 50 m). We note in passing, that these values over the Sofala Bank are in fact the biggest anywhere within MC, and are among the biggest anywhere in the world.

Figure 14.5 shows the overall maximum of the depth-integrated body force within a complete semi-diurnal tidal cycle for the Sofala Bank area (with the tidal currents computed for 4 December 2009, and shown in Fig. 14.4). Together with this 2D body force, a composite map of ISWs observations is also shown for ease of interpretation, in terms of the hotspot regions associated with observations. The dataset used in the composite map contains a total of 6 images acquired from 2003 to 2006, during the southern spring and summer (a complete description may be found in da Silva et al. 2009; see their Table 1). Note that the present body force map shown in Fig. 14.5 was computed for a different day of the ISWs presented in this figure, but results would remain fairly the same for any other choice of dates. A closer inspection of the body force in Fig. 14.5 shows that there are localized regions of maximum forcing coincident with steep slopes and focused along a narrow line (with depths approximately between 100 and 500 m) that is generally along the edge of the continental shelf. The tidal forcing is actually very weak everywhere else (in general), meaning that the body force values decay rapidly to zero in the shallower

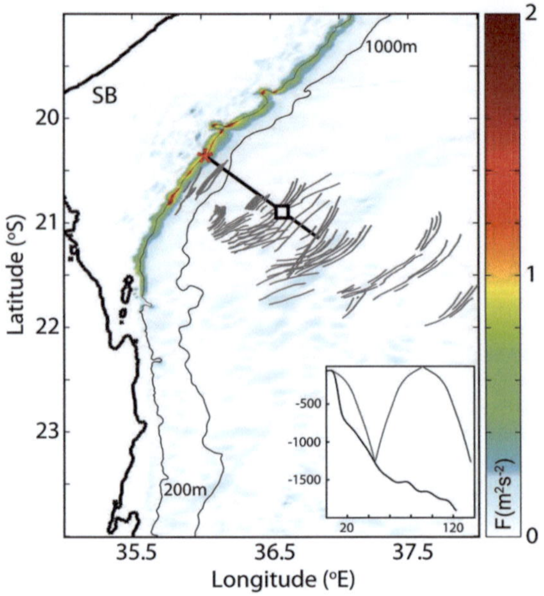

Fig. 14.5 Maxima of Baines (1982) depth-integrated body force (F) over a complete tidal cycle, calculated using Eq. (14.2) for 4 December 2009 (same date as Figs. 14.2 and 14.4). The area shown corresponds to the *black rectangle* in Fig. 14.1. A reference location for the Sofala Bank is labelled *SB*. The coastline is marked by a *black contour* and the *thin lines* are the 200 and 1,000 m isobaths. A composite map of ISW observations (six in total, see text for details) propagating in MC is also shown in *grey thick lines*. The inset on the *bottom-right corner* is a vertical profile corresponding to the section in a *black line*, which goes from the region of maximum F towards the southeast, along the ISWs propagation path. The vertical and horizontal axes are depth in meters and distance in kilometres (measured from the northwest end). The vertical section shows a ray-tracing diagram for the M_2 tidal constituent and assuming stratification typical for the summer in the MC. The *red star* in the map represents the critical bathymetry from where IT beams are generated, and the *black square* is the corresponding reemergence at the thermocline (see text for more details)

areas above 100 m, and in the deep ocean basin that extends farther away into the MC.

The section shown by a black line in Fig. 14.5, which extends in an approximately perpendicular direction to the shelf-edge (from the northwest to the southeast), is representative of a vertical profile for IT ray-paths (see the inset shown in the lower right corner). The red star marks the position of critical bathymetry along that same profile for the M_2 tidal constituent and using the October stratification used earlier (the same used to calculate F; see Fig. 2a in da Silva et al. 2009). The black square is the corresponding reemergence at the thermocline. It can be seen that there are internal tidal beams originating from critical topography coincident with elevated regions of F that initially slope downward into the ocean interior. The IT beam then reflects upward from the seafloor between 1,100 and 1,300 m, and reemerges later at the thermocline at about 90 km offshore.

Considering the relative positions of the ISWs signatures and the geometry of the IT beam reflection, a region of low activity can be identified near the shelf-break that lies behind the re-emergence of IT energy. This region is followed by another area of ISW activity that closely coincides with the surfacing positions of the IT rays. Similarly to the case in the Bay of Biscay, this suggests that there is a region of elevated forcing near the shelf edge that is simultaneously generating ISWs directly above, and IT rays that are propagating to the deeper ocean, reflecting from the sea floor, and locally generating ISWs some 90 km farther offshore.

14.3 Mode 2 ISWs

14.3.1 Introduction

IWs propagation is naturally bounded, both above and below by the ocean's surface and bottom—thus confining them within a natural waveguide. However, the ocean's pycnocline can also be seen as another form of waveguide, since it confines the higher frequency oscillations into interfacial waves (e.g. the ISWs seen in SAR imagery). The vertical confinement imposed by these boundaries results in horizontally propagating waves, which nonetheless have a standing character in the vertical. In this case, a classical Sturm-Liuville problem arises if the linear equations of motions (under the Boussinesq and Traditional approximations) are solved with boundary conditions for zero vertical velocities at the surface and bottom of the ocean. Therefore, IW solutions with an infinite number of modes can either exist in nature on their own or in any number of combinations (see e.g. Gerkema and Zimmerman 2008 for a detailed description on IW dynamics). The first case, usually results in low mode propagation of IWs (like interfacial ITs or ISWs), while the second may result in an increasing superposition of modes, which makes the spatial appearance of the IW field to resemble a beam structure (like IT beams). In the particular case of SAR imagery, ISWs are generally described as interfacial mode 1 waves of depression (propagating along the pycnocline, and depressing it) that are usually ridding on the troughs of the IT (which is also an interfacial disturbance of the pycnocline). However, recent attention has been given to a different set of observations, where SAR surface signatures are not consistent with mode 1 ISWs, but rather suggest mode 2 solitary-like waves (for simplicity referred to in this chapter as mode 2 ISWs). In fact, despite that first-mode internal waves are most commonly observed in the ocean, higher-mode waveforms also occur under the proper conditions. A mode-2 wave is best thought of in terms of a three-layer stratified ocean, when the middle-layer thickness is relatively thinner than the upper and lower layers. In this configuration, the wave travels as a bulge in the middle layer, displacing isopycnals upward into the upper layer and downward into the bottom layer (see e.g. Shroyer et al. 2010c).

Cases of satellite observations and in situ measurements of mode 2 oceanic ISWs are seldom, and they are hardly ever found in the literature. In fact, despite that they have long been predicted in theory, it was not until recently that they were seen to

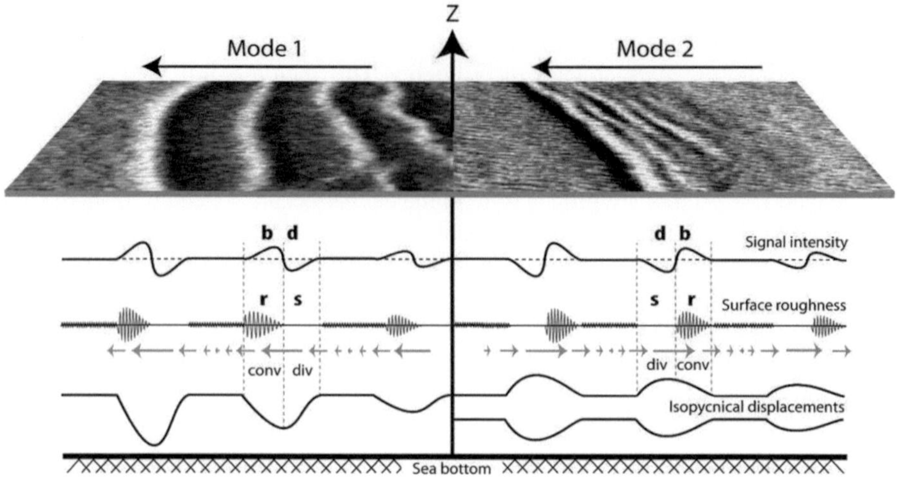

Fig. 14.6 Examples of SAR images with sea surface signatures of mode 1 and 2 ISWs are shown at the top (*left and right*, respectively). The ISWs are assumed to be moving from *right to left*. From *top to bottom* the horizontal profiles represent the following features: SAR intensity profile along the ISWs, with bright enhanced backscatter (*b*) preceding dark reduced backscatter (*d*) in the direction of propagation for mode 1 waves; surface roughness representation indicating how rough (*r*) and smooth (*s*) the surface is along an ISW wave packet; surface current variability induced by ISWs (note indication of convergence and divergence fields near the surface); isopycnical displacements produced by ISW propagation. Note how the SAR signatures, surface roughness, current fields and convergence and divergence patterns are all reversed between mode 1 (*on the left*) and mode 2 (*on the right*) ISWs

exist in nature using in situ and satellite image data (see e.g. Shroyer et al. 2010c; da Silva et al. 2011). However, SAR observations now suggest that mode 2 ISWs can be more frequent than previously acknowledged. In this Section it will be shown that mode 2 ISWs can also be observed in MC, and are generated by IT beams with semi-diurnal frequency.

Sea surface signatures of mode 1 and mode 2 ISWs can be unambiguously identified in SAR images provided the waves propagate in relatively deep water (when the lower layer is significantly deeper than mixed layer depth). If interfacial waves (in a two layer system) travelling along the pycnocline are waves of depression (which is often the case in deep waters), then mode 1 ISWs will be revealed in SAR images by bright bands preceding darker ones in their direction of travel (see Alpers 1985). However, mode 2 ISWs will have exactly the opposite contrast in the SAR, since their radar signature consists of dark bands preceding bright bands in their direction of propagation (see Fig. 14.6).

This is because surface velocity fields induced by travelling ISWs of different modes can create different convergence and divergence patterns, which then modulate the surface roughness and thus the intensity of the radar backscatter signal. In particular, a mode 2 ISW travelling along the thermocline will generate a divergence pattern at the surface, followed by a convergence pattern in the propagation direction (see Fig. 14.6). It is this dynamical feature (reversed with respect to mode 1 ISWs) that produces the characteristic dark bands ahead of the bright bands in their direction of travel.

Research studies on mode 2 ISWs generated by IW rays hitting a pycnocline have been recently presented using numerical models and laboratory experiments (Grisouard et al. 2011 and Mercier et al. 2012, respectively). These studies were focused on the local generation mechanism (see Sect. 14.2), and have also addressed the generation of higher mode waves by IW rays. Grisouard et al. (2011) used a (non-linear and non-hydrostatic) numerical model and concluded that higher mode ISWs (particularly mode 2) may also be generated by an IW beam hitting a pycnocline at an angle. In their studies, high mode waves are favoured by a strongly stratified pycnocline or, equivalently, by short-wavelength IT beams. Laboratory measurements of ISWs generated by an internal wave beam were presented by Mercier et al. (2012), and are in agreement with the results of Grisouard et al. (2011). Following these numerical and laboratory studies, we will now show satellite observations of mode 2 ISWs, which are consistent with local generation by an IW beam impinging on the ocean pycnocline from below.

14.3.2 SAR Observations of Mode 2 ISWs

Figure 14.7 shows a typical example of a SAR image where two wave trains can be seen to propagate towards the shore (approximately to the west-northwest, and in a near perpendicular direction to the local isobaths) in the southern end of MC (near parallel 24°S). The left panel shows an ERS-2 SAR image dated 24 September 2001 (acquired at 7:39 UTC), which is centred approximately at 23.75°S and 35.75°E. Note that land and coast lines are also marked in gray areas outlined by a black contour, respectively, and that bathymetry levels for 100, 200, 500 and 1,000 m are also shown. The black filled circle on the image indicates the predicted position where an IT beam would impinge on the thermocline from below (assumed at an average depth of 50 m), and the dashed line indicates its trajectory coming from deeper bathymetry to the east (passing through the centre and taken perpendicular to the wave crests). The right hand side of Fig. 14.7 shows a vertical profile, which is aligned and partially coincident with the black dashed line on the left, together with a simulated IT beam of semi-diurnal frequency based on local stratification. The arrow indicates the direction of energy propagation, and the inset shows the geographic location of the observation (the star in the inset denotes the geographic location of the image centre).

The SAR image shown in Fig. 14.7 reveals that the ISWs in the deeper wave packet are characterized by the usual rank-ordered bright and dark bands of radar intensity, which decrease in amplitude from the front to the rear. However, while these signatures look like many other typical SAR signatures of ISWs, it can be seen that in this case the contrast pattern is reversed in relation to the general pattern (of mode 1 ISWs). Here, dark bands precede bright bands in the propagation direction,

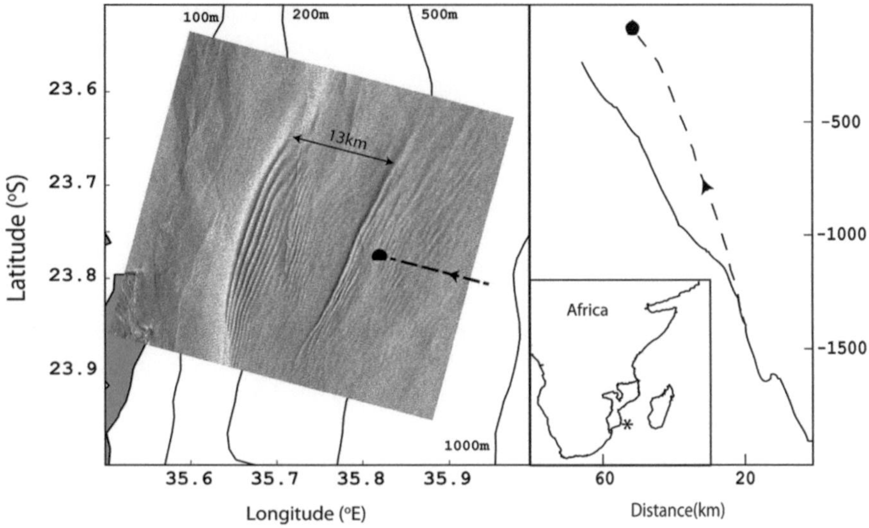

Fig. 14.7 ERS2-SAR image dated 24 September 2001 and acquired at 7:39 UTC, showing two packets of ISWs consistent with a mode 2 vertical structure (separated by ~ 13 km). Isobaths for depths of 100, 200, 500, and 1,000 m are also shown. On the *right hand side* a vertical profile corresponding to the dashed section on the *left* (perpendicular to the ISW crests) is shown with a simulated ray trajectory (*dashed line*) coming from critical bathymetry and based on local stratification (the *arrows* indicate the direction of energy propagation). The *black circle* on both panels indicates the predicted position where an IT beam impinges on the thermocline from below. The inset shows the geographic location of the observation in the Mozambique Channel

indicating that these radar signatures are consistent with mode 2 ISWs (see schematics shown in Fig. 14.6 for mode 2 ISWs). Another important characteristic of this image is that it shows the ISWs to appear close to (but in front of) the surfacing position of the predicted IT rays of semi-diurnal frequency (with a M_2 tidal constituent), generated at the critical slopes farther east. This strongly suggests that these mode 2 ISWs are generated locally after the impact of an IT beam in the pycnocline from below. We note that, this is in agreement with the numerical and laboratory studies done by Grisouard et al. (2011) and Mercier et al. (2012), respectively, since they too predicted the possibility of ISWs with higher mode structures being "locally generated".

14.3.3 Comparison of Average Propagation Speeds with Theory

To reinforce the hypothesis that the observed ISWs in Fig. 14.7 are of mode 2, we will also estimate an average phase speed from the SAR image, and compare it with a theoretical model. To do so the Taylor–Goldstein (TG) model is used, which is solved numerically to investigate the propagation speeds of the lowest IT modes (see e.g. Kundu and Cohen 2004). For the MC case, we have assumed a semi-diurnal frequency for the IT with a M_2 tidal constituent (with $\sigma = 1.4 \times 10^{-4}$ rad s^{-1}), and a mean density profile typical for the southern winter stratification (namely October, see Fig. 2a of da Silva et al. 2009). Based on Fig. 14.7, we also considered an average ocean depth of 300 m to the west of the ISWs along their apparent propagation path (region between the two consecutive packets that are observed in the SAR). Additionally, we assumed that the two packets in Fig. 14.7 were generated at the same phase of the tide and at the same position—that is near the impact point of the IT beams with the pycnocline. Under those assumptions, and considering that the two ISW packets in the SAR are separated along their apparent propagation path by an average distance of 13 km (see black arrow on top of left panel in Fig. 14.7), the estimated phase speed of the mode 2 ISWs in the SAR is 0.3 ms^{-1}. On the other hand, the TG model assuming an average depth of 300 m predicts linear phase speeds of 0.4 and 1.0 ms^{-1} for mode 2 and mode 1 linear IWs of tidal frequency, respectively. This further suggests that the waves in Fig. 14.7 are indeed mode 2 waves, since both signatures (discussed in Sect. 14.3.2) and phase speeds are consistent with a mode 2 vertical structure.

14.4 Eddy-Like Structures and IW Refraction

Da Silva et al. (2009) reported that, on some occasions, ISWs in MC (off the Sofala Bank; see Fig. 14.1 for location) were observed farther south than the average pattern. In their study, the seasonal variability of currents during the southern winter was the plausible explanation for this ISWs propagation anomaly. To justify this seasonal variability they referred to the work done by Maltrud et al. (1998), who believed that the Mozambique Current (which flows southward along the western slopes of the basin) was much stronger during the austral winter. Based on the knowledge of the Mozambique Current available at that time, da Silva et al. (2009) then proposed that the increased southward flow in the southern winter months could explain why some ISWs were refracted (or advected) towards more southerly pathways at that time of year. However, more recent research indicates that the MC currents are in fact highly variable, which is mainly due to the presence of eddy-like features that originate in the Indian Ocean and travel southwards throughout MC (Ridderinkhof et al. 2010; Ullgren et al. 2010). In this Section, we will show an example of ISWs propagating through eddy-like structures (which are observed in altimetry data) and undergoing dramatic refraction patterns.

Figure 14.8 shows a map of ISWs based on our interpretation of an Envisat ASAR image (WS mode), acquired on 4 December 2009 at 20:07 UTC, that is centred approximately at 21.0°S and 36.3°E. For clarity, two types of wave packets (discussed earlier in Sect. 14.2) were marked with ‘SG’ for shelf-generation and ‘LG’ for local generation), until they reach 21.5°S.

This is because further south of 21.5°S along their trajectory, the interpretation of the sea surface signatures starts to become ambiguous, mainly due to dissipation,

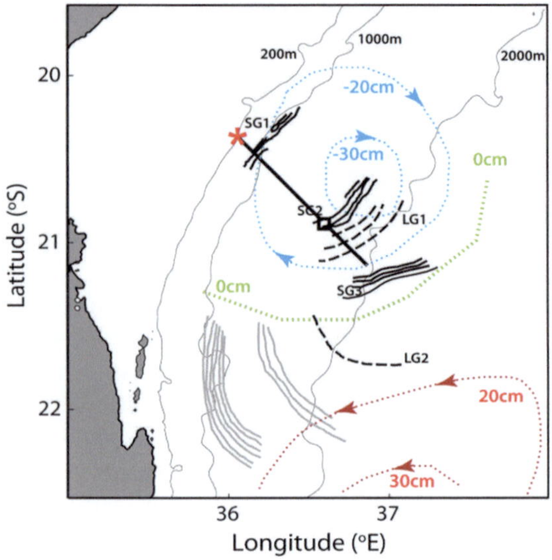

Fig. 14.8 Location map for ISWs observed to propagate in an Envisat-ASAR image (WS mode) over the Mozambique Channel, dated 4 December 2009 and acquired at 20 h 07 m UTC. Land is shown in *grey areas* and isobaths for 200, 1,000, and 2,000 m are marked in thin *black lines*. ISWs interpreted to be generated directly over the shelf (*SG*) are in solid black lines, while locally generates waves (*LG*) are in dashed *black lines*. Each group is numbered chronologically from their generation to later stages of development. ISWs that could not be discriminated (for their generation mechanism) are marked in *grey* (see text for details). The *red star*, the *black square*, and the *black section* are the same as in Fig. 14.5, and are shown here for a reference location of critical bathymetry and the re-emergence of IT rays at the thermocline. The *blue and red dashed lines* mark the locations of eddy-like features detected in altimetry data, and whose rotation is indicated by *arrows*. The green line represents a reference level of neutral sea surface height

interactions with currents and winds, and interaction among different ISW packets. We note that, this interpretation is further reinforced if the average propagation speeds of the several packets are taken into account. Indeed, according to da Silva et al. (2009), the average phase speeds of these waves were found to be around 1.5 ms^{-1}, and the same values can also be found for the sequences SG1, SG2, and SG3, and LG1 and LG2—based on distances measured in the SAR. The map in Fig. 14.8 also displays Sea Surface Height (SSH) anomalies taken from altimetry data[1], where different colours and dashed quasi-circular lines are used to indicate the presence of eddy-like features observed at approximately the same time as the SAR image (4 December 2009 at 00:00 UTC, the closest product available in time). The blue and red curves refer to negative and positive SSH anomalies, respectively, and a green line has been drawn for reference indicating the boundary between them (an apparent reference level).

[1] Namely, from Topex-Poseidon data products. The altimeter products were produced by Ssalto/Duacs and distributed by Aviso, with support from Cnes (http://www.aviso.oceanobs.com/duacs/).

There are tangential surface currents associated to these SSH anomalies, whose rotation directions are indicated by arrows in Fig. 14.8. These currents can be rather large in the MC (as large as 0.5 ms^{-1} according to Swart et al. 2010 and thus of the order of the ISWs propagation speeds) and their influence can extend in diameter for some 200 km. In terms of vertical sections the SSH anomalies are associated with isopycnical deformations, which are in general in the opposite way to the surface displacements. From Fig. 14.8, it is clear that the eddy-like features associated to the SSH anomalies interfere with the propagation of the ISWs, apparently causing an overall refraction pattern (to be discussed in the next paragraphs).

In the southern hemisphere eddies with an anti-clockwise rotation are warm eddies, whose SSH anomalies are positive at the surface but are characterized by bowl-like (or bulge) downward displacements of their isopycnicals at depth (isopycnicals can be displaced some 100 m downwards or more, at the core of a warm eddy near the pycnocline). On the other hand, cold eddies spin in the clockwise direction, which are characterized by negative SSH anomalies, and have their isopycnicals displaced upwards near the core.

This means that the eddy-like features identified in Fig. 14.8 are those of a cold eddy just off the Sofala Bank (in blue), and a warm eddy somewhat farther south and farther offshore (in red). It can be seen that the ISW crests showed in Fig. 14.8 are clearly interacting with the eddy structures that were measured in the SSH data. First, the cold eddy closer to the shelf and afterwards the warm eddy farther offshore (and southwards of the cold eddy) are located directly along the ISWs path. Note that the sequence of ISW crests shown in Fig. 14.8 also exhibits what appears to be a clockwise rotation pattern.

In fact, the packets labelled SG1, SG2, LG1, SG3 and LG2 (forming a chronological sequence) progressively propagate more to the south and less to the east, until they reverse the component of east-west propagation velocity and return back on to the shore, undergoing a near 135 clockwise rotation. In this interpretation, we considered that the ISWs sequence mentioned above is progressively older in time, with respect to their generation sites and propagation paths. We believe that the pattern of ISWs rotating clockwise in their direction of propagation is essentially caused by interaction with the near-surface currents produced by eddies. However, refraction may also be caused (at least to some extent) by different stratifications and mixed layer depths resulting from the relative positions of eddies along the ISWs propagation pathways. The later effect has been explained successfully by Sherwin et al. (2002) off the northern Portuguese shelf (in the west Iberian Coast), where refraction of ISWs occurs basically due to an offshore downward tilt of the isopycnicals. In the Iberian case, similarly to the MC, ISWs were first found to propagate in an off-shelf direction and then refract back onto the coast some 50 km north of their generation site (roughly the same distance that ISWs need in MC to refract back to shore).

Fig. 14.9 Envisat-MERIS image dated 16 September 2004 and acquired at 07 h 15 m UTC. The image covers an area of about 580 × 660 km^2 and is centred near 20.4°S and 38.0°E. The *black compass* overlaid on the *top-left corner* indicates north, and the *black circle* has been drawn to have approximately 70 km in diameter (for a scale reference). Two waves can be seen in the southwest part of the Mozambique Channel, one large-scale AGW (*L*) going southwest and a smaller AGW (*S*) going southeast

14.5 AGWs in the MC and Oceanic ISW Look-Alikes

In the previous Sections we have clearly shown that MC is a hotspot for oceanic IWs, which adds to the idea that these features represent ubiquitous phenomena in the world's oceans. However, MC is also a region where atmospheric IWs are frequently observed, being amongst the largest (in terms of spatial scales and crestlengths) observed anywhere in the world (see da Silva and Magalhaes 2009).

A particular case of large-scale AGWs propagating in MC can be found in Fig. 14.9, where we present a RGB composite from an Envisat-MERIS image dated 16 September 2004, and acquired at 07:15 UTC. The image covers an area of about 580 × 660 km^2 in the southwest part of MC, and it is centred near 20.4°S and 38.0°E. A large-scale AGW can be seen in the centre of the image (labeled L), and travelling southwest towards the south end of MC. The crestlength of the leading wave, in this case, extends for more than 500 km and its crest-to-crest spatial scale is approximately 10 km on average. It is interesting to note that several (but not all) of these individual waves are made visible by characteristic cloud bands, which have considerable lateral extensions and uniform cross-sections.

Satellite imagery (mainly SAR, visible and near infrared) exhibiting these large-scale AGWs are a regular occurrence in MC, particularly between July and October when they are more frequently observed. Satellite data (such as MODIS, MERIS and Envisat-ASAR in WS mode) have been used to characterize the horizontal structure of these AGWs, which have mean dispersive wavelengths between 3.5 and 8.5 km, mean crestlengths of 443 km, and propagation speeds that can exceed 10 ms^{-1}. Two preferential pathways were found to dominate the AGWs, which propagate mainly to northeast (from the south end of the MC) and to southwest (coming from Madagascar). It was also found that high pressure systems born in the Polar Regions (known as Moving Polar Highs, MPHs), which are associated with air subsidence in MC, can play a major influence on the propagation of these waves. This is because MPHs change the atmosphere's stability, and influences its vertical structure to form suitable waveguides, necessary for AGWs to propagate, and can even trigger their own generation (see da Silva and Magalhaes 2009 for a detailed study).

The sea surface signatures of the AGWs in MC can usually be distinguished from their oceanic counter parts (discussed in Sect. 14.2 to 14.4) due to their large dimensions (wavelength and crestlength). However, in many cases, other smaller-scale AGWs can also be found to propagate across the MC, including those areas where oceanic ISWs are frequently present (e.g. see composite map in Fig. 14.5). A good example can be seen in Fig. 14.9, where a small-scale AGW (labeled S, and much smaller than the largest packet labeled L) is seen to travel to the southeast, in an almost perpendicular direction to the larger-scale AGW (labeled L) that is travelling southwest. In fact, this observation is very similar to the oceanic ISWs presented in Sect. 14.2, since it has the same characteristics (namely, spatial scales, crests orientation, propagation direction and geographic location).

Care should be taken when interpreting sea surface manifestations of IWs in SAR (as mentioned in the Introduction Section). On the one hand, SAR images cannot detect the presence of cloud structures, which would clearly indicate the atmospheric nature of AGWs. On the other hand, clouds may not form at all in the presence of large amplitude AGWs, since the ascending branches of the waves may not meet the necessary moisture for condensation to occur. In fact, Fig. 14.9 shows that some of the solitary waves in the trailing edge of the large packet (labeled L) are only partially recognizable as clouds—the remaining segments are merely seen as sea surface patterns of banded roughness on a smoother background. In such cases, alternative methodologies should be taken under consideration to discriminate between oceanic and atmospheric IWs. A detailed discussion of these issues can be found in the work presented by Alpers and Huang (2011), where lists of discriminating criteria were presented based on soliton and SAR theory, and further complemented by additional information on the ocean and atmosphere.

14.6 Summary

This chapter describes remote sensing data from SARs and image spectrometers to clearly show that MC is a hotspot for observing ISWs in the ocean and in the atmosphere. Oceanic ISWs were found to be generated directly off the shelf by the steepening of an interfacial tidal wave, and through a different generation mechanism by which the impact of a tidal beam from below the thermocline generates ISWs (known as local generation of ISWs). Both mechanisms are associated with an elevated region of body forcing near the Sofala Bank, which results from strong barotropic tidal currents interacting with critical slopes. SAR imagery has also revealed that the MC is a preferential region for observing other IW phenomena such as mode 2 ISWs and strong refraction patterns. The data suggests that the mode 2 ISWs have also been generated by tidal beams, similarly to some of the mode 1 waves observed in the SAR. This confirms some conclusions drawn from other independent studies using modeling and laboratory experiments (Grisouard et al. 2011 and Mercier et al. 2012, respectively) about the origin of mode 2 solitary-like waves by tidal beams. The refraction patterns in SAR revealed that it is likely that ISWs in MC can frequently interact with oceanic features such as eddy-like structures, which propagate year-round through the MC and affect the local vertical structure of the ocean. Large-scale AGWs are also a ubiquitous feature in MC, especially from July to October, and care should be taken to avoid miss-interpretations in SAR with oceanic ISWs.

References

Akylas TR, Grimshaw RHJ, Clark SR, Tabaei A (2007) Reflecting tidal wave beams and local generation of solitary waves in the ocean thermocline. J Fluid Mech 593:297–313. doi:10.1017/S0022112007008786

Alpers W (1985) Theory of radar imaging of internal waves. Nature 314:245–247. doi:10.1038/413245a0

Alpers W, Huang W (2011) On the discrimination of radar signatures of atmospheric gravity waves and oceanic internal waves on synthetic aperture radar images of the sea surface. IEEE Trans Geosci Remote Sens 49(3):1114–1126

Alpers W, Stilke G (1996) Observations of a nonlinear wave disturbance in marine atmosphere by synthetic aperture radar aboard the ERS 1 satellite. J Geophys Res 101:6512–6525

Azevedo A, da Silva JCB, New AL (2006) On the generation and propagation of internal waves in the southern Bay of Biscay. Deep-Sea Res Part I 53:927941. doi:10.1016/j.dsr.2006.01.013

Baines PG (1982) On internal tides generation models. Deep-Sea Res Part I 29:307–338. doi:10.1016/0198-0149(82)90098-X

Bedard AJ, Canavero F, Einaudi F (1986) Atmospheric gravity waves and aircraft turbulence encounters. J Atmos Sci 43(23):2838–2844

Colosi JA, Beardsley RC, Lynch JF, Gawarkiewicz G, Chiu CS, Scotti A (2001) Observations of nonlinear internal waves on the outer New England continental shelf during the summer Shelfbreak Primer study. J Geophys Res 106(C5):9587–9601. doi:10.1029/2000JC900124

Christie DR, Muirhead KJ (1983) Solitary waves: a hazard to aircraft operating at low altitudes. Aust Meteorol Mag 31:97–109

Crook AN (1986) The effect of ambient stratification and moisture on the motion of atmospheric undular bores. J Atmos Sci 43(2):171–181

da Silva JCB, Helfrich KR (2008) Synthetic aperture radar observations of resonantly generated internal solitary waves at Race Point Channel (Cape Cod). J Geophys Res 113:C11016. doi:10.1029/2008JC005004

da Silva JCB, Magalhaes JM (2009) Satellites observations of large atmospheric gravity waves in the Mozambique Channel. Int J Remote Sens 30(5):1161–1182

da Silva JCB, Ermakov SA, Robinson IS, Jeans DRG, Kijashko SV (1998) Role of surface films in ERS SAR signatures of internal waves on the shelf. 1. Short-period of internal waves. J Geophys Res 103(C4):8009–8031. doi:10.1029/97JC02725 da Silva JCB, New AL, Srokosz M, Smyth T (2002) On the observability of internal tidal waves in remotely-sensed ocean color data. Geophys Res Lett 29(12):1569–1572. doi:10.1029/2001GL013888

da Silva JCB, New AL, Azevedo A (2007) On the role of SAR for observing "local generation" of internal solitary waves off the Iberian Peninsula. Can J Remote Sens 33:388–403. doi:10.5589/m07-041

da Silva JCB, New AL, Magalhaes JM (2009) Internal solitary waves in the Mozambique Channel: observations and interpretation. J Geophys Res 114:C05001. doi:10.1029/2008JC005125

da Silva JCB, New AL, Magalhaes JM (2011) On the structure and propagation of internal solitary waves generated at the Mascarene Plateau in the Indian Ocean. Deep-Sea Res Part I 58:229–240. doi:10.1016/j.dsr.2010.12.003

Egbert GD, Erofeeva SY (2002) Efficient inverse modeling of barotropic ocean tides. J. Atmos. Oceanic Technol 19:183–204. doi:10.1175/1520-0426(2002)019<0183:EIMOBO>2.0.CO;2

Gerkema T (2001) Internal and interfacial tides: beam scattering and local generation of solitary waves. J Mar Res 59:227–255. doi:10.1357/002224001762882646

Gerkema T, Zimmerman JTF (2008) An introduction to internal waves. Lecture notes, Royal NIOZ, pp. 207

Grisouard N, Staquet C, Gerkema T (2011) Generation of internal solitary waves in a pycnocline by an internal wave beam: a numerical study. J Fluid Mech 676:491–513. doi:10.1017/jfm.2011.61

van Haren H, Groenewegen R, Laan M, Koster B (2005) High sampling rate thermistor string observations at the slope of Great Meteor Seamount. Ocean Sci 1:17–28. doi:10.5194/os-1-17-2005

Jackson CR (2007) Internal wave detection using the moderate resolution imaging spectroradiometer (MODIS). J Geophys Res 112:C11012. doi:10.1029/2007JC004220

Jackson CR, da Silva JCB, Jeans G (2012) The generation of nonlinear internal waves. Oceanography 25(2):108–123. doi:http://dx.doi.org/10.5670/oceanog.2012.46

Kundu PK, Cohen IM (2004) Fluid Mechanics. Third Edition. Academic Press. pp. 759

Lamb KG, Farmer D (2011) Instabilities in an internal solitary-like wave on the Oregon Shelf. J Phys Oceanogr 41:67–87. doi:10.1175/2010JPO4308.1

Magalhaes JM, Araújo IB, da Silva JCB, Grimshaw RHJ, Davis K, Pineda J (2011) Atmospheric gravity waves in the Red Sea: a new hotspot. Nonlinear Proc Geoph 18:71–79. doi:10.5194/npg-18-71-2011

Maltrud ME, Smith RD, Semtner AJ, Malone RC (1998) Global eddy-resolving ocean simulations driven by 1985–1995 atmospheric winds. J Geophys Res 103:30825–30853. doi:10.1029/1998JC900013

Mercier MJ, Mathur M, Gostiaux L, Gerkema T, Magalhaes JM, da Silva JCB, Dauxois T (2012) Soliton generation by internal tidal beams impinging on a pycnocline: laboratory experiments. J Fluid Mech 704:37–60

Merrifield MA, Holloway PE (2002) Model estimates of M2 internal tide energetics at the Hawaiian Ridge. J Geophys Res 107(C8):3179. doi:10.1029/2001JC00996

Moum JN, Farmer DM, Smyth WD, Armi L, Vagle S (2003) Structure and generation of turbulence at interfaces strained by internal solitary waves propagating shoreward over the continental shelf. J Phys Oceanogr 33:2093–2112. doi:10.1175/1520-0485(2003)033<2093:SAGOTA>2.0.CO;2

Nash JD, Moum JN (2005) River plumes as a source of large-amplitude internal waves in the coastal ocean. Nature 437:400–403. doi:10.1038/nature03936

New AL, da Silva JCB (2002) Remote-sensing evidence for the local generation of internal soliton packets in the central Bay of Biscay. Deep-Sea Res Part I 49:915–934. doi:10.1016/S0967-0637(01)00082-6

New AL, Pingree RD (1992) Local generation of internal soliton packets in the central Bay of Biscay. Deep-Sea Res Part A 39:1521–1534. doi:10.1016/0198-0149(92)90045-U

Niwa Y, Hibiya T (2004) Three-dimensional numerical simulation of M2 internal tides in the East China Sea. J Geophys Res 109(C4) C04027. doi:10.1029/2003JC001923

Pingree RD, New AL (1989) Downward propagation of internal tidal energy into the Bay of Biscay. Deep-Sea Res Part A 36:735–758 doi:10.1016/0198-0149(89)90148-9
Pingree RD, New AL (1991) Abyssal penetration and bottom reflection of internal tidal energy in the Bay of Biscay. J Phys Oceanogr 21:28–39. doi:10.1175/1520-0485(1991)021<0028:APABRO>2.0.CO;2
Pingree RD, Mardell GT, New AL (1986) Propagation of internal tides from the upper slopes of the Bay of Biscay. Nature 312:154–158. doi:10.1038/321154a
Pinkel R (2000) Internal solitary waves in the warm pool of the western equatorial pacific. J Phys Oceanogr 30:2906–2926. doi:10.1175/1520-0485(2001)031 < 2906:ISWITW > 2.0.CO;2
Reeder MJ, Christie DR, Smith RK, Grimshaw RHJ (1995) Interacting morning glories over northern Australia. Bull Amer Meteor Soc 76:1165–1171
Ridderinkhof H, van der Werf PM, Ullgren JE, van Aken HM, van Leeuwen PJ, de Ruijter WPM (2010) Seasonal and interannual variability in the Mozambique Channel from moored current observations. J Geophys Res 115:C06010. doi:10.1029/2009JC005619
Sandstrom H, Elliott JA (1984) Internal tide and solitons on the Scotian Shelf: a nutrient pump at work. J Geophys Res 89(C4):6415–6426. doi:10.1029/JC089iC04p06415
Sherwin TJ, Vlasenko VI, Stashchuk N, Jeans DRG, Jones B (2002) Along-slope generation as an explanation for some unusually large internal tides. Deep-Sea Res Part I 49:1787–1799. doi:10.1016/S0967-0637(02)00096-1
Shroyer EL, Moum JN, Nash JD (2010a) Vertical heat flux and lateral mass transport in nonlinear internal waves.Geophys Res Lett 37:L08601. doi:10.1029/2010GL042715
Shroyer EL, Moum JN, Nash JD (2010b) Energy transformations and dissipation of nonlinear internal waves over New Jersey's continental shelf. Nonlinear Proc Geoph 17:345–360. doi:10.5194/npg-17-345-2010
Shroyer EL, Moum JN, Nash JD (2010c) Mode 2 waves on the continental shelf: ephemeral components of the nonlinear internal wavefield. J Geophys Res 115:C07001. http://dx.doi.org/10.1029/2009JC005605
Smith WHF, Sandwell DT (1997) Global sea floor topography from satellite altimetry and ship depth soundings. Science 277(5334):1956–1962. doi:10.1126/science 277.5334.1956
Swart NC, Lutjeharms JRE, Ridderinkhof H, de Ruijter WPM (2010) Observed characteristics of Mozambique Channel eddies. J Geophys Res 115:C09006. doi:10.1029/2009JC005875
Ullgren J, van Aken H, Ridderinkhof H (2010) Unique measurements of ocean currents east of Africa. NIOZ Annual Report 2010, 21–22
Vlasenko V, Alpers W (2005) Generation of secondary internal waves by the interaction of an internal solitary wave with an underwater bank. J Geophys Res 110:C02019. doi:10.1029/2004JC002467

Chapter 15
Ship Surveillance with High Resolution TerraSAR-X Satellite in African Waters

Susanne Lehner, Andrey Pleskachevsky, Stephan Brusch, Miguel Bruck, Matteo Soccorsi and Domenico Velotto

Abstract Ship detection is an important application of monitoring of environment and security or safety issues in African Waters. In order to overcome the limitations by other monitoring systems, e.g. coastal radar, surveillance with satellite Synthetic Aperture Radar (SAR) is used because of its potential to detect ships at high resolution over wide swaths and in all weather conditions and independent from sun illumination. TerraSAR-X (TS-X) is an X-band polarimetric SAR capable of imaging up to 1 m resolution in Spotlight mode. TS-X can be used for a wide variety of applications and methods of analysis including visual interpretation, mapping, digital-elevation-model creation, disaster monitoring, and oceanography. Results on the combined use of TS-X ship detection, Automatic Identification System (AIS), and satellite AIS (SatAIS) are presented. Using AIS is an effective terrestrial method for tracking vessels in real time typically up to 40 km off the coast. SatAIS is a space-based system with nearly global coverage for monitoring of AIS equipped ships. Since not all ships operate their AIS and smaller ships are not equipped with AIS, space borne SARs provide complimentary means for ship monitoring. As cases, images were acquired over the Somali Coast Area, South African Coast and Gibraltar in Stripmap mode with a resolution of 3 m at a coverage of 30 km × 50 km. The rapid tasking performance as well as the short response time of the TS-X data acquisition of the ground segment DLR-BN (Ground Station Neustrelitz, Germany), are very helpful to monitor hotspot areas such as the Gulf of Aden. For ascending orbits the delivery time of ship detection products is less than 20 min. Along with the detected ship positions, estimated wave heights and wind fields derived from large-area TS-X imagery can be used to get a detailed maritime picture of the situation.

15.1 Introduction

It is well known that the state of the world's oceans is deteriorating due to human activities. As noted by the African Maritime Safety and Security Agency (AMSSA) the consequences of these activities is that the world's oceans and coasts will be

S. Lehner (✉) · A. Pleskachevsky · S. Brusch · M. Bruck · M. Soccorsi · D. Velotto
Institut für Methodik der Fernerkundung, SAR-Signalverarbeitung,
Deutsches Zentrum für Luft- und Raumfahrt e.V. (DLR),
Bremen Flughafen, Bremen, Germany
e-mail: Susanne.Lehner@dlr.de

V. Barale, M. Gade (eds.), *Remote Sensing of the African Seas*,
DOI 10.1007/978-94-017-8008-7_15,

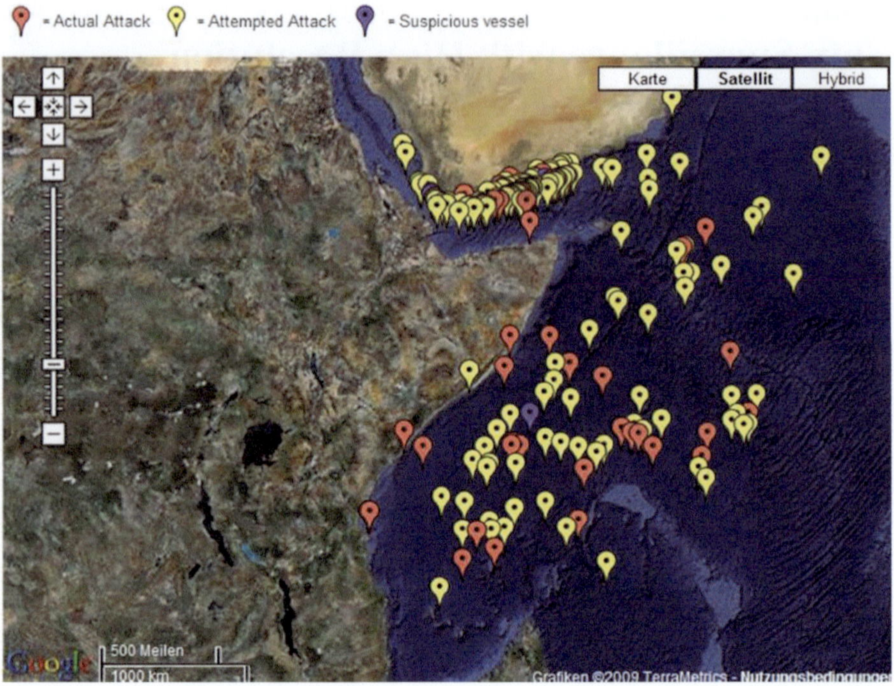

Fig. 15.1 Pirate attacks in Somalia waters and Golf of Arden in 2009 (Piracy Reporting Centre, Live Piracy Report)

increasingly stressed and move to a situation where they not able to maintain the ecosystems. Fishing and maritime sectors, in particular maritime transport systems, oil production have a specific interaction with the marine environment. It is necessary to ensure that these meeting requirements do not damage the marine environment. Threats and vulnerability due to natural disasters and environmental degradation, environmental crimes, illegal fishing, oil theft, money laundering, illegal arms and drug trafficking, trafficking humans, terrorism and maritime piracy and armed robbery (on this account see e.g. Fig. 15.1) are thus monitored by the authorities.

Space-based information provides a powerful tool platform to monitor both environment and security and safety concerns. Security comprises a broad set of policy concerns, e.g. border surveillance and piracy. Safety topics are traffic safety, environmental protection, fisheries control and Search and Rescue (S&R).

A new generation of radar satellites allows the acquisition of the area of interest with high resolution and quality. The scanning of the ocean surface using remote sensing instruments provides an opportunity not only to detect and monitor ships, but also the surrounding environmental processes like turbulent ship wakes or wave breaking. Due to daylight, weather independency and global coverage radar sensors onboard TerraSAR-X (TS-X) and its twin TanDEM-X, short for TerraSAR-X add-on for Digital Elevation Measurement (TD-X) are suitable for practical support of ships security and safety. Ship position, length and speed can be identified, compared to Automatic Identification System (AIS) messages and provided to integrating

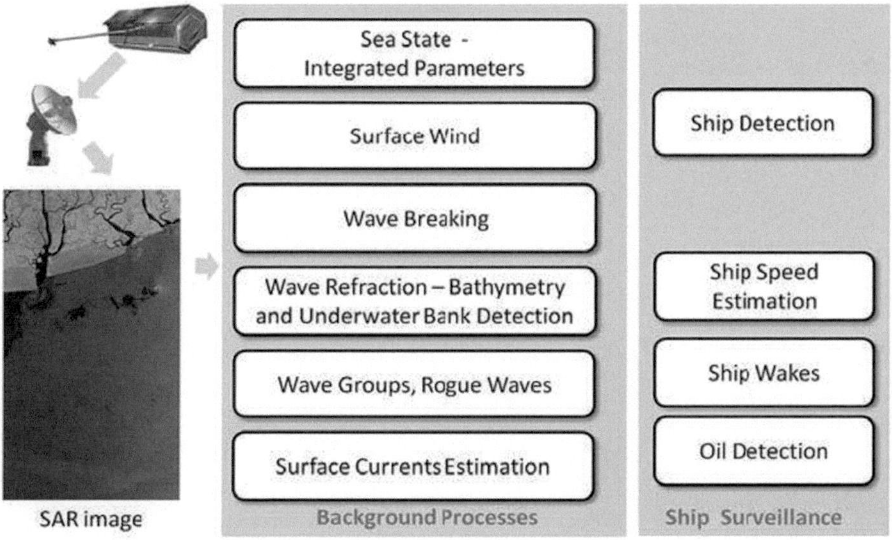

Fig. 15.2 Using remote sensing data for ship surveillance. A wide spectrum of features is involved besides ship detection like waves, surface wind effects, etc. The knowledge of such background geophysical processes and their SAR imaging mechanism is important for an assessment of services in terms of safety and security issues

services. Cases of oil discharge from the ships and platforms can also be detected (Velotto et al. 2011) and reported in Near Real Time (NRT). In addition to ship detection, the surrounding marine and meteorological parameters can be estimated for operational oceanographic services.

The development of algorithms and methods for remote sensing data processing are carried out along two lines. On one hand products for NRT services (e.g. Schwarz et al. 2010), such as ship detection are provided, on the other hand satellite images taken by space borne sensors are used to explain atmospheric as well as oceanographic geophysical processes and features, e.g. wind field and sea state. The whole set of such information contained in the images has to be extracted (see Fig. 15.2).

15.2 Ship Surveillance in African Waters

At present, vessel tracking in African waters depends on the combination of coastal monitoring and airborne or surface craft patrol. Coastal monitoring includes the use of coastal surveillance radar, automatic identification system receivers and local vessel traffic monitoring systems (e.g. for ports).

The main restriction of these systems is their limited coverage. In addition, some fishing vessels must also carry identification systems that transmit location data via a satellite telecommunications link. Maritime Patrol Aircraft can detect vessels over

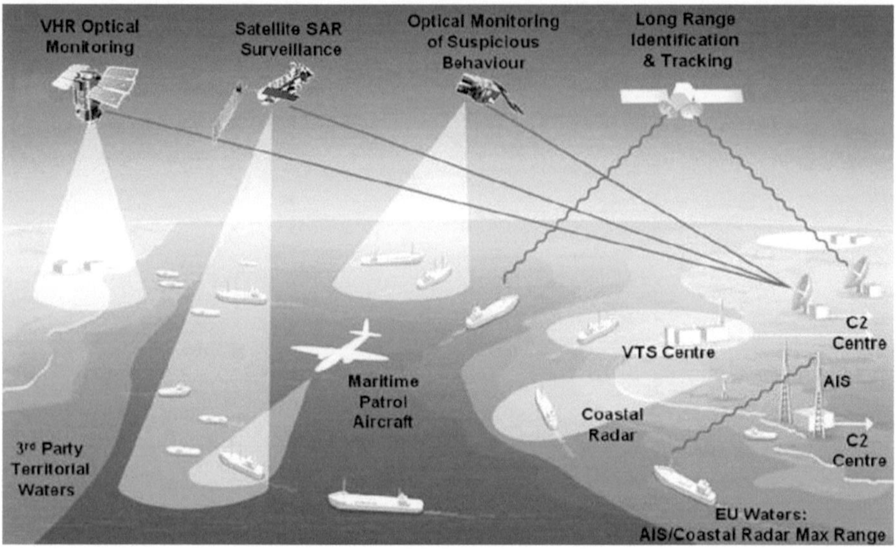

Fig. 15.3 The integration of satellite based surveillance and monitoring for enhanced operational maritime border control and maritime domain awareness. (Source: ESA MARISS)

a wider area than coastal systems but in most cases are still not capable of providing total coverage over all national waters. In addition, such platforms have only limited time on station.

In the framework of the European Space Agency (ESA) project Maritime Security Information Services (MARISS) satellite-based Synthetic Aperture Radar (SAR) is used for detection of vessels in areas beyond the range of conventional identification systems. In addition, areas that constitute potential sources of threat can be regularly monitored using satellite based high resolution radar and optical imagery. Within MARISS, the recognised maritime picture can be enhanced through the integration of satellite based vessel detection with data streams, such as coastal radar, Automatic Identification System (AIS), Vessel Monitoring System (VMS) and Vessel Traffic System (VTS).

Satellite based SARs enables the detection of vessels in areas beyond the range of coastal monitoring systems for support of tactical monitoring. Like coastal radar, satellite radar information is independent of on the transmission of identification data by vessels. By combining the cooperative identification systems with non-cooperative monitoring and detection, a more comprehensive Maritime Picture can then be generated. For strategic monitoring and threat identification, satellite imagery can complement traditional sources of intelligence by providing regular routine monitoring of potential threat areas as well as supporting the construction of statistics on heavily utilised trafficking routes. The synergetic surveillance system with use of space-based and ground based monitoring systems is illustrated in the following Fig. 15.3.

Fig. 15.4 "MV York" and a TerraSAR-X Stripmap (reduced resolution). Snapshot

The next study shows the importance of high resolution SAR data to detect and recognize ships and to monitor abnormal behaviour. The hijacking of the gas tanker "MV York" (see Fig. 15.4) in 2010 by Somali pirates initiated a case study to test the ability of SAR data for monitoring the situation. "MV York" was hijacked in the Somali Basin an October 23, 2010 after leaving Mombasa en route to the Seychelles. One of four motherships was used by pirates to extend their range into the Indian Ocean. With the detected ship positions derived from a TS-X image acquisition on November 14, 2010, the follow up data take were planned. The second TS-X product from November 18, 2010 did not only yield clear detections of hijacked ships but provided furthermore details of small boats shipping between the coast and the ships (see Fig. 15.5). These anomaly detections can be performed by SAR satellite data several times daily if more than one sensor is in use.

15.3 TerraSAR-X and TanDEM-X Satellites

The German X-Band SAR satellite TS-X[1] was launched in June 2007. Since January 2008, data and products are available for researchers and commercial customers. TS-X operates from a 514 km height sun-synchronous orbit. The TS-X ground speed is 7 km/s (15 orbits/day). It operates with a wavelength of 31 mm and frequency of 9.6 GHz. The repeat-cycle is 11 days, but the same region can be imaged with different incidence angles after 3 days dependent on latitude. The resolution of TS-X images is improved in comparison to earlier SAR missions (about 1 m for Spotlight mode against 25 m for the ASAR, onboard the ENVISAT orbital platform of the ESA). Typical incidence angle range for TS-X is between 20° and 60° (Breit et al. 2007, 2010).

[1] See www.dlr.de/TerraSAR-X

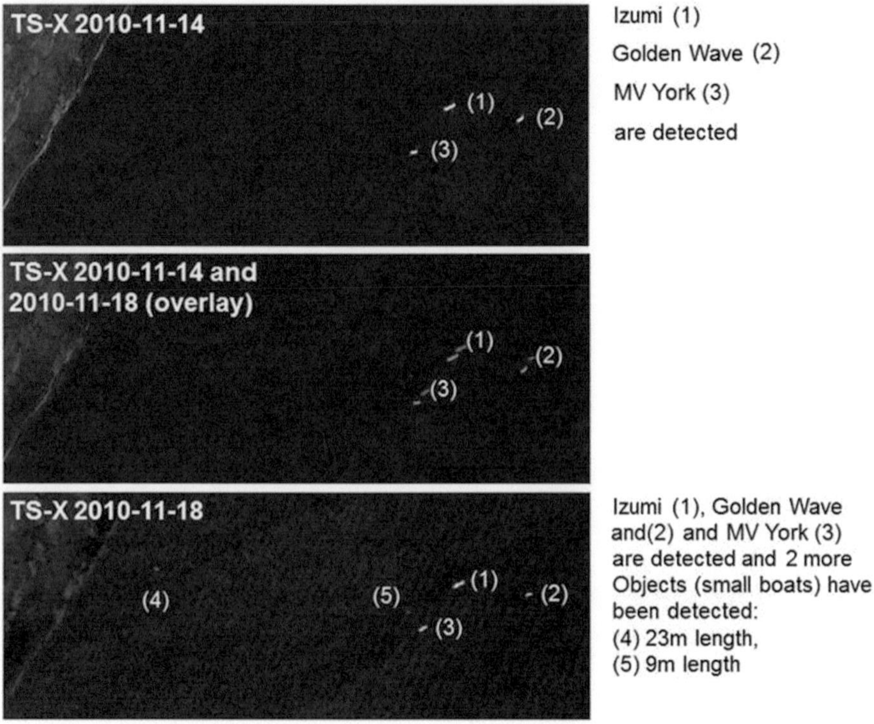

Fig. 15.5 "MV York" detections and anomaly detections on two TerraSAR-X scenes (2010-1114; 2010-11-18). In scenes overlay (*middle*) moving of ships "Izumi"(*1*), "Golden Wave"(*2*) and "MV York" (*3*) is visible. The main interest are the detections of small boats commuting between hijacked ships and land. (See the TS-X 2010-11-18)

TerraSAR-X operates four different basic imaging modes with different spatial resolution, technical parameters of the modes are provided in Fig. 15.6. The analysis of TerraSAR-X images is used for different applications: ship, ice, oil and underwater shallow area (bank and bars) detection as well as measuring the wind field, coast line and underwater topography fields. Compared to earlier SAR missions like ENVISAT ASAR, TerraSAR-X offers besides higher resolution a number of further advantages: e.g. the Doppler-shift of a scatterer, moving with velocity *ur* toward the sensor (radial velocity) at distance *Ro* (slant range) is reduced. For instance, for the same incidence angle 22° and $ur = 1$ m/s the target's displacement in azimuth direction $D_x = (u_r/V_{sar})R_0$ (Lyzenga et al. 1985) is reduced by a factor of about two and results in $\sim$ 73 m for TerraSAR-X ($\sim$ 115 m for ENVISAT) due to different platform velocity V_{sar} (7,55 km/s for ENVISAT) and slant range R_0 (ENVISAT altitude is 800 km). Thus, the azimuth smoothing of the wave crests moving is noticeably reduced. Due to this, the imaging of the ocean surface is more stable. The twin satellite TanDEM-X was launched on June 21, 2010. TerraSAR-X and TanDEM-X are orbiting in a close formation with typical distances between satellites of 250 m–500 m.

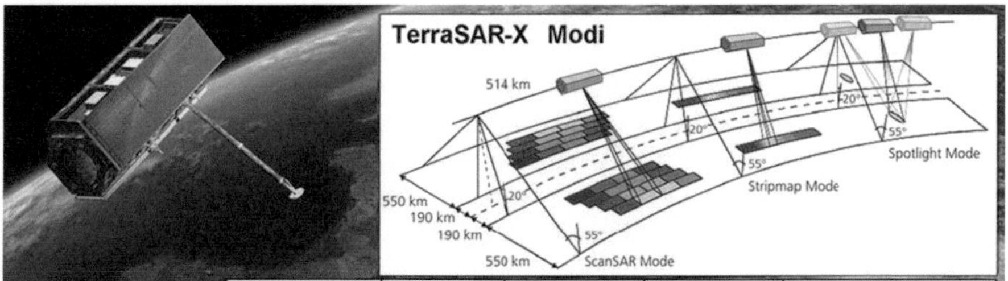

	ScanSAR	StripMap	Spotlight	High-Resolution Spotlight
Swath Width (ground range)	100km	30 km	10km × 10km	5km × 10km
azimuth resolution	18m	3.3 m	2 m	1.7m
range resolution	1.7m–3.5 m	1.7m–3.5m	1.5m–3.5 m	1.5m–3.5m

Fig. 15.6 TerraSAR-X satellite and characteristics of four TerraSAR-X modes: ScanSAR, Stripmap, Spotlight and High-Resolution Spotlight

15.4 Ship Surveillance with Synthetic Aperture Radar

In 2004 a benchmark test for SAR ship detection was performed, which consisted of eight different system analyses, the same set of 17 SAR images and a comparison of detection results. The outcome is reported in (Greidanus et al. 2004). Of the eight detectors, six were operational (i.e. being used for customers) and two experimental.

The operational systems rely on detection of pixels with a high value compared to the local background statistics—Constant False Alarm Rate approach (CFAR)—sometimes employing a template to match expected target shape. Experimental systems further use approaches based on wavelets and subaperture cross-correlation. It was found that the operational systems, although performing markedly different on individual images, have an overall comparable performance.

However, in reality, the radar echo includes apart from ship targets signals due to different noise sources and distortions from environmental processes. For instance, so called "white capping" (crushing of upper wave crest part by a wind speed higher than 10–12 m/s) and wave breaking produces turbulence at the sea surface causes a high radar backscatter that can be falsely detected as a ship. An optimum detector for this situation should maximize the probability of detection while minimizing the probability of false alarm. The methods are developed and signal filters are adopted aiming at maximizing the detection of ship targets. Validation of ship such SAR detection methods can be performed by using AIS or SatAIS data (Brusch et al. 2011).

15.4.1 Automatic Identification System, Terrestrial and Satellite AIS

AIS is a ship-borne broadcast system by which ships inform each other about their position, their course, speed, name, and many other ship-specific parameters. This is used in navigation for maritime safety and in particular for collision avoidance. AIS messages from ships are collected by coastal receivers too. The range of coastal AIS receivers is typically 40 km offshore. For ships over 300 t in international waters, and over 500 t in national waters, the use of AIS is mandatory.

Several institutions and companies have demonstrated that AIS signals can be effectively received using space born systems. This fact makes SatAIS a powerful data source to provide global situational awareness of ship positions, vessel and fleet movements and traffic density mapping. AIS signal detection is possible in the entire radio visibility range of a satellite equipped with an AIS payload. For a payload in Low Earth Orbit (LEO) at an altitude of 650 km the distance from message source to receiver varies between 2,800 km (very low elevation angle) and 650 km (space-based receiver is directly over the ship in zenith). This means that the average field of view of the payload is well above 20 million km^2. The advantage of SatAIS is clearly the possibility of complete global coverage and the availability in areas without land based means of vessel detection. Moreover, SatAIS features the availability of all information contained in the actual AIS messages as transmitted by ships, providing a complete and extensive information source on all the parameters with respect to individual ships. This enables a convenient and effective way of analysing maritime traffic around the globe. The data received by SatAIS have demonstrated the applicability of AIS data in maritime surveillance scenario's underlining the advantages as previously listed.

15.4.2 Automatic Identification System, Case Study 'MV Sirius Star'

The hijacking of the super tanker "MV Sirius Star" by Somali pirates initiated a case study to test the response time of the TS-X data acquisition as well as the rapid tasking performance of the ground segment DLR-BN (Ground Station Neustrelitz, Germany) for monitoring the situation. The supertanker was hijacked close to Somalia (Africa) on November 18, 2008 and was released on January 9, 2009. With the detected ship positions derived from large area ScanSAR and Stripmap search acquisitions, follow up data takes in high resolution but at smaller coverage modes Spotlight and High-Resolution Spotlight (HS) were planned.

The first two ScanSAR products from November 22, 2008 did not only yield a clear detection at 4° 35′35″N, 48°06′14″E but provided already enough details for identification (Fig. 15.7). The subsequent HS data take on the 24th was placed at this location, but showed no sign of the ship since it had moved in the meantime. This initiated a new search acquisition revealing on November 27th, the new anchoring site at 4° 48′59″N, 48°15′52″E approximately 30 km North-East of the old position. Several high resolution data takes were acquired of this area in the following weeks and the ship was found to remain within 1000 m of this position until it was released on January 9, 2009. The test case proved that the TS-X system is suited to task and executes acquisitions based on the analysis of preceding products within 2–3 days

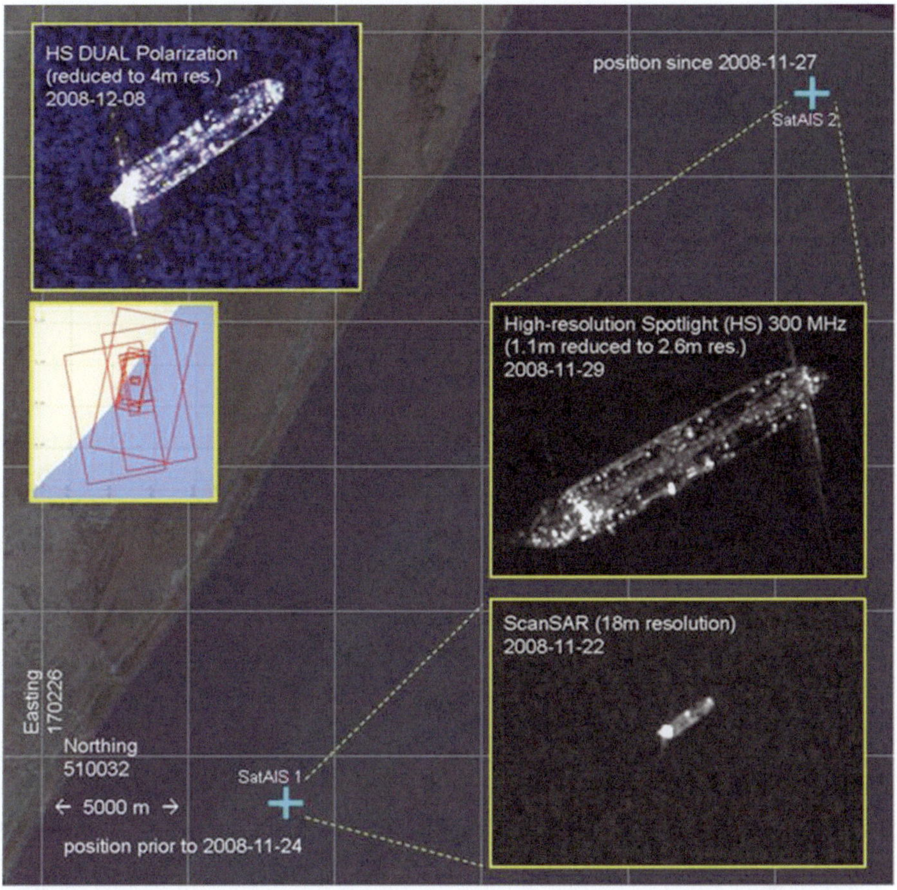

Fig. 15.7 Colour composite of three TerraSAR-X ScanSAR acquisitions (*red*: 2008-1127T15:04, *green*: 2008-11-22 T03:02, *blue*: 2008-11-22 T14:56) showing the position of the "Sirius Star" off Somali coast. During the two acquisitions of 22 November, the ship was anchoring at the exact same position (yielding a cyan colour). The blow-ups are from three of the 16 single acquisitions with the TerraSAR-X footprint are indicated on the map plot. The AIS Coordinates retrieved later on are indicated by white crosses. (Source LuxSpace)

even close to the equator, using the nominal ordering, processing and delivery chains. Due to applicable German laws, the highest resolution data including the 1.1 m HS 300 MHz products are not approved for publication and hence available only in Radiometrically Enhanced (RE) product variants with a resolution coarser than 2.5 m. Well visible are the bridge, the pipelines in the middle and the two helicopter decks near the bow. SatAIS reports are superimposed by red crosses. The positions derived from TS-X agree perfectly in the spatial domain with the TS-X positions. Temporal agreement of the information depends on the satellite orbit. Coastal AIS is not available in this region. A synergetic use of SatAIS and SAR data allows assessing the maritime situation around hijacked ships.

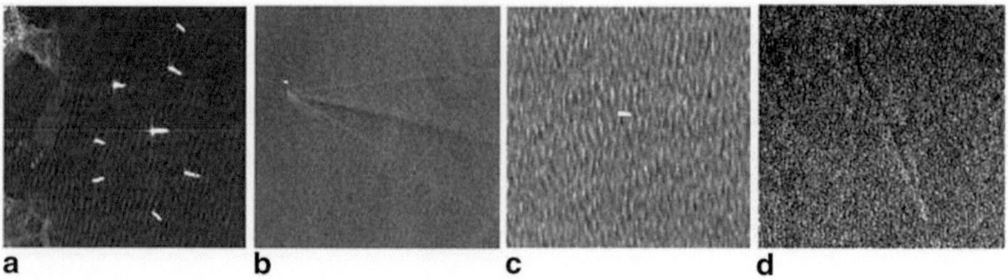

Fig. 15.8 Examples of wakes and ships in SAR images

15.5 Ship surveillance with TerraSAR-X

In the following, the appearance of ships and their wakes is discussed. A detailed account of the TS-X ship detection process and error analysis is also introduced.

15.5.1 Ships and Wakes in SAR images

The detection of ships through the wake is limited by the ship motion and the visibility of the wake. Examples on TS-X images of ships and wakes are shown in Fig. 15.8. In Fig. 15.8a, a group of ships close to the harbour is visible. Their vicinity and orientation make the automatic detection hard as well as the determination of the heading is difficult because they are not moving. On the other hand, a simple thresholding would work in this example because of the glassy sea state. In Fig. 15.8b a moving ship is visible together with the wake. The typical displacement which affects moving target is visible because the ship is not imaged in its real location but shifted in the azimuth proportional to the line-of-sight along-range speed vector component. The occurrence presented in Fig. 15.8b is a favourable case, because the turbulent wake and the Kelvin envelope are clearly visible and they provide the heading of the vessel. The opposite case is shown in Fig. 15.8c, where the wake is not visible for a moving vessel. This phenomenon, probably due to the destructive interference with the sea waves, has often been observed in some cases. In the last case, shown in Fig. 15.8d, only the shape of the wake is visible but not the vessel which causes it.

The challenge of automatic detection of ships and wakes has to be approached by different methods and data. The former is better with HH and HV channels because the sea clutter appears reduced. In opposite, the environmental like parameters (e.g. wind and swell waves) are better provided using VV channel. Due to sufficient backscattering of short surface capillary waves, these processes can be imaged. All these cases make the automatic target detection a hard task. A comparative study of the wake from sonar and SAR data has been proposed in (Soloviev et al. 2010). In Fig. 15.9 the wake model is shown. The simplified ship reflection model illuminated from a SAR sensor is shown in Fig. 15.10. In order to go from target detection to

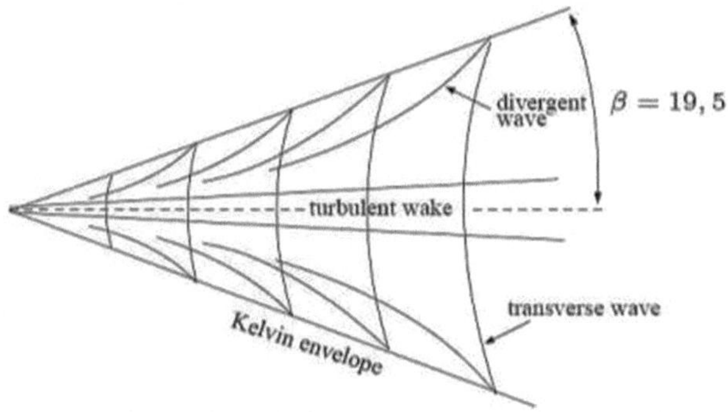

Fig. 15.9 Wake model characterised by the Kelvin envelope and the turbulence wake. The angle is constant while the frequency of the transverse waves depends on the speed of the source

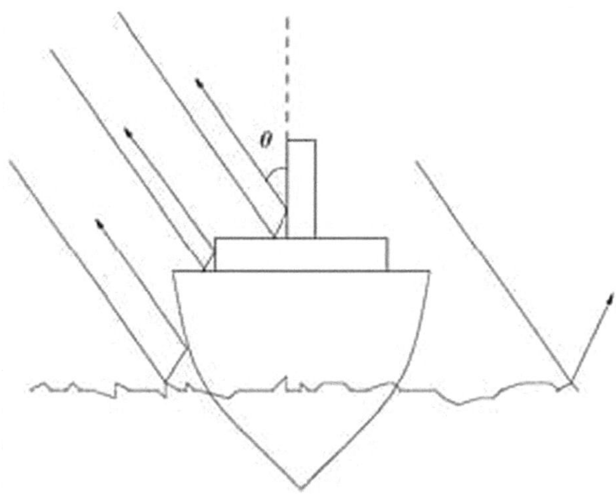

Fig. 15.10 Ship reflection model for transversal profile. It is characterized by several double bouncing due to the structures (according to Crisp 2004)

target recognition more complicated 3D models have to be taken into account or other techniques, e.g. the Inverse Synthetic Aperture Radar (ISAR) technique (Cooke et al. 2006).

15.5.2 TerraSAR-X Ship Detection Method

Ship detection systems (work flow) generally consist of five stages for standard ship detection (see Fig. 15.11):

- *Preprocessing:* SAR products are calibrated and geo-referenced. The expected pixel location accuracy for TS-X is around 1–2 m.
- *Land masking:* using shore-line databases with an eligible buffer zone included to limit the processing of ship detection to sea area or automatic detection of land zones. Land masking is important because ship detectors can produce high numbers of false alarms when applied to land areas. Accurate land masking is

Fig. 15.11 Ship detection work flow scheme

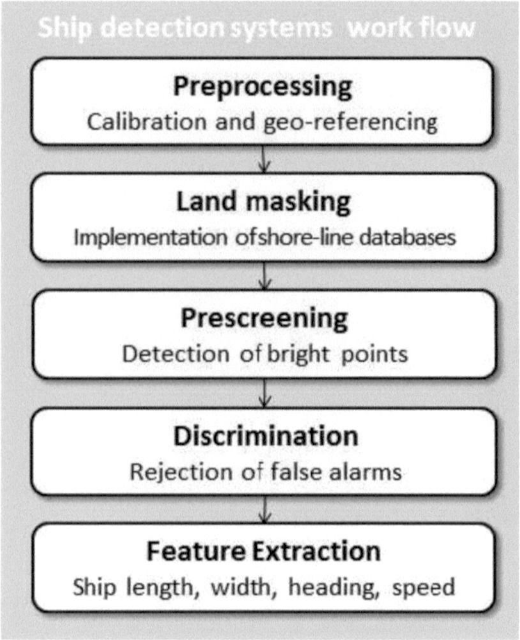

generally difficult due to inaccuracy of recorded coastline, tidal variations, and coastal constructions.

- *Prescreening:* applying a simple moving window adaptive threshold algorithm to detect bright points. In general classical CFAR algorithms are chosen due to its robustness. Details of other ship detection methods can be found in (Crisp 2004).
- Discrimination: rejecting some false alarms using target measurements or characterization of specific oceanographic or meteorological phenomena.
- *Feature Extraction:* Based on standard SAR intensity images ship length, width, and heading is automatically extracted. If the Displacement between ship wake and the ship is detectable, this feature can be used as a direct measurement for ship speed retrieval.

Several algorithms have been developed to detect targets in SAR imagery. CFAR algorithms have been applied widely among them. However, CFAR and its improved algorithms have to build a statistical model for the clutter. Since there are a number of factors influencing the clutter, for example, the type of the ground, it is difficult to build an accurate model to simulate the clutter.

The CFAR is an adaptive threshold algorithm. Its aim is to search for pixel values which are unusually bright compared to those in the surrounding area. This is done by setting a threshold which depends on the statistics of the area under analysis. The pixels which lie above the threshold are selected as target samples. If the threshold is chosen so that the percentage of background pixel values which lie above the threshold is constant, thus the method is called constant false alarm rate detector.

One method for CFAR detection is to work directly with the histogram of the background windows and set the threshold at the appropriate point in the tail of the distribution. Choosing a parametric distribution model for the background is equivalent to specifying the associated parametric probability density function *f(x)*, where x gives possible pixel values. Once *f(x)* has been chosen and its parameters are estimated from the background samples, the probability of false alarm (PFA) for the threshold T is given by

$$PFA = 1 - \int_{-\infty}^{T} f(x)dx = \int_{T}^{\infty} f(x)dx \qquad (15.1)$$

Designing a CFAR detector involves solving the Eq. (15.1) for the threshold T in terms of the specified PFA and the estimated parameters of the *pdf f(x)*. An analytic solution to this problem is not always possible and numerical methods may be needed. One approach is to search for the correct value of T by trial and error.

A commonly used statistical model is the Gaussian distribution. Gaussian distribution is the best approximation due to an increasing of the signal to noise ratio by resizing SAR images (less time consuming). In this case, the detector is

$$x_t > \mu_b + \sigma_b t \Leftrightarrow TARGET \qquad (15.2)$$

where x_t is the pixel value under test, μ_b is the background mean, σ_b is the background standard deviation, and t is a detector design parameter which controls the *PFA* (or equivalently the false alarm rate). The Gaussian distribution is not an accurate model of radar imagery unless the data have been averaged by a large number of looks. More appropriate models for radar intensity are the negative exponential for single look imagery and the Gamma distribution for multi-look image. SAR sea clutter is well modelled by $K\nu$ distribution. It is characterised by fast and local fluctuation (capillary waves) and slower spatial fluctuations due to gravity waves (swell). A comparison between Gaussian model and *K*-distributed model on RADARSAT data is provided in (Brekke et al. 2010). The authors show that the number of false alarms is reduced by using the K-distribution model.

The following example shows the synergetic use of high resolution earth observation data and satellite based AIS and terrestrial AIS data. Figure 15.12 shows a TS-X Stripmap mode image in the southern Atlantic Ocean near Cape Town (SA).

Vessels detected by SAR are superimposed and marked with red rectangles. Collocated ships with terrestrial AIS messages are marked by green rectangles. Satellite AIS is superimposed with yellow rectangles. SAR as well as terrestrial and Satellite AIS are reporting the same position of targets as can be observed in Fig. 15.12a. Figure 15.12b shows the general limitation of terrestrial AIS, covering only an area up to 40 km of the Cape Town area. Three small vessels are detected by SAR. One of them is reporting an AIS signal via SatAIS. Furthermore, a moving target is visible. The SAR detected ship reported by SatAIS is moving south.

In the next example, the synergetic use of high-resolution EO data and terrestrial AIS data is demonstrated in the Street of Gibraltar (Fig. 15.13). Vessels detected

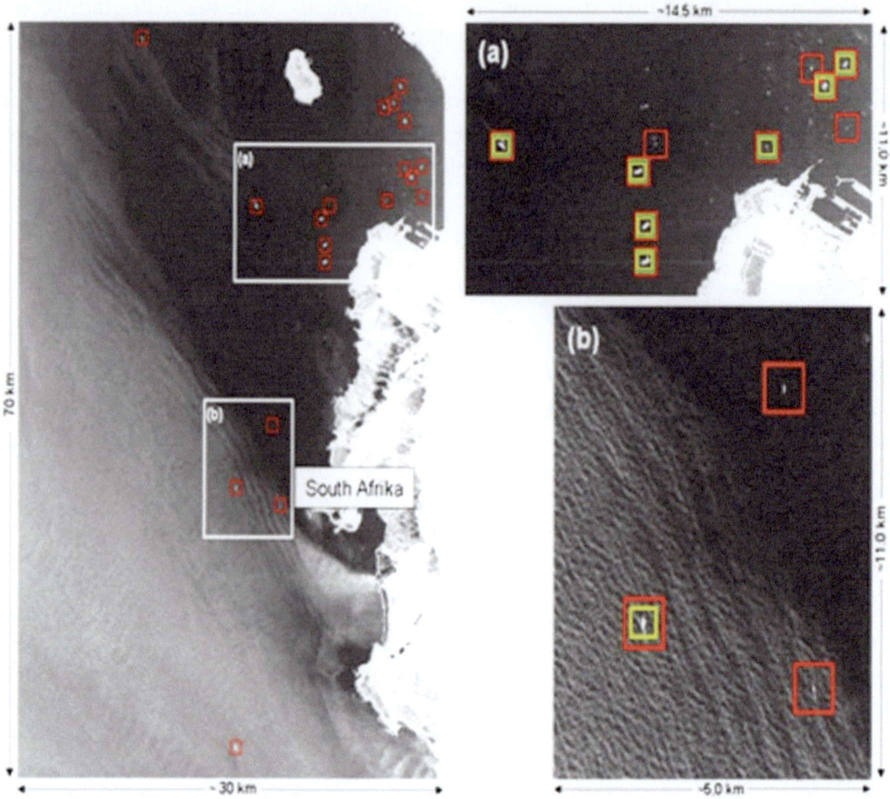

Fig. 15.12 TerraSAR-X Stripmap sub-image (HH-Pol, 3 m resolution) acquired on 16 December 2008 over the southern Africa near Cape Town with ships detected by the DLR ship detection algorithm (*red* superimposed AIS (source: HIS Fairplay) data with *green rectangles*, and Satellite AIS reports (*yellow rectangles*, source: Luxspace)

by SAR are superimposed and marked by red rectangles. Collocated ships with terrestrial AIS messages are marked by green rectangles.

Bright point detection is sometimes not sufficient and when used itself not able to avoid detections related to oceanographic features, e.g. breaking waves or meteorological events like rain cells (Brusch et al. 2008). Further errors can arise from SAR image artefacts, such as side lobes, azimuth ambiguities from land or strong point targets over the open sea. The distance of azimuth ambiguities Δx to point targets is calculated by:

$$\Delta x = \Delta t_{ambi}{}^{*} f_0 \quad (15.3)$$

where f_0 is the Zero Doppler Velocity and

$$\Delta t_{ambi} = \frac{PRF}{FM} \quad (15.4)$$

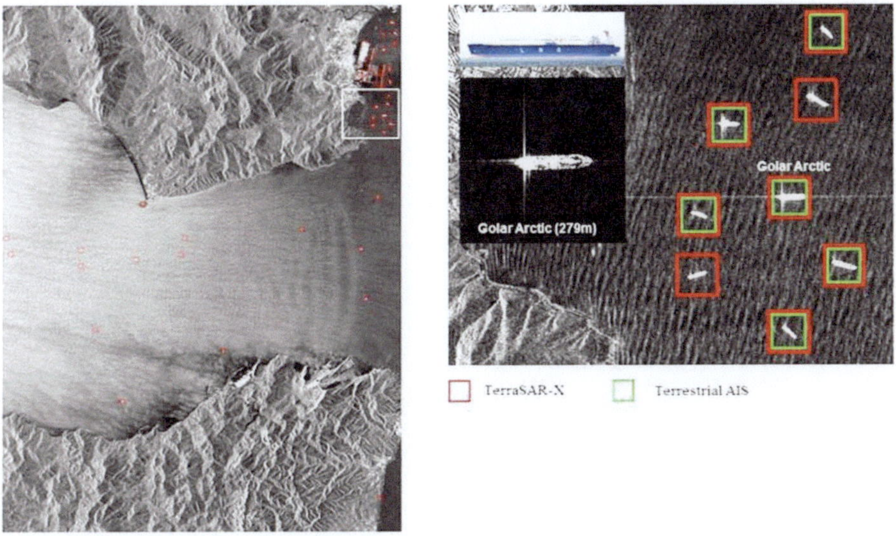

Fig. 15.13 Street of Gibraltar. Ship detected (*red*), Terrestrial AIS (*green*)

where *PRF* is the pulse repetition frequency and *FM* is the derivative of the Doppler frequency in time (Mittermayer and Runge 2003). Azimuth ambiguities in TS-X Stripmap mode images of strong point targets are visible at around 5.2 km offset.

Detection algorithms are designed to exclude azimuth ambiguities of ship targets. For strong scatterers over land a couple of SAR images are used to create look up tables of scatterers and ambiguities. These tables are used to avoid false alarms. Azimuth ambiguities from strong point targets like ships in front of the Cape Town harbour and point targets over land are shown in Fig. 15.14.

Another methodology for azimuth ambiguities removal for ship detection purposes is the exploitation of fully Polarimetric SAR (PolSAR) data combined with the high resolution capability of X-Band sensor. For the purposes, fully polarimetric X-band SAR data acquired by the German satellite TS-X (TS-X) during the experimental Polarimetric SAR (PolSAR) campaigns are accounted for (Velotto et al. 2012).

Imaging is possible in single- and dual-polarisation for SL and SM basic products. Carrying a redundant receiver chain, the measurement of the complete scattering matrix (quadpolarisation) is achievable creating two receiver channels by splitting the antenna electrically in two halves and hence the Pulse Repetition Frequency (PRF) is doubled, leading to half single polarization range extent, as for dual-polarisation products. This configuration is called Dual Receive Antenna (DRA) and is operated in the SM mode only. DRA configuration allows also creating two distinct images from the two halves enabling Along Track Interferometry (ATI) capability (Mittermayer and Runge 2003).

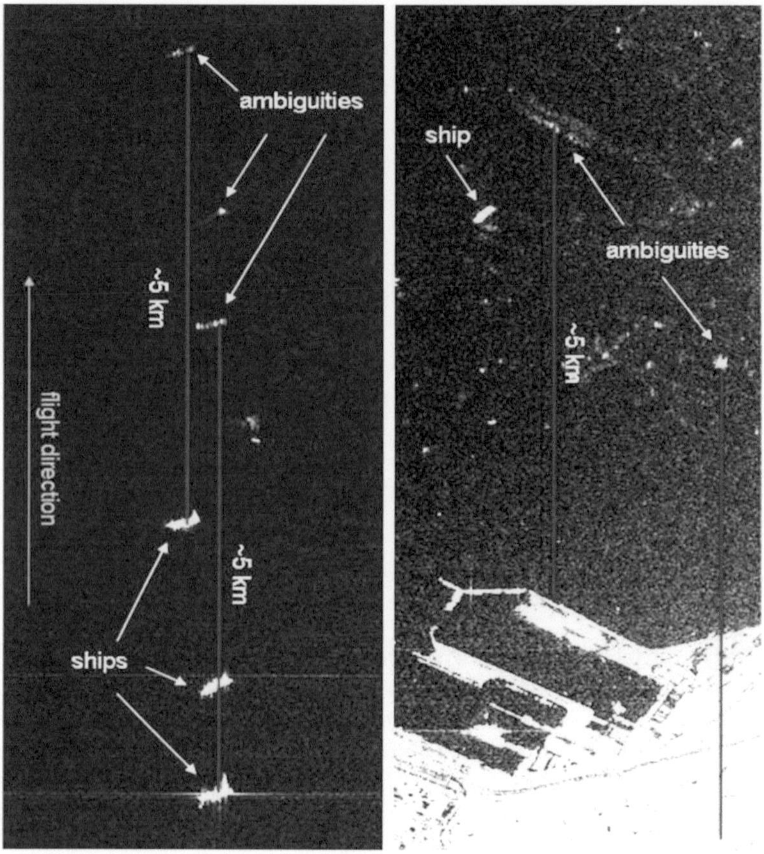

Fig. 15.14 *Left*: example of azimuth ambiguity at around 5.2 km above the real vessel. This kind of ambiguity can easily lead to a false detection. *Right*: strong azimuth ambiguity at around 5 km above the real target over land. Most of the bright spots inside the ambiguity could be detected and confused as vessels

The methodology proposed in (Velotto et al. 2012) addresses both the discrimination of real targets from non-trivial false positives, namely those due to azimuth ambiguities and the detection itself using a Generalized-K (GK-) distribution approach. According to the flowchart in Fig. 15.11, the preprocessing and land masking step are basically the same, while the discrimination is performed before the prescreening and detection itself. This is because exploiting the fact that, for targets, HV and VH channels are approximately equal in magnitude and phase, i.e. the reciprocity theorem apply, but opposed in phase for azimuth ambiguities, is possible to create an ambiguity free HV intensity image.

It is important to note that with this methodology the native full resolution of the sensor is kept because it acts on pixel by pixel basis without any adaptive filtering of the area polluted by the ambiguities. Being the detector in Eq. 15.2 appropriate for HH

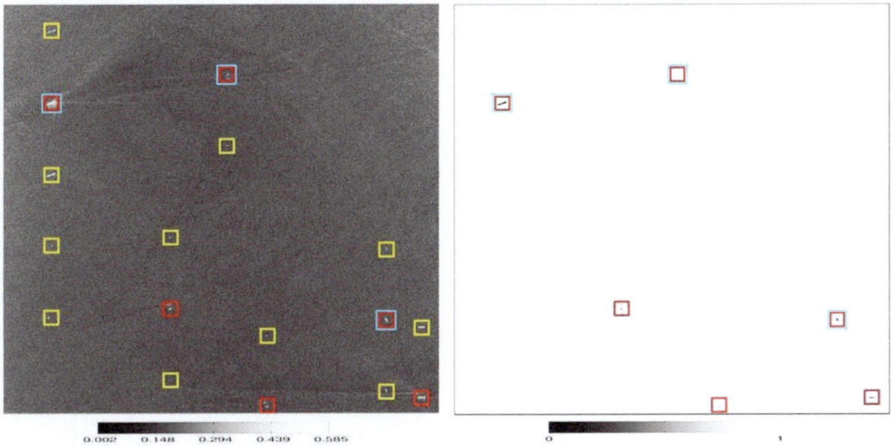

Fig. 15.15 Example of TerraSAR-X DRA Mode data; (*left*) HH slant range amplitude projection image, (*red rectangle*) ships/targets, (*yellow rectangle*) azimuth ambiguities caused by ships/targets, (*cyan rectangle*) available AIS data; (*right*) comparison of available AIS info and logical true-false output

images, a more suitable detector for HV images has been adapted for the prescreening step. GK distribution is an extension of the well-known two parameters K-distribution for the case of weak scattering regime in which a non-uniform distribution of the phase is considered. In (Migliaccio et al. 2007) the GK distribution has been proposed to characterize the speckle of full-resolution SLC single-polarization SAR marine scenes, showing a physically consistent transition among different scattering scenario present over the ocean surface.

The validation of the complete processing chain developed for single-look Slant Range Complex (SRC) quad-pol TS-X product has shown that false positives caused by targets or land azimuth ambiguities over the ocean surface can be reduced to zero exploiting PolSAR data (Velotto et al. 2012). Besides, the proposed methodology shows that even small targets are observed and efficiently detected and a good detection consistency, i.e. none of the ambiguities is misguided as target, when compared with ground truth data is achieved (Fig. 15.15). The results obtained in (Velotto et al. 2012) strongly support the use of X-band PolSAR data for marine targets detection purposes, especially where the problem of ambiguities might be critical, i.e. harbours and coastal areas.

15.5.3 Error Analysis of CFAR Algorithm with TerraSAR-X

The first experiment was to compare TS-X detection products with AIS and/or SatAIS data. It was found that time gaps between the acquisition and AIS data may appear. Other discrepancies occur also due to incorrect reporting from AIS geo-location

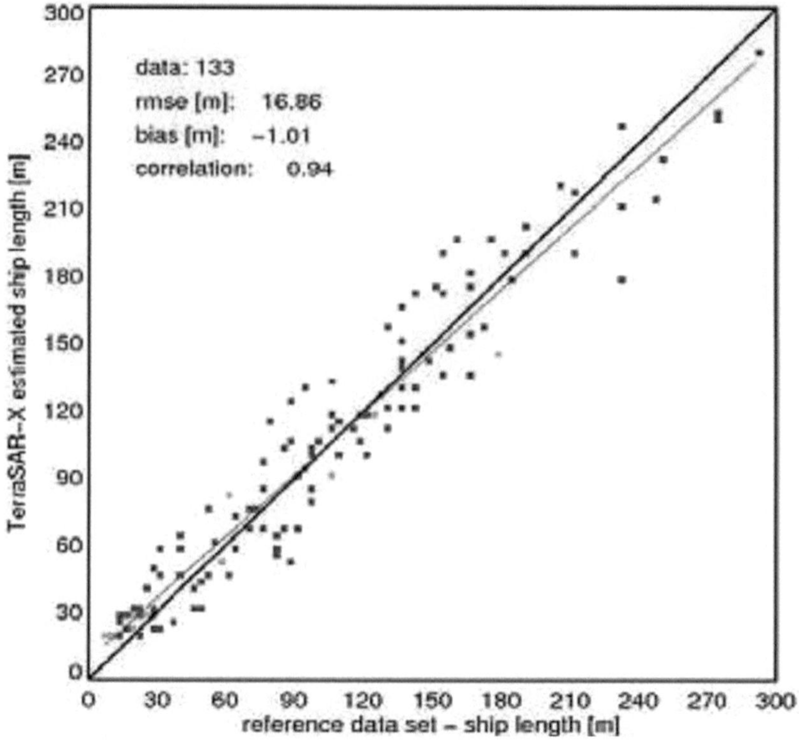

Fig. 15.16 Ship length measured automatically in SAR images against length reported by terrestrial AIS (133 data pairs [reference data set])

inaccuracies. A validation study based on 133 out of 2243 AIS reports for 134 TS-X images (including several meteorological situations) show that the estimation of length has an error of about 16 m and a small bias of ~ -1 m (see Fig. 15.16). Results show that AIS-carrying larger vessels (length above 60 m) were detected in around more than 92 % of cases. For smaller vessels (length < 30 m), the assessment of quality is more difficult due to increasing false alarms for small vessels in high-sea state situations with breaking waves. Results of an additional manual validation are summarised in the pie chart shown in Fig. 15.17, where the true positive rate of 92 % is confirmed over the database. The remaining detections are buoys, ambiguities and 2 % of targets in all TS-X images of the validation data set were not detected automatically.

15.5.4 Possible Error Sources in Ship Detection

Sources of errors in the ship detection processors are due to land masking methods, tidal changes, breaking waves, ambiguities (as described above) and spurious targets,

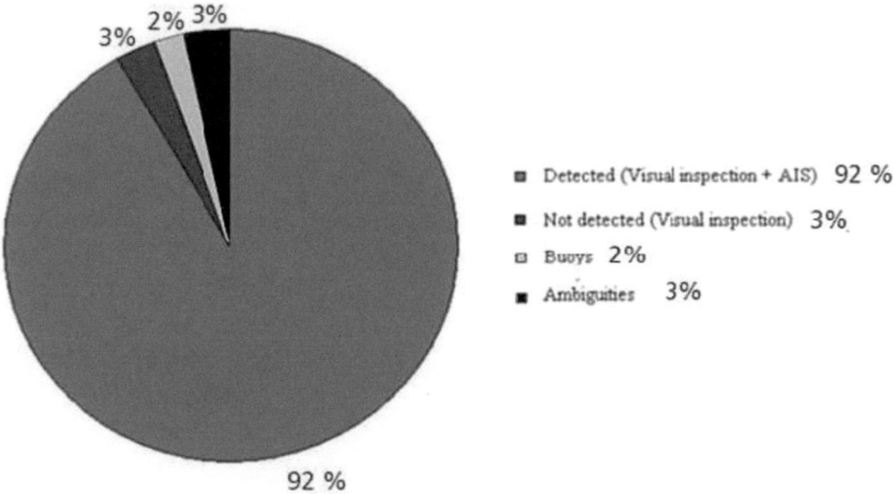

Fig. 15.17 Pie chart of the detected ship: detected ships 92 %, not detected 3 %, buoys 2 % and ambiguities 3 %

e.g. buoys or oil platform. Land: Coast line variations due to human constructions or tidal variations. A possible solution is an adaptive land masking estimated from the image or the creation of a varying coast line database depending on the tidal status.

Breaking waves, caused e.g. by underwater topography, appear as bright pixel streaks. They can be distinguished from the ships by shape and intensity which is the goal of applying texture filters.

The spurious targets might be separated by ship shape filters and/or by mapping their positions. Shape and the size are the primary characteristics which allow separating them. Different ships have different profiles which theoretically might allow going from detection to recognition. However, classification is a difficult task due to the dependencies of the backscattering from several parameters, i.e. incidence angle, orientation, motion, material, etc.

15.5.5 Ship Velocity Estimation on TerraSAR-X Complex data

The speed of a moving object can be retrieved from one single channel complex image exploiting the spectral properties of the coherent SAR imaging system. This potential has multiple applications: control of harbour areas and maritime traffic, security issue, etc. Furthermore the retrieved information might be integrated to AIS and Long Range Identification and Tracking (LRIT) information.

One effect of moving objects in SAR images is that the target (ship) is imaged in another position and not in its real location. The real location of the ship is indicated by the wake which can be taken as reference. The across track motion

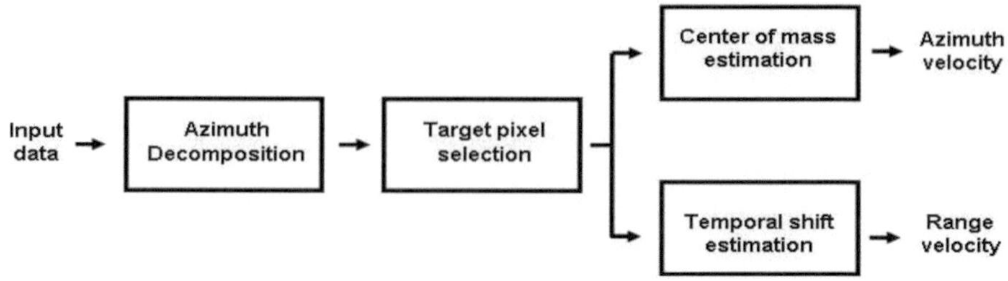

Fig. 15.18 Flow chart of speed estimation

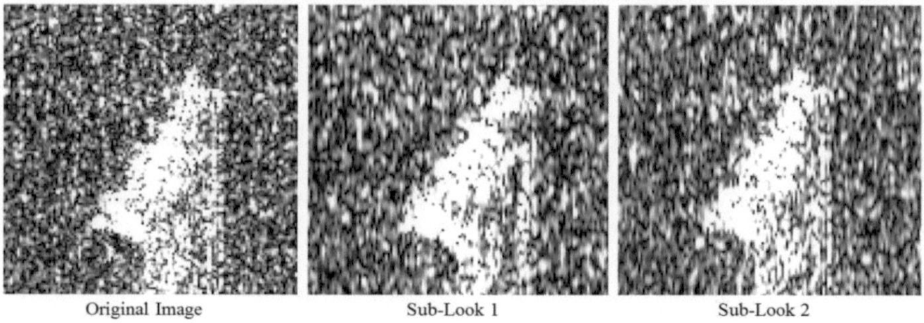

Fig. 15.19 sequence of two single-look images based on a TerraSAR-X complex image

(ship movement not parallel to the satellite flight direction) effect can be used to estimate the range component of speed of a moving target. The offset of the ship and its wake is proportional to the motion speed towards the radar. The method is not always applicable cause of the invisibility of ship wakes due to strong winds or high sea state. To overcome this limitation, a method that does not rely on the detectability of the ship wake is used instead.

The technique introduced in (Kirscht 1998), is here extended to marine application. The process to reconstruct the two components (azimuth and range) of the ship's velocity vector consists of three main steps (Fig. 15.18).

In the first stage, two complex SAR sub-scenes at reduced resolution are obtained from the full resolution complex SAR data by applying the azimuth-split decomposition, i.e. spectrum bandpass filtering.

The absolute value of the original data and of the two obtained sub-scenes of an excerpt of SAR data in Fig. 15.19 shows the effect of the reduced resolution.

In the second step a change detection analysis of the target in the two sub-scenes is applied (in Fig. 15.18 called Target pixel selection). In fact, decomposing the original data in two sub-scenes is like creating two synthetic aperture images with two different aspect angles (or more precisely each look corresponds to a different integration time), therefore the moving ship will appear in different position between the two sub-scenes (Fig. 15.19). The third step consists in the estimation of the centre of mass and of the temporal shift. The first one is related to the azimuth velocity component, while the second is related to the range velocity component as will be described in the following.

The azimuth velocity component is accessible directly by evaluating the displacement vector of the moving target in the two consecutive sub-scenes. The relationship between the azimuth velocity component and the displacement vector of the moving target is given by Eq. 15.5:

$$v_{Ta} \approx -\frac{\Delta x \cdot dx \cdot v_s^2}{\Delta f \cdot \lambda \cdot R_T} \tag{15.5}$$

where Δx is the displacement vector, dx is the pixel spacing, v_s^2 is the spaceborne velocity, Δf is the distance between the centre frequency of the two sub-scenes, λ is the radar wavelength, and R_T is the distance between the SAR sensor and the target. It is clear from Eq. 15.5 that v_{Ta} depends on the estimation accuracy of the displacement vector Δx. In order to obtain an accurate estimation of Δx the distance between the centres of mass of the azimuth amplitude distributions (averaging in range the ship's pixels) is calculated. The displacement vector is calculated according to $\Delta \mathrm{x} = |c_1\text{-}c_2|$ where c_1 and c_2 are the centres of mass for sub-look, respectively. Each centre of mass is computed by weighting the time position with the amplitude value in the following way

$$c_i = \frac{\sum_j m_{ij} x_{ij}}{\sum_j m_{ij}} \tag{15.6}$$

where the sum is over the number of pixels belonging to the target of the i-th look, m_{ij} represents the amplitude value and x_{ij} the position of the i-th look at time position j. The range velocity component is estimated by evaluating the time shift centre Δt, of the centre of mass in the sequence. Δt is given by:

$$\Delta t \approx -\frac{f_{DT} \cdot \lambda \cdot R_T}{2 \cdot v_s^2} = \frac{v_{Tr} \cdot \sin\theta \cdot R_T}{v_s^2} \tag{15.7}$$

where f_{DT} is the Doppler frequency and the other parameters have been already defined. Taking into account that:

$$f_{DT} = -\frac{2 \cdot v_{Tr} \cdot \sin\theta}{\lambda} \tag{15.8}$$

after manipulation of Eq. 15.7 the following equation can be obtained:

$$\Delta t \approx \frac{v_{Tr} \cdot \sin\theta \cdot R_T}{v_s^2} \tag{15.9}$$

where θ is the local incident angle. The local incident angle θ counts for the projection of the range velocity component in the look direction of the sensor. The estimation of the temporal shift Δt is done by weighting the signal amplitude in each image with the time t_i of this image. The azimuth profiles of the target after masking and averaging in range are shown in Fig. 15.20. The estimated velocity of the ship imaged in Fig. 15.18 is 12.2 knots with a dispersion of 1.6 knots when compared to the 11.3 knots available from the AIS information. However, the time lag between the SAR acquisition and the AIS may allow for the difference.

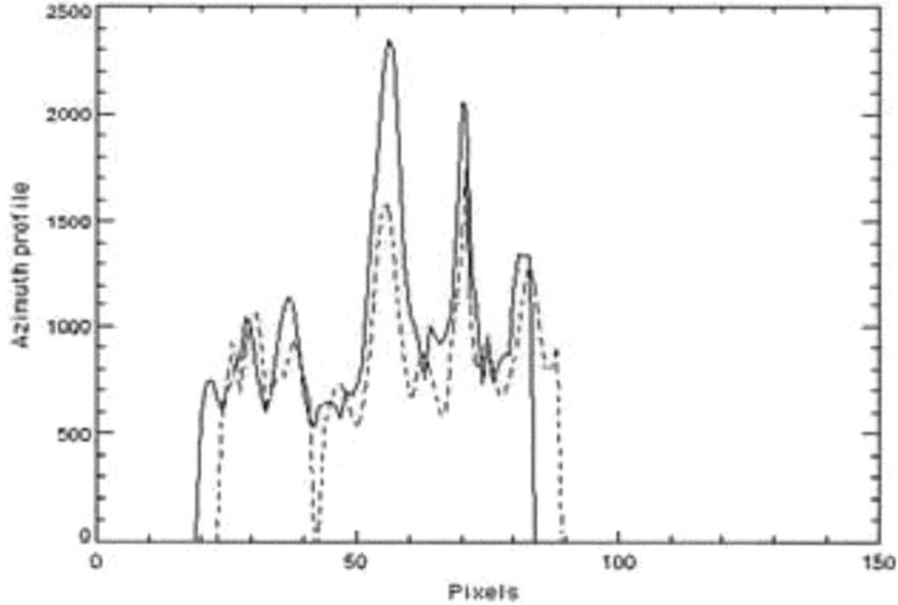

Fig. 15.20 Profiles of the target in sub-look 1 (*solid line*) and sub-look 2 (*dashed line*). The profile is obtained by range averaging and the pixels not belonging to the target are masked

15.5.6 Ship Velocity Estimation on TerraSAR-X (ATI) data

Ground moving target indication is a very difficult problem, due to the difficulty of separating the signal returned from a moving target from the stationary background (clutter). A method that can be used to detect ground moving target is Along Track Interferometry (ATI). Along Track Interferometric Synthetic Aperture Radar systems use more than one SAR antenna (typically two), mounted on the same platform and displaced along the platform moving direction. TS-X can acquire images with a split antenna. The so called Dual Receive Antenna (DRA) Mode allows the use of ATI methods. The information about the radial velocity of the moving target is estimated from the interferometric phase of the images using statistical estimation techniques based on the statistical distribution of the measured phases. The phase is proportional to the speed, which has zero mean for non-moving target:

$$\phi_{ATI} = \frac{4\pi}{\lambda} \frac{B_{ATI}}{v_s} v_{gr} \sin\theta \tag{15.10}$$

B_{ATI} is half of the distance between the two antennas, λ is the wavelength, v_s is the sensor velocity, v_{gr} is the target velocity to be estimated and θ is the local incidence angle. In Fig. 15.21 an example of retrieved ship speed with TS-X ATI Mode data over Gibraltar is shown.

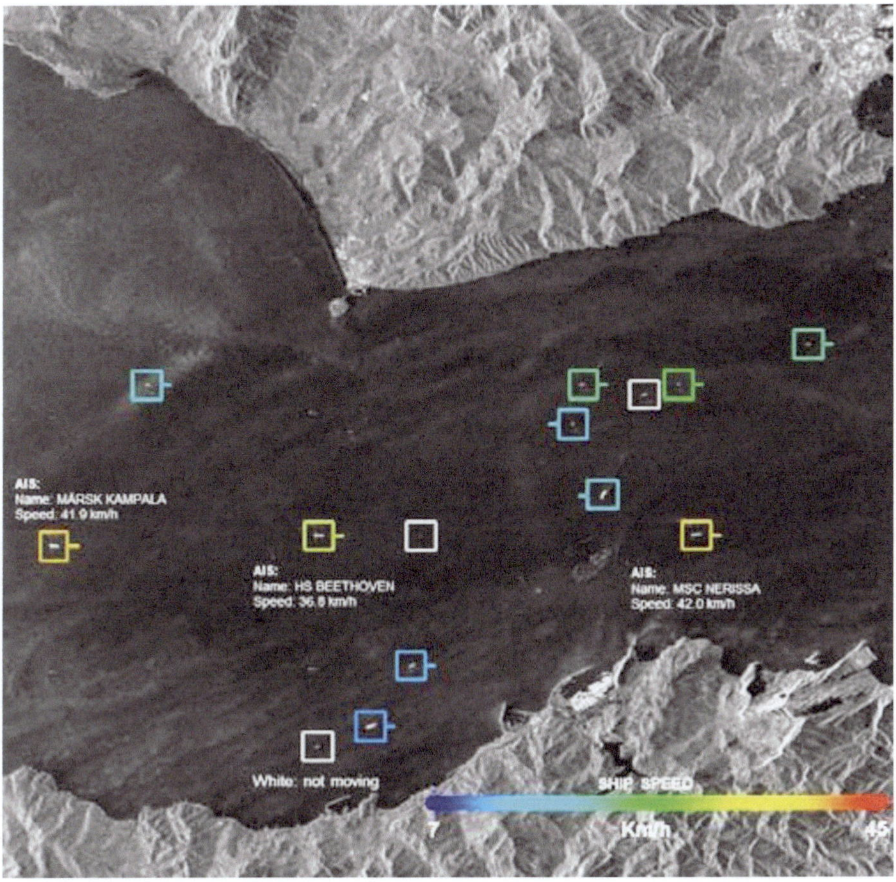

Fig. 15.21 Ship velocity estimation with TerraSAR-X DRA (Dual Receive Antenna) mode data in comparison to AIS data

15.6 Meteo-Marine Parameters

The detection capabilities of ship targets are influenced by the wind speed and wind direction compared to the track of the satellite and thus in change of sea state. In the following the main meteo-marine parameter influencing ship detection will be discussed.

15.6.1 Wind

Synthetic aperture radar can provide information on wind speed by measuring the roughness of the sea surface. An X- Band Geophysical Model Function (GMF)

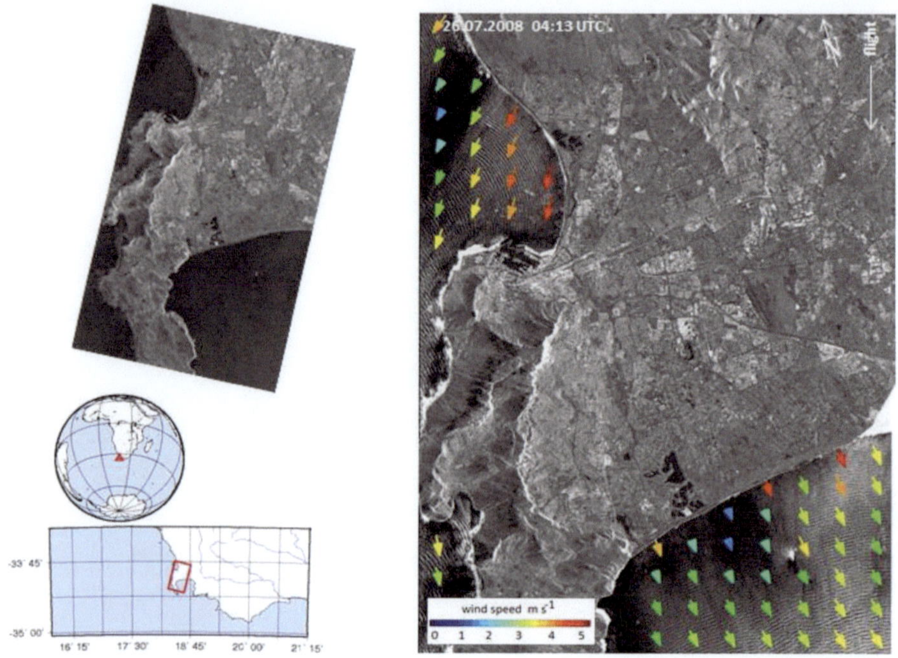

Fig. 15.22 Wind field derived from TerraSAR-X Stripmap image acquired over southern Africa near Cape Town on July 26, 2008 at 04:13 UTC

algorithm (XMOD) was established for VV and HH polarized data to obtain the wind fields (Ren et al. 2012). The relationship between X-band radar cross section and wind speed, wind direction and incidence angle is applied. Wind field parameters can be retrieved to an accuracy of about 2 m/s in absence of precipitation for incidence angles θ of $20 \sim 60\,^{\circ}$ and wind speed in the range from about 2 to 25 m/s. The X-band linear geophysical model function XMOD given in Ren et al. (2012). However, the XMOD linear algorithm yields reasonable measurement for wind up to about 15 m/s. A new algorithm XMOD-2 has been developed that takes the full nonlinear physical model function into account (Li et al. 2011). An example of the wind estimation is shown in Fig. 15.22.

15.6.2 Sea state

Several retrieval algorithms to derive the two-dimensional ocean wave spectrum or integral wave parameters from SAR data have been developed based on a first guess model output of the 2-D wave spectrum. As the ocean surface gravity waves are moving targets, the mechanism of their SAR imaging consist of the linear transformation of tilt and hydrodynamic modulation, as well as the non-linear distortion induced by the radial wave motions generally (Hasselmann et al. 1985). This leads, among other effects, to image smearing and to a loss of information beyond the so-called azimuth cut-off wavelength (Alpers and Brüning 1986). For ERS and ENVISAT SAR, this corresponds typically to wavelengths shorter than about 200 m in the along track. As motioned in introduction, the improved parameters of TS-X results in cut-off of about 100 m (azimuth) and 30 m (range) direction, the nonlinear effects are significantly reduced.

First guess information, needed for algorithms to derive sea state from SAR image spectra like PARSA (Schulz-Stellenfleth 2003), are not easily available. The use of empirical functions is here needed. The empirical X-WAVE.2 model for obtaining integrated wave parameters (wave height and period) has been developed for X-Band data (Lehner et al. 2011, 2012) without needing a priori information. The algorithm is based on analysis of image spectra and uses parameters fitted with collocated hind cast wave model results of the German Weather Services (DWD). For the validation and tuning of the X-WAVE algorithm, different data sets were used:

- Collocated buoy measurements,
- WaMoS II—Wave and Current Monitoring System (Borge et al. 2004, Dankert et al. 2005),
- Radar altimeter data,
- DWD wave hind cast model resulTerraSAR (Bruck and Lehner 2010, Pontes et al. 2010).

Comparison of TS-X derived significant wave height with the significant wave height obtained by the buoy located at Ekofisk oil platform shows a correlation of 0.83, for National Data Buoy Center (NDBC) buoy 44066 the scatter index is 0.19, correlation is 0.95, mean square error 0.89. More details for parameter estimation and comparisons are given in (Bruck et al. 2010, 2011; Lehner et al. 2012). Figure 15.23 shows an example of sea state parameter estimation carried out near Nigerian coast from TanDEM-X Stripmap image acquired on February 19, 2012 with an incidence angle of 23.9 ° at centre coordinates.

15.6.3 *Surface Current obtained from TerraSAR-X Dual Receive Antenna Mode*

A method to estimate the sea surface speed of the SAR cross-track velocity component from TerraSAR-X image has been developed. The technique exploits the phase of the Along Track Interferometry (ATI) acquisition. The ATI image pair is multiplied in order to obtain the interferogram which is processed in order to obtain the speed. The application of several filters, linear and non-linear is needed to get rid of the SAR image noise and to deal with the surface scattering which moves. The measured currents are derived by using TerraSAR-X Dual Receive Antenna (DRA) Stripmap data. Due to the Pulse Repetition Frequency (PRF) a pre-filtering is needed to remove the ambiguities which strongly affect these data. The demonstration on DRA data is

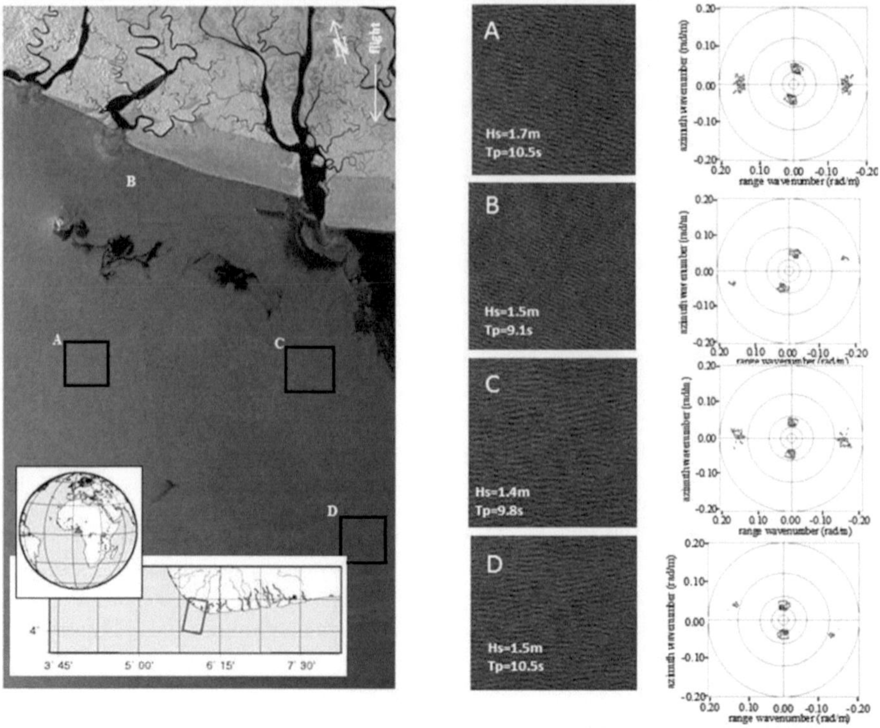

Fig. 15.23 Tandem-X VV polarized image acquired, taken on February 19, 2012 at 5:46 UTC over the Nigerian coast (*left*). For sub-scenes *A, B, C, D* the estimated parameters and spectra are shown (*right*)

new as well as the proposed method which is an extension to DRA data. The filters are appropriately tuned for DRA configuration.

The images acquired over Gibraltar (Fig. 15.24) are characterised by a surface speed flowing from West to East with an average absolute value of 1.1–1.3 m/s. Although wind features are visible in the image acquired over Gibraltar on April 12, 2010, it is characterised by low wind speed according to the Deutscher Wetterdienst (DWD) model (0.8–1.2 m/s). Some strong ambiguities from land and from vessels are still visible in both Gibraltar acquisitions; they are phase error sources which causes the estimated current to flow in the opposite direction. This effect is especially visible in the image of Gibraltar on 12th of April close to the coast. The procedure estimation accuracy is of ~0.5 m/s and the spatial resolution ~1 km, which are acceptable values for this application.

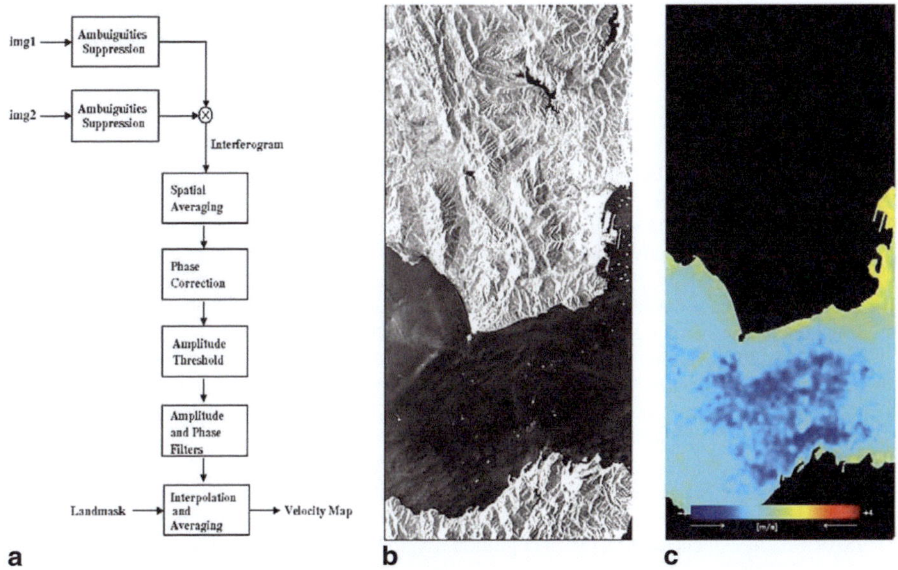

Fig. 15.24 Block diagram to estimate the sea surface motion (a), Strait of Gibraltar on Mai 12, 2010, Interferogram from TerraSAR-X ATI image pair (b), and derived current field (c)

15.7 Summary

The algorithm has been integrated to the DLR Toolbox (SAR AIS Integrated Toolbox, or SAINT) for automatic detection of both large and small vessels. Results show in case of the AIS-carrying larger vessels (length above 60 m) were detected in around more than 92 % of cases. For smaller vessels (length < 30 m) the assessment of quality is more difficult due to increasing false alarms for small vessels in high sea state situations with breaking waves. For moving and non-moving ships terrestrial AIS or Satellite AIS show good agreement with detected ships by SAR. The limitation of terrestrial AIS with coverage of about 40 km off the coast makes Satellite AIS and SAR a perfect candidate to fill this gap of ship traffic information over open oceans. Further, algorithms to derive the meteo-marine parameters from the surrounding sea surface like wind speed, sea state and currents have been developed.

References

Alpers W, Brüning C (1986) On the relative importance of motion-related contributions to the SAR imaging mechanism of ocean surface waves. IEEE Trans Geosci Remote Sens 24:873–885

Borge JN, Rodriguez G, Hessner K, González P (2004) Inversion of marine radar images for surface wave analysis. J Atmos Oceanic Technolo 21:1291–1300

Breit H, Fritz T, Schattler B, Borner E, Lachaise M, Niedermeier A, Eineder M, Balss U (2007) Terrasar-X Payload data processing—first experiences. In: Geosciences and remote sensing symposium, IEEE International, 2007 Igarss 2007

Breit H, Fritz T, Balss U, Lachaise M, Niedermeier A, Vonavka M (2010) The TerraSAR-X satellite. IEEE Trans Geosci Remote Sens 48(2):615–622

Brekke C, Anfinsen SN, Eltorf T (2010) Marine Target Detection Based On Dual Channel SAR Images, In Proc. 8th European Conference On Synthetic Aperture Radar, Aachen, Germany, Jun. 2010, pp 786–789

Bruck M, Lehner S (2010) Extraction of wave field from Terrasar-X data. Proc. Oceanography Work Shop SEASAR 2010, advances in SAR oceanography from ENVISAT, ERS and ESA third party missions, ESA ESRIN, Frascati/Rom, Italy, 25–29 January 2010, ESA, ISBN 9789292212438

Bruck M, Pontes MT, Azevedo E, Lehner S (2011) Study of sea-state variability and wave groupiness using terrasar-x synthetic aperture radar data. Ewtec 2011 On Cd, Pp. 1–7. 9th Ewtec 2011, 5–9 September 2011, Southampton, Uk. Peer Reviewed Proceeding

Brusch S, Lehner S, Schulz-Stellenfleth J (2008) Synergetic use of radar and optical satellite images to support severe storm prediction for offshore wind farming. IEEE J Sel Top Appl Earth Obs Remote Sens 1(1):57–66

Brusch S, Lehner S, Fritz T, Soccorsi M, Soloviev A, Van Schie B (2011) Ship surveillance with TerraSAR-X. IEEE Trans Geosci Remote Sens 49(3):1092–1103 (ISSN: 0196–2892)

Cooke T, Martorella M, Haywoodc B, Gibbinsa D (2006) Use Of 3D ship scatterer models from ISAR image sequences for target recognition. Digital Signal Process 16(5):523–532

Crisp DJ (2004) The state-of-the-art in ship detection in synthetic aperture radar imagery. Intelligence, Surveillance and Reconnaissance Division Information Sciences Laboratory, Department Of Defence Defence Science And Technology Organisation, Dsto-Rr-0272

Dankert H, Horstmann J, Rosenthal W (2005) Wind and wave field measurements using marine X-Band radar image sequences. IEEE J Oceanic Eng 30(3):534–542

Greidanus H, Clayton P, Indregard M, Staples G, Suzuki N, Vachon P, Wackerman C, Tennvassas T, Mallorquí J, Kourti N, Ringrose R, Melief H (2004) Benchmarking operational SAR ship detection. In Proc. Igarss 2004, Alaska, 20–24 Sept 2004

Hasselmann K, Raney RK, Plant WJ (1985) Theory of synthetic aperture radar ocean imaging: a marsen view. J Geophys Res 90:4659–4686

Kirscht M (1998) Detection, velocity estimation and imaging of moving targets with single-channel SAR. Proc Eusar'98, pp 587–590

Lehner S, Pleskachevsky AL, Bruck M, Li X, Brusch S (2011) Validation of coastal wind and wave fields by high resolution satellite data. Proc. 12th international workshop on wave hindcasting and forecasting

Lehner S, Pleskachevsky A, Bruck M (2012) High resolution satellite measurements of coastal wind field and sea state. Int J Remote Sens 33(23):7337–7360 (Special issue: Pan Ocean remote sensing: connecting regional impacts to global environmental change)

Li XM, Lehner S, Brusch S, Ren YZ (2011) "Sea surface wind measurement over offshore wind farm using TerraSAR-X data". Proc. of SPIE 2011 Conference—Remote Sensing Of The Ocean, Sea Ice, Coastal Waters, And Large Water Regions 2011

Lyzenga DR, Shuchman RA, Lyden JD (1985) SAR imaging of waves in water and ice: evidence for velocity bunching. J Geophys Res 90:1031–1036

Migliaccio M, Ferrara G, Gambardella A, Nunziata F, Sorrentino A (2007) "A physically consistent speckle model for marine SLC SAR images". IEEE J Ocean Eng 32(4):839–847

Mittermayer J, Runge H (2003) "Conceptual studies for exploiting the Terrasar-X dual receive antenna". In Proc. IGARSS, pp 2140–2142

Pontes M, Bruck M, Lehener S, Kabuth A (2010) Using satellite spectral wave data for wave energy resource characterization. In: 3rd International Conference On Ocean Energy, 6 October, Bilbao

Ren YZ, Lehner S, Brusch S, Li XM, He MX (2012) An algorithm for the retrieval of sea surface wind fields using X-band TerraSAR-X data. Int J Remote Sens 33(23):7310–7336

Schulz-Stellenfleth J (2003) Ocean wave measurements using complex synthetic aperture radar data. Ph.D. Thesis, Universität Hamburg

Schwarz E, Lehner S, Brusch S (2010) Ship detection service. Geoforum Mv 2010 Vernetzte Geodaten: Vom Sensor Zum Web, Seiten 115–118. Gito-Verlag Berlin. Geoforum Mv 2010-Vernetzte Geodaten: Vom Sensor Zum Web, 26–27 Apr. 2010, Rostock-Warnemünde. ISBN 978-3-942183-01–7

Soloviev A, Gilman M, Young K, Brusch S, Lehner S (2010) Sonar measurement in ship wakes simultaneous with TerraSAR-X overpasses. IEEE Trans Geosci Remote Sens 48(2):841–851. doi:10.1109/Tgrs.2009.2032053. ISSN 0196-2892

Velotto D, Migliaccio M, Nunziata F, Lehner S (2011) Dual-polarized TerraSAR-X data for oil-spill observation. IEEE Trans Geosci Remote Sens 49(12):4751–4762, part 1. doi:10.1109/Tgrs.2011.2162960

Velotto D, Soccorsi M, Lehner, S (2014) Azimuth Ambiguities Removal for Ship Detection Using Full Polarimetric X-Band SAR Data. IEEE Transactions on Geoscience and Remote Sensing, 52 (1), Seiten 76–88. IEEE Xplore. ISSN 0196-2892

Part IV
Marginal and Enclosed Water Bodies

Chapter 16
Remote Sensing of Coral Reefs and Their Environments in the Red Sea and Western Indian Ocean: Research and Management

Serge Andréfouët

Abstract The Red Sea and the Western Indian Ocean regions offer a wide variety of coastal configurations, including large areas of coral reefs. In regions that include both developed and developing countries, coral reefs and the services they provide are critical for the livelihoods, economies and well being of local communities, while they are at risk due to both anthropogenic and natural threats. Coral reefs also offer to scientific research a unique array of ecological and geological processes specific to these regions. Both applied and fundamental aspects motivate conservation and management projects. Satellite and airborne remote sensing has contributed to scientific and management projects with a variety of sensors and products, including some of the earliest coral reef applications worldwide 30 years ago. High spatial resolution geomorphology, bathymetry and habitat maps provide a biophysical description of the reefs themselves. Oceanic fields, such as sea surface temperature regime and their anomalies, quantify how coral reefs are exposed to weather and climate forcing. The combined local and large scale spatial information help understand how coral reefs status and processes vary across space and time. In particular, data products are increasingly combined to map the resilience of coral reef and associated habitats, and to design conservation plans at different spatial scales. However, both high-tech state-of-the-art and user-friendly remote sensing are still not used as systematically as it could be, while most countries still need basic spatial information consistent at national and international scales, such as habitat maps. Future applied developments will need to address these gaps in priority products. At the same time, on-going research will continue to refine local and regional products relevant for the science and the management of the Red Sea and the Western Indian Ocean coral reefs and their environments.

S. Andréfouët (✉)
Institut de Recherche pour le Développement, Nouméa,
New Caledonia, France
e-mail: serge.andrefouet@ird.fr

V. Barale, M. Gade (eds.), *Remote Sensing of the African Seas,*
DOI 10.1007/978-94-017-8008-7_16,

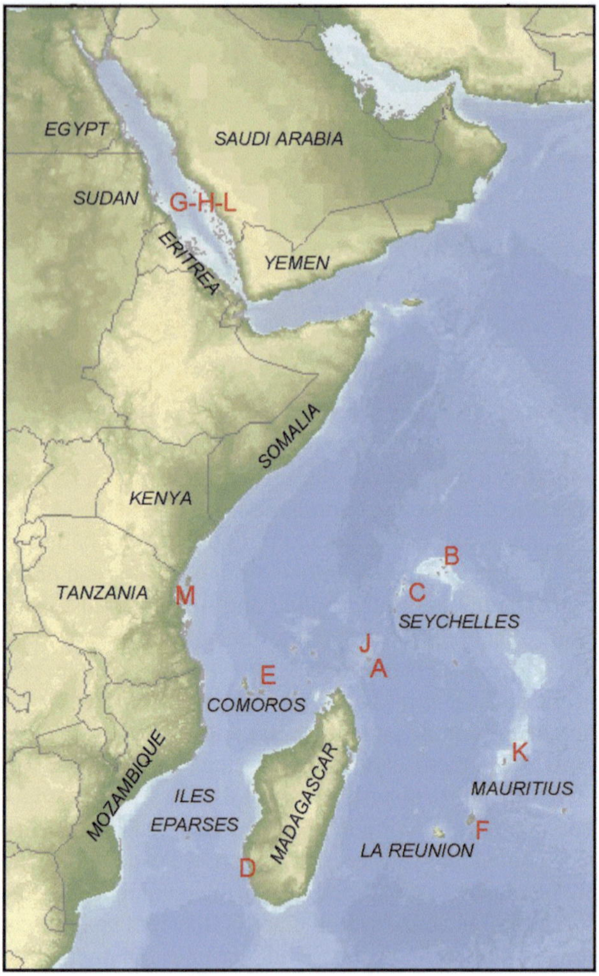

Fig. 16.1 Map of the focal area of this review, which includes the Red Sea and the Western Indian Ocean (background image from ReefBase, http://www.reefbase.org). Red letters refer to reef sites shown in Fig. 16.2

16.1 Introduction

The Red Sea, the East African Coast from Somalia to South Africa, Madagascar and the Western Indian Ocean islands offer a vast array of diverse coastal configurations stretched across both a wide latitudinal gradient and a rich coastal-oceanic gradient (Fig. 16.1). The Red Sea is a deep semi-enclosed water body bounded by an arid environment; the Mozambique Channel is a inner sea sandwiched between Madagascar and Mozambique, sheltering the high Comoros Islands and the low Iles Eparses; the Seychelles, Mauritius and La Reunion are made of isolated and remote islands, atolls and banks; finally the eastern shores of Madagascar Tanzania, Kenya and Somalia are directly exposed to the Indian Ocean and its monsoon driven-climate and to land influences as well (Sheppard 2000).

These different configurations create significant contrasts in terms of exposure of coastal habitats and their benthic and pelagic living communities to physical factors (*e.g.* temperature, circulation, wave climate, cloud cover, sedimentation) and create a large diversity of ecological and functional niches. Among the main coastal habitats, coral reefs, seagrass meadows, and mangroves are of high importance given their daily use by human populations. The amount of ecosystem services (fishery and aquaculture, shoreline protection, nutrient recycling and so forth) provided by this trio of coastal habitats is essential to the livelihood of islanders and coastal continental populations (Gullstroem et al. 2002). Increasing human populations and climate change threaten ecosystem resilience and the sustainability of these services in the long term (Hughes et al. 2003). Conservation projects at various scales intend to mitigate these impacts by implementing a range of management actions, some purely local at the scale of villages, others at national and trans-boundary scales (Wells et al. 2007).

This extremely brief panorama of the Red Sea and Western Indian Ocean context identifies the potential use of remote sensing for coastal research and management. First, the large ocean-scale domain and the variety of configurations naturally call for the use of synoptic low to moderate resolution (50 km down to 1 km) observations of the atmosphere, ocean and land dynamics along the coast and near the reefs. Second, the ecological and economical importance of coral reefs, seagrass meadows and mangroves call for high to very high resolution (30 to 1 m) observations in order to map the distribution of these habitats and monitor their changes. Third, climate change-induced threats promote the combined use of low resolution and high resolution data in synergy to observe (and predict) the trajectories of habitats under the observed (or forecasted) physical forcing. Fourth, management and conservation activities need accurate habitat and physical stress maps to design adequate conservation plans and assess the amount and sustainability of services provided by habitats at a variety of scales, from transboundary to local, from the entire region to few tens of kilometre square.

This review compiles representative remote sensing studies to illustrate these different applications in the region of interest. The review is not necessarily completely exhaustive and does not cite all existing papers in peer reviewed, conference proceedings and grey literature, but it points to gaps and perspectives. When interesting remote sensing applications have not yet been conducted in the Red Sea and Indian Ocean, pilot studies from other regions are cited. Also, this author assumes that the reader is familiar with basic remote sensing principles and vocabulary and with the most commonly used remote sensing sensors to date. The following chapters do not enter methodological details but rather emphasize the applications. Finally the main focus is on coral reef ecosystems, but seagrass and mangroves studies are also frequently cited. In agreement with the terminology used by Andréfouët and Riegl (2004), *direct* remote sensing refers to the use of remote sensing to study the habitats themselves, while *indirect* remote sensing implies that the ocean and the atmosphere surrounding the habitats are studied.

16.2 Direct Remote Sensing of Coral Reefs

In a coastal remote sensing context, habitats are often defined by a number of attributes which generally include a geomorphological, a depth and a benthic description (Andréfouët et al. 2003). These attributes have been mapped with remote sensing, afterwards leading to the study of habitat-related processes. This section reviews Red Sea and Western Indian Ocean habitat mapping and functional studies achieved through direct use of remote sensing.

16.2.1 Geomorphology

Modern geomorphology is the result of the convolution of a number of processes acting at different time-scales (Camoin et al. 1997; Hopley 2011). A coral reef geomorphological unit can be created by the combination of geological processes occurring in the area, type of climate and weather forcing, variation of sea levels and level of accretion and erosion due to the respective activity of dominant calcifying communities (corals, coralline algae) and bioeroders. Once mapped, modern geomorphological features may bring clues pointing to the past geological events and climate regime that have shaped the reefs present on the studied sites. For instance, Purkis et al. (2010) have explained the occurrences in the Red Sea of karstic reticulated structures identified on high resolution Quickbird images by a Quaternary signature of paleohumidity in the now hyperarid Red Sea.

Except for numerous past studies that have used aerial photographs (color and black & white) to help geomorphological interpretation (*e.g.* Stoddart et al. 1971) and Purkis et al. (2010), no other satellite borne geomorphological maps have been directly used to infer the geological processes that have shaped coral reefs in the Red Sea and Western Indian Ocean. Instead, remotely sensed geomorphological maps have rather remained descriptive, as a first coarse layer of habitat description. For instance, in the course of the Millennium Coral Reef Mapping project (Andréfouët et al. 2006; Andréfouët 2011; Andréfouët et al. (2009) have compiled in one single atlas geomorphological maps for Madagascar and every Western Indian Ocean Islands. The maps were derived from Landsat 7 images at 30 m resolution. They depicted the main coral reef geomorphological units down to approximately 25–30 m depth on average (Fig. 16.2) following a hierarchical typology of units that is used globally. The value of this work was to provide maps that were regionally consistent from one country to another. Often, these maps yielded the first accurate inventory of the surface area covered by coral reefs and associated lagoons, at national scale and per individual island. These inventories are required to design national scale conservation plans obeying international conservation targets (Wabnitz et al. 2010).

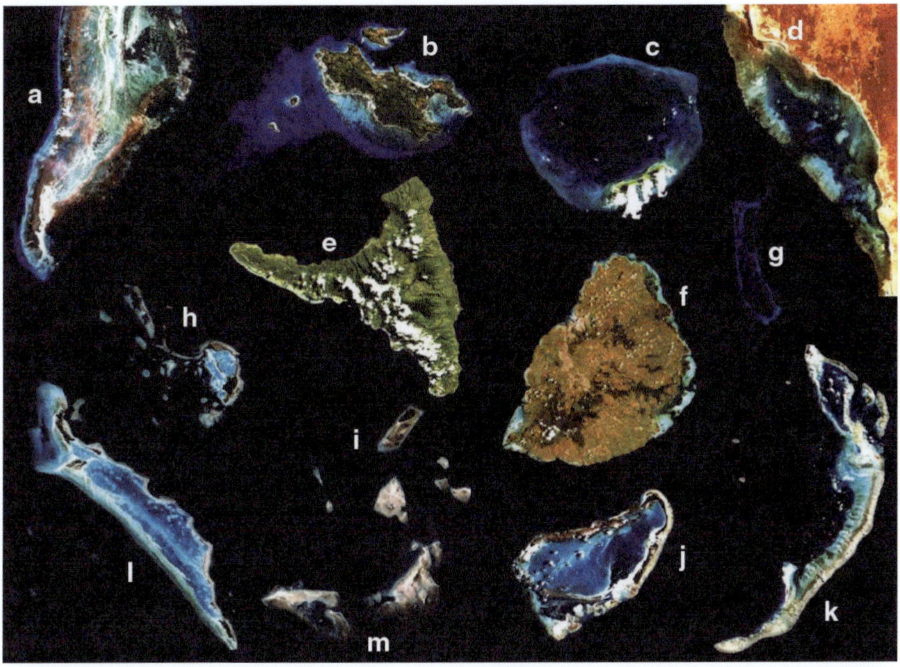

Fig. 16.2 Diversity of coral reefs geomorphology found in the focal area. Examples as seen with the Landsat 7 ETM + sensor (scales are different for each image). See Fig. 16.1 for site location. Barrier reefs are represented here by D (Madagascar), L (Saudi Arabia). Fringing reefs are represented by D (Madagascar, in the lagoon), B (Seychelles), E (Comorros) and F (Mauritius). Different types of atolls are illustrated with C (Seychelles), G (Saudi Arabia) and J (Seychelles). Patch reef systems are represented by M (Tanzania) and H (Saudi Arabia). Finally, large oceanic banks are A (Seychelles) and K (Mauritius)

16.2.2 Bathymetry

Bathymetry (water depth) is an important habitat variable since it is directly correlated to light availability, water temperature, water motion and other critical physical variables that define the range of conditions suitable for organisms to thrive.

Bathymetric algorithms have been used since the emergence of optical airborne and satellite remote sensing (Lyzenga 1985), and have been continually refined afterwards (Lyzenga et al. 2006). These semi-empirical algorithms are based on a number of assumptions related principally to the curve of light attenuation with depth and the seafloor reflectance and their ratios between different spectral bands (Maritorena et al. 1994). With multispectral imagery, few algorithmic options are possible, generally using the available green and blue spectral bands. Active bathymetric mapping can be achieved using lidar (for Light Detection and Ranging) sensors (Brock and Purkis 2009). However, to the best of knowledge, bathymetry lidar data have been acquired only in La Reunion and Iles Eparses in our region of interest, and no reports seem available yet.

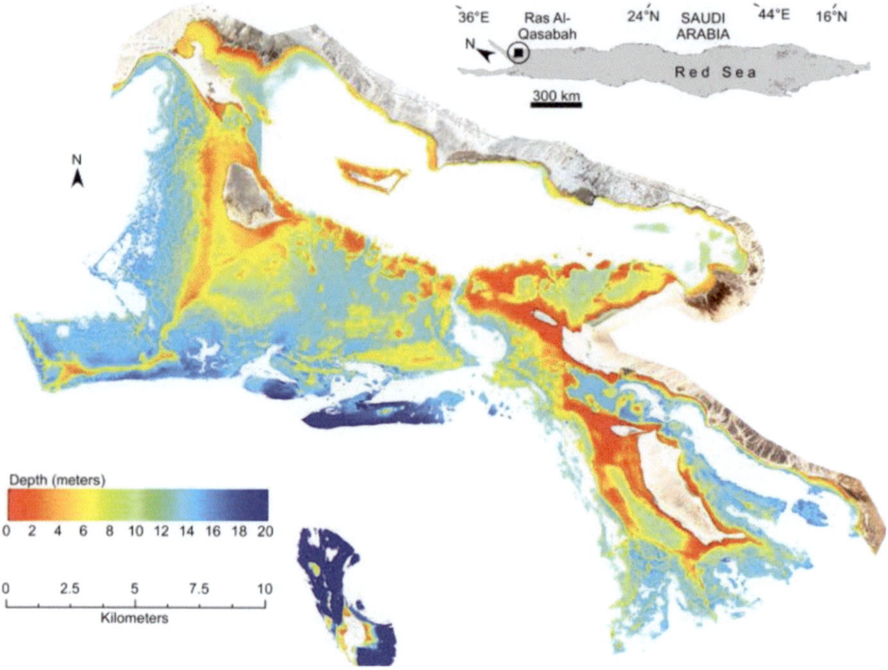

Fig. 16.3 Shallow water bathymetry for Ras Al-Qasabah, Saudi Arabia. (location map on the *upper right corner*). The example is derived from QuickBird satellite imagery and represents a small sub-set of a vast assessment of the Saudi Arabian coastal resource carried out under the auspices of the Khaled bin Sultan Living Oceans Foundation and the Saudi Wildlife Commission. Further map examples, and image processing methodology can be found in an atlas of Saudi Arabian Red Sea marine habitats (Bruckner et al. 2011). Geological and ecological applications of this or similar bathymetric data are found in Purkis et al. (2010) and Rowlands et al. (2012)

It is now fairly common practice to derive bathymetric maps from optical imagery (Fig. 16.3), especially with the growing combined availability of acoustic sonar (for Sound Navigation and Ranging) data that can be used to calibrate the process and assess the accuracy.

Benny and Dawson (1993); Purkis and Pasterkamp (2004); Vanderstraete et al. (2005); Knudby et al. (2010a); Hamylton and Spencer (2011); Bruckner et al. (2011); Rowlands et al. (2012) have all created bathymetry maps from their imagery—using Landsat Thematic Mapper^{TM} and Enhanced Thematic Mapper (ETM +), IKONOS, Compact Airborne Spectrographic Imager (CASI), Quickbird—for their applications in Egypt, Saudi Arabia, Tanzania and Seychelles. However, no large-scale data sets have been created to map entire regions and countries. The most systematic production occurred for the Saudi coast of the Red Sea (see Fig. 16.3), as described in Bruckner et al. (2011) and Rowlands et al. (2012).

Accuracy of modelled optical bathymetry is deemed acceptable especially for ecological and management applications, but it would not be the case for marine charts and navigation.

16.2.3 Benthic Communities

Beyond the geomorphological and depth attributes, habitats are also defined by the type of benthic cover found on the seafloor. The high diversity of substrate types (*e.g.* terrestrial and carbonate sand, rubble, boulders rocks, dead eroded corals, pavement) and benthic organisms (hard and soft corals, macroalgae, seagrass, sponges, ascidians, microbial mats, etc) results in a virtually infinite number of mosaics of communities. Most of the time, these communities are organized according to local energy (due to wave and wind exposure) and depth gradients. Obviously, corals are the focus of many coral reef remote sensing studies, but in the field, coral dominated areas may be extremely narrow compared to vast expanses of sedimentary areas where other organisms, such as seagrass, dominate. The importance of inter-connections and resulting fluxes between the different habitats are now fully acknowledged in ecological studies and in management plans (*e.g.* see, for Zanzibar Island, Dorenbosch et al. 2005).

The reflectance spectral properties of coral organisms from the Red Sea has been studied by Minghelli-Roman et al. (2002) in the course of a hyperspectral mapping project. The radiometric discriminant functions obtained from these local studies were confirmed by measurements that have characterized globally the spectral signatures of a variety of coral reef end-members. Hochberg et al. (2003, 2004) have measured the spectral differences between benthic end-members from all coral reef regions worldwide, including samples from Mayotte Island in the Mozambique Channel.

Measuring the spectral reflectance of individual organisms is not necessarily immediately useful for mapping communities and habitats with images because communities are made of the intricate spatial assemblages of organisms. These assemblages can also change with time, according to seasons and disturbances. Using directly spectral signatures differences and scaling discriminant functions from organisms to communities and habitats to achieve a map is not a trivial task (Andréfouët et al. 2004; Hedley et al. 2012). In the Red Sea, Purkis and Pasterkamp (2004) have described the benefits of using *in situ* reflectance for habitat mapping, but spectral measurements were achieved for meter-scale communities, which is already a step ahead in terms of scaling and mixing compared to using organism signatures. As a result, when mapping habitats, photo-interpretation, purely statistical classification and object-based segmentation approaches often all appear more suitable especially when using multispectral images limited in spectral resolution.

The number of satellite-based habitat mapping applications in the Red Sea and Western Indian Ocean is quite high. We can cite the early Red Sea SPOT study by Courboulès et al. (1987, 1988); Courboulès and Manière (1992) and Manière and Jaubert (1985). Later, Landsat TM has been used by Purkis et al. (2002) and Purkis and Pasterkamp (2004) also in the Red Sea, by Chapman and Turner (2004) in Rodrigues Island and by Klaus et al. (2003) in Socotra Island. New generation of very high resolution sensors such as IKONOS and Quickbird have been used in Mayotte, Zanzibar and the Red Sea (Andréfouët et al. 2003; Knudby et al. 2010a, Rowlands et al. 2012). Specifically with a focus on seagrass meadows, we can cite the Landsat TM processing by Dahdouh-Guebas et al. (1999) in Kenya, the Landsat TM/ETM + change detection work by Gullstroem et al. (2006) and Knudby et al. (2010b) both around Zanzibar Island, Tanzania, and the IKONOS mapping by Knudby and Nordlund (2011) also next to Zanzibar.

From the digital airborne hyperspectral realm, Despinoy and Naim (1997) have mapped reefs of La Reunion Island and Borstad et al. (1997) looked at Mauritius reefs and lagoons. Red Sea reefs have been investigated by Hamylton (2011a) and Rowlands et al. (2012). Possibly the largest hyperspectral mapping exercise was conducted in Amirantes, Seychelles. This work yielded a well-documented atlas of the reefs and islands of this archipelago (Spencer et al. 2009). All the above hyperspectral work relied on a CASI.

In addition to "classic" satellite and airborne missions, an ultra-light aircraft has been used in La Reunion Island to acquire very high resolution vertical images of coral bleaching events (Pennober and Borius 2010). Finally, hand-held digital photographs taken by astronauts from the International Space Station served as a background image for mapping reefs in Iles Eparses (Samani 2005).

Methodologically, multispectral image processing techniques for habitat mapping have significantly evolved during the two last decades. A number of the aforementioned studies had a strong innovative methodological goal (Purkis and Pasterkamp 2004), whereas others were more thematically oriented and relied on standard processing methods (Andréfouët et al. 2003; Knudby et al. 2010b). Part of the habitat mapping challenge is to maintain both high thematic richness (number of mapped habitat classes) and high classification accuracy. Andréfouët (2008) discusses the relative merits of producers and users approaches in addressing this challenge, and devises the benefits of combining simple visual interpretation methods with traditional image processing methods, especially in a capacity building context. Examples of habitat maps produced with this user-oriented approach are shown Fig. 16.4.

16.2.4 Processes and Functional Studies

Habitat maps, once created, should be used for some type of applications. A map is not necessarily an end in itself. However, few studies go the extra-step, beyond the habitat mapping exercise. Fundamental science and resource management benefited poorly from habitat maps. Thus far, functional studies looking at processes controlling coral reef communities remain scarce. They are limited to the characterization of hydrodynamic regime, geological processes (see Purkis et al. 2010 as discussed above), and fish community modelling. Finally, there is an increasing effort to map coral reef resilience potential at local and regional scale.

The spatial distribution of living communities is partly constrained by the hydrodynamic regime. Water movements around reefs can often be inferred on visible and near infra-red images, especially when significant sunglint can offer an image of the sea surface. The patterns obviously only capture snapshots in time if only one image is used. Nevertheless, surface features visible on SPOT images have been used to infer the relative difference of hydrodynamic exposure among different reef sections in the

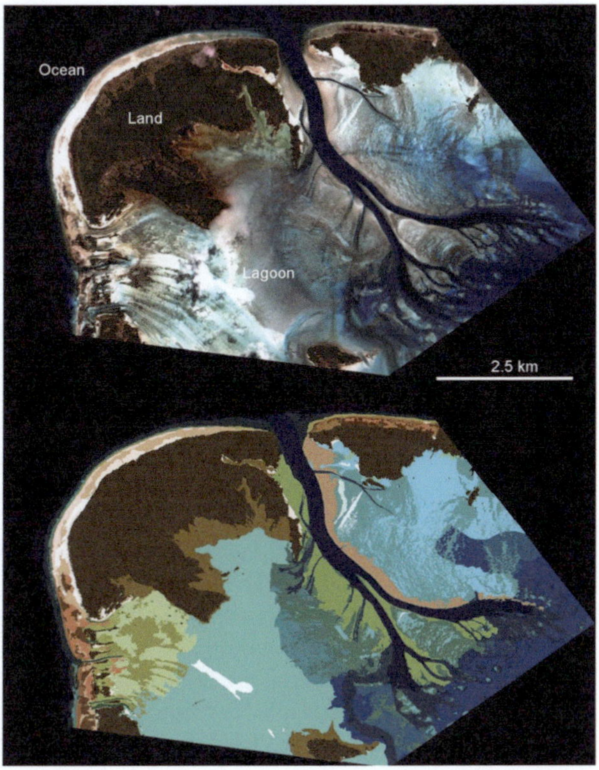

Fig. 16.4 Illustration for Aldabra atoll (Seychelles) of a geomorphological map derived from a WorldView 2 image (2 m spatial resolution), mapped in 53 classes. The map also included benthic information on seagrass meadows (detailed caption is not show given the total number of classes, but see Andréfouët 2012). The total processed area represents 64 km^2 of a fairly complex reef system, in particular with the passes and channels that dissect the lagoonal platform. Processing took 5 days to complete as per the user-oriented methods outlined in Andréfouët (2008)

Red Sea (Courboulès and Manière 1992). Relationships between hydrodynamics and morphology of seagrass beds have been quantified using both images and numerical models, allowing time integration. Hamylton and Spencer (2011) characterized in Alphonse Atoll in Seychelles the combined influence of spur and groove amplitude and incident wave power in the variation of sand and seagrass patch linearity across a habitat map derived from CASI hyperspectral imagery. Hamylton (2011b) also used CASI-derived maps and wave models to integrate hydrodynamic energy for a period of 5 years and relate this forcing to spatial patterning of Alphonse reef flat surface cover.

Mapping biodiversity is a major challenge in tropical coastal environments. Since biodiversity censuses are difficult and costly to conduct in the field and require taxonomic expertise, surrogates have been proposed to do so indirectly. This includes habitats mapped with remote sensing (Dalleau et al. 2010). Thus far, no true surrogacy analysis measuring the linkages between species diversity and habitat diversity for instance has been done in the Red Sea and the Western Indian Ocean. However, as a different type of biodiversity application, reef fish communities variables (richness and abundance) have been modelled in Zanzibar by deriving from IKONOS images the variables (depth, structural complexity of the substrate, and live coral cover) that were related to fish communities according to ancillary field studies (Knudby et al. 2010b).

Coral reefs and associated ecosystems such as seagrass meadows are at risk given the cumulative stress induced by human activities and climate change. Understanding ecosystem resilience is now a priority and a challenge (Nyström et al. 2008). Resilience capacity results from a combination of numerous processes related to biodiversity, functional redundancy, connectivity, heterogeneity, and their dynamics under human pressure (*e.g.* fisheries) and climate change. Mapping resilience is in its infancy given our ignorance on many aspects of reef resilience, the high complexity at stake and the amount of new data needed. However, we identified two relevant remote sensing approaches to study resilience.

In its simpler form, resilience can be measured by multidate image analysis and change detection. Across decades, it may be possible to observe cycles of loss of important communities and their recovery which suggests resilience, as in Scopélitis et al. (2009) for Saint Leu Reef in La Réunion Island. Long term observations may also suggest steady degradation and continuous loss of critical habitats, due to chronic disturbances, precluding the return to previous functional states, *e.g.* Toliara, Madagascar (Bruggeman et al.). The temporal dynamics of coral reefs, seagrass and mangroves habitats have been studied in the Western Indian Ocean with a variety of image sensors (Makota et al. 2004; Wang et al. 2003; Gullstroem et al. 2006; Ferreira et al. 2009; Knudby et al. 2010b), sometimes with a multi-sensor, very high resolution, approach (Scopélitis et al. 2009; Andréfouët et al. 2013) (Fig. 16.5).

Using a more complex modelling design, resilience has been tentatively mapped throughout Saudi Arabia west coast by Rowlands et al. (2012). They identified first a number of key variables and indices required to characterize resilience. These indices were then inferred and mapped from remote sensing using a variety of assumptions, images and processing. Both "landscape" (related to coral, framework, depth) and "stress" (related to fishery, temperature and distance to urban areas) factors and indices were mapped to be merged into an integrated, spatially explicit, resilience index. The result is a consistent modelled regional view of the capacities of individual reefs to cope with climate change and human induced disturbances.

16.3 Indirect Remote Sensing of Coral Reefs

This section addresses remote sensing of meso-scale hydrodynamic and atmospheric physical processes around coral reefs. The goal is to characterize how these physical processes influence the functioning and structures of local communities and habitats. This idea has been already expressed in the previous section. For instance, Hamylton (2011b) coupled a wave model with habitat maps to characterize community patterning. The resilience study by Rowlands et al. (2012) also included meso-scale forcing with an index of thermal stress computed with Moderate Resolution Imaging Spectroradiometer (MODIS) 1 km Sea Surface Temperature (SST) data. It would be possible to include here the land and riverine processes that also significantly affect coastal reefs through run-offs, and even remote reefs at long distance through river plumes dispersal.

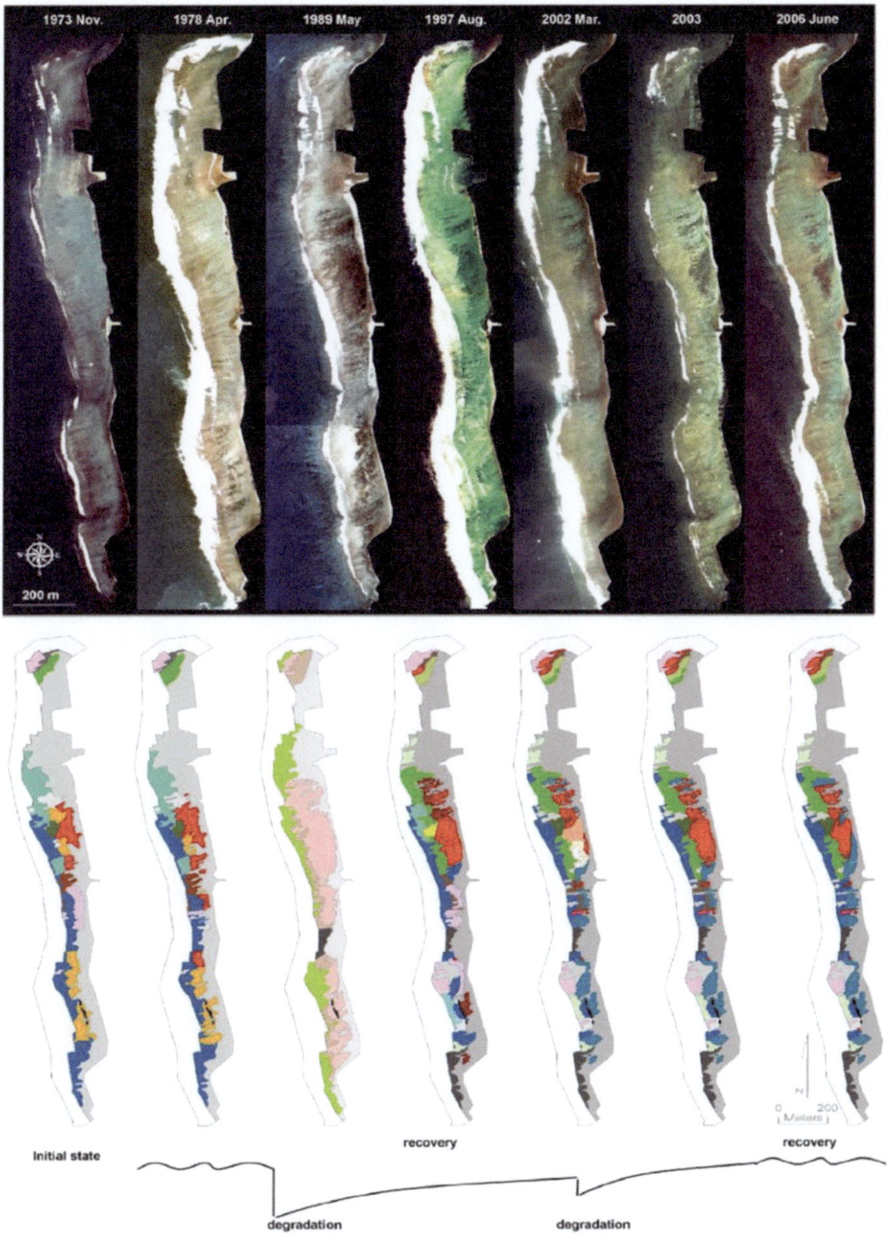

Fig. 16.5 Example of time-series of different images (including aerial photographs from 1973 to 1998, and in 2003, and Quickbird images in 2002 and 2006) for the reef flat of Saint Leu Reef in La Reunion Island. The maps highlight the distribution of different coral communities (one for each color), and their dynamics. Just before the 1989 image, hurricane Firinga stroke the reef leading to the quasi complete loss of coral communities. Despite a minor bleaching event in 2002, the reef returned in 2006 to a situation comparable to the 1973 situation. This example illustrates the value of long term, multi-sensor, time series to understand the trajectory of reef communities, and their resilience to severe disturbances. More details in Scopélitis et al. (2009)

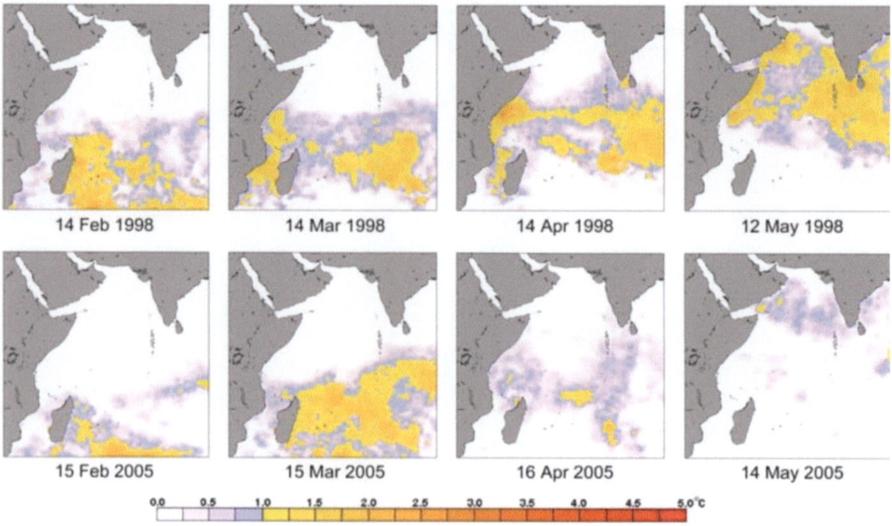

Fig. 16.6 From McClanahan et al. (2007), the above panels show the spread of SST positive anomalies, in degrees Celsius) in the Western Indian Ocean in 1998. As a result, the summer 1998 has experienced the worst mass coral bleaching event to date. In contrast, the 2005 summer (below) has seen moderate thermal stress and bleaching. Anomalies are computed at 50 × 50 km resolution against SST climatologies. Data from http://www.coralreefwatch.noaa.gov/satellite

Climate change and its consequences are obvious drivers of most recent indi rect remote sensing studies. Projected and observed trends in SST are of particular concerns. Changes in temperature regimes and associated positive anomalies have driven local and regional coral bleaching events worldwide.

Bleaching is a physiological process where coral polyps expel their symbiotic zooxanthellae micro-algae, which results in the loss of the coral autotrophic capacities and eventually death (Baker et al. 2008). The impacts of massive global bleaching events have been catastrophic especially in 1998 and 2002. The consequences had significant effects on reefs and many did not recover, impacting on the long term structural complexity, community structures, trophic networks and other functional processes (Graham et al. 2008). These events have promoted the refinement of SST-based early warning systems that use Advanced Very High Resolution Radiometer (AVHRR) data to monitor thermal fluctuations in the tropics. SST around Red Sea coral reefs have also been studied with MODIS by Blythe et al. (2011). Globally, a number of bleaching-specific indices of cumulative thermal stress have been proposed (*e.g.* Degree Heating Weeks), after examining empirically at large scale the spatial relationships between SST and bleaching intensity. This is still a very active field of research (Teneva et al. 2012).

In the Indian Ocean, McClanahan et al. (2007) have measured the reliability of SST based predictions of bleaching when using 50 km resolution Degree Heating Weeks data with a large regional coral data sets (Fig. 16.6). They demonstrated that cumulative temperature anomalies identified general areas with bleaching but with both large over- and underestimates of bleaching intensity. They pointed out the necessity to couple SST observations with field data on coral communities, especially where acclimation and reorganization of the coral community have occurred due to past bleaching events.

To date the only multi-factor analysis of bleaching events for the Indian Ocean, Maina et al. (2008) combined a variety of remotely sensed data to model the susceptibility of coral reefs to environmental stress. Specifically, they used surface currents, wind velocity; SST, UV radiation, photosynthetically active radiation (PAR), and chlorophyll a concentration to develop predictors of thermal stress. In addition, fairly original (in an oceanography context) soft computing methods, such as fuzzy logic algorithms, have been used to integrate these environmental variables into the measure of susceptibility. The Maina et al. (2008) approach will certainly pave the way to more integrated interdisciplinary, multisensory and multiscale study of complex regional coral reef processes in the Red Sea, the Indian Ocean and elsewhere (Maina et al. 2011).

Beside coral bleaching, ocean warming can impact coral reef ecosystems through the widening of the tropical belt. The previous regions of marginal coral growth and marginal communities should shift southward and northward at the latitudinal extremes. As pointed out by Lloyd et al. (2012) these consequences are largely unexplored to date. These authors have compared changes in fish community structures paralleled by a 0.46 °C increase of average sea surface temperature between the time periods 1989–97 and 2002–2007. SST monthly observations came from 4 km resolution AVHRR products. They concluded that the relative abundance of temperate species as a whole decreased whereas that of tropical species increased, and broadly distributed species showed little change. These results confirmed the expected poleward shift in species ranges, and the increase in species richness and diversity with increasing sea temperature along the East Africa coastline.

A number of key processes for coral reef resilience could be studied with indirect remote sensing data. This includes physical connectivity between reefs through the modelling of a tracer that simulates larvae dispersal, and ocean color observations. Conducted elsewhere (Soto et al. 2009), these connectivity applications are yet to be conducted in the Western Indian Ocean. However, Acker et al. (2008) observed in the northern Red Sea with Sea-viewing Wide Field-of-view Sensor (SeaWiFS) and MODIS data the geochemical consequences of water enrichment by the nearby coral reefs. Their data indicated that large coral reef complexes could export either nutrients or chlorophyll-rich detritus and sediment, enhancing chlorophyll a concentration in normally oligotrophic waters adjacent to the reefs.

16.4 Remote Sensing for Management and Coral Reef Conservation

In principles, a number of the aforementioned direct and indirect applications are relevant for coral reef management and conservation. Throughout the planet, remote sensing derived products are increasingly used to map habitats and monitor the physical environment. These products are then used to provide habitat inventories, design better conservation plans, optimize sampling strategies for monitoring and for biological species inventories, assess fishery stocks, model their sustainability, and identify areas vulnerable to climate change (Andréfouët 2008, 2011).

We remind that the vast majority of biological coral reef information come from field surveys and monitoring programs. All data for model parameterization and validation, as well as data used to interpret remote sensing images, will come from the field. Designing sampling strategy that represents exhaustively the diversity found in an ecological system is difficult without remote sensing information and some sort of clear protocol for allocation of efforts. However, to date, few monitoring sampling strategies seem to have taken advantage of remote sensing data to optimize the representativity of the sampling as advocated by Scopélitis et al. (2010). High resolution images offer a powerful mean to select representative monitoring sites and to design adaptive monitoring strategies but this remain underused.

In the Red Sea and Western Indian Ocean, spatial applications dedicated to coastal management remain scarce, irrespective of the origin of the spatial data (remote sensing or interpolation of field data, as in Riegl and Piller 2000). No mentions of coral reef fishery stock assessments conducted using remotely sensed habitat maps could be found in the peer reviewed literature (Hamel and Andréfouët 2010). However, finfish, sea cucumber and several other species important for subsistence and commercial fisheries are targeted throughout the region. Also, there are marine protected areas in both the Red Sea and Western Indian Ocean, and traditional fisheries in Kenya, Tanzania and Madagascar are under the scrutiny of scientists (McClanahan et al. 2009; Barnes and Rawlinson 2009; McClanahan et al. 2011). Yet, no spatial resource management plans have been enhanced with remote sensing products to date, but see work in progress in the French Iles Eparses in the Mozambique Channel (Grellier et al. 2012).

Beyond what may be considered best as long term management perspectives, several direct remote sensing studies had immediate potential for enhanced management, such as the demonstration of human impacts on reefs due to urban developments and other activities. Change detection analysis makes a powerful diagnostic tool if changes visible on images can be correlated to human activities. Moufaddal (2005) demonstrated the human infringements on local coral reefs on the Egyptian coast with a time series of Landsat TM and ETM+. Furthermore, Vanderstraete et al. (2005) proposed reef risk indices after mapping changes with Landsat and potential sources of disturbances.

Recently, to support the plans of the Government of Madagascar to increase marine protected area coverage by over one million hectares, Allnut et al. (2012) compared four methods for marine spatial planning of Madagascar's west coast. Among the four methods, two used spatial conservation target-based optimization software (Marxan and Zonation), now a typical approach for marine conservation planning. Input data was drawn from the following variables: fishing pressure, exposure to climate change, and biodiversity (habitats, species distributions, biological richness, and biodiversity value). Habitat data came from the geomorphological Landsat-derived coral reef atlas provided by Andréfouët et al. (2009) and exposure to climate index was draw from the Maina et al. (2008) database (mentioned in the indirect remote sensing section above). This work integrated products issued from both direct and indirect remote sensing data towards one management goal. The authors emphasized that choosing an appropriate approach ultimately depends on scientific and political factors including representation targets, likelihood of adoption, and persistence goals. The portfolio created by Allnut et al. (2012) using a variety of remotely sensed data offered a quantitative integrated decision support tool. Similar approaches will likely rise in the future to support the sound implementation of international and national conservation targets.

16.5 Conclusion and Perspectives

The review of applications presented here highlights a fair diversity of valuable applications useful to tackle science and management questions at a variety of spatial and temporal scales. However, as in 2012, remote sensing still remains a tool poorly used routinely for coastal management in the Red Sea and Western Indian Ocean. This statement holds in fact for most regions supporting coral reefs (Andréfouët 2008). Applications remain often limited to academic studies, while a number of applications could be performed routinely in the real world. Thus, the full potential of space technology for the monitoring and management of coral reefs, seagrass and mangroves of these regions is not achieved. Even technically simple priority tasks, such as habitat mapping, have not been fully accomplished.

One of the first priorities remains ecosystem inventory and change detection. Accurate ecosystem inventories remain scarce in most countries. To date, no complete and consistent seagrass or mangroves inventories are available for the entire region. While geomorphological coral reef maps exist from Landsat data, the number of benthic habitat maps achieved at very high resolution (both thematic and spatial) remains extremely limited. The immediate benefits of these inventories would be to enhance management plans with proper estimates and optimization of costs and benefits related to the loss or preservation of ecosystem services. Obviously, conservation planning is a complex task where political, social and economic imperative are often more important than the background environmental analysis. Yet, the provision of accurate, updated and spatially explicit data is a key to the effectiveness of a conservation planning project. Furthermore, accurate inventories are essential when resilience is at stake. Indeed, basic requirement to maximize resilience within a network of protected areas is to have enough patches of habitats and enough replications spaced adequately. Without accurate inventories, this is impossible to achieve realistically (Mora et al. 2006; Wabnitz et al. 2010).

Another priority is certainly to continue developing indices of relevant climate stressors at adequate spatial resolution (1–4 km) and optimize them locally. Thus far, the environmental characterizations seen in the literature remain too coarse for local use in coral reef management plans and these need to be addressed. Although SST remains of high importance for coral stress, local variations are explained by other factors and processes as well (*e.g.* local hydrodynamics). The multi-variable work presented by Maina et al. (2008; 2011) and Rowlands et al. (2012) combined with adequate field data, highlights possible multisource fusion schemes. In addition to pointing to vulnerable areas for climate change and catastrophic events, injecting environmental information in the design of conservation and management trans-boundary plans would be extremely valuable (Allnut et al. 2012). With this goal in mind, we predict future developments of new multi-scale multi-disciplinary studies combining remote sensing data, field data, soft computing and modelling techniques.

Acknowledgments Figure 16.3 was provided by the Khaled bin Sultan Living Oceans Foundation (courtesy G. Rowlands, S. Purkis, A. Bruckner, P; Renaud). The Aldabra 2011 Worldview-2 image used for Figure 16.4 was provided through a partnership between IRD and Collecte Localisation Satellites (CLS) when evaluating Earth Observation Information Products/Services for World Bank projects (grant ESRIN/Contract No. 4000103101/11/I-IW).

References

Acker J, Leptoukh G, Shen S, Zhu T, Kempler S (2008) Remotely-sensed chlorophyll a observations of the northern Red Sea indicate seasonal variability and influence of coastal reefs. J Marine Syst 69:191–204

Allnutt TF, McClanahan TR, Andréfouët S, Baker M, Lagabrielle E, McClennen C, Rakotomanjaka AJM, Tianarisoa TF, Watson R, Kremen C (2012) Comparison of marine spatial planning methods in Madagascar demonstrates value of alternative approaches. PLoS ONE 7:e28969

Andréfouët S (2008) Coral reef habitat mapping using remote sensing: a user *vs* producer perspective. Implications for research, management and capacity building. J Spatial Sci 53:113–129

Andréfouët S (2012) Mozambique channel coral reef habitat mapping. Report to collecte localisation satellites, earth observation information products/services. World Bank projects Project A, Mozambique Channel Mapping IRD Nouméa January, p 56

Andréfouët S, Riegl B (2004) Remote sensing: a key-tool for interdisciplinary assessment of coral reef processes. Coral Reefs 23:1–4

Andréfouët S et al (2003) Multi-sites evaluation of IKONOS data for classification of tropical coral reef environments. Remote Sens Environ 88:128–143

Andréfouët S, Payri C, Hochberg EJ, Hu C, Atkinson MJ, Muller-Karger F (2004) Use of *in situ* and airborne reflectance for scaling-up spectral discrimination of coral reef macroalgae from species to communities. Mar Ecol Prog Ser 283:161–177

Andréfouët S, Chagnaud N, Kranenburg C (2009) Atlas des récifs coralliens de l'Océan Indien Ouest. Atlas of Western Indian Ocean Coral Reefs. Centre IRD de Nouméa, Nouméa, Nouvelle-Calédonie, p 102

Andréfouët S, Guillaume M, Delval A, Rasoamanendrika F, Blanchot J, Bruggemann J (2013) Fifty years of changes in reef flat habitats of the Grand Récif of Toliara (SW Madagascar) and the impact of gleaning. Coral Reefs 32:757–768

Baker AC, Glynn PW, Riegl B (2008) Climate change and coral reef bleaching: an ecological assessment of long-term impacts, recovery trends and future outlook. Estuar Coast Shelf Sci 80:435–471

Barnes D, Rawlinson K (2009) Traditional coastal invertebrate fisheries in south-western Madagascar. Journal of the Marine Biological Association of the United Kingdom. Barnes DKA, Rawlinson KA 89:1589–1596

Benny A, Dawson GJ (1993) Satellite imagery as an aid to bathymetric charting in the Red Sea. Cartogra J 20:5–16

Blythe JN, da Silva JCB, Pineda J (2011) Nearshore, seasonally persistent fronts in sea surface temperature on Red Sea tropical reefs. Ices J Mar Sci 68:1827–1832

Borstad G, Brown L, Cross W, Nallee M, Wainwright P (1997) Towards a management plan for a tropical reef-lagoon system using airborne multispectral imaging and GIS, 4th Int. Conf. on Remote Sensing for Marine and Coastal Environments, Orlando, USA

Brock JC, Purkis SJ (2009) The emerging role of Lidar remote sensing in coastal research and resource management. J Coastal Res 25:1–5

Bruckner A, Rowlands GP, Riegl B, Purkis S, Williams A, Renaud P (2011) Khaled bin Sultan living oceans foundation Atlas of Saudi Arabian Red Sea Marine Habitats. Panoramic Press, Phoenix, p 262

Bruggemann H, Rodier M, Guillaume M, Andréfouët S, Arfi R, Cinner J, Pichon M, Ramahatratra F, Rasoamanendrika F, Zinke J, McClanahan TR (2012) Wicked social-ecological problems forcing unprecedented change on the latitudinal margins of coral reefs: the case of southwest Madagascar. Ecology and Society, 17:47

Camoin GF, Colonna M, Montaggioni LF, Casanova J, Faure G, Thomassin BA (1997) Holocene sea level changes and reef development in the southwestern Indian Ocean. Coral Reefs 16:247–259

Chapman B, Turner JR (2004) Development of a geographical information system for the marine resources of Rodrigues. J Nat Hist 38:2937–2957

Courboulès J, Manière R (1992) Apport de la télédétection à l'étude de la relation entre l'hydrodynamisme de surface et les récifs coralliens. Int J Remote Sens 13:1911–1923

Courboulès J, Manière R, Bouchon C (1988) Systèmes d'informations géocodés et télédétection à haute résolution. Exemples d'application aux *cô*tes jordaniennes. Oceanol Acta 11:337–351

Courboulès J, Manière R, Jaubert J, Mandura AS, Khaffaji AK, Eshky AA, Saifullah S (1987) Cartographie d'une zone littorale en Mer Rouge par traitements numériques des données HRV de SPOT 1. Photo-Interprétation 87:35–38

Dahdouh-Guebas F, Coppejans E, Van Speybroeck D (1999) Remote sensing and zonation of seagrasses and algae along the Kenyan coast. Hydrobiologia 400:63–73

Dalleau M, Andréfouët S, Wabnitz CCC, Payri C, Wantiez L, Pichon M, Friedman K, Vigliola L, Benzoni F (2010) Use of habitats as surrogates of biodiversity for efficient coral reef conservation planning in Pacific Ocean Islands. Conserv Biol 24:541–552

Despinoy M, Naim O (1997) Etude sur l'océanographie cotière à l'ile de la Réunion par télédétection multibande. In: Dubois J, Bernier M, Fortin J, Boivin F (eds) VII Journées scientifiques du réseau Télédétection de l'AUPELF-UREF. Agence Universitaire de la Francophonie, Québec, pp 29–36

Dorenbosch M, Grol MGG, Nagelkerken I, van der Velde G (2005) Distribution of coral reef fishes along a coral reef-seagrass gradient: edge effects and habitat segregation. Mar Ecol-Prog Ser 299:277–288

Ferreira MA, Andrade F, Bandeira SO, Cardoso P, Mendes RN, Paula J (2009) Analysis of cover change (1995–2005) of Tanzania/Mozambique trans-boundary mangroves using Landsat imagery. Aquat Conserv-Marine and Freshwater Ecosystems 19:S38–S45

Graham NAJ et al (2008) Climate warming, marine protected areas and the ocean-scale integrity of coral reef ecosystems. Plos One 3:e3039

Grellier M, Nicet JB, Ringelstein J (2012) Etude d'identification des zones de conservation marines prioritaires dans le cadre de la mise en place d'une réserve naturelle dans les Iles Eparses: cas de l'île d'Europa. EngineerThesis. Ecole Nationale du génie de l'Eau et de l'Environnement de Strasbourg, MNHN, Ifrecor, TAAF, PARETO, 99 p

Gullstroem M, de la TCastroM, Bandeira SO, Bjoerk M, Dahlberg M, Kautsky N, Roennbaeck P, Oehman MC (2002) Seagrass ecosystems in the Western Indian Ocean. Ambio 31:588–596

Gullstroem M, Lunden B, Bodin M, Kangwe J, Ohman MC, Mtolera MSP, Bjork M (2006) Assessment of changes in the seagrass-dominated submerged vegetation of tropical Chwaka Bay (Zanzibar) using satellite remote sensing. Estuar Coast Shelf S 67:399–408

Hamel M, Andréfouët S (2010) Using very high resolution remote sensing for the management of coral reef fisheries: review and perspectives. Mar Pollut Bull 60:1397–1405

Hamylton S (2011a) Estimating the coverage of coral reef benthic communities from airborne hyperspectral remote sensing data: multiple discriminant function analysis and linear spectral unmixing. Int J Remote Sens 32:9673–9690

Hamylton S (2011b) The use of remote sensing and linear wave theory to model local wave energy around Alphonse Atoll, Seychelles. Estuar Coast Shelf S 95:349–358

Hamylton SM, Spencer T (2011) Geomorphological modelling of tropical marine landscapes: optical remote sensing, patches and spatial statistics. Cont Shelf Res 31:S151–S161

Hedley JD, Roelfsema C, Phinn S, Mumby PJ (2012) Environmental and sensor limitations in optical remote sensing of coral reefs: implications for monitoring and sensor design. Remote Sens 4:271–302

Hochberg EJ, Atkinson MJ, Andréfouët S (2003) Spectral reflectance of coral reef community-types worldwide and implications for coral reef remote sensing. Remote Sens Environ 85:159–173

Hochberg E, Atkinson M, Apprill A, Andréfouët S (2004) Spectral reflectance of coral. Coral Reefs 23:84–95

Hopley D (2011) Encyclopedia of modern coral reefs. Structure, form and process. Springer, Berlin, p 1206

Hughes TP et al (2003) Climate change, human impacts, and the resilience of coral reefs. Science 301:929–933

Klaus R, Jones D, Turner J, Simoes N, Vousden D (2003) Integrated marine and coastal management: a strategy for the conservation and sustainable use of biological resources in the Socotra Archipelago. J Arid Environ 54:71–80

Knudby A, Nordlund L (2011) Remote sensing of seagrasses in a patchy multi-species environment. Int J Remote Sens 32:2227–2244

Knudby A, LeDrew E, Brenning A (2010a) Predictive mapping of reef fish species richness, diversity and biomass in Zanzibar using IKONOS imagery and machine-learning techniques. Remote Sensing of Environment 114:1230–1241

Knudby A, Newman C, Shaghude Y, Muhando C (2010b) Simple and effective monitoring of historic changes in nearshore environments using the free archive of Landsat imagery. Int J Appl Earth Observ Geoinformation 12:S116–S122

Lloyd P, Plaganyi EE, Weeks SJ, Magno-Canto M, Plaganyi G (2012) Ocean warming alters species abundance patterns and increases species diversity in an African sub-tropical reef-fish community. Fish Oceanogr 21:78–94

Lyzenga D, Malinas N, Tanis F (2006) Multispectral bathymetry using a simple physically based algorithm. IEEE T Geosci Remote 44:2251–2259

Lyzenga DR (1985) Shallow water bathymetry using combined lidar and passive multispectral scanner data. Int J Remote Sensing 6:115–125

Maina J, Venus V, McClanahan MR, Ateweberhan M (2008) Modelling susceptibility of coral reefs to environmental stress using remote sensing data and GIS models. Ecol Modell 212:180–199

Maina J, McClanahan TR, Venus V, Ateweberhan M, Madin J (2011) Global gradients of coral exposure to environmental stresses and implications for local management. Plos One 6:e23064

Makota V, Sallema R, Mahika C (2004) Monitoring shoreline change using remote sensing and GIS: a case study of Kunduchi area, Tanzania. Western Indian Ocean J Mar Sci 3:1–10

Manière R, Jaubert J (1985) Traitement d'image et cartographie de récifs coralliens en Mer Rouge (Golfe d'Aqaba). Oceanol Acta 8:321–328

Maritorena S, Morel A, Gentili B (1994) Diffuse reflectance of oceanic shallow waters: influence of water depth and bottom albedo. Limnol Oceanogr 39:1689–1703

McClanahan TR, Ateweberhan M, Sebastian CR, Graham NAJ, Wilson SK, Bruggemann JH, Guillaume MMM (2007) Predictability of coral bleaching from synoptic satellite and *in situ* temperature observations. Coral Reefs 26:695–701

McClanahan TR, Cinner JE, Graham NAJ, Daw TM, Maina J, Stead SM, Wamukota A, Brown K, Venus V, Polunin NVC (2009) Identifying reefs of hope and hopeful actions: contextualizing environmental, ecological, and social parameters to respond effectively to climate change. Conserv Biol 23:662–671

McClanahan TR, Graham NAJ, MacNeil MA, Muthiga NA, Cinner JE, Bruggemann JH, Wilson SK (2011) Critical thresholds and tangible targets for ecosystem-based management of coral reef fisheries. P Natl Acad Sci USA 108:17230–17233

Minghelli-Roman A, Chisholm JRM, Marchioretti M, Jaubert JM (2002) Discrimination of coral reflectance spectra in the Red Sea. Coral Reefs 21:307–314

Mora C, Andréfouët S, Costello M, Kranenburg C, Rollo A, Veron J, Gaston K, Myers R (2006) Coral reefs and the global network of marine protected areas. Science 312:1750–1751

Moufaddal WM (2005) Use of satellite imagery as environmental impact assessment tool: a case study from the NW Egyptian Red Sea coastal zone. Environ Monit Assess 107:427–452

Nyström M, Graham NAJ, Lokrantz J, Norström A (2008) Capturing the cornerstones of coral reef resilience. Linking theory to practice. Coral Reefs 27:795–809

Pennober G, Borius A (2010) Détection à très haute résolution spatiale du blanchissement corallien sur les récifs de l'île de La Réunion. Télédétection 9:1–3

Purkis S, Pasterkamp R (2004) Integrating *in situ* reef top reflectance spectra with Landsat TM imagery to aid shallow tropical benthic habitat mapping. Coral Reefs 23:5–20

Purkis S, Kenter JAM, Oikonomou EK, Robinson IS (2002) High resolution ground verification, cluster analysis and optical model of reef substrate coverage on LANDSAT TM imagery (Red Sea, Egypt). Int J Remote Sensing 23:1677–1698

Purkis SJ, Rowlands GP, Riegl BM, Renaud PG (2010) The paradox of tropical karst morphology in the coral reefs of the arid Middle East. Geology 38:227–230

Riegl B, Piller WE (2000) Mapping of benthic habitats in northern Safaga Bay (Red Sea, Egypt): a tool for proactive management. Aquat Conserv Marine Freshwater Ecosystems 10:127–140

Rowlands G, Purkis S, Riegl B, Metsamaa L, Bruckner A, Renaud P (2012) Satellite imaging coral reef resilience at regional scale. A case-study from Saudi Arabia. Mar Pollut Bull 64:1222–1237

Samani D (2005) Une démarche SIG appliquée aux récifs coralliens: Le cas des Iles Glorieuses et Juan de Nova. Msc thesis. Université de La Réunion et ARVAM

Scopélitis J, Andréfouët S, Phinn S, Chabanet P, Naim O, Tourrand C, Done T (2009) Changes of coral communities over 35 years: integrating *in situ* and remote-sensing data on Saint-Leu Reef (la Reunion, Indian Ocean). Estuar Coast Shelf S 84:342–352

Scopélitis J, Andréfouët S, Phinn S, Arroyo L, Dalleau M, Cros A, Chabanet P (2010) The next step in shallow coral reef monitoring: combining remote sensing and *in situ* approaches. Mar Pollut Bull 60:1956–1968

Sheppard CRC (2000) Coral Reefs of the Western Indian Ocean: an overview. In: McClanahan T, Sheppard CRC, Obura DO (eds) Coral reefs of the Indian Ocean. Oxford University Press, New York, pp 3–34

Soto I, Andréfouët S, Hu C, Muller-Karger FE, Wall CC, Sheng J, Hatcher BG (2009) Physical connectivity in the Mesoamerican Barrier Reef System inferred from 9 years of ocean color observations. Coral Reefs 28:415–425

Spencer T, Hagan A, Hamylton S, Renaud P (2009) Atlas of the Amirantes. University of Cambridge, p 66

Stoddart DS, Taylor JD, Fosberg FR (1971) Geomorphology of Aldabra Atoll. Philos Trans R Soc Lond B 260:31–65

Teneva L, Karnauskas M, Logan C, Bianucci L, Currie J, Kleypas J (2012) Predicting coral bleaching hotspots: the role of regional variability in thermal stress and potential adaptation rates. Coral Reefs 31:1–12

Vanderstraete T, Goossens R, Ghabour TK (2005) Realization of a local coral reef monitoring system in the Red Sea using Landsat 7 ETM + data, New Strategies for European Remote Sensing. Millpress Science Publishers, Rotterdam, pp 479–487

Wabnitz CCC, Andréfouët S, Muller-Karger FE (2010) Measuring progress toward global marine conservation targets. Front Ecol Environ 8:124–129

Wang Y, Bonynge G, Nugranad J, Traber M, Ngusaru A, Tobey J, Hale L, Bowen R, Makota V (2003) Remote sensing of mangrove change along the Tanzania coast. Mar Geod 26:35–48

Wells S, Burgess N, Ngusaru A (2007) Towards the 2012 marine protected area targets in Eastern Africa. Ocean Coast Manage 50:67–83

Chapter 17
Basic Ecosystem Dynamics in the Red Sea as Seen by Sundry Remote Sensing Techniques

Vittorio Barale and Martin Gade

Abstract Patterns of algal blooming are considered to be indicators of basic ecosystem dynamics in marginal and enclosed seas. A time series of chlorophyll-like pigment concentration (*chl*)–which can be interpreted as a proxy of biomass and, under certain circumstances, productivity–derived from SeaWiFS data, from July 1999 to June 2009, was considered to explore the space–time variability of algal blooming in the Red Sea. The comparison with concurrent surface wind speed (*ws*) and direction, derived from QuikSCAT data, allowed to correlate *chl* variability with atmospheric forcing. The observed *chl* seasonality is essentially bimodal, with a *fall-winter* period of extended blooming, progressing from south to north and back, followed by a *spring-summer* period of reduced blooming, at least in the northern sub-basin. This annual cycle seems to match the climatic characteristics of the basin, the monsoon-driven wind regime in particular, and the ensuing thermohaline circulation. The correlation with *ws* suggests that the Red Sea behaves like a classical sub-tropical basin, where production is never limited by sunlight, but always limited by nutrient availability–a condition relaxed only in the colder season, when (wind-driven) convection processes can enrich the euphotic zone with nutrients from deeper layers. However, at the same time, it appears that other blooming episodes, in the southern sub-basin in particular, are not driven directly by the wind field, but rather by other factors such as the exchange of water with the Arabian Sea, via the Gulf of Aden and Bab-el-Mandeb.

V. Barale (✉)
European Commission, Joint Research Centre,
Institute for Environment and Sustainability,
Ispra, Italy
e-mail: vittorio.barale@jrc.ec.europa.eu

M. Gade
Institut für Meereskunde, Universität Hamburg,
Hamburg, Germany
e-mail: martin.gade@uni-hamburg.de

V. Barale, M. Gade (eds.), *Remote Sensing of the African Seas*,
DOI 10.1007/978-94-017-8008-7_17,

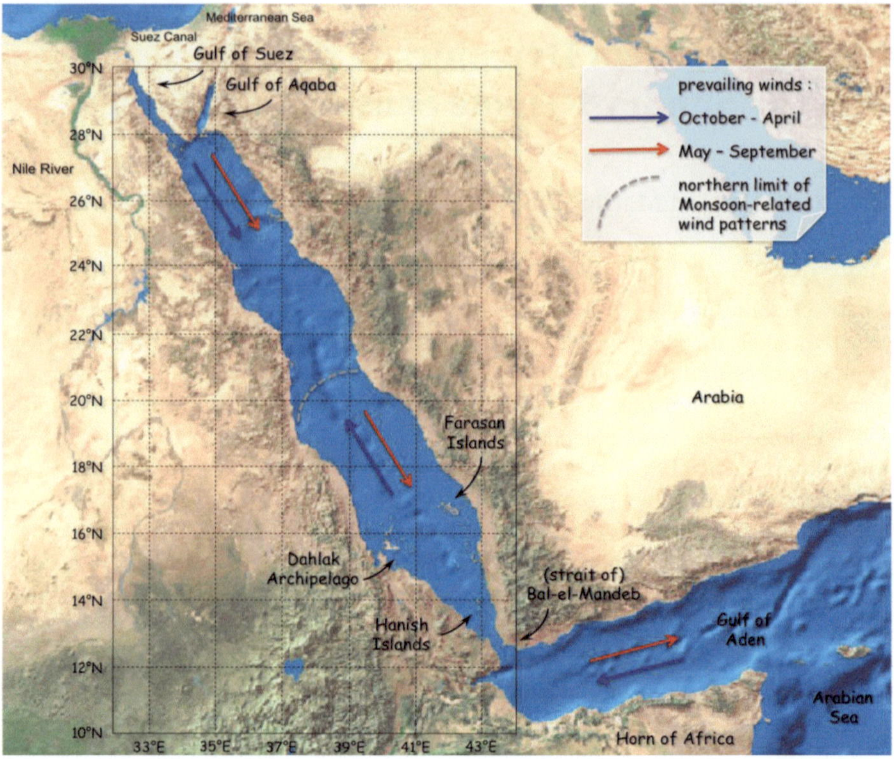

Fig. 17.1 Red Sea, geographical boundaries, nomenclature and prevailing wind patterns. The framed area to the left corresponds to the one covered by the SeaWiFS and QuikSCAT images shown in the following

17.1 Introduction

Patterns of algal blooming, described by variations in the abundance of planktonic agents, are commonly adopted as indicators for ecological balance in coastal and marine environments. Remote Sensing (RS) from Earth orbit can help to assess these patterns over a wide range of space and time scales, by describing basic ecosystem dynamics of entire marginal or enclosed seas. Optical RS, in particular, allows monitoring of the large-scale, long-term concentration of chlorophyll-like pigments (in the following referred to as *chl*), which can be interpreted as a proxy of biomass and, under certain circumstances, productivity (Barale 1994). In the present case, an extended time series of RS-derived *chl* statistical maps was used to explore the space–time variability of algal blooming in the Red Sea (see Fig. 17.1 for geographical boundaries and basin nomenclature). The comparison with statistical assessments of surface wind speed (in the following referred to as *ws*), also obtained from RS techniques, allowed to correlate such heterogeneity with patterns of atmospheric forcing.

Given the peculiar name of the basin (where "Red"—a term coming from antiquity and recurring in all languages spoken in the region—may refer to the recurrence of red-coloured cyanobacteria blooms[1] or, according to other sources, to the ancient designation of cardinal points according to a geo-chromatic code, red being associated with south), it would seem rather appropriate to investigate the Red Sea ecosystem by virtue of optical observations. However, althought widely applied for the characterization of other Eurasian enclosed seas (see e.g. various authors in Barale and Gade 2008), satellite observations, particularly in the visible and near infrared spectral range, have not been used extensively for basin-scale studies of the Red Sea. Kirby et al. (1993) considered Coastal Zone Colour Scanner (CZCS) data over a 2-year period (1979–1980) to describe the basin surface optical properties and to determine their seasonality due to the monsoon cycle. Barale (2007) used a 9-year time series (1998–2006) generated by the Sea-viewing Wide Field-of-view Sensor (SeaWiFS) to look at the space-time monthly variability of *chl* patterns, as well as interannual anomalies and other irregular features of the pigment field, occurring over monthly to seasonal scales.

More frequently, it has been the northern part of the basin, and primarily the Gulf of Aqaba, that has been the subject of in-depth investigations of bio-optical properties, from both satellite and *in situ* measurements (Labiosa et al. 2003; Stambler 2005), or of site-specific algorithms for their derivation from RS radiance measurements (Iluz et al. 2003; Sokoletsky et al. 2003, 2004). Acker et al. (2008), finally, used seven years of SeaWiFS data (1998–2004), coupled to concurrent 2003–2004 Sea Surface Temperature (SST) data collected by the Moderate Resolution Imaging Spectroradiometer (MODIS), on the Aqua orbital platform, to observe seasonal phytoplankon dynamics, local impacts of coastal features such as coral reefs, and surface circulation patterns described by optical tracers, primarily in the northern Red Sea, between 21.5°N and 27.5°N.

In the limited scientific literature available, much attention is devoted to the link between sea surface bio-optical environment and atmospheric processes, as related chiefly to the monsoon cycle (see Fig. 17.1 for a schematic representation of wind patterns). Marked seasonal variations of the local wind field, controlled by the large-scale monsoon wind system, are indeed the key features of the basin (Edwards 1987). During the Southwest (SW) Monsoon, in boreal summer, winds over the Arabian Sea are southwesterly, changing to north-northwesterly over the southern Red Sea. During the Northeast (NE) Monsoon, in boreal winter, winds over the Arabian Sea are northeasterly, changing to south-southeasterly over the southern Red Sea. Conversely, in the northern Red Sea, the prevailing winds (linked to the eastern Mediterranean weather systems) are north-northwesterly year-round (Morcos 1970). Thus, the region north of 20°N is not markedly affected by the monsoons, while the region south of 20°N is subject to a monsoon-driven seasonal wind reversal (Sheppard et al. 1992).

[1] see e.g. the spatial and temporal distribution of *Trichodesmium* spp. in the Gulf of Aqaba as reported by Post et al. (2002).

The basin's location in a hot and arid climate zone, where evaporation far exceeds precipitation, and where fluvial runoff is negligible, results in a negative hydrological budget (Morcos 1970). This deficit is compensated by a net water inflow from the Gulf of Aden, via the Strait of Bab-el-Mandeb (Eshel and Naik 1997). The input of fresher water from the Gulf of Aden and high evaporation rate result in a gradual northward increase in salinity (in excess of 40 pps, in the northern basin) and density (Sofianos et al. 2002). Conversely, SST increases from north to south, only to decrease again in the southermost part of the basin.

The Red Sea surface currents are somewhat irregular and can vary significantly on a seasonal scale. In the northern part of the basin, thermohaline forcing generates a cyclonic circulation all around the year, in contrast to the prevailing wind forcing (Quadfasel and Baudner 1993). The central part of the Red Sea, again around 20°N, is dominated by recurring or persistent anticyclonic eddies. The formation mechanisms and temporal variability of which are still unclear, but which are thought to be an indication of a very strong response of the Red Sea to atmospheric forcing (Sofianos and Johns 2003). In the southern part, the circulation is very complex (due to the exchanges with the Gulf of Aden, the peculiar bathymetry with wide continental shelves on both sides of the basin, and the high space/time variability of atmospheric forcing) and results in strong mixing of the different water masses involved (Sofianos and Johns 2007).

In general, two distinct seasonal situations prevail. During the NE monsoon (winter), the direction of net surface transport is northward. Following the typical inverse-estuarine circulation of concentration basins, fresher surface water enters the Red Sea from the Gulf of Aden. It moves northward, mixing and sinking in the northern basin, and then returning south as a saltier subsurface current that ultimately outflows at depth over the Hanish Sill (Smeed 2000). During the SW monsoon (summer), the direction of net surface transport is southward. A three-layer system develops in the Bab-el-Mandeb, as a consequence of opposing wind stress and thermohaline forces, composed of Red Sea water outflowing at the surface and at depth, and of Gulf of Aden intermediate water, upwelled by the monsoon winds, inflowing between these two layers (Sofianos and Johns 2007). In between these different seasonal states, transitional periods occur around April and October, when the winds reverse direction (in the southern basin), surface currents are weakest and variable, and the flow in the Bab-el-Mandeb reaches a minimum (Patzert 1974; Murray and Johns 1997).

In the Red Sea, the water column permanent stratification, only somewhat weaker in winter than in summer, is the main factor hindering nutrient renewal from deeper layers, particularly in the northern part of the basin. The persistent thermocline (and halocline), extending from about 50 m to 300 m of depth, isolates a homogeneous deep layer with uniform salinity of 40.6 pps (Woelk and Quadfasel 1996). Consequently, the biological production of surface waters depends on the nutrient concentration found in the upper water column, and in particular on the nutrients received through the surface (winter) or intermediate (summer) water flow from the Gulf of Aden and the ensuing northward overall drift (Edwards 1987). Vertical mixing induced by winter surface cooling, due to a negative heat flux at the air-sea interface, can create a deeply mixed layer (normally around 300 m, but mixing depths in excess of 800 m have been measured in unusually cold winters) and increase nutrient concentration in surface waters (Genin et al. 1995). At the basin scale, nutrient concentration generally increases from north to south, and so does, consistantly with the nutrient distribution, primary production (Weikert 1987).

According to the environmental traits outlined above, the Red Sea can be divided in various geographical provinces bounded by bathymetry, physical properties and dynamics, and wind patterns (as e.g. in Siedler 1969). In the following, the general trends of the Red Sea pigment field were analyzed at the basin scale, as well as in the northern and southern regions separated by the critical area around 20°N, where the seasonal monsoon wind shift reaches its limit and the surface circulation appears to be dominated by anticyclonic eddies (Johns et al. 2000, Acker et al. 2008). Similarly, the relationship to wind speed was considered over the same scales (full, northern and southern basin), in order to try and couple bio-optical patterns to their potential atmospheric forcing.

17.2 SeaWiFS and QuikSCAT Historial Data Record

The present work is based on a time series collected by the SeaWiFS, a dedicated ocean colour imager that operated between 1997 and 2010, acquiring Earth radiance data in 8 spectral bands (ranging from 402 to 885 nm), with a spatial resolution of ~ 1.1 km at nadir. Concurrent data collected by the MODIS-Aqua, a multispectral imager active since 1999 (on board the Aqua orbital platform) and equipped with 36 spectral bands, 9 of which (ranging from 405 to 877 nm) dedicated to acquire ocean radiance data at ~ 1 km resolution, were also used for comparison. In order to avoid incomplete coverage due to cloud cover, in particular during the summer monsoon, or to desert dust aerosols, rather common in this region–or again to processing algorithm failure due to unaccounted air or water constituents–only monthly composites (and climatologies) of the original data were considered. As the number of valid pixels increases with larger and longer binning intervals, the compositing process generates complete, cloud-free images. However, the variability due to highly dynamical events in a given area is averaged out too, and the composites retain only those patterns that persist over significant space and time scale.

The monthly composite (and climatological) data were obtained directly from web-based services[2], which provide a suite of data products that includes SeaWiFS

[2] i.e. the Goddard Earth Sciences (GES) and Data Information Services Center (DISC) Interactive Online Visualization ANd aNalysis Infrastructure, (GIOVANNI), available for public access and use at http://disc.sci.gsfc.nasa.gov/giovanni/overview/, by the National Aeronautic and Space Administration (NASA); as well as the Global Marine Information System (GMIS), at http://gmis.jrc.ec.europa.eu/, by the Joint Research Centre (JRC) of the European Commission (EC). Both web services allow for the extraction of fully processed data, from multiple sensors, and the use of various data analysis and visualization options.

and MODIS-Aqua imagery, fully processed to correct top-of-the-atmosphere radiances from atmospheric noise; to derive normalized water-leaving radiances; and to compute from these a score of derived parameters. In the web services' version used in the present case, the parameter *chl* is calculated using the OC4 empirical algorithm for SeaWiFS data, and the similar OC3M algorithm for MODIS-Aqua data (O'Reilly et al. 2000). Composite data products are obtained, for both SeaWiFS and MODIS, by re-mapping on a common equal-area grid and then averaging with a grid cell, or "bin", ranging in size from $2 \times 2\,\text{km}^2$ to $9 \times 9\,\text{km}^2$, depending on the web service used, at weekly or monthly intervals. Annual, multi-annual and climatological composites can also be obtained, over the same grids. For each space/time bin, the mean value (weighted by the number of valid pixels used in the binning process) is accompanied by related data quality statistics (Campbell et al. 1995). The resulting data products should be considered with caution, owing to the impact of various coloured water constituents, or to direct bottom reflection, which may significantly alter the *chl* signal derived from the sensor measurements, particularly in shallow coastal areas. In fact, the presence of optically active materials other than phytoplankton and related pigments, with partially overlapping spectral signatures—i.e. dissolved organic matter and suspended inorganic particles—can prevent the computation of reliable *chl* absolute values (Sathyendranath et al. 2000). However, when these limitations are accounted for, the data can provide significant—if qualitative, at worst—information on recurrent or anomalous algal blooms, and related boundary conditions.

SeaWiFS data from ten consecutive full-year cycles, from July 1999 to June 2009, were used to compute *chl* climatologies and *chl* average basin values (chl_{ABV}) from the monthly composite images. Three different regions of interest were selected for the spatial averaging: the entire basin (10°N–30°N, 32°E–44°E), the southern sub-basin (10°N–20°N, 32°E–44°E) and the northern sub-basin (20°N–30°N, 32°E–44°E). When the resulting SeaWiFS historical record proved to be incomplete (i.e. January, February, March and July 2008; May 2009), the chl_{ABV} time series was filled in using MODIS-derived estimates. A statistical comparison of SeaWiFS and MODIS-Aqua data, for the 7-year period of overlap between the two time series (i.e. July 2002 to June 2009), over the three regions considered, produced very high correlation coefficients (about 0.9, for the entire basin case) of the chl_{ABV} series generated by each sensor. However, the linear regression of the two data sets indicated also that the MODIS chl_{ABV} are systematically larger of those computed from SeaWiFS data by almost a factor of 2 (slope of 1.75, with a negative offset of 0.2, for the entire basin case; see Fig. 17.2). Although this apparently systematic inconsistency was not investigated further, it must be recalled that such uncertainties and biases are not uncommon, when considering concurrent ocean colour products generated by different orbital sensors (Zibordi et al. 2012).

Data collected by the microwave scatterometer SeaWinds, which was launched on the QuikBird satellite in 1999 and operated until 2009, were used to derive wind direction and speed statistics. The QuikBird SeaWinds, later dubbed QuikSCAT, was the third in a series of National Aeronautic and Space Administration (NASA) scatterometers, operating at 13.4 GHz (Ku-band). Scatterometers transmit microwave

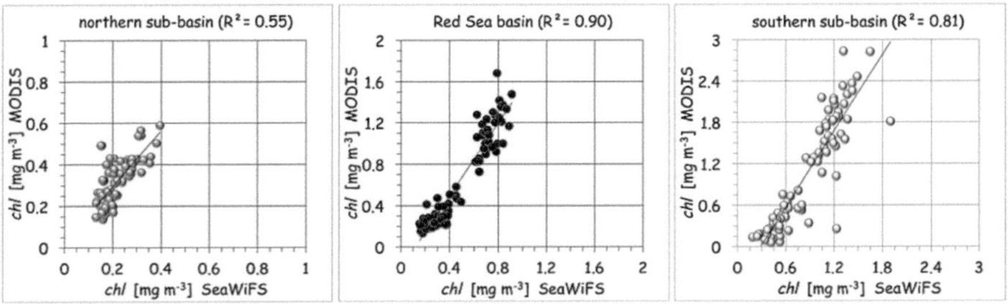

Fig. 17.2 Linear regressions of SeaWiFS and MODIS-Aqua chl_{ABV} [mg m^{-3}], for the 7-year period from July 2002 to June 2009, in the Red Sea entire basin (*central panel*), the northern sub-basin (*left panel*) and the southern sub-basin (*right panel*)

pulses down to the Earth's surface and then measure the power that is scattered back to the sensor and is related to surface roughness. In the sea, surface roughness is highly correlated with the near-surface wind that generated it (Schroeder et al. 1982; Donelan and Pierson 1987). Hence, wind speed and direction (at 10 meters over the water surface) can be retrieved from the scatterometer backscattered power. The 10-year QuikSCAT data set, collected from July 1999 to June 2009, used here for comparison with the SeaWiFS data, was mapped to a 0.25 degree grid. Once again, monthly means of *ws* were obtained directly from dedicated web-based services[3] and then processed to derive *ws* (and direction) climatologies and *ws* average basin values (ws_{ABV}) analogous to those computed for *chl*, over the same areas, i.e. the entire basin, the southern sub-basin and the northern sub-basin.

17.3 Climatological *chl* Field and Inter-annual Variability

The SeaWiFS-derived "climatological" annual mean (based on the 1999–2009 time series) shows a *chl* field differentiated between successive latitudinal zones (Fig. 17.3, left panel). The two northern Gulfs appear to be quite different from the remainder of the basin, and from one another as well, with the Gulf of Suez presenting mean concentration a factor 2 higher than that of the Gulf of Aqaba (i.e. 0.4 mg m^{-3} *vs* 0.2 mg m^{-3}). The northern part of the Red Sea proper displays a decisively oligotrophic character, with mean values on the order of 0.2 mg m^{-3}. The central transition area (20°N $\pm$ 2°) presents higher concentrations, ranging from 0.4 mg m^{-3} to 0.6 mg m^{-3} in the near-coastal zones and up to 1.0 mg m^{-3} or more at selected coastal sites. In the southern part of the Red Sea, the wide, shallow, near-coastal areas display concentrations larger than 1.0 mg m^{-3}, in particular around the Dahlak Archipelago and the Farasan Islands, while lower concentrations, ranging from 0.6 mg m^{-3} to 1.0 mg m^{-3}, appear in the middle part of this sub-basin, north of the Hanish Islands.

[3] The QuikSCAT *ws* data are produced by Remote Sensing Systems (RSS) and sponsored by the NASA Ocean Vector Winds Science Team. The entire historical data set is available at http://www.remss.com/.

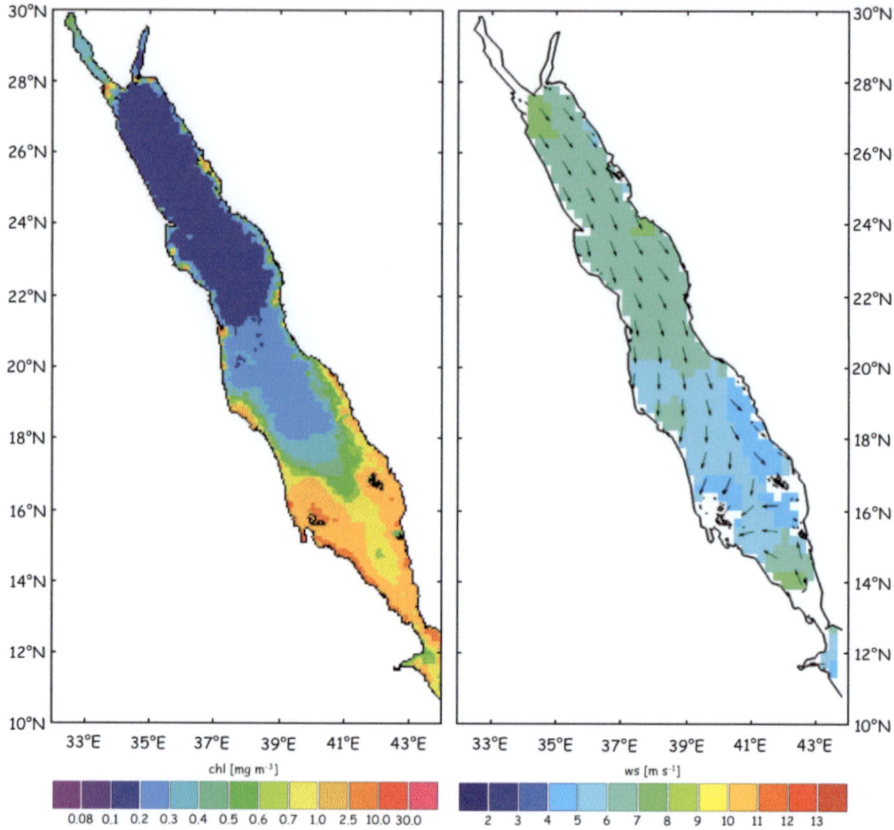

Fig. 17.3 Red Sea "climatological" annual means of SeaWiFS-derived *chl* [mg m^{-3}] (*left panel*) and of QuikSCAT-derived *ws* [m s^{-1}] *(right panel)*. Both composite images were derived from the historical data sets collected between 1999 and 2009

These *chl* values are well in line with *in situ* measurements reported in the literature (at least in the northern sub-basin), even though the surface estimates from RS data cannot take into account the deep chlorophyll maximum, 2 to 3 times larger than surface values, which commonly occurs at 30 m–60 m of depth (Stambler 2005).

The QuikSCAT-derived "climatological" annual mean (again, based on the 1999–2009 time series) shows that *ws* (and direction) are also differentiated between successive latitudinal zones (Fig. 17.3, right panel). Strong northerly winds prevail in the northern sub-basin, and southerly winds in the southernmost part of the Red Sea. An intermediate zone of converging wind patterns appears from 18°S to 16°S. The temporal evolution of the climatological *chl* annual mean, described by the chl_{ABV} computed from the SeaWiFS climatological monthly means, is shown in Fig. 17.4 (left panels). The monthly sequence is arranged from July to June, so that the plots are centered around the winter maxima. In July, August and September, the basin-wide *chl* estimates start to increase slowly from the summer minimum.

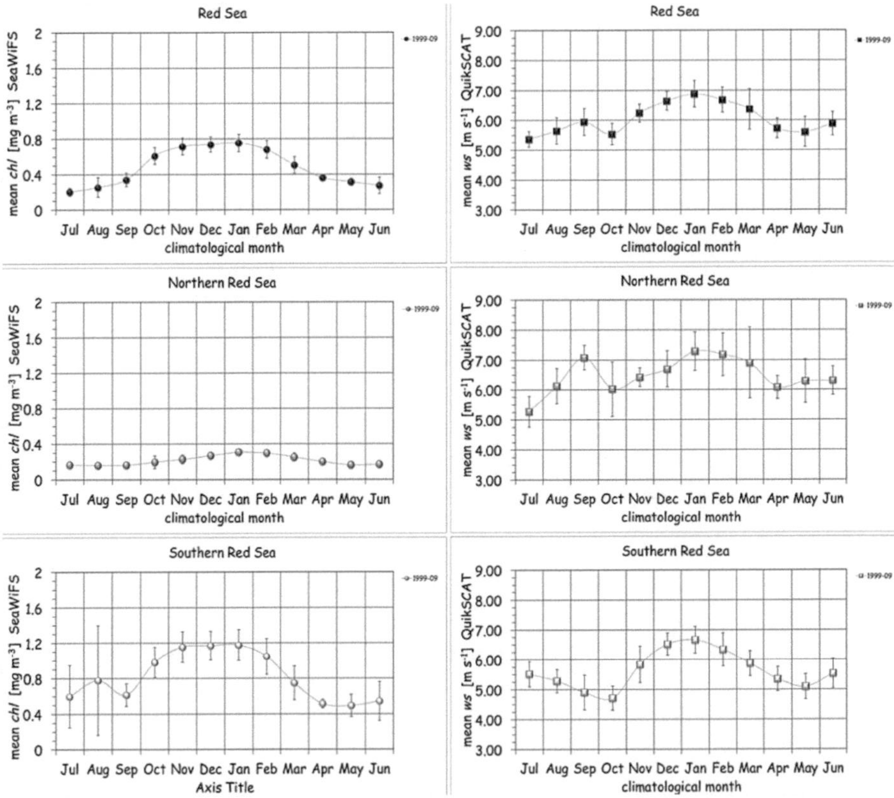

Fig. 17.4 Red Sea climatological (monthly scale) SeaWiFS-derived chl_{ABV} [mg m^{-3}] (*left panels*) and QuikSCAT-derived ws_{ABV} [m s^{-1}] (*right panels*), for the entire basin (*upper row*), the northern sub-basin (*middle row*) and the southern sub-basin (*lower row*). The error bars represent $\pm$ 1 standard deviation

However, while the northern sub-basin exhibits the lowest *chl* values of the whole sequence, the southern sub-basin registers a relative maximum in August. A blooming period follows in fall and winter, from October to March. Higher chl_{ABV} start to appear already in October for the southern sub-basin, but grow at a slower pace and reach a peak only in January for the northern sub-basin. A steady decline of chl_{ABV} in April, May and June leads again to the summer minimum (anticipated to May, in the south) from which the sequence had started.

The corresponding ws_{ABV} are also shown in Fig. 17.4 (right panels). These plots present the same seasonal characteristics as the chl_{ABV}, i.e. a summer minimum and a winter maximum that can be associated respectively with a less productive, stratified, upper water column and more productive, vertically mixed, water column conditions. This general pattern is reminiscent of the classical seasonal variation of phytoplankton biomass in sub-tropical basins (see e.g. Cushing 1959), where producion is never limited by sunlight, but is always limited by nutrient availability—a condition relaxed only in the colder season, when (wind-driven) convection processes can enrich the euphotic zone with nutrients from deeper layers. However, some of the details appearing in the *ws* pattern—in particular the late summer (September) maximum in the northern sub-basin, and both early fall (October) and spring (April) relative minima in the southern sub-basin—can not be easily associated with this general scheme. Instead, they seem to conincide rather well with the monsoon reversal periods, when the wind over the southern sub-basin, first aligned with that over the northern basin (late summer), starts blowing from the opposite, southerly, direction (early fall), and then again when the wind field over both sub-basins aligns once more from the same, northerly, direction (spring).

The spatial heterogeneity of this seasonal sequence above is illustrated in some detail by the *chl* and *ws* climatological (1999–2009) monthly mean images, derived from the SeaWiFS and QuikSCAT historical data sets, and shown in Figs. 17.5 and 17.6 respectively. These composites (also arranged from July to June) present a highly structured pigment field and wind field, exhibiting a large degree of variability (both in space and time) for all sub-basins already identified in the annual mean images.

In July, August and September, low *chl* values characterize most of the Red Sea, in particular in the interior of the northern sub-basin, while the southern basin maintains higher values only in near-coastal areas. Note that, in summer, the climatic conditions of the southern sub-basin are such that some areas remain permanently covered by clouds (mostly in July), for the entire SeaWiFS period of coverage, so as to prevent the computation of a mean value. Furthermore, note that the highest values recorded in August for the southern sub-basin appear to be located in the Bab-el-Mandeb and in the Gulf of Aden, along the southeastern coastline, possibly having more to do with local oceanographic conditions, than with those typical of the Red Sea proper. Sofinaos and Johns (2007) suggest that in summer the basic structure of the three-layer flow, in the southernmost part of the basin, involves banking toward the coasts. In particular, the Red Sea Outflow Water should be banked toward the western coast, and the Gulf of Aden Intermediate Water, and surface waters, toward the eastern coast. Indeed, such an inflow from the Gulf of Aden, composed of nutrient-rich waters upwelled by the summer monsoon wind pattern over the Arabian Sea, would provide the apt process for generating the enhanced patterns observed in the southeastern coastal zone in August.

The corresponding *ws* images show that, in this period, northerly winds prevail over the entire Red Sea, with mean speed increasing over the northern sub-basin and decreasing over the southern sub-basin. The summer *chl* minimum is followed by increasing values in October, November and December, starting from the south and progressing northward.

Note also the persistent traces of intense eddy-like circulation features, around 20°N, in the October transition period. This corresponds to a period of wind reversal over the southern sub-basin, from northerly to southerly winds, as well as increasing mean wind speed. After reaching their maximum extent in January and February,

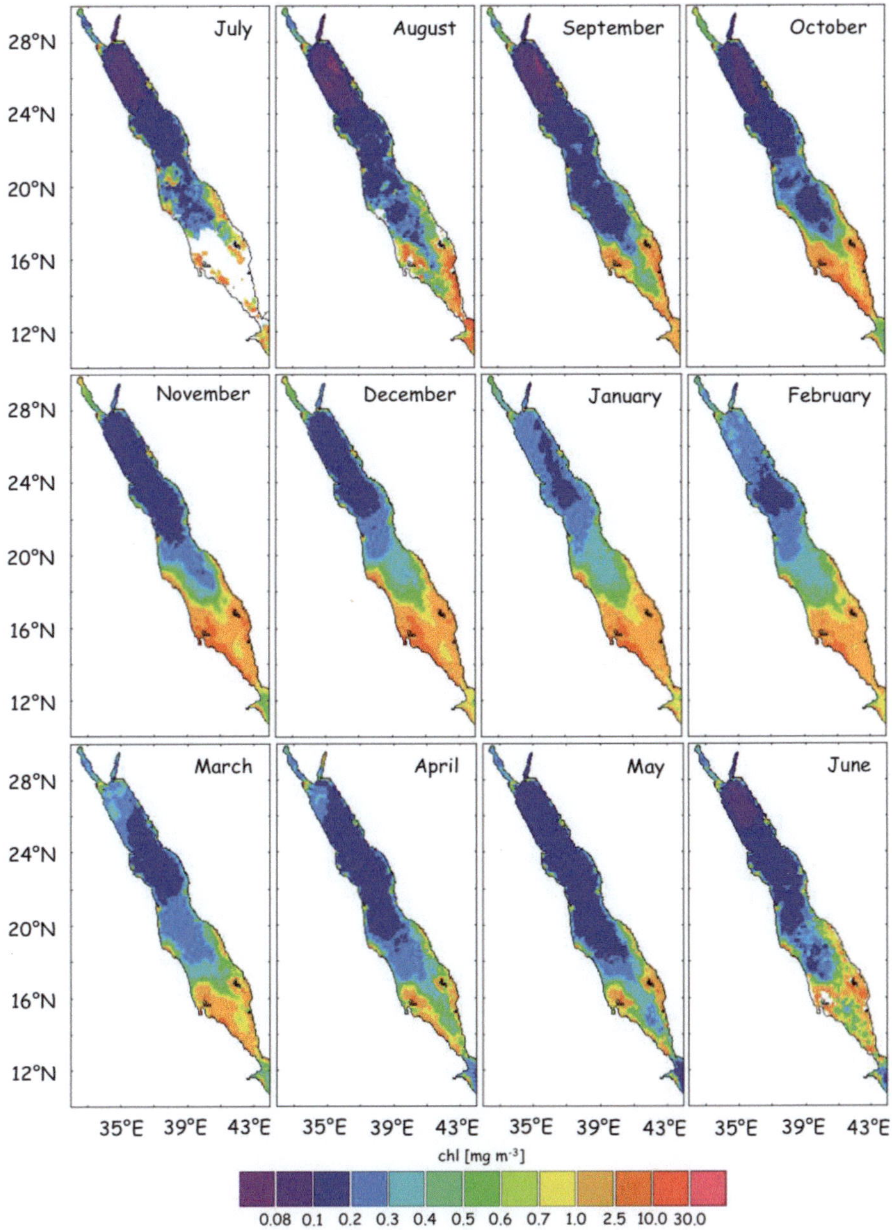

Fig. 17.5 Red Sea *chl* [mg m^{-3}] climatological monthly means (SeaWiFS data set)

higher *chl* values start to recede from the north and progressively move southward in March and April. The same pattern (maximum followed by a decrease) applies also for *ws* values, and is concluded by another reversal of wind direction.

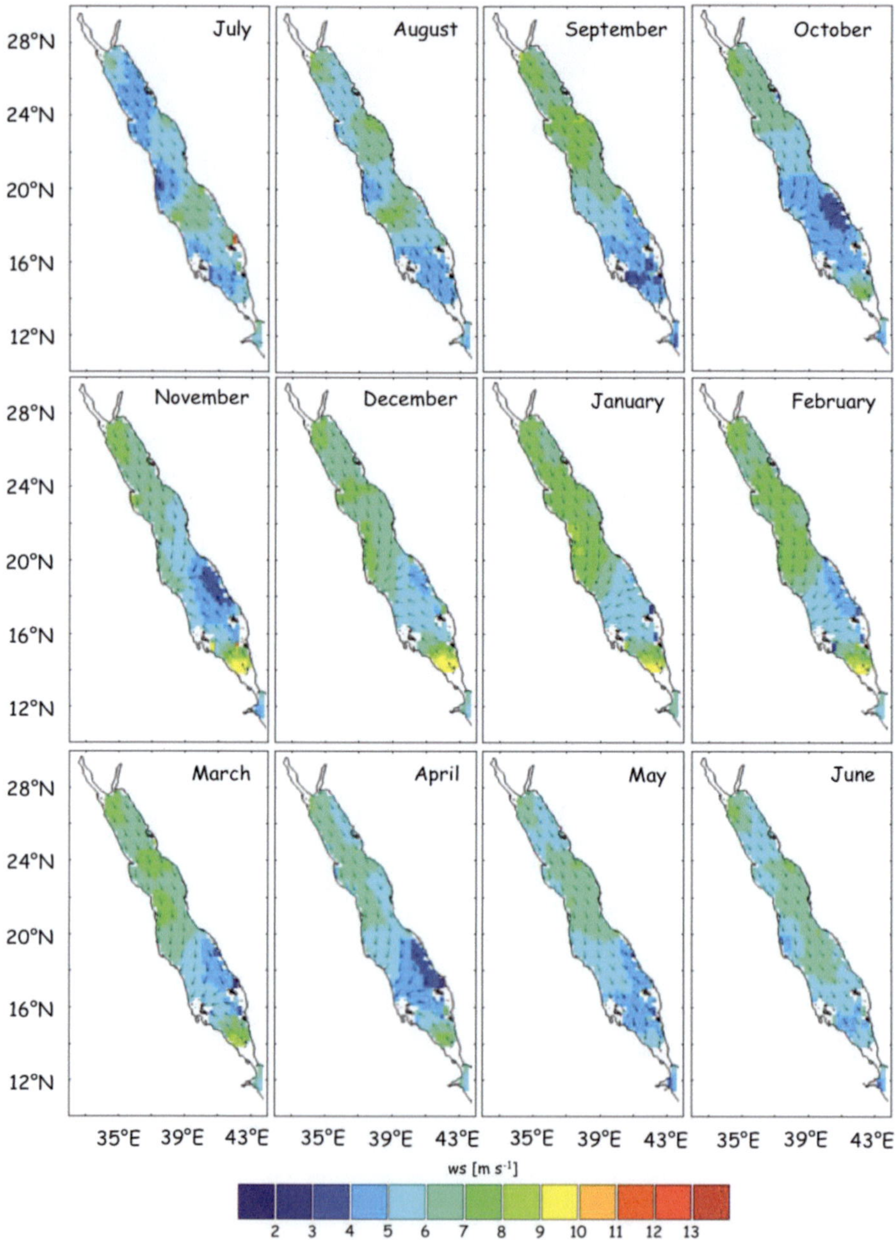

Fig. 17.6 Red Sea *ws* [m s^{-1}] climatological monthly means (QuikSCAT data set)

Most *chl* values remain low, in particular for the basin interior, from May to June, until the same summer spatial distribution seen at the beginning of this seasonal sequence is reached once again. Similarly, the prevailing swinds continue to

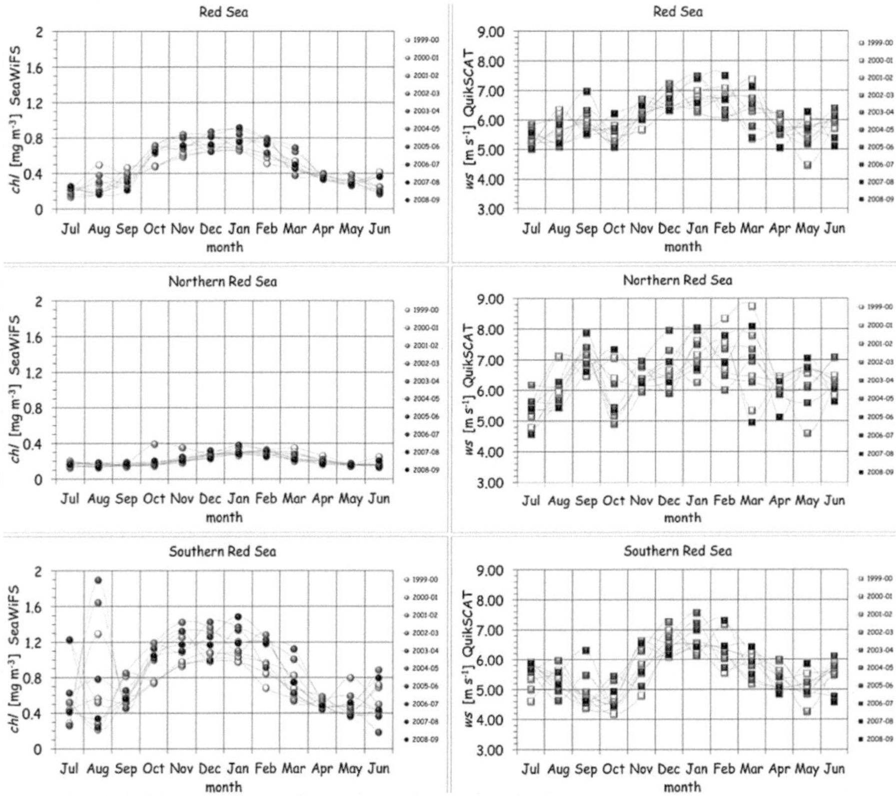

Fig. 17.7 Red Sea inter-annual variability of the SeaWiFS-derived chl_{ABV} [mg m^{-3}] (*left panels*) and QuikSCAT-derived ws_{ABV} [m s^{-1}] (*right panels*), for the entire basin (*upper row*), northern (*middle row*) and southern (*lower row*) sub-basins

be northerly, with rather low *ws* values. The inter-annual variability of the *chl* climatological monthly sequence, as bescribed by the chl_{ABV} computed from the SeaWiFS monthly means, is shown in Fig. 17.7 (left panels). Again, the general scheme, for all years considered, is that of spring-summer minima opposed to fall-winter maxima (see also Barale 2007), with notable exceptions. At the scale of the entire basin (upper left panel), the spread of monthly mean values appears to be limited and rather constant, over the whole year, except for an expansion in August and a reduction in April. The largest variability in the basin-wide time series is driven by the increased inter-annual spread observed in the southern sub-basin for August (lower left panel), but starting already in June-July (and pointing once more to the sustained, though highly variable, impact of nutrient-rich upwelling waters entering the southern Red Sea from the Gulf of Aden in summer).

The smallest variability in the basin-wide time series is due to the reduced range appearing in all sub-basins in April. Conversely, the sizeable spread appearing in fall, from October to December, and even more so in winter, from January to March,

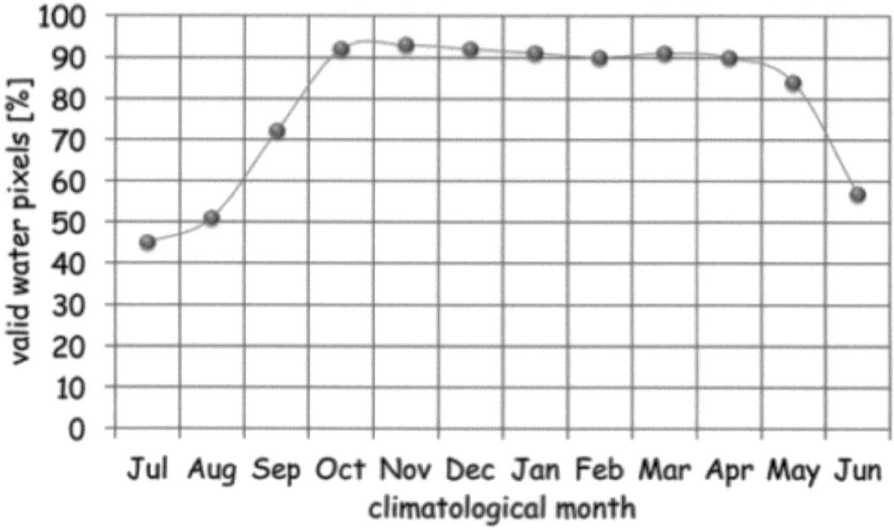

Fig. 17.8 Percentage of valid (water) pixels used for computing the Red Sea *chl* climatological monthly means (shown in Fig. 17.5) from the SeaWiFS data set

is driven primarily by the southern sub-basin (lower left panel). The limited spread of monthly mean values in the northern sub-basin hardly contributes to this picture.

The corresponding ws_{ABV} computed from the QuikSCAT monthly means are also shown in Fig. 17.7 (right panels). The plot of *ws* basin-scale values (upper right panel) presents similar seasonal characteristics as the chl_{ABV}, i.e. a summer minimum and a winter maximum, already discussed in the climatological case. As noted earlier, the recurring September maxima, in the northern sub-basin (central right panel), as well as the October and April minima, in the southern sub-basin (lower right panel), do not seem to have direct links with the dynamics of the *chl* field, within the more general seasonal scheme above. In the north, the inter-annual variability is always very high, and seems to increase even further in winter, reaching a peak in March. In the south, the inter-annual variabilty is somewhat lower, and follows a bimodal pattern of summer and winter highs, interposed by lows in the transitional months of October and April.

17.4 Seasonal Patterns and *chl-ws* Correlation

Two main seasonal periods, in line with the dominant monsoon cycle, seem to emerge from the SeaWiFS (and QuikSCAT) historical data set. This periodicity can be easily illustrated using the percentage of valid (water) pixels (see Fig. 17.8) used for computing the Red Sea *chl* climatological monthly means—and reflected in the error bars width appearing in Fig. 17.4. A period that will be referred to as

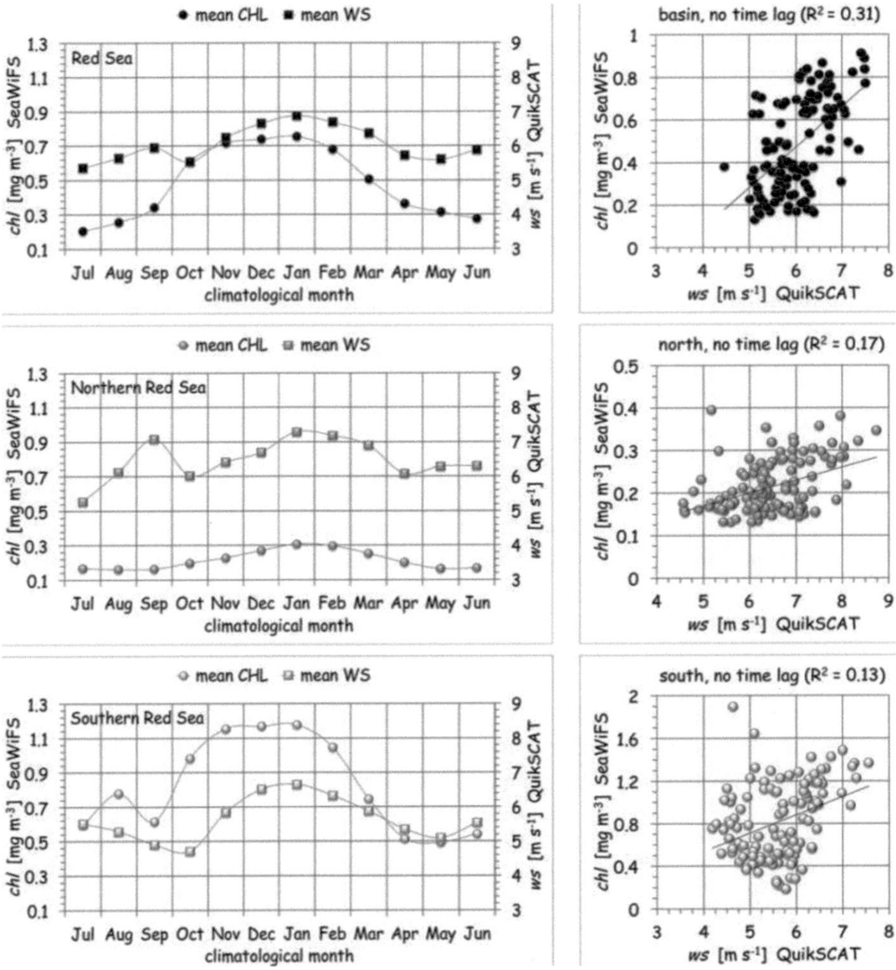

Fig. 17.9 Red Sea climatological (monthly scale) SeaWiFS-derived chl_{ABV} [mg m^{-3}] and QuikSCAT-derived ws_{ABV} [m s^{-1}] (*left panels*). Linear regressions of the monthly mean chl_{ABV} and ws_{ABV} (*right panels*), for the entire basin (*upper row*), the northern sub-basin (*middle row*) and the southern sub-basin (*lower row*)

fall-winter, going roughly from October to March (i.e. that with the largest number of valid pixels), is characterized by clear skies—as well as by higher *chl* and *ws* values (and by prevailing southerly winds in the southern-sub-basin). A period that will be referred to as *spring-summer*, going roughly from April to September (i.e. that with a smaller number of valid pixels), is characterized by persistent cloud cover—as well as by lower *chl* and *ws* values (and by prevailing northerly winds in the southern-sub-basin).

As seen in Fig. 17.9 (left panels), *fall-winter* starts with a first phase, from October to December, in which *chl* values are increasing everywhere. Starting from the relative minimum of October, a maximum is reached already in November, in the southern sub-basin, and then in December, elsewhere. A second phase follows, from January to March, with *chl* values decreasing everywhere. Starting from the absolute maximum of January, a relative minimum for the period is reached again in March. In *spring-summer*, again a first phase, from April to June, shows the *chl* values continue their decline, reaching a minimum around May and perhaps starting to increase again in June, in both the northern and southern sub-basins. In a second phase, from July to September, the increase seems to continue, with a relative maximum in August, for the southern sub-basin.

The *ws* values (still Fig. 17.9, left panels) seem to follow the same general pattern of recurrent maxima in fall-winter followed by minima in spring-summer, at least at the basin scale. The winds of the northern and southern sub-basins, however, tend to behave in rather different ways. During *spring-summer*, when they are aligned in the SW monsoon (i.e. northerly everywhere), *ws* is rather constant at first, only to reach a relative maximum in September, in the north, and a relative minimum in October, in the south. In this period, northerly winds are reinforced in the northern sub-basin, but will eventually die off in the southern sub-basin, before reversing their direction. During *fall-winter*, when the NE monsoon induces winds of opposite direction (i.e. southerly) in the southern sub-basin, *ws* increases from October on, and then decreases to a minimum in April. While the general *fall-winter vs spring-summer* bimodal scheme can be readily associated with the wind forcing required to have increased vertical mixing, and a consequent enhanced fertility of surface waters, in *fall-winter*, the diverse sub-basin pattern followed by *ws* in *spring-summer* does not seem to have any significant impact on the *chl* field.

The linear regression of the monthly mean chl_{ABV} and ws_{ABV} (Fig. 17.9, right panels) indicates that the correlation is positive, but very small ($R^2=0.43$), at the basin scale. Conversely, the correlation is extremely small ($R^2=0.19$) or virtually non existent ($R^2=0.06$) in the northern and southern sub-basins, respectively. Introducing a time lag between chl_{ABV} and ws_{ABV} (i.e. taking into account that there might be a delay of one or more months in the *chl* response to changes in *ws*), the correlation does not improve, indeed it decreases further. Once again, the general trend of fall-winter maxima followed by spring-summer minima suggests that the general blooming pattern of the Red Sea is indeed related to the classical "tropical sea" scheme, driven by wind forcing in wintertime. But, at the same time, it appears that specific, recurrent blooming episodes, in the southern Red Sea in particular, are not related directly to wind forcing (rather to other factors such as the periodic exchange of water with the Arabian Sea, via the Gulf of Aden and Bab-el-Mandeb).

17.5 Conclusions

Space and time heterogeneity of algal blooming in the Red Sea was studied using a SeaWiFS-derived (and in part MODIS-derived) time series of *chl* data. Further, a QuikSCAT-derived time series of *ws* data was also considered, to assess the impact of monsoon-driven atmospheric forcing on the *chl* variability. Data from ten consecutive full-year cycles, from July 1999 to June 2009, were used to compute chl_{ABV} and ws_{ABV} from monthly composite images. A correlation between these parameters has been attempted, to explore the basic ecosystem dynamics of the Red Sea.

The SeaWiFS imagery details the diverse characteristics of the northern (oligotrophic) sub-basin and the southern (mesotrophic) sub-basin, between which a central, transitional area presents more variable environmental conditions. Interestingly, hardly any subdivision between a strictly pelagic region and a coastal zone appears anywhere in the Red Sea, with the exception of that occurring between the basin interior and the broad shelf areas, dotted by numerous islands, of the south.

The observed seasonal pattern is essentially bimodal, with a *fall-winter* period of extended blooming, which progresses from south to north, followed by a *spring-summer* period of much reduced blooming, at least in the northern sub-basin. Overall, this annual cycle seems to be governed by the climatic characteristics of the basin, the monsoon-driven wind regime in particular, and by the ensuing thermohaline circulation.

Given that fertilization in the Red Sea occurs mainly *via* the inflow from the Gulf of Aden and the southern part of the basin, *chl* values should indeed start to increase with the inflow of surface waters from the Gulf of Aden throughout phase one (October to December) of the *fall-winter* period. Thus, relative *chl* maxima appear first in the southern sub-basin, between October and November, and then progress northward in the northern sub-basin, between November and December. From December to January, Ekman transport due to the NE monsoon would contribute to spread surface waters toward the northern region, thus fostering the *chl* absolute maximum observed in the north during phase two (January to March) of the *fall-winter* period. The flux of nutrients of southern origin would then stop, or be much reduced, during the spring inter-monsoon, from March to May. Hence the decreasing *chl* values toward the end of the *fall-winter* period, and the continuing decrease in phase one (April to June) of the *spring-summer* period.

Following the May absolute minimum, starting in June and through phase two (July to September) of the *spring-summer* period, the intrusion of intermediate waters from the Gulf of Aden, driven by the SW monsoon, would favour the mixing of deep nutrients into the upper layer of the southern sub-basin. This could fertilize the southern Red Sea and result in the *chl* relative maxima recurring in summer months, between July and September. At the end of this phase, following the September minimum, the conventional anti-estuarine regime of the Red Sea would take over again, with the Red Sea overflow coupled to the inflow from the Gulf of Aden leading once again to the *fall-winter* situation detailed above.

Acknowledgments Analyses and visualizations used in this paper were produced with the online data services featured by the Goddard Earth Sciences (GES) and Data Information Services Center (DISC) Interactive Online Visualization ANd aNalysis Infrastructure (GIOVANNI), of the National Aeronautic and Space Administration (NASA); by the Global Marine Information System (GMIS), of the Joint Research Centre (JRC), European Commission (EC); and by Remote Sensing Systems (RSS), sponsored by the NASA Ocean Vector Winds Science Team. We acknowledge the Mission

Scientists and Principal Investigators who provided the data used in this research effort. Thanks are also due to the reviewers of this paper, who contibuted substantially to the improvement of its content and graphical presentation.

References

Acker J, Leptoukh G, Shen S, Zhu T, Kempler S (2008) Remotely-sensed chlorophyll a observations of the northern Red Sea indicate seasonal variability and influence of coastal reefs. J Mar Sys 69:191–204

Barale V (1994) Ocean colour, planktonic pigments & productivity. Memoires de l'Institut Oceanographique Monaco 18:23–33

Barale V (2007) Marine and coastal features of the Red Sea. European Commission, EUR 23091 EN, p 56

Barale V, Gade M (eds) (2008) Remote Sensing of the European Seas. Springer, Heidelberg, p 514

Campbell JW, Blaisdell JM, Darzi M (1995) Level-3 SeaWiFS Data Products: spatial and Temporal Binning Algorithms. NASA Technical Memorandum 104566, vol 32, SeaWiFS Technical Report Series, Hooker SB, Firestone ER, Acker JG (eds) NASA Goddard Space Flight Center, Greenbelt, MD, p 73

Cushing DH (1959) The seasonal variation in oceanic production as a problem in population dynamics. J du Cons CIEM Cons Int pour l'Explor de la Mer/ICES Int Counc Explor Sea 24(3):455–464

Donelan A, Pierson WJ (1987) Radar scattering and equilibrium ranges in wind-generated waves with application to scatterometry. J Geophys Res 92:4971–5029

Edwards FJ (1987) Climate and oceanography. In: Edwards AJ, Head SM (eds) Key environments: Red Sea. Pergamon Press, Oxford, pp 45–69

Eshel G, Naik N (1997) Climatological coastal jet collision, intermediate water formation and the general circulation of the Red Sea. J Phys Ocean 27(7):1233–1257

Genin A, Lazar B, Brenner S (1995) Vertical mixing and coral death in the Red Sea following the eruption of Mount Pinatubo. Nature 377:507–510

Iluz D, Yacobi YZ, Gitelson A (2003) Adaptation of an algorithm for chlorophyll-a estimation by optical data in the oligotrophic Gulf of Eilat. Int J Remote Sens 24:1157–1163

Johns WE, Jacobs GA, Kindle JC, Murray SP, Carron M (2000) Arabian Marginal Seas and Gulfs. University of Miami RSMAS Technical Report 2000–01, pp 60

Kirby CM, Parmeter MC, Arnone RA, Oriol RA (1993) Optical Properties of the Red Sea. Technical Note TN 03–93, Naval Oceanographic Office, Stennis Space Center, MS, p 37

Labiosa RG, Arrigo KR, Genin A, Monismith SG, van Dijken G (2003) The interplay between upwelling and deep convective mixing in determining the seasonal phytoplankton dynamics in the Gulf of Aqaba: evidence from SeaWiFS and MODIS. Limnol Oceanogr 48(6):2355–2368

Manasrah R, Badran M, Lass HU, Fennel WG (2004) Circulation and winter deep-water formation in the northern Red Sea. Oceanologia 46:5–23

Morcos SA (1970) Physical and chemical oceanography of the Red Sea. Oceanogr Mar Biol Annu Rev 8:73–202

Murray SP, Jones W (1997) Direct observations of seasonal exchange through the Bab al Mandab Strait. Geophys Res Lett 24:2557–2560

O'Reilly JE, 24 co-authors (2000) Ocean color chlorophyll a algorithms for SeaWiFS, OC2, and OC4: version 4. NASA Technical Memorandum 206892, vol 11, SeaWiFS Postlaunch Calibration and Validation Analyses, Part 3. Hooker SB, Firestone ER (eds) NASA Goddard Space Flight Center, Greenbelt, MD, pp 9–23

Patzert WC (1974) Seasonal reversal in the Red Sea circulation. In: L'oceanographie Physique de la Mer Rouge, Centre National pour l'Exploitation des Oceans, Paris, pp 55–89

Post AF, Dedej Z, Gottlieb R, Li H, Thomas DN, El-Absawi M, El-Naggar A, El-Gharabawi M, Sommer U (2002) Spatial and temporal distributions of Trichodesmium spp. in the stratified Gulf of Aqaba, Red Sea. Mar Ecol Prog Ser 239:241–250

Quadfasel D, Baudner H (1993) Gyre-scale circulation cells in the Red Sea. Oceanol Acta 16:221–229

Sathyendranath S et al (ed) (2000) Remote Sensing of Ocean Colour in Coastal, and other Optically-Complex, Waters. IOCCG Report, vol 3, IOCCG Project Office, Dartmouth, Nova Scotia. p 140

Schroeder C, Boggs H, Dome G, Halberstam M, Jones L, Pierson J, Wentz J (1982) The relationship between wind vector and normalized radar cross section used to derive Seasat-A satellite scatterometer winds. J Geophys Res 87:3318–3336

Sheppard C, Price A, Roberts C (1992) Marine ecology of the Arabian Region, Patterns and processes in extreme tropical environments. Academic Press, London, p 359

Siedler G (1969) General circulation of the water masses in the Red Sea. In: Degens ET, Ross DA (eds) Hot Brines and Recent Heavy Metals Deposits in the Red Sea. Springer, Berlin, pp 131–137

Smeed DA (1997) Seasonal variations of the flow in the strait of Bab el Mandab. Oceanol Acta 20:773–781

Smeed DA (2000) Hydraulic control of three-layer exchange flows: application to the Bab el Mandab. J Phys Oceanogr 30:2574–2588

Sofianos SS, Johns WE (2003) An Oceanic General Circulation Model (OGCM) investigation of the Red Sea circulation: 2. Three- dimensional circulation in the Red Sea. J Geophys Res 108(C3), 3066, doi:10.1029/2001JC001185

Sofianos SS, Johns WE (2007) Observations of the summer Red Sea circulation. J Geophys Res 112:C06025. doi:10.1029/2006JC003886

Sofianos SS, Johns WE, Murray SP (2002) Heat and freshwater budgets in the Red Sea from direct observations at Bab el Mandab. Deep-Sea Research Part II 49:1323–1340

Sokoletsky L, Dubinsky Z, Shoshany M, Stambler N (2003) Estimation of phytoplankton pigment concentration in the Gulf of Aqaba (Eilat) by in situ and remote sensing single-wavelength algorithms. Int J Remote Sens 24:1–25

Sokoletsky L, Dubinsky Z, Shoshany M, Stambler N (2004) Single-wavelength algorithms for in situ or remote sensing estimation of mean pigment concentration. Int J Remote Sens 25:1517–1525

Stambler N (2005) Bio-optical properties of the northern Red Sea and the Gulf of Eilat (Aqaba) during winter 1999. J Sea Res 54:186–203

Weikert H (1987) Plankton and the pelagic evironment. In: Edwards AJ, Head SM (eds.) Key environments: Red Sea. Pergamon Press, Oxford, pp 90–111

Woelk S, Quadfasel D (1996) Renewal of deep water in the Red Sea during 1982–1987. J Geophys Res 101(C8):18155–18165

Zibordi G, Melin F, Berthon JF (2012) Intra-annual variations of biases in remote sensing primary ocean color products at a coastal site. Remote Sens Environ 124:627–636

Chapter 18
Eddies in the Red Sea as seen by Satellite SAR Imagery

Svetlana S. Karimova and Martin Gade

Abstract We present the results of our observations of mesoscale and sub-mesoscale eddies in the Red Sea based on Synthetic Aperture Radar (SAR) imagery. The dataset used includes about 500 Envisat Advanced SAR (ASAR) images obtained in 2006–2011 over the Red Sea. We found more than 1,000 sub-mesoscale eddies, which manifest in the SAR imagery both due to surfactant films ("black" eddies) and wave/current interactions ("white" eddies), depending on the local wind speed. Sub-mesoscale eddies in the Red Sea seem to be more innumerous than in other inner seas, presumably due to a relatively deep upper mixed layer in this basin. Moreover, more than 50 meso- and basin-scale eddies were found, whose rotation was mostly anti-cyclonic and whose diameters ranged up to approximately 200 km. Most of the basin-scale eddies were found between 21 and 24°N, which is in agreement with earlier observations and with numerical modeling.

18.1 Introduction

On a basin scale the water circulation in the Red Sea has been subject of several studies (Barale and Gade 2014, and literature therein). Most of the efforts undertaken focused either on analyses of large-scale gyres and were based on conductivity-temperature-depth (CTD) (Woelk and Quadfasel 1996; Manasrah et al. 2004) and expendable bathythermograph (XBT) measurements (Quadfasel and Baudner 1993) or on numerical models (Sofianos and Johns 2003). It was shown that the general circulation in the Red Sea is of inverse estuarine type (Sofianos and Johns 2003), with an additional wind-driven upper layer during summer (June–September). The presence of some recurring or persistent large-scale gyres in the northern and central parts of the basin was reported, though there were disagreements on the gyres'

S. S. Karimova (✉)
Space Research Institute, Russian Academy of Sciences, Moscow, Russia
e-mail: svetlana.karimova@hzg.de

Institut für Küstenforschung, Helmholtz-Zentrum Geesthacht, Zentrum für Material- und Küstenforschung GmbH, Geesthacht, Germany

M. Gade
Institut für Meereskunde, Universität Hamburg, Hamburg, Germany
e-mail: martin.gade@uni-hamburg.de

V. Barale, M. Gade (eds.), *Remote Sensing of the African Seas*,
DOI 10.1007/978-94-017-8008-7_18,

number, location, and sign of vorticity (cyclonic or anti-cyclonic). Ultimately, there is still a lack of systematic analyses of the occurrence of mesoscale and even sub-mesoscale eddies (i.e. turbulent vortical structures in the upper water layer with diameters smaller than, or up to, the baroclinic Rossby radius).

This paper aims at providing some statistical information on the spatial and temporal distribution of sub-mesoscale to basin-scale eddies in the Red Sea, as gained from high-resolution satellite data. Images obtained at infrared (IR) and visible (VIS) ranges of the electromagnetic spectrum can be used for the detection of mesoscale eddies (Alpers et al. 2013). Moreover, synthetic aperture radar (SAR), not only because of its high ground resolution, but also because of its independence of daylight and cloud conditions, offers great opportunities for studies of the surface ocean circulation at meso- and sub-mesoscales. Consequently, Karimova (2012) used a large number of Synthetic Aperture Radar (SAR) data of the Baltic, Black, and Caspian Seas to derive statistics on the occurrence of sub-mesoscale eddies. Those investigations are adopted herein and are extended towards the analysis of Red Sea surface dynamics based on satellite imagery.

18.2 Data and Methods

The present study is mainly based on the analysis of 492 Envisat Advanced SAR (ASAR) Wide Swath (WS) imagery acquired over the Red Sea during the whole years 2006–2011. The spatial resolution of the ASAR WS imagery is about 150 m, thus allowing for a detection of features, whose sizes are on the order of a few kilometers and below. Figure 18.1 shows the areal coverage of the Red Sea region with ASAR images for the whole observation period. For the northern part of the Red Sea (north of 20°N) more than 100 ASAR scenes are available at any given location, whereas some areas in the southern part were imaged less than 20 times in 2006–2011.

In general, sub-mesoscale eddies manifest in SAR imagery due to two main mechanisms, namely the accumulation of surfactants and wave/current interactions (Karimova 2012). At low to moderate wind speeds (of 3–5 m/s) the former is the main mechanism and eddies usually appear in SAR images due to the presence of natural films on the sea surface (Espedal et al. 1998). Surfactants dampen small-scale surface waves, thereby reducing the radar backscattering from the sea surface (Alpers and Hühnerfuss 1989), and they often accumulate along the shear lines and convergence zones, which in turn are associated with the spiral flow (Eldevik and Dysthe 2002). Thus, the eddies become visible on SAR imagery (Gade et al. 2013). Since this effect causes vortices to be marked dark, for shortness sake, eddies visualized due to slicks are hereinafter referred to as "black" eddies. In Fig. 18.2 examples of "black" eddies in the Red Sea are presented. The image (shown area is 130 × 95 km) was acquired on June 29, 2006 (i.e. during summer) and shows the north-western coast of the Red Sea, north of Cape Banas. Starting on the right bottom of the image, and reaching up into the upper middle, a chain of (cyclonic) "black" eddies can be inferred, which may result from the interaction of the general flow in this area with

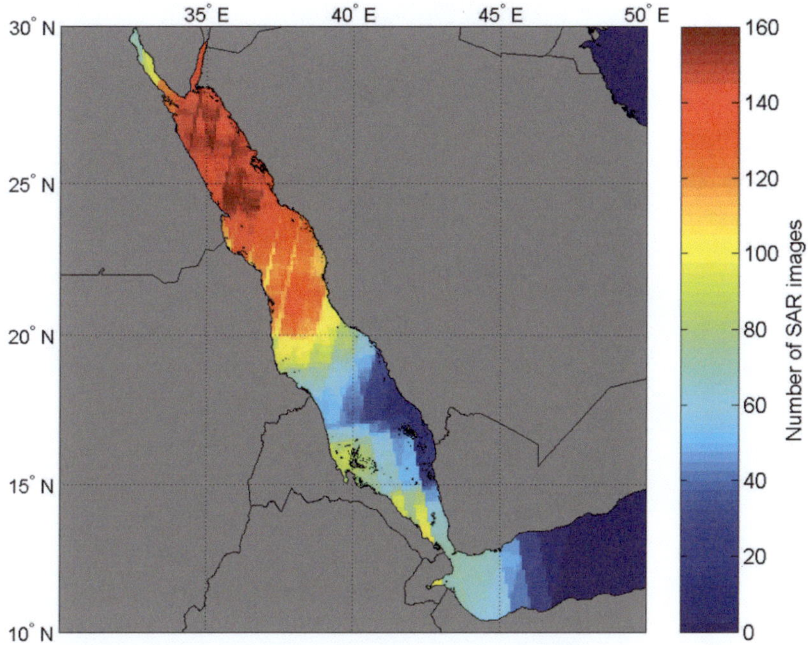

Fig. 18.1 Envisat ASAR WS coverage of the Red Sea in 2006–2011

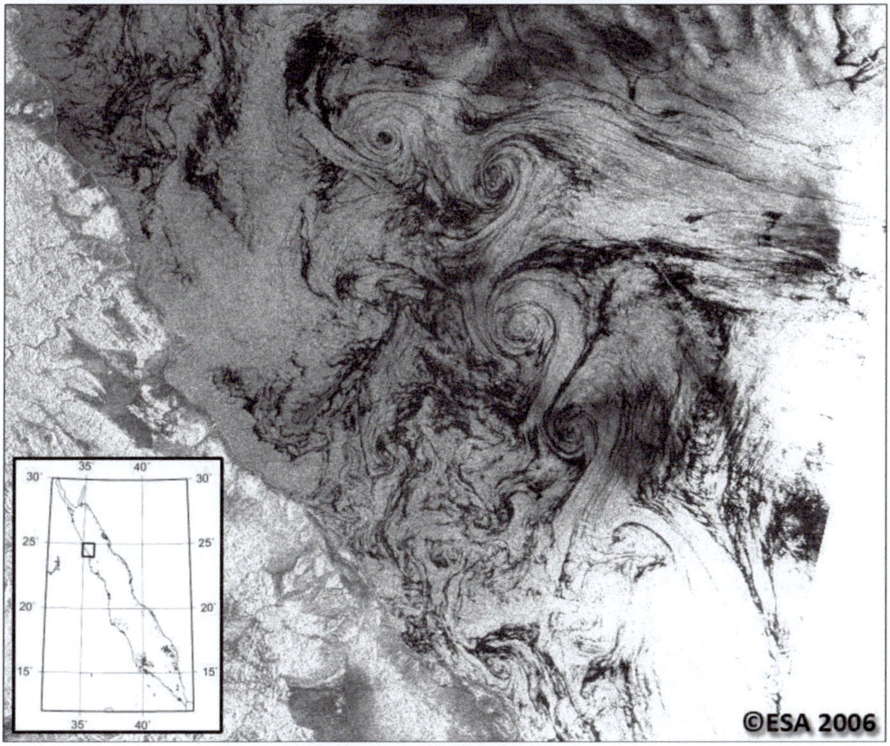

Fig. 18.2 Manifestation of "*black*" eddies in an Envisat ASAR WS image acquired on June 29, 2006 at 07:42 UTC over the north-western coast of the Red Sea, north of Cape Banas. The dimensions of the shown area are 130 × 95 km

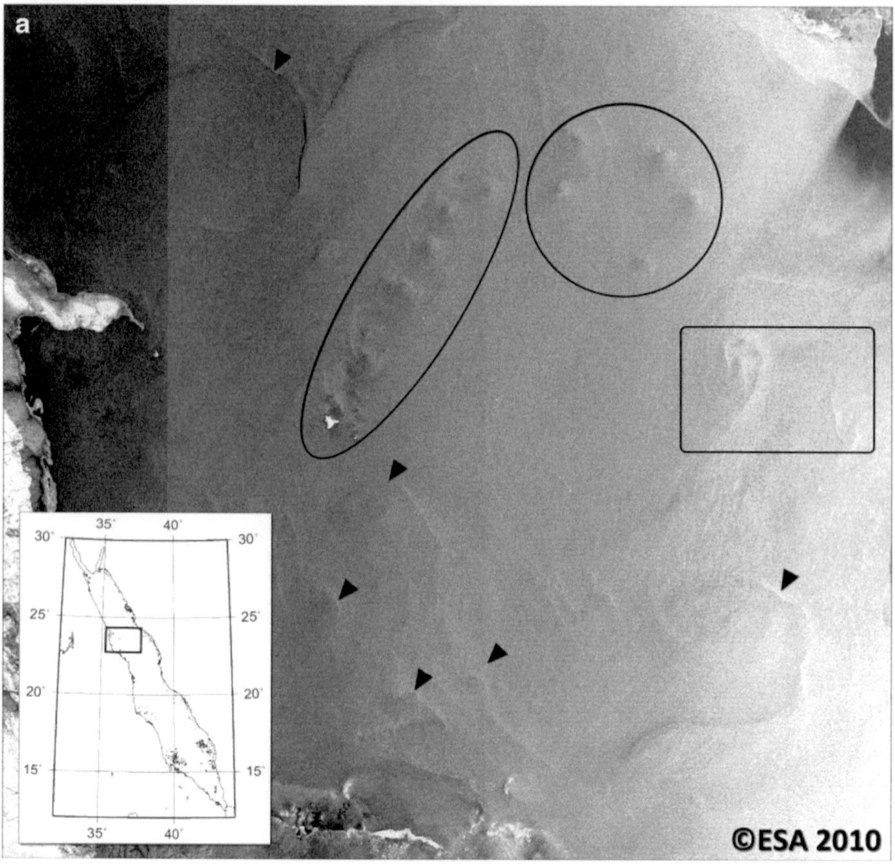

Fig. 18.3 **a** Manifestations of "*white*" eddies (*black marks* and *ellipse/circle/rectangular*) in an Envisat ASAR WS image (shown area is 220 × 215 km) acquired on December 18, 2010 at 07:29 UTC, between Cape Banas (*left*) and the eastern Red Sea coast (*upper right*). **b** Zoomed in areas marked with an *ellipse*, *circle*, and *rectangular* in the Envisat ASAR image shown in Fig. 18.3a

the local wind field, which is usually oriented towards south-west (Barale and Gade 2014). While those sub-mesoscale eddies have diameters of 10 km and more, several eddies with diameters of less than 5 km can be found in the image center and in the upper left corner.

At wind speeds of 5–7 m/s surfactant films start to disrupt, and as a result those dark spiral lines representing "black" eddies disappear (Dokken and Wahl 1996). At higher wind speeds eddies appear in SAR imagery only as a result of wave/current interactions along the lines of current shear, which manifest as bright curved lines. Dokken and Wahl (1996) reported that due to this mechanism eddies can be visualized in SAR imagery at wind speeds as high as 12 m/s. Since the radar backscattering is locally enhanced due to this effect, eddies of this type are hereinafter referred to as "white" eddies. Examples of such eddy manifestations are provided in Fig. 18.3a, b.

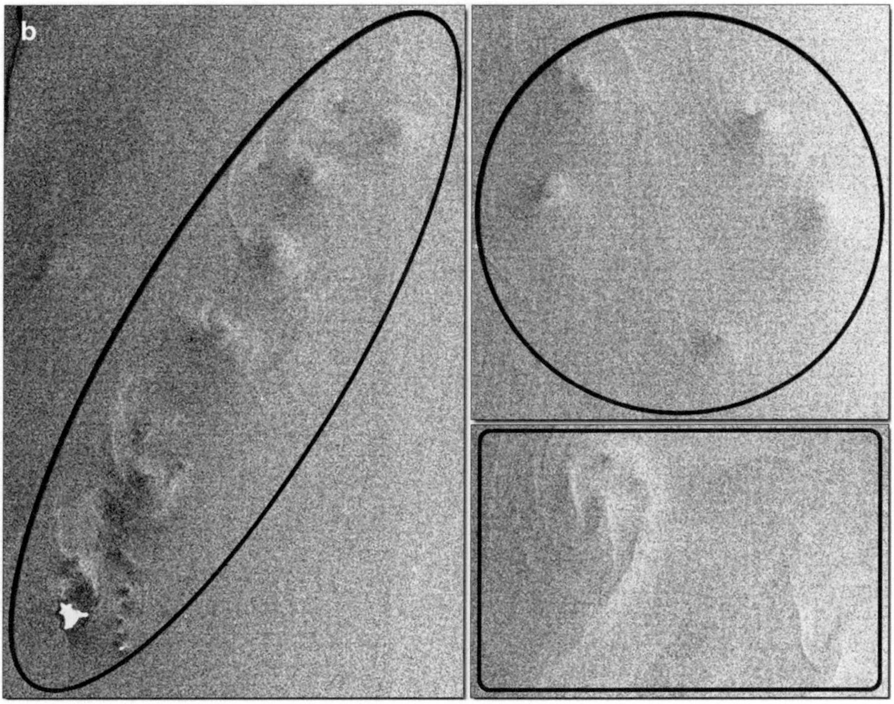

Fig. 18.3 (continued)

This Envisat ASAR WS image was acquired on December 18, 2010 (i.e. during the winter period). The "white" eddies, which are of interest herein, can be inferred from bright spiral signatures, which practically appear all over the image and some of which are marked with arrows. Two vortex streets generated behind, i.e. north-east of, the island of Zapargad (the bright triangular island in the left image center) can be seen in the central part of the image (marked with an ellipse in Fig. 18.3a and zoomed in Fig. 18.3b). Similar signatures of cyclonic vortices were already found by Fu and Holt (1982) on a Sea SAR image of the Atlantic Ocean south of Grand Bahama Island. Given a homogeneous wind field the circular (spiral) flow of the vortices enhances or reduces the effective wind speed sensed by the sea surface, which in turn enhances or reduces the sea surface roughness, respectively, and thus the radar backscatter. Another group of small eddies is arranged further east, in a circular pattern (marked with a circle in Fig. 18.3a and zoomed in Fig. 18.3b). Note that the shear lines and convergence zones inside the eddies, in combination with wave/current interactions and the local wind field, causes a combination of increased and (slightly) decreased radar backscattering, thus resulting in a combination of darker and brighter lines.

The main problem concerning the observation of "white" eddies is that they should be discriminated from the manifestations of small-scale atmospheric processes, such as vortices, rain and convective cells, wind field inhomogeneities, etc.

(Ufermann and Romeiser 1999). As a criteria for the discrimination between "white" eddies and atmospheric phenomena we used the presence of thin shear lines typical for "white" oceanic eddies. Such shear lines can be found, e.g., in Fig. 18.3a, b inside the manifestations of "white" eddies marked with the ellipse and with a rectangle.

All SAR images used for the present investigation were visually inspected and for each detected eddy, its sign of rotation (cyclonic or anti-cyclonic), center coordinates and diameter were recorded using the BEAM Visat software. The rotational sign was defined based on the spiral structure of the eddies: if the spiral was swirling counter-clockwise (clockwise) the eddy was regarded as cyclonic (anti-cyclonic). The results of our statistical analyses of sub-mesoscale eddies (with diameters less than approximately 30 km) are provided in the following section, while those of meso- and basin-scale eddies (with larger diameters) are discussed in Sect. 4 and 5, respectively.

18.3 Sub-mesoscale Eddies

In total, 582 "black" and 440 "white" sub-mesoscale eddies were detected in the 492 SAR images analyzed. It is noteworthy that the number of eddies per SAR image is much smaller than that discovered for the Baltic, Black, and Caspian seas, where about 14,000 eddies were detected in approximately 2,000 images (Karimova 2012). Presumably this is due to the greater depth of the upper mixed layer, which is typically about 100 m for the Red Sea, compared to 10–40 m (depending on the season) for the above mentioned basins. Following Tragou and Garrett (1997) the baroclinic Rossby radius in the Red Sea is approximately 30 km; therefore, we chose this length to discriminate between sub-mesoscale (smaller) and meso- to basin-scale (larger) eddies. The vast majority ($> 90\,\%$) of eddies found in the ASAR imagery are cyclonical, which is in line with theoretical considerations and numerical modeling (Kleppin 2012).

We have to note here that the very mechanisms of sub-mesoscale eddy generation, especially in case of spatial densely-packed groups of such eddies, are still under investigation. In some particular cases, the mechanisms of eddy generation are quite obvious and coincide with those of bigger eddies, such as lateral friction of the near-coastal flow, influence of a coastline inhomogeneity, baroclinic instability at the hydrological fronts, and barotropic (shear) instability in the shear current zones. For a general case, when none of the hypotheses mentioned can be applied, eddy origination was explained by current's baroclinic instability (Eldevik and Dysthe 2002; Akitomo 2010; Kleppin 2012), quasi-periodic impulsive impact on the water surface (Voropayev and Afanasyev 1992), and convection in the near-surface layer (Boubnov and Golitsyn 1995). The results on sub-mesoscale eddy spatio-temporal distribution provided here and in (Karimova 2012) seem to support the influence of water baroclinicity on sub-mesoscale eddy activity.

In the following subsections we will analyze spatial scale and spatial distribution of sub-mesoscale eddies detected.

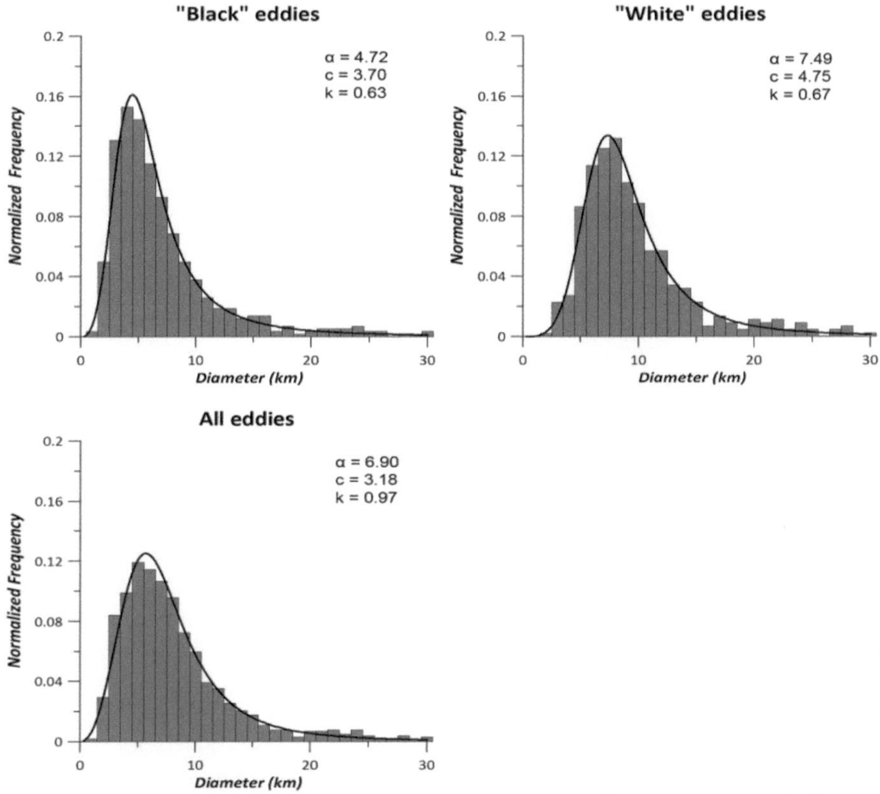

Fig. 18.4 Diameter distributions of "black" (*upper left*), "white" (*upper right*), and all (*lower*) submesoscale eddies found on Envisat ASAR WS imagery of the Red Sea from 2006–2011. Added to each histogram is the best-fit Burr distribution with scaling parameters given in the diagrams

18.3.1 Spatial Scale

For each eddy detected, its diameter (or the distance between the two most remote parts of the eddy) was measured (manually) and recorded. Figure 18.4 shows normalized histograms of those diameters, separated by the visualization type ("black" and "white", see above). Burr (Type XII) distributions (Tadikamalla 1980), which were found to provide the best fit to the data, are added to each histogram. The mean diameter (i.e. the first moment of the respective best-fit Burr distribution) of "black" eddies is 7.4 km, while that of "white" eddies is 9.8 km. The reason for this difference in size might be the fact that "black" and "white" eddies are usually observed under different wind conditions: at lower wind speeds (generating lower surface drifts) eddies with smaller diameters can survive, while at higher wind speeds only bigger and stronger eddies still exist. Since the local wind speed is generally lower when "black" eddies appear on SAR imagery, as a consequence, the diameter of the turbulent rotational flow should be smaller, which was indeed observed. The histogram for all detected eddies (lower panel in Fig. 18.4) is broader, and the total mean eddy diameter was found to be 8.3 km.

18.3.2 Spatial Distribution

Figure 18.5 shows the spatial distributions of the detected "black" eddies (Panel a) and "white" eddies (Panel b) for the entire period 2006–2011. In order to avoid a confusion with the small circles (denoting eddy locations) we have removed all islands from the underlying map.

"Black" eddies were mostly found in the northern part of the Red Sea, with highest densities along the coasts. Again, since lower wind speeds are needed for the visualization of "black" eddies, more of them should be found in coastal zones, where the interaction with, and the sheltering by, the land reduces the mean wind speed. Moreover, in addition to the constraints given by the SAR imaging of surface films, both the shallow bathymetry in coastal zones and the closer distance to the shore should have an impact on the generation of small eddies and, thus, on their size distribution in those areas. In this respect, it is interesting to note that the overall availability of chlorophyll-like pigments in the northern part of the Red Sea is lower than in its southern part (Barale and Gade 2014). Therefore, a simple relationship between the occurrence of "black" eddies and the biological productivity (and, thus, the availability of surfactants at the sea surface) cannot be inferred.

In contrast to the "black" eddies we found "white" eddies mostly offshore, i.e. in the open part of the Red Sea (Fig. 18.5b). Again, the overall higher (mean) wind speed on the open sea supports the visualization of "white" eddies, since the visibility of surfactants on SAR imagery is reduced. Moreover, most of the "white" eddies were found in the northern part of the Red Sea. At a first glance, this may be due to the higher availability of SAR images from the northern part (see Fig. 18.1); however, we also found only very few "white" eddies at the southern tip of the Red Sea, close to the Strait of Bal-el-Mandeb, where more SAR imagery was available again. Therefore, it is unlikely that the heterogeneous distribution of the "white" eddies (and of the "black" eddies as well) is just a manifestation of the heterogeneous coverage of the total area by SAR imagery. Instead, hydrodynamic differences, in combination with a different bathymetry and wind climatology in either parts of the basin, must be responsible for the observed differences in eddy density.

In order to support the statement that the spatial eddy distribution does not simply depend on the SAR image coverage, we have calculated normalized densities, *i.e.* the number of (detected) eddies per resolution cell of size $0.2° \times 0.2°$, for convenience multiplied by 100. The respective maps are shown in Fig. 18.6 and demonstrate that the general distribution of both "black" and "white" eddies is looking the same, with or without normalization. We note, however, that the local density of "black" eddies

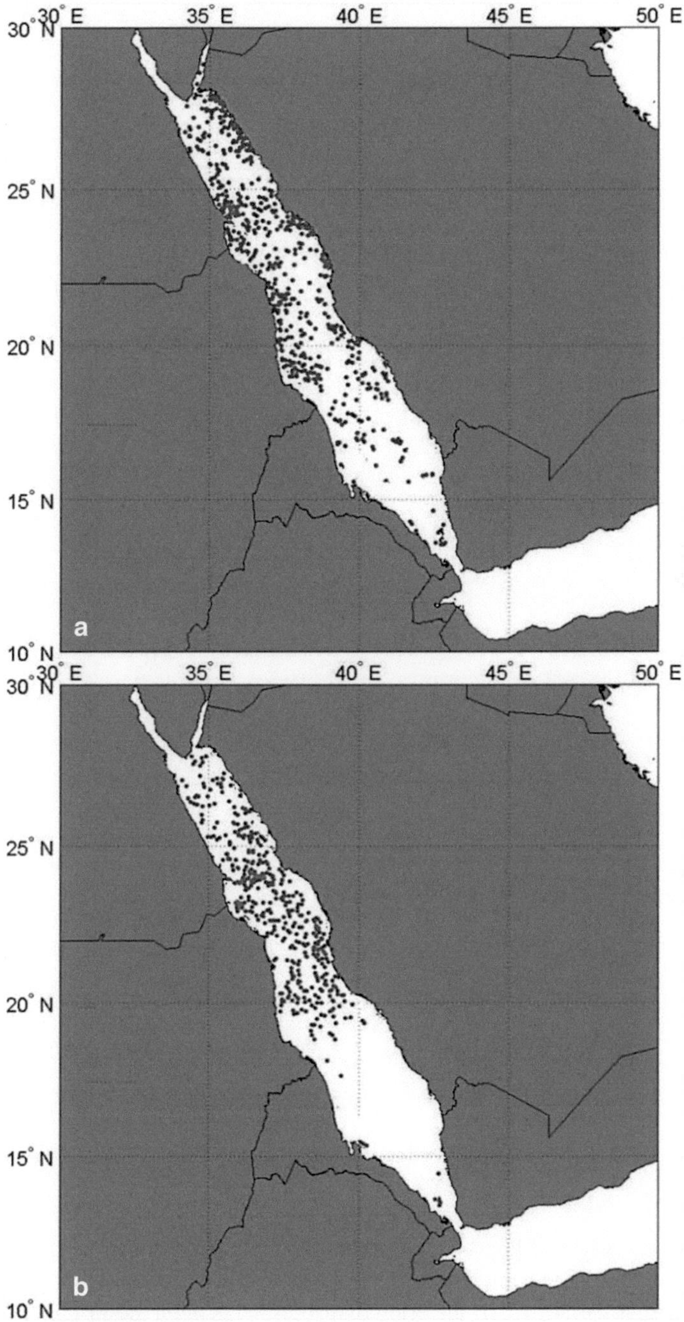

Fig. 18.5 Spatial distribution of "*black*" and "*white*" sub-mesoscale eddies (Panels a and b, respectively). Islands were removed from the maps, as not to be confused with the circles denoting the locations of the detected eddies

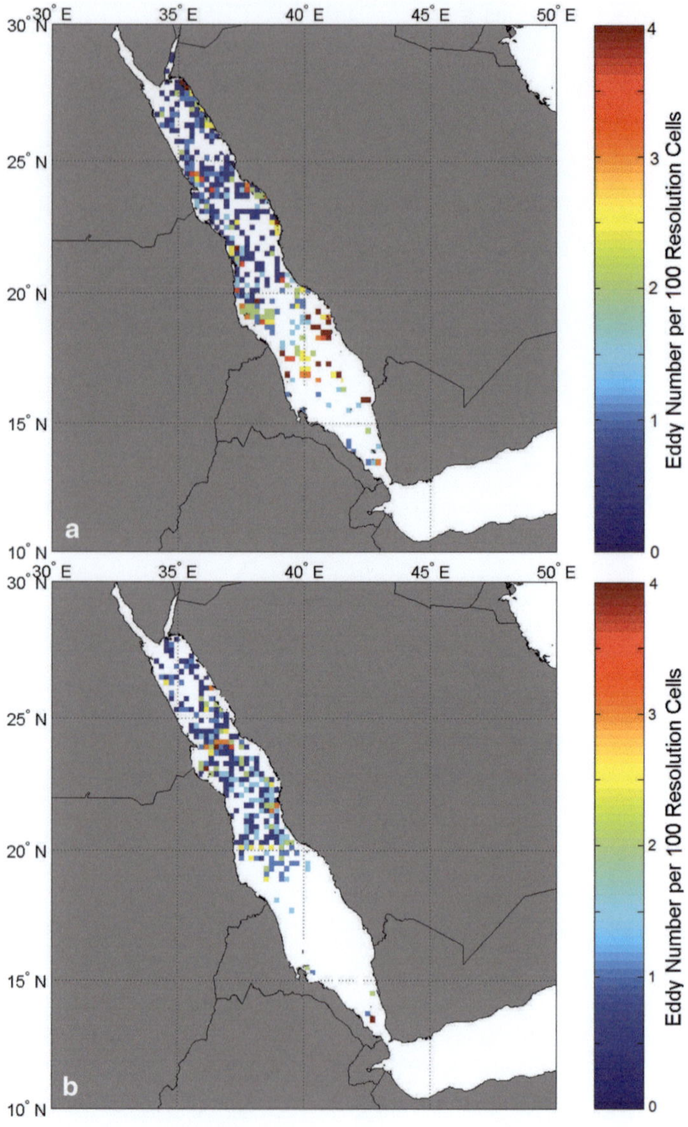

Fig. 18.6 Normalized eddy density (i.e. the number of eddies per image and per resolution cell of size 0.2° × 0.2°) of "*black*" eddies (Panel a) and "*white*" eddies (Panel b)

in the southern Red Sea is even higher than in the northern Red Sea, which in fact might be only due to the limited number of SAR images available.

The seasonal variability of the eddy distributions is depicted in Fig. 18.7a, b, the former showing distributions in winter and spring, the latter in summer and autumn. In all panels, again, islands have been removed as to not being confused with the dots representing eddies. Though six years is a rather too short time to derive robust

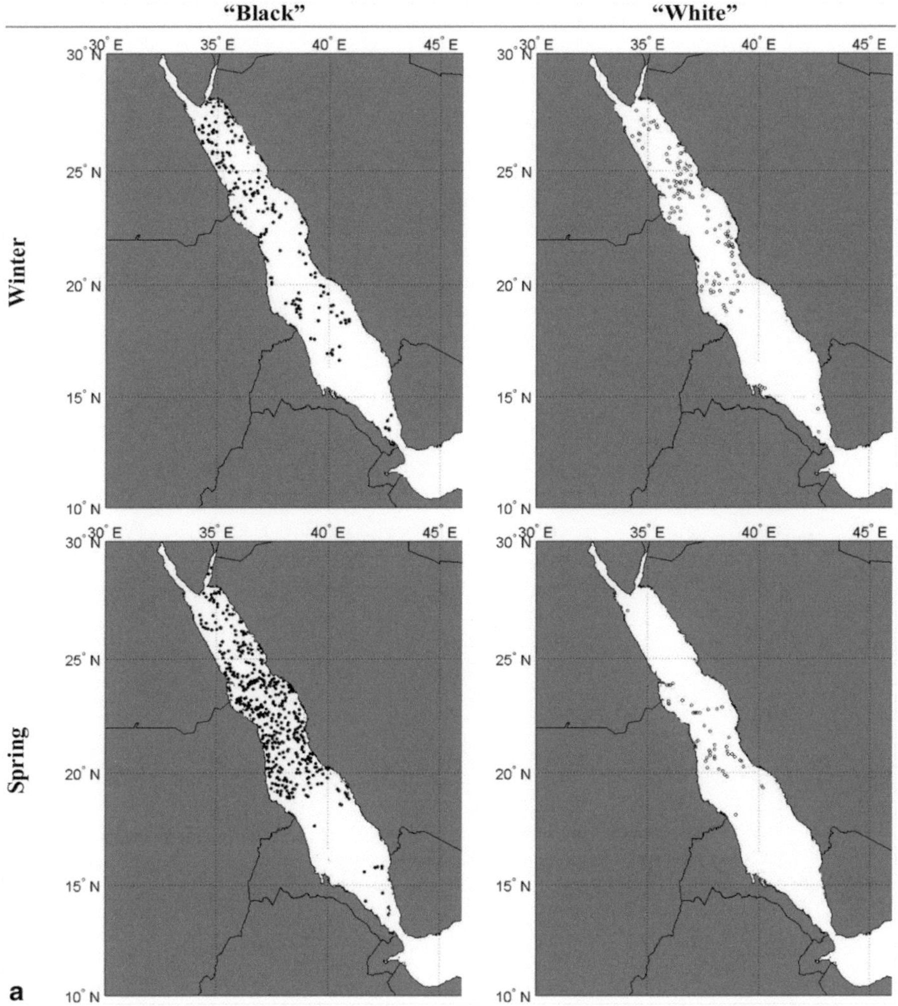

Fig. 18.7 **a** Spatial distribution of sub-mesoscale eddies at different seasons. The *left* column shows the distribution of "black" eddies and the *right* column that of "white" eddies, the *upper row* corresponds to winter (Jan–Mar), the *lower row* to spring (Apr–Jun). In order not to be confused with eddies, small islands were erased from the maps. **b** Same as Fig. 18.7a, but for summer (Jul–Sep, *upper row*) and autumn (Oct–Dec, *lower row*)

statistics on the eddy climatology, our results already show that there are some general trends in the appearance of sub-mesoscale eddies in the Red Sea, and in their visualization by SAR sensors:

- The density of "black" eddies is highest in the warm season, i.e. during spring and summer, and in the northern part of the Red Sea, i.e. north of 20°N.

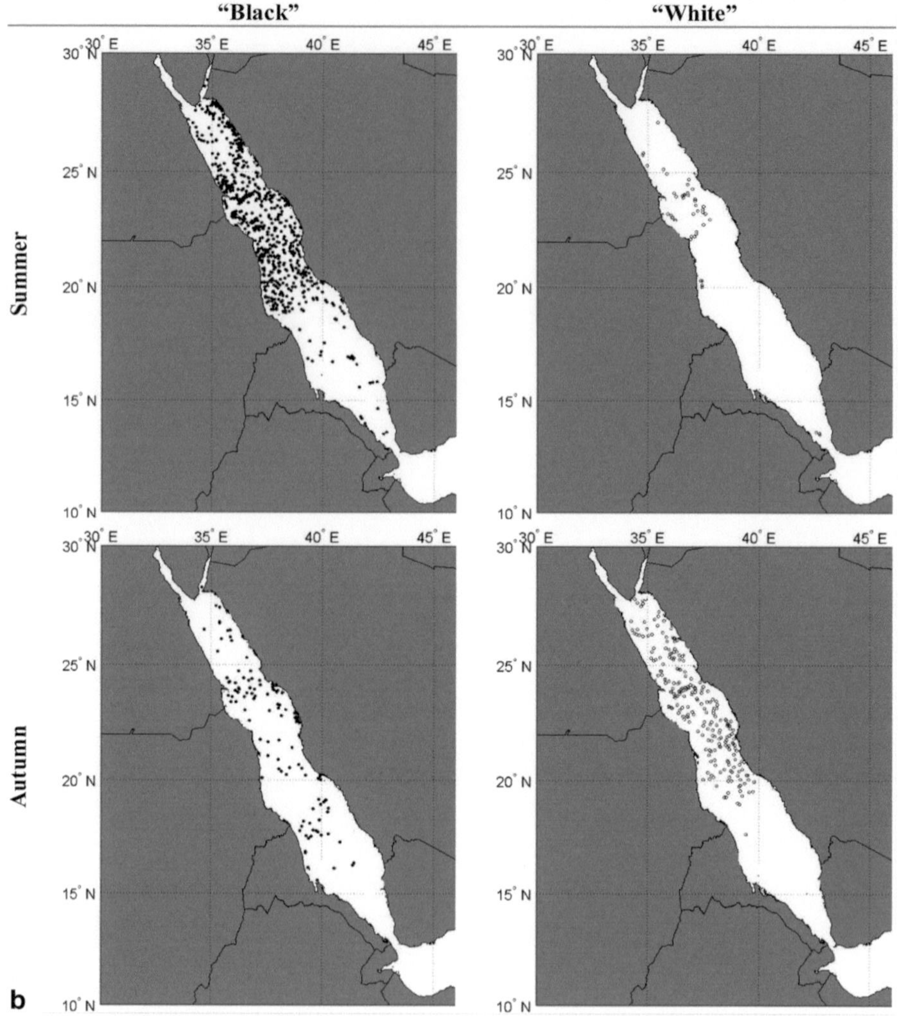

Fig. 18.7 (continued)

- In the cold and windy season, i.e. during autumn and winter, the number of detected eddies, and thus their respective densities, is about the same for "black" and "white" eddies.
- During autumn and winter, but also during the warm season (albeit not that pronounced), highest densities of "white" eddies were measured in the central basin around 24–25°N and along the eastern coast at 22°N. Those areas coincide with abysses (Laughton 1970).

Our results also indicate that the wind has an effect on the generation of eddies in the Red Sea, as already suspected by Tragou and Garrett (1997).

18.4 Mesoscale Eddies

In total we detected 39 mesoscale eddies, whose diameters were ranging from about 30 km to almost 100 km. Like the smaller, sub-mesoscale eddies they were visualized in the SAR imagery both through surfactant films (23 eddies) and through wave/current interactions (16 eddies). 24 eddies were cyclonically rotating, while the rest, 15, were anti-cyclonical. The locations of all anti-cyclonic and cyclonic mesoscale eddies are shown in Fig. 18.8.

Both anticyclonic and cyclonic mesoscale eddies can be found in different parts of the Red Sea, but most of them (especially the cyclonic eddies) are located along the longer axis of the basin where probably there are more favorable conditions for cyclonic eddy generation/evolution. For instance, mesoscale cyclonic eddies in the central part of the sea can be a result of the (preceding) evolution of sub-mesoscale eddies. Eddies in the near-coastal zone (both cyclonic and anti-cyclonic) can appear as a result of a lateral friction of a general flow.

Spatial distributions of both meso- and basin-scale eddies at different seasons are given in Fig. 18.9, separated by their vorticity sign (left column: anti-cyclonic and right column: cyclonic) and by the season (Fig. 18.9a: winter and spring, i.e., Jan–Mar and Apr–Jun, respectively, and Fig. 18.9b: summer and autumn, i.e., Jul–Sep and Oct–Dec, respectively). The sizes of the circles correspond to the sizes of the eddies, and according to our nomenclature, "black" eddies are marked by a solid circle and "white" eddies by an open circle.

From Fig. 18.9a, b we draw the following conclusions:

- In general, the detected anti-cyclonic eddies are larger than the cyclonic eddies.
- Most of the eddies were found in the central Red Sea.
- During the warm period (spring and summer), the anti-cyclonic eddies are dominating, while the cyclonic eddies dominate during the cold period (autumn and winter).
- Mostly "black" eddies were found during the warm period, i.e. during spring and summer.
- During the warm period (spring and summer), the "black" eddies dominate, while the "white" eddies dominate during the cold period (autumn and winter).

18.4.1 *Basin-Scale Eddies*

Eddies with a diameter greater than 100 km were regarded in the present paper as basin-scale. The maximum diameter of such eddies reached almost 200 km. Totally, there were 15 basin-scale eddies detected. Again, they were manifested due to both surfactant films (8 eddies) and wave/current interactions (7 eddies). Only two of them were cyclonically rotating.

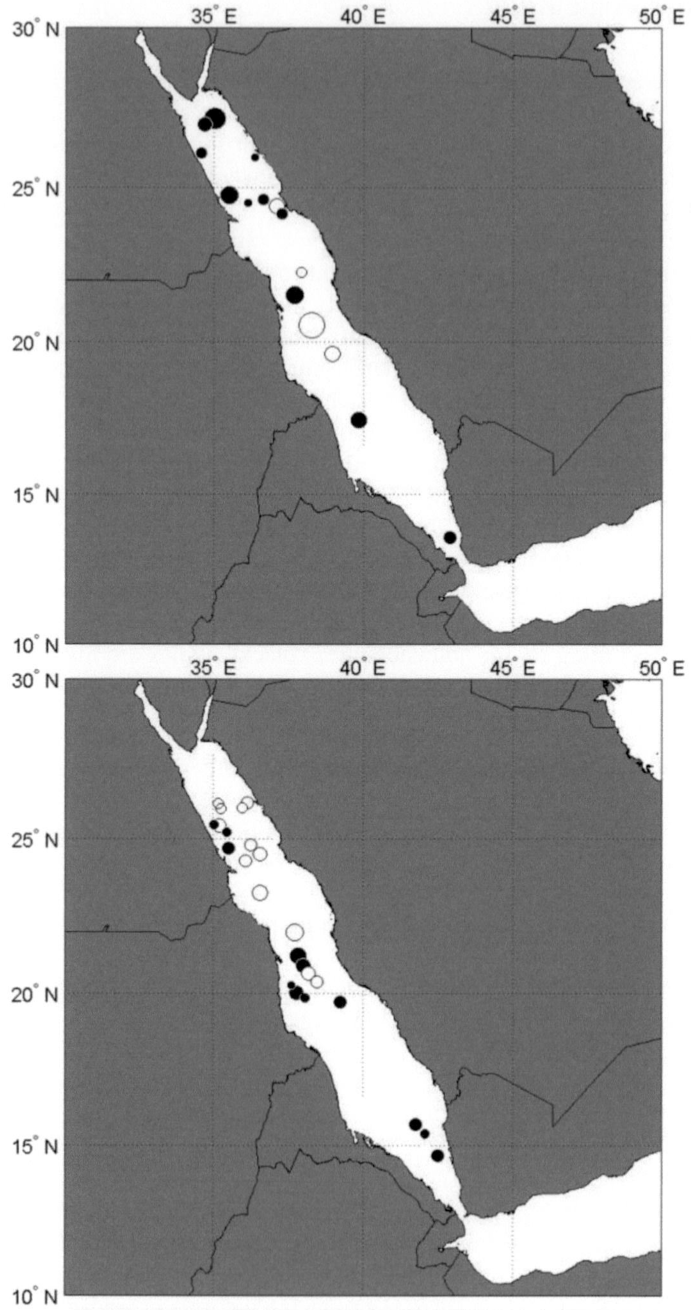

Fig. 18.8 Spatial distribution of anti-cyclonic (*upper panel*) and cyclonic (*lower panel*) mesoscale eddies detected in 2006–2011. Solid circles mark "*black*" eddies, open circles mark "*white*" eddies. The diameters of the circles correspond to those of the eddies

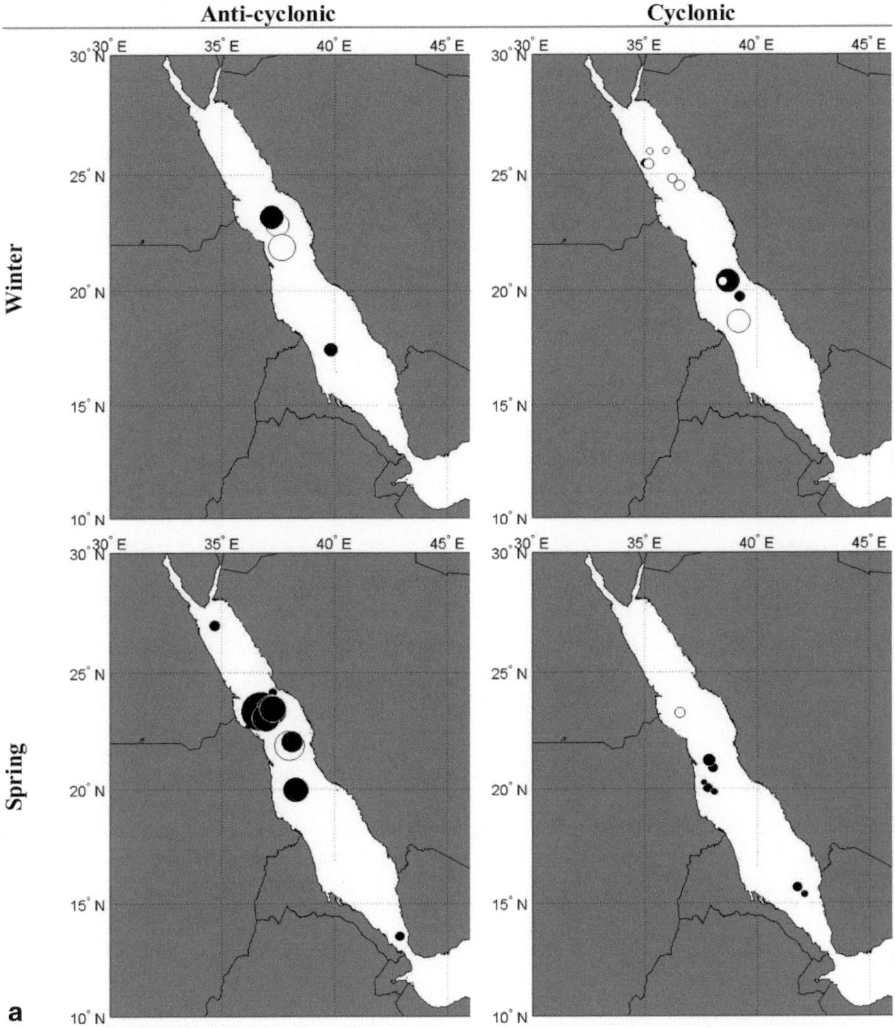

Fig. 18.9 **a** Spatial distribution of basin- and mesoscale eddies detected in winter (Jan–Mar, *upper row*) and spring (Apr–Jun, *lower row*). The *left* column contains all anti-cyclonical eddies, the right column all cyclonical. Solid circles mark "*black*" eddies, open circles mark "*white*" eddies. The diameters of the circles correspond to those of the eddies. **b** Same as Fig. 18.9a, but for summer (Jul–Sep, *upper row*) and autumn (Oct–Dec, *lower row*). Islands were removed from the maps

In order to define those places where basin-scale eddies were most frequently observed Fig. 18.10a, b show the spatial distribution of the largest eddies detected. It is obvious that the majority of them were found in a relatively small area, roughly between 21 and 24°N, i.e. in the center part of the basin. Apparently, in most cases we found manifestations of the same gyre, which was recently observed by Sofianos and Johns (2007). This gyre, as well as the adjacent ones, is generated through wind

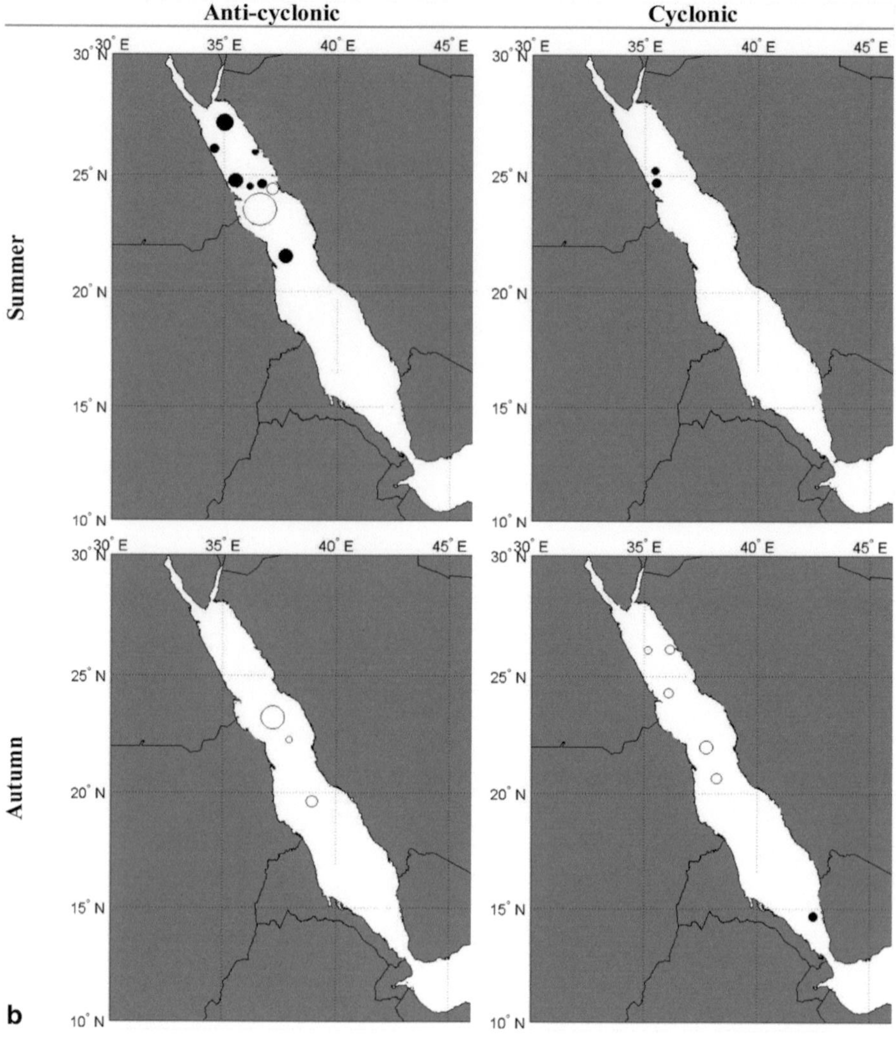

Fig. 18.9 (continued)

forcing and the general current's interaction with the local topography (Quadfasel and Baudner 1993). We further note that basin-scale eddies were mainly detected during spring season.

More quantitative information on the basin-scale eddy manifestations is summarized in Table 18.1 in chronological order. Again, we can note that only the "smallest" two eddies were cyclonically rotating. This table also supports our finding that the largest eddies detected in all SAR images (with diameters exceeding 120 km) used for the present study were mainly found during spring.

Fig. 18.10 **a** Spatial distribution of anti-cyclonic basin-scale eddies detected in 2006–2011. Solid circles mark "*black*" eddies, open circles mark "*white*" eddies. The diameters of the circles correspond to those of the eddies. **b** Spatial distribution of cyclonic basin-scale eddies detected in 2006–2011. Solid circles mark "*black*" eddies, open circles mark "*white*" eddies. The diameters of the circles correspond to those of the eddies

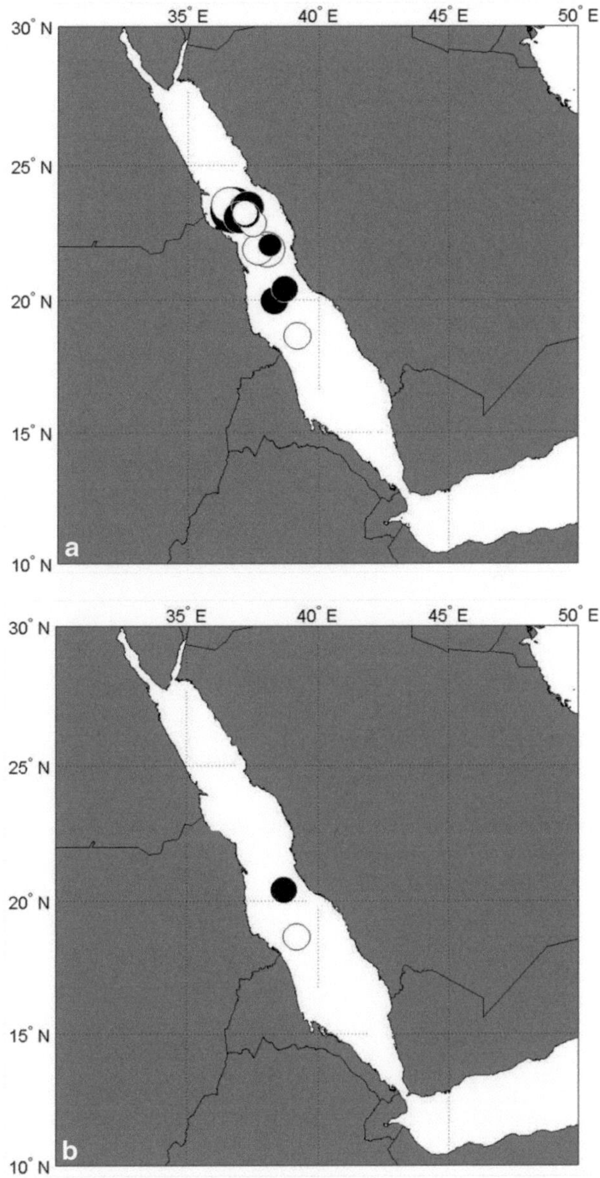

On May 19 and 24, 2009, two Envisat ASAR WS images were acquired over the same part of the Red Sea, and they show manifestations of basin-scale "black" and "white" eddies, respectively. A closer look at the observed signatures reveals that apparently the same eddy was imaged twice, but visualized in two different ways, i.e. as "black" eddy (May 19, Fig. 18.11a) and as "white" eddy (May 24, Fig. 18.11b). The time in between the two SAR acquisitions is 5 days, but eddies with diameters of

Table 18.1 Basin-scale eddies detected in the Red Sea in 2006–2011

Date	Time (UTC)	Longitude	Latitude	Diameter (km)	Visualization	Rotation
Jan 20, 2008	07:29	37.47	22.88	115.5	w	AC
May 4, 2008	07:29	38.01	21.90	142.2	w	AC
Jun 11, 2008	19:29	37.19	23.21	114.1	w	AC
Feb 2, 2009	07:18	39.18	18.67	111.8	w	C
Feb 5, 2009	07:23	38.69	20.42	113.6	b	C
Mar 26, 2009	19:29	37.67	21.88	129.9	w	AC
Apr 14, 2009	19:32	36.97	23.09	139.3	b	AC
May 19, 2009	19:32	37.24	23.40	147.1	b	AC
May 24, 2009	07:28	37.15	23.23	110.2	w	AC
Jun 4, 2009	19:29	36.74	23.34	194.2	b	AC
Jun 12, 2009	07:28	38.13	22.06	103.5	b	AC
Sep 17, 2009	19:30	36.56	23.55	153.3	w	AC
Feb 23, 2010	19:32	37.21	23.19	113.7	b	AC
May 9, 2010	07:28	37.29	23.46	133.4	b	AC
Jun 24, 2010	19:29	38.30	19.99	122.8	b	AC

b black, *w* white, *AC* anti-cyclonic, *C* cyclonic

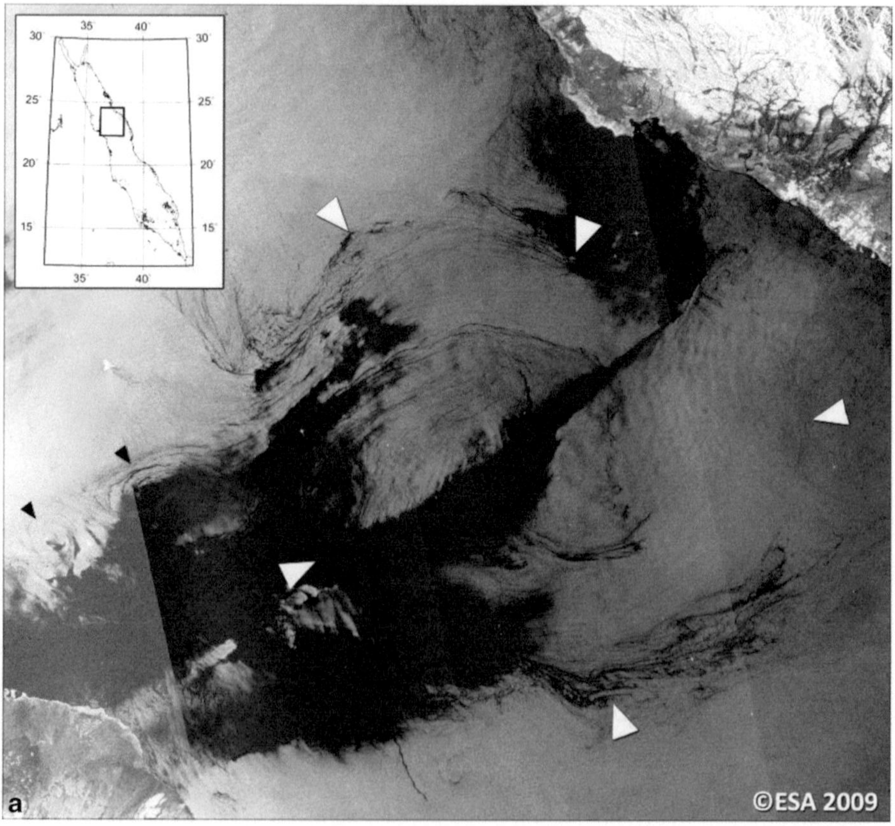

Fig. 18.11 **a** Manifestation of an anti-cyclonic eddy in an Envisat ASAR WS image acquired on May 19, 2009 at 19:32 UTC. The shown area is 245 × 225 km. **b** Manifestation of an anti-cyclonic eddy in an Envisat ASAR WS image acquired on May 24, 2009 at 07:28 UTC. Shown area is 300 × 240 km

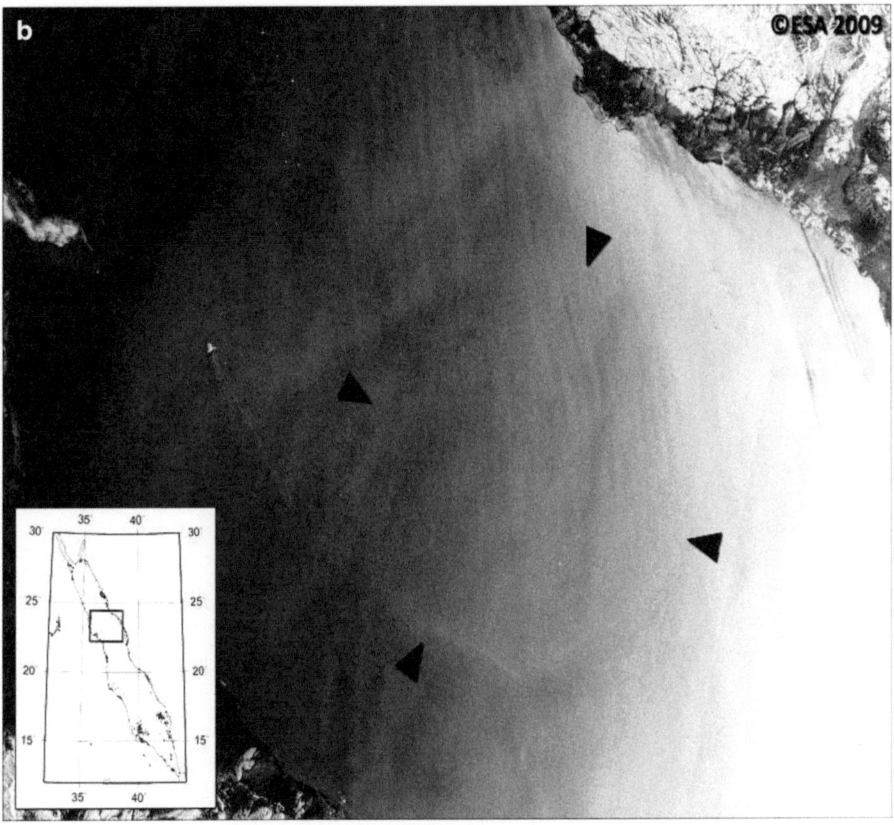

Fig. 18.11 (continued)

more than 40 km were already found to have lifetimes of about 5 weeks (Zhan 2013). The eddy's diameter was found to be 147 km on May 19, and 110 km on May 24; however, because of the different visualization mechanisms, care should be taken when directly comparing those diameters.

Obviously, in the former (earlier) case, the wind speed was lower: QuikScat wind scatterometer measurements (for more details see Barale and Gade 2014) from May 19, 2009, revealed that the mean wind speed in that area was 2–3 m/s (evening passes), while the same sensor measured 10–12 m/s on May 24, 2009 (morning passes). These data, together with the fact that sub-mesoscale eddies were found only in the earlier SAR image, clearly show the impact of the local wind field on the appearance of sub-mesoscale eddies in SAR imagery.

The same basin-scale eddy was also observed in an Aqua MODIS true color image shown in Fig. 18.12. The circular bright pattern in the image center is caused by higher concentrations of phytoplankton advected along the outer current lines of this gyre. Those higher phytoplankton concentrations may have given rise to the

Fig. 18.12 Manifestation of the same basin-scale (anti-cyclonic) eddy shown in Fig. 18.11 in an Aqua MODIS true color image (shown area is 440 × 530 km) acquired over the central Red Sea on May 20, 2009, at 11:05 UTC

higher amounts of surfactants, which in turn caused the visualization of the "black" eddy on the day before.

Basin-scale eddies in the central Red Sea have already been found in-situ (Sofianos and Johns 2007), in altimeter data (Zhan 2013), and in ocean color data (Barale 2007). They differ in diameter only slightly, but their exact location does vary along the basin axis.

18.5 Conclusions

We used more than 500 Envisat ASAR WS images acquired in 2006–2011 over the Red Sea to study the spatial and temporal distribution of eddies at sub-meso-, meso-, and basin-scale. Most of the images covered the northern part of the basin and therefore, most of the eddies detected were also located there.

However, the normalized densities (i.e. the numbers of detected eddies per image resolution cell) show evidence that the observed heterogeneity is not simply due to the inhomogeneous coverage by SAR imagery, but that a general trend exists of a greater number of sub-mesoscale eddies in the northern part of the basin (north of 20°N).

In total more than 1,000 sub-mesoscale eddies were found, which is about two eddies per SAR image. This is generally less than previous studies revealed for the Baltic, Black, and Caspian seas. We hypothesize that this finding is linked to the greater depth of the upper mixed layer in the Red Sea (about 100 m), but also to the overall shape of the (narrow) basin and to the different circulation within the basin.

"Black" eddies, i.e. eddies manifesting in SAR images through the accumulation of surfactants along the shear lines, were found mostly in coastal areas, while "white" eddies, i.e. manifesting through wave-current interactions, were found further offshore. In general, "black" eddies visualize in SAR images at lower wind speeds, since the surfactants start to disrupt when the wind speed increases.

We also found about 50 meso- and basin-scale eddies with diameters up to approximately 200 km. Their rotation was both cyclonic and anti-cyclonic; however, most of the basin-scale eddies (with diameters exceeding 100 km) were found to be anti-cyclonic, which is in contrast to the smaller, sub-mesoscale eddies. Those basin-scale eddies were mainly found in SAR images acquired in spring, and they were located between 21 and 24°N, which supports the hypothesis of Quadfasel and Baudner (1993) that wind forcing and interactions with the local topography in that area are the main generation mechanisms.

SAR imagery of six consecutive years is certainly not sufficient for the derivation of complete climatologies, particularly if the entire basin was not covered homogeneously. However, our results indicate very well that the systematic analyses of SAR imagery with respect to the detection of sub-mesoscale, mesoscale, and basin-scale eddies has great potential to further the knowledge about the hydrodynamics in certain sea areas or in enclosed or semi-enclosed seas.

Acknowledgments This work was supported within the framework of the Federal Target Program "Scientific and scientific-pedagogical personnel of innovative Russia" in 2009–2013. The data used for the present study were kindly provided by ESA under project ID 14120 (SESAMeSEA).

References

Akitomo K (2010) Baroclinic instability and submesoscale eddy formation in weakly stratified oceans under cooling. J Geophys Res 115:C11027. doi:10.1029/2010JC006125

Alpers W, Hühnerfuss H (1989) The damping of ocean waves by surface films: a new look at an old problem. J Geophys Res 94(5):6251–6265

Alpers W, Brandt P, Lazar A, Dagorne D, Sow B, Faye S, Hansen M, Rubino A, Poulain PM, Brehmer P (2013) A small-scale oceanic eddy off the coast of West Africa studied by multi-sensor satellite and surface drifter data. Remote Sens Environ 129:132–143

Barale V (2007) Marine and coastal features of the Red Sea. European Commission, EUR 23091 EN, p 56

Barale V, Gade M (2014) Basic ecosystem dynamics in the Red Sea as seen by sundry remote sensing techniques. In: Barale V, Gade M (eds) Remote sensing of the african seas. Springer, Heidelberg, this issue

Boubnov BM, Golitsyn GS (1995) Convection in rotating fluids. Kluwer Academic Publishers, Dordrecht

Dokken ST, Wahl T (1996) Observations of spiral eddies along the Norwegian Coast in ERS SAR images. FFI Rapport 96/01463

Eldevik T, Dysthe K (2002) Spiral eddies. J Phys Oceanogr 32:851–869

Espedal HA, Johannessen OM, Johannessen JA, Dano E, Lyzenga D, Knulst JC (1998) COASTWATCH '95: A tandem ERS-1/2 SAR detection experiment of natural film on the ocean surface. J Geophys Res 103:24969–24982

Fu LL, Holt B (1982) Seasat views the oceans and sea ice with synthetic-aperture-radar. JPL Publication, California

Gade M, Byfield V, Ermakov SA, Lavrova OYu, Mitnik LM (2013) Slicks as indicators for marine processes. Oceanography 26(2):138–149

Karimova S (2012) Spiral eddies in the Baltic, Black and Caspian seas as seen by satellite radar data. Adv Space Res 50(8):1107–1124

Kleppin H (2012) Analysing spiral eddies as an example of baroclinic Mixed Layer Instabilities. MSc thesis, Dept. Geosci., University of Hamburg, Germany

Laughton AS (1970) A new bathymetric chart of the Red Sea. Phil Trans Roy SocLondA 267:21–22

Manasrah R, Badran M, Lass HU, Fennel WG (2004) Circulation and winter deep-water formation in the northern Red Sea. Oceanologia 46:5–23

Quadfasel D, Baudner H (1993) Gyre-scale circulation cells in the Red Sea. Oceanologica Acta 16: 221–229

Sofianos SS, Johns WE (2003) An Oceanic General Circulation Model (OGCM) investigation of the Red Sea circulation: 2. Three-dimensional circulation in the Red Sea. J Geophys Res 108(3):3066. doi:10.1029/2001JC001185

Sofianos SS, Johns WE (2007) Observations of the summer Red Sea circulation. J Geophys Res 112:C06025. doi:10.1029/2006JC003886

Tadikamalla PR (1980) A look at the burr and related distributions. Int Stat Rev 48:337–344

Tragou E, Garret C (1997) The thermohaline circulation of the Red Sea. Deep-Sea Res I 44(8):1355–1376

Ufermann S, Romeiser R (1999) Numerical study on signatures of atmospheric convective cells in radar images of the ocean. J Geophys Res 104. doi:10.1029/1999JC900224

Voropayev SI, Afanasyev YaD (1992) Two-dimensional vortex dipoles interactions in a stratified fluid. J Fluid Mech 236:665–689

Woelk S, Quadfasel D (1996) Renewal of deep water in the Red Sea during 1982–1987. J Geophys Res 101(8):18155–18166

Zhan P (2013) Sea surface height variability and Eddy statistical properties in the Red Sea. MSc thesis, King Abdullah Univ of Science and Technology, Thuwal, Saudi Arabia p 66

Chapter 19
Satellite Surveys of Lagoon and Coastal Waters in the Southeastern Mediterranean Area

Mahmoud Ahmed and Vittorio Barale

Abstract Satellite observations provide elements about the environmental status of the major lagoons in northern Egypt (i.e. Lakes Maryut, Edku, Burullus, Manzala and Bardawil) and the adjacent marine area off the Nile River delta, in the southwestern Mediterranean region. Examples are shown of data collected by various sensors onboard the Landsat, Système Pour l'Observation de la Terre (SPOT) and EgyptSat-1 satellite series, and by the Sea-viewing Wide Field-of-view Sensor (SeaWiFS). The remote sensing-derived parameters (e.g. lagoon size, aquatic vegetation patterns and other water quality parameters) allow to evaluate evolution and degradation of the near-coastal water bodies in the last four decades. The basic ecosystem indicators delivered by the satellite surveys are considered important contributions for the establishment of a sound decision-making process for environmental management purposes, aiming to ensure protection of natural systems and sustainability of human activities.

19.1 Introduction

The coast of northeastern Africa, in the southern Levantine Basin of the Mediterranean Sea, is contiguous to and influenced by vast desert areas. However, it is also bordered by a number of wetlands, marshes and lagoons, particularly in and around the Nile river delta area. Suitable combination of satellite data and ground observations can be used to draw elements for the systematic monitoring of these water bodies at large, regional, space scales and over long, decadal, time periods. The present review—focusing on the major lagoons of northern Egypt (see the next section, and Fig. 19.1 in particular, for a description of the main water bodies in this region) and the adjacent marine area—will cover the key issues involved in using

M. Ahmed (✉)
Marine Sciences Department, National Authority Remote Sensing and Space Sciences, Cairo, Egypt
e-mail: mahahmed.narss@yahoo.com

V. Barale
European Commission, Joint Research Centre, Institute for Environment and Sustainability, Ispra, Italy

V. Barale, M. Gade (eds.), *Remote Sensing of the African Seas*,
DOI 10.1007/978-94-017-8008-7_19,

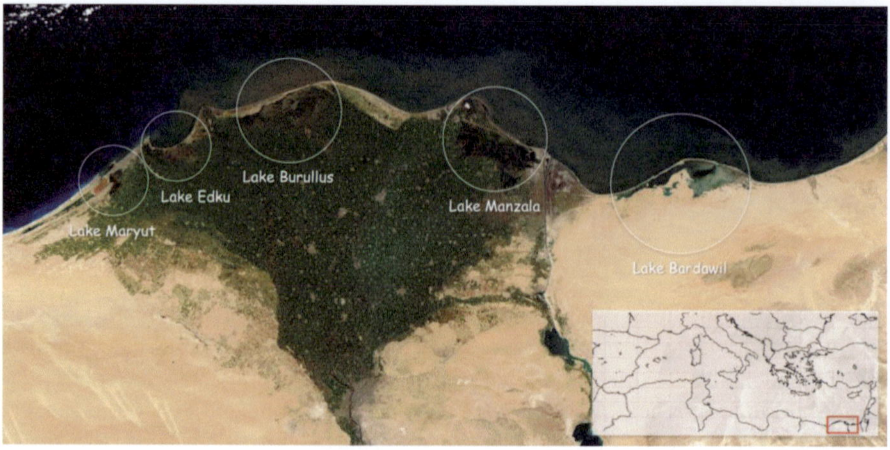

Fig. 19.1 Lagoons and coastal waters of the Mediterranean southeastern area. Simulated true colour image, from Moderate-resolution Imaging Spectrometer (MODIS) data collected on 6/02/2003

Remote Sensing (RS) for the appraisal of enclosed and coastal waters, and use sample satellite images, collected during the last four decades, to examine some key aspects of their environmental status.

The basic requirement for environmental monitoring by RS is the availability of coherent time series of data, in which the same area can be observed repeatedly. A variety of satellite systems, mostly equipped with multi-spectral instruments, have been available since the 1970s to detect environmental patterns on the Earth surface. Consequently, today sizeable historical archives—generated by a number of sensorns, performing measurements at various spectral, spatial and temporal resolutions—provide more data than have ever been available at any other time. In spite of their varying technical characteristics, and their uneven time coverage, such datasets constitute a powerful tool to detect, follow and evaluate ecosystem changes, as well as to develop management strategies for ecosystem resources, over a wide range of time and space scales (Graetz 1990).

In the coastal domain, satellites can sample interacting physical and bio-geo-chemical processes at scales impractical to cover with conventional *in situ* data gathering techniques. RS of marginal and enclosed seas from the vantage point offered by the Earth's orbit allows to observe at a glance the dynamical relationships of natural setting, water exchange, basic ecological relations and their main driving forces, as well as the problems faced by coastal and marine waters (see e.g. various authors in Barale and Gade 2008). Complementary techniques, exploiting the potential of RS in the visible, thermal infrared and microwave parts of the spectrum, can be used in the context of environmental status and marine pollution investigations (as exemplified by Barale 2005, for the Mediterranean basin). Optical assessments have been particularly successful in describing basic ecosystem dynamics, and related space/time anomalies, in both the coastal and pelagic Mediterranean environments

(Bosc et al. 2004; Volpe et al. 2007; Barale et al. 2008). Analogous applications, using high-sensitivity sensors that collect data at low spatial resolution (at km scale) but medium spectral resolution (~10's of nm), geared to measure indicators like concentration of chlorophyll-like pigments (CHL) or total suspended matter (TSM) in marine waters, have been developed also for lake studies (Zilioli 2001).

Low-sensitivity sensors working in the visible and near-infrared parts of the spectrum, with higher spatial resolution (at dm scale) but lower spectral resolution (~100's of nm), are primarily designed for detecting land features. However, current instruments and processing algorithms provide better performance for aquatic studies than ever previously available (Kloiber et al. 2002; Sawaya et al. 2003). Thus, monitoring programs of inland waters have been conducted using land observation sensors as well, with the obvious advantage of data on multiple small lakes being collected quickly and relatively inexpensively, reducing the effort and cost of field sampling. Such programs have shown good correlations between RS data and *in situ* measurements of Secchi Disk Transparency (SDT), Colored Dissolve Organic Matter (CDOM), or CHL and TSM (Lathrop 1992; Dewider and Khedr 2001; Brezonik et al 2005). However, it is in the assessment of structural properties, e.g. varying water surface area or aquatic vegetation cover, that sensors borrowed from land-observation programs can best exploit their combined spatial and spectral potential (Hess et al. 2003; Ozemi and Bauer 2004; Cozar et al. 2005; Ahmed et al. 2009).

As for the region of interest here, Barale and Folving (1996) gave early examples of studying coastal runoff by satellite RS, looking at the fluvial and marine interactions that occur within the area off the Nile river delta. More recently, Dewidar and Khedr (2005) combined satellite and surface measurements to map surface water parameters, and multiple linear regression models to prepare digital cartographic products depicting water quality over large areas. Over the past decade, Ahmed et al. (2000, 2001, 2003, 2007, 2009) used RS techniques for repeated bio-geo-chemical assessments of the Egyptian lagoons, and Geographic Information System (GIS) techniques to produce spatial distribution maps of their properties. In particular, Ahmed (2003) applied change analysis methods to sequences of satellite images, in order to identify modifications that have occurred both inside and outside the coastal lagoons, documenting variations of water areas, as well as erosion and accretion processes along the lagoon openings.

In the following, the use of RS to identify key environmental features of enclosed and near-coastal water bodies of northern Egypt, in the southeastern Mediterranean region, will be reviewed. Examples will be given of sensors, techniques, parameters and applications available to monitor the status and evolution of lagoon and coastal waters, and to describe some of their basic ecological problems. An evaluation of changes that have occurred in the lagoons, as well as the recent trends along the adjacent coastlines, will be proposed. Finally, GIS descriptions of lagoon size, vegetation patterns and land cover/use, will be presented, to show how elements for environmental management purposes can be derived from RS data.

19.2 Lagoon and Coastal Waters of Northeastern Africa

The coastal waters of northern Egypt are under the direct impact of the Nile river outflow (Barale and Larkin 1998). Although the river plume is unlikely to have maintained the same impact on the pelagic zone it must have had, in the era preceding the 1963 closure of the Aswan High dam, its influence on the coastal zone is still (or perhaps is again, according to Nixon 2003) a major feature of the area. The lagoons bordering the Nile river delta are all located by the plume impact area, and represent hotspots of great environmental and socio-economic significance. Four of them, i.e. the Lakes Maryut, Edku, Burullus and Manzala, are deltaic water bodies, while a fifth one is the non-deltaic Lake Bardawil (Fig. 19.1).

The four lagoons situated in the Nile river delta are shallow basins (depth < 3 m, with an average depth of $\sim$1 m) of brackish waters. Lake Maryut is artificially enclosed and has long been without a major connection to the sea. Lakes Edku, Burullus and Manzala display typical lagoon characteristics, are separated from the Mediterranean by low-lying, narrow coastal sand barriers, and are connected to the sea by protected channels. The latter are vital for dozen of species of fish that depend on the lakes for at least part of their life cycle, and constitute the only inlet/outlet available for thousands of commercial fishing boats that operate in the lakes. The deltaic lakes receive much of their freshwater input from irrigation drains located along the southern, eastern, and western margins. The fifth lagoon, the non-deltaic Lake Bardawil, is situated away from the Nile delta region, in northern Sinai. Contrary to the other lagoons of the area, this large, somewhat deeper, hypersaline water body does not have any freshwater inflow, and communicates only with the Mediterranean Sea.

Like other Mediterranean lagoons, those of northern Egypt originated as a result of changes in sea level, combined with neotectonic land subsidence. In the Nile river delta lagoons, bottom sediments are composed of plant- and shell-rich muds and sandy silts, whereas the Bardawil lagoon displays sand-size sediments mixed with evaporates (Levy 1974). In general, the coarsest sediments are found near the openings toward the sea, where current velocities are largest, and the finest sediments are in the innermost reaches, where drains are located and current velocities approach zero. Biological processes contribute significantly to production of carbonate sediments through formation of mollusk shells, ostracod and foraminiferal tests (Bernasconi and Stanley 1994). In fact, according to the trophic state index suggested by Carlson (1977), the lagoons can be considered eutrophic during the entire year. The lagoon margins are bordered by extensive marshes rich in aquatic macrophytic plants.

The ecosystem in these coastal water bodies is controlled by the interaction of natural and anthropogenic factors. Ongoing natural processes, including barrier erosion, channel siltation, land subsidence and rising sea level, as well as the prevailing sedimentary processes, have induced substantial changes in the lagoons' environment (Ahmed et al. 2000). Human activities, such as inlet engineering works, reclamation of water areas, wastewater discharge, salt paning, fishing, hunting of water birds and recreation, aggravated by the lack of effective and consolidated management practices, continue to have a major impact on the lagoons' ecological balance

(Ahmed et al. 2009). A coastal highway has been built along the Mediterranean coast, from El Salum in the west to Rafah in the east. The road is parallel to the shoreline, from which it maintains a minimum distance of about 1 km (Mohsen 1992), and crosses lagoon openings by bridges that permit navigation of fishing boats as well as the water exchanges with the sea. This highway has improved both access to and control of the Nile delta coastal zone and communications between adjacent communities, residential centers, agriculture and industrial areas. However, it has also stimulated the growth of urban and economic activities around the lagoons, causing even more pressure on their ecosystems.

19.2.1 Key Features of the Egyptian Coastal Lagoons

Lake Maryut, a completely enclosed body of water with depths ranging from 0.6–2.7 m, is separated from the Mediterranean Sea by the narrow isthmus on which the city of Alexandria was built. It covered approximately 200 km^2 at the beginning of the twentieth century, but has now reduced to just 70 km^2. The lagoon has a strategic importance at the local and regional level. It plays an important role in the water balance of the western Nile Delta, regulating drainage to the sea and preventing flooding of the coastal zone. Its shores are home to local fisheries (one of the major activities in the area, with 2,000 fishing boats, owned by about 5,000 fishermen, sustaining a community of about 25,000 individuals) and saltworks. The fish catch is about 4,000 t annually, declining steadily due to the deterioration of water quality and the draining of vast areas for land reclamation (GAFRD 2008). In addition, due to the scarcity of land for new development in Alexandria, the lagoon and its surroundings are viewed as a prime area for urban expansion (Ahmed and Elaa 2003). A vast expanse of reclaimed land borders the lagoon, harboring various residential, commercial, industrial and recreational activities. The spreading of settlements, often uncontrolled, and the combined discharges of the activities above are the main sources for Lake Maryut's environmental problems, related to shrinking of the water surface, deterioration of water quality, overgrowth of aquatic vegetation, and declining biological diversity.

Lake Edku is located about 40 Km east of Alexandria and 18 Km west of the Nile River Rosetta branch. This shallow lagoon, with depths varying from 0.1 to 1.4 m, maximum depths being in the central and eastern parts, is connected to the adjacent Abu Qir Bay through the opening of Boughaz El Maadia, a short shallow channel, 300 m in length and 2.5–6 m deep. After the closure of its connection to the Nile system, and until the beginning of the twentieth century, the lagoon had acquired predominantly marine characteristics. Its size has decreased substantially since 1964, to about 60 km^2, due to the reclamation of a large area on the eastern side. Today, the lagoon receives large quantities of drainage water, from about 1,260 km^2 of farmland in the Beheirah province. Two main drains, named El-Khayry and Barsik, supply agricultural, industrial and domestic wastewaters, as well as the drainage of more than 300 fish farms. Most of the time, the freshwater inflow, moving from both west and south to the north, creates a seaward flow through the lagoon, the mean surface of which is about 16 cm higher than sea level. Hence, during the period of high discharge, there is an outflow of fresh water from the lagoon, while seawater influx occurs only during the high tide. The marine water influence is limited to the areas near the Boghaz E1 Maadia (Okbah and El-Gohary 2002).

Lake Burullus, located east of the Nile River Rosetta branch, is the second largest of the coastal lagoons in northern Egypt, and covers an area of more than 400 km^2. It is about 54 km long and has a maximum width of 12 km. Water depth ranges from 0.1–2.4 m. Its longer axis lies parallel to the coastline and is separated from the Mediterranean Sea by a sand barrier. The other lagoon margins are irregular and bordered by wetlands considered to be a site of international importance for migratory birds. The lagoon is divided into several sub-basins by natural and artificial barriers, which includes numerous mud and/or sand islands. These are important paleo-geographic indicators of relict deltaic features, such as beach ridges, dunes and riverbanks of former distributaries (Ahmed et al. 2001). Several water plants, reeds in particular, spread all over the lagoon, affecting water dynamics and exchange, but playing an important role in the preservation of wildlife along the internal coastal marshes. The only connection of the lagoon with the Mediterranean Sea is the Burullus opening. A large amount of freshwater (carrying mainly agriculture waste, but relatively free from industrial waste) enters the lagoon from the south, through several drains. This causes a rise of the water level above that of the sea, with consequent net outflow from the lagoon and dilution of incoming seawater. When the outflow towards the sea weakens, sediments accumulate at the outlet, which must be periodically cleared of deposits.

Lake Manzala, located east of the Nile River Damietta branch, is the largest of the coastal lagoons in northern Egypt (although its size decreased from $> 1{,}400\,km^2$ to $< 700\,km^2$, in the last three decades). This shallow water body, with depth between 0.7 and 2.0 m, is about 50 km long and has a maximum width of 30 km. It accounts for the most productive fishing grounds in Egypt, having contributed nearly 50 % of the total country yield in the 1970s and 35 % in the 1980s, and is rich in natural resources, ranging from wildlife refuge areas to table salt production sites. The sand barrier that separates the lagoon from the sea suffers from erosion. As a result of this, the coastal road between Damietta and Port Said had been replaced by a new road farther inland connecting a number of islands inside the lagoon. The El Gamil channel is the only opening allowing the exchange of water with the sea. Two jetties have been constructed to protect this channel from siltation and longshore sediment migration. The lagoon is further divided into sub-basins by natural and artificial barriers. The western and southern sections are supplied with drainage water from several sources, and are bordered by extensive marshes. The southeastern Ginka sub-basin is heavily polluted by nutrients and heavy metals from agricultural runoff, municipal sewage and industrial effluents from the Bar El Baqar drain (GEF 1992).

Lake Bardawil is located along the northern coast of Sinai, 45 km east of Port Said and 20 km west of El-Arish. This complex lagoon system—which includes a number of bays, about 50 islets and a few enclosed ponds with water depth of only a few cm—has a total area of almost 600 km^2 and a maximum depth of about 3 m. It extends for about 80 km in the east-west direction, with a long narrow arm of about 50 km to the west and a main water body of about 30 km to the east, ending with Zaranik pond. A long, arc-shaped, 300–1,000 m wide, low sand bar, occasionally covered by water, separates the lagoon from the Mediterranean Sea. An extensive sabkha[1] opens to the south, interrupted by series of sand dunes, with ridges running parallel to the coast (Shaheen 1998). The Tineh Sabkha to the west constitutes the eastern margin of the Nile Delta plain. There is little evidence that the lake is a remnant of the estuary created by the easternmost Pelusiac branch of the Nile, but undoubtedly the Nile supplied the large quantities of sand that formed the bar separating the lagoon depression from the sea (Neev and Friedman 1978). At present, seawater enters the lagoon through two artificial tidal inlets (i.e. Boughaze, in the western and eastern parts of the lagoon, about 1 km long, 250–300 m wide, and 4–7 m deep), kept open by periodic dredging, and the natural inlet of Zaranik, at the eastern end, occasionally closed by silting. Being a concentration basin, with no freshwater inflow, Lake Bardawil is the most saline of the northern Egyptian lagoons. Its ecosystem depends on exchanges with the sea, which regulate salinity and fish recruitment. It is considered to be one of the top fishing areas in Egypt, although its production (1.500–2.500 t/year) accounted for less than 1 % of national fish production in 2005 (GAFRD 2005).

19.2.2 Environmental Problems of the Egyptian Coastal Lagoons

The coast of northern Egypt, like all low-lying deltaic regions, is vulnerable to even minor changes in shoreline processes. Any variation in sea level, due e.g. to climatic deviations, is expected to have a pronounced impact on the Nile delta area and on the lagoon ecosystems (El Fishawi and Fanos 1989; Frihy 1992; Stanley and Warne 1993). At present, the sand barriers separating the coastal lagoons from the sea are undergoing intense erosion, due to the prevailing current and wave effects, and to the absence of a steady sediment supply (resulting from the construction of the Aswan High Dam in 1964). This shoreline retreat causes the lagoon barriers to reduce their function as a natural protection line from the sea (Orlova and Zenkovich 1974; Fanos et al. 1995). However, the coastal lagoons are also experiencing severe environmental problems due to increasing human pressure and interventions. Tampering with coastal morphology, via the construction of artificial inlets associated with long jetties, to connect the lagoons with the Mediterranean Sea, has worsened local erosion on the down drift side of adjacent beaches. Similar impacts are due to dredging of the inlets, to prevent or remedy siltation taking place as a result of sand transport in both longshore and cross-shore direction (Fanos et al. 1995).

In addition to natural processes, anthropogenic alterations have caused deep physical and ecological transformations in the lagoons. The rapid and uncontrolled urban expansion in the whole region has changed shoreline stability and altered water quality. Large parts of the deltaic lagoons have been drained for land reclamation

[1] The word "sabkha" (a transliteration of the corresponding Arabic word) indicates a supratidal coastal plain, normally covered by a salt crust.

purposes. Land reclamation rates range from less than 1 km^2/year, in Lake Bardawil, to 2–3 km^2/year, in Lake Edku and Lake Burullus, and up to 7–8 km^2/year, in certain parts of Lake Manzala (Flower et al. 2009). Untreated domestic, industrial and agricultural wastes are dumped into the Nile delta drainage system, which in turn discharges locally into the lagoons (Oczkowski et al. 2011). The degree of pollution varies from high in the eastern part of Lake Maryut and in the southern eastern part of Lake Manzala, to minor in Lake Burullus (Saad et al. 1981; Siegel et al. 1994). The high concentrations of nutrients in organic waste, together with the accumulation of pesticides and heavy metals in the water and sediments, have already altered the lagoons' ecosystem profoundly. The increase of nutrients load, in particular, has lead to overgrowth of macrophytes, plants that have adapted to aquatic environments in either freshwater or saltwater, and to widespread eutrophication.

19.3 RS and GIS Applications

Satellite data from several different sources have been used, by a number of studies to be reviewed here, for repeated observations of the coastal lagoons of northern Egypt. These include data originated by the Landsat Multi Spectral Scanner (MSS), Thematic Mapper^{TM} and Enhanced Thematic Mapper Plus (ETM+); the Système Pour l'Observation de la Terre (SPOT); and EgyptSat −1. The Landsat sensors—i.e. the MSS, TM and ETM+ (see e.g. Markham et al. 2004 and reference therein)—have been the workhorse of land-oriented RS for 4 decades, offering observations in a wide visible and infrared spectral range, with a relatively high spatial resolution (from about 80 m to 30 m and 15 m, in the visible and panchromatic bands, for the successive versions of the instruments) that allows precision mapping in space, and with a temporal resolution (16 days) that allows systematic monitoring over time, when the local climate is good enough to ensure repeated cloud-free conditions. The series of SPOT sensors—i.e. the High-Resolution Visible (HRV), on SPOT-2/3, High-Resolution Visible-Infrared (HRV-IR), on SPOT-4, and the High Resolution Geometry (HRG), on SPOT-5 (see e.g. Clark et al. 2010, and references therein)—has also privileged visible and near-infrared observations, in both panchromatic and spectral modes, with high spatial resolution (20 m, and 10 m in panchromatic mode, originally, and then 10 m, and 5 m in panchromatic mode) and specific tasking capabilities. Finally, EgyptSat-1 was the first in a series of orbital sensors planned in Egypt, jointly built by Egypt's National Authority for Remote Sensing and Space Sciences together with the Yuzhnoye Design Bureau of Ukraine (see e.g. Elshehaby and Taha 2010, and references therein). The sensor provided data in the visible (green and red bands) and near infrared, with 7.8 m spatial resolution and a 50 days revisit cycle.An additional band was also available in the mid-infrared, but at a coarser spatial resolution (∼40 m).

In general, the processing of RS data includes image registration and referencing to a common coordinate system, nearest-neighbor resampling and compositing (when required), as well as image classification, based on a supervised or unsupervised approach (see e.g. Ahmed et al. 2009). When the observations address primarily structural parameters (distances, areas, morphological changes), typical of land features, an accurate spatial registration of the images used is essential for the application of any change-detection method. This demands the use of geometric rectification algorithms, which register the images to each other and/or to a standard map projection. The change-detection algorithms that generated the results shown here are based upon an unsupervised classification technique (Olmanson et al. 2001). The basic premise in multi-spectral classification is that the objects being observed manifest sufficiently different reflectance properties (commonly indicated as spectral signatures) in different regions of the electromagnetic spectrum. Based on these spectral signatures, natural and artificial surface features can be discriminated and a new output image is created, identifying a specific number of classes or categories.

The same approach is taken when the observations address parameters related to water surfaces, by performing first a land-water classification (as described e.g. in Olmanson et al. 2002). Since water has spectral characteristics different from those of land features, water pixels can be easily placed into one or a few distinct classes (notwithstanding the possible classification errors generated by bottom reflection or floating vegetation). Then, proper masking of terrestrial features allows creating a water-only image, in which a second classification can be performed on the water pixels alone. Average brightness values from this second classification can be graphed to show the spectral signature of each class. The various signatures, along with the location where the pixels occur, can be used to differentiate classes containing clear, turbid, or shallow water. Based on this information, classes are redefined so that vegetation, bottom and other land effects can be identified, when selecting water pixels within the lagoon area. In the case of the images shown here, steps were also taken to identify the type of aquatic vegetation present in the lagoons. The first step was to categorize the lagoon surface in terms of emerging and submerged vegetation features. The second consisted of two main procedures for aquatic plant discrimination: separation of image features into discrete units and recognition of the pixels in each unit. This was accomplished by performing yet another unsupervised classification (see Olmanson et al. 2008 for details). Vegetation classes were spectrally analyzed to select 7 different plant species, later identified and named according to field survey information. The amount and percentage of each plant type was also identified.

The results of the above classifications have been interpreted with the aid of supporting data from topographic maps, aerial photography, local knowledge, and the spectral and/or temporal properties of the classes themselves, obtained by virtue of field spectroscopy campaigns. In addition, several geospatial ancillary data sets (e.g., digital terrain models, geological maps, land use maps, urban plans, housing census, population distribution, native vegetation species, agricultural production) were also used to guide data analysis, collect ground truth and validate the results obtained. Further, *in situ* sampling at fixed stations, covering the entire coastal lagoons area, allowed obtaining complementary data on lake water, sediments and vegetation. The final database was completed with information on the nature and location of anthropic activities in the region, urban/suburban runoff from all drainage areas, diffuse pollution sources and runoff volume based on rainfall rates and canal discharges, as well as erosion and sedimentation sites. This information was finally assembled in

GIS form, to determine the spatial and temporal patterns that characterize the coastal lagoons region (Ahmed et al. 2009).

Finally, a regional assessment of water properties in the near-coastal area adjacent to the Egyptian coastal lagoons was performed using a time series of Sea-viewing Wide Field-of-view Sensor (SeaWiFS) data (McClain et al. 2004), covering 10 full annual cycles, from September 1997 to August 2007. In this case, the analysis focused on water quality parameters such as CHL, in order to highlight the possible impact of exchanges between enclosed and near-coastal water bodies. Climatological and monthly composite images were obtained directly from web-based services[2], which provide a suite of data products that includes SeaWiFS imagery, fully processed to correct top-of-the-atmosphere radiances from atmospheric noise; to derive normalized water-leaving radiances; and to compute from these a score of derived parameters (including CHL). In the web services' version used in the present case, the parameter CHL was computed using the OC4 empirical algorithm for SeaWiFS data (O'Reilly et al. 2000). The compositing was obtained by re-mapping on a common equal-area grid and then averaging with a grid cell, or "bin", ranging in size from 2×2 to $9 \times 9\,\text{km}^2$, for various time intervals. For each space/time bin, mean values (weighted by the number of valid pixels used in the binning process) are coupled to data quality statistics (Campbell et al. 1995). The resulting data products should be considered with caution, owing to the impact of various colored water constituents, or to direct bottom reflection, which may significantly alter the CHL derived estimates, particularly in shallow coastal areas. The presence of optically active materials (other than phytoplankton and related pigments) with partially overlapping spectral signatures—i.e. dissolved organic matter or suspended inorganic particles—can prevent the computation of reliable CHL absolute values (Sathyendranath et al. 2000). However, when these limitations are accounted for, the analysis of such data can provide interesting, if qualitative, information on recurrent or anomalous features, and related environmental boundary conditions.

19.4 Observed Features of Lagoon and Coastal Waters

19.4.1 Lake Maryut

A series of Landsat MSS/TM, SPOT HRV and EgyptSat-1 data, collected over the past four decades, was assembled to monitor costal margins and water surface properties of Lake Maryut. The time series, provided by the different sensors in 1972, 1985, 1999 and 2007 (see samples in Fig. 19.2), allowed the identification of changes in the lagoon shape and extension.

[2] i.e. the Goddard Earth Sciences (GES) and Data Information Services Center (DISC) Interactive Online Visualization ANd aNalysis Infrastructure, (GIOVANNI), available at the address http://disc.sci.gsfc.nasa.gov/giovanni/overview/, by the National Aeronautic and Space Administration (NASA). This web services portal allows for the extraction of fully processed data, from multiple sensors, and the use of various data analysis and visualization options.

Fig. 19.2 Evolution of the main Lake Maryut sub-basins (i.e. Main Basin, Fisheries Basin, Northwest Basin, Southwest Basin), as seen in MSS (1972), TM (1985), Spot (1999), and EgyptSat-1 (2007) data. Water-covered areas appear the *darkest* in these arbitrarily colour-coded images

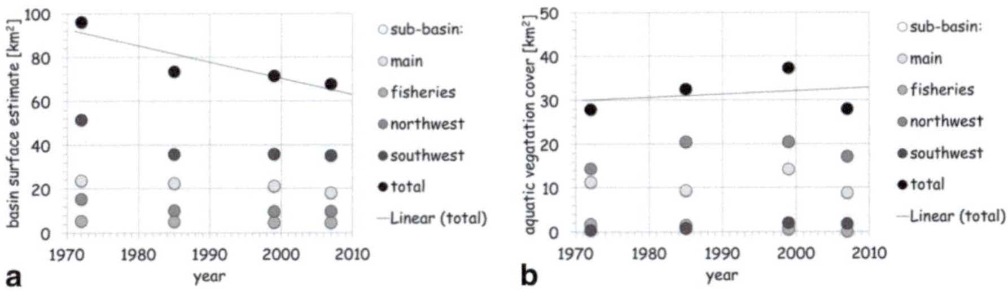

Fig. 19.3 Changes in **a** basin surface estimate and **b** aquatic vegetation cover, for Lake Maryut and its principal sub-basins (i.e. Main Basin, Fisheries Basin, Northwest Basin, Southwest Basin), from MSS (1972), TM (1985), Spot (1999), and EgyptSat-1 (2007) data. The *solid lines* represent the linear regression of **a** total area estimates and **b** total aquatic vegetation cover

The results indicate that the main sub-basins of Lake Maryut (i.e. Main Basin, Fisheries Basin, Northwest Basin and Southwest Basin; see Fig. 19.2, left panel) underwent multiple stages of draining, possibly finalized to the development of urban settlements and various economic activities. The evolution of environmental conditions in the main sub-basins, over the last four decades, can be quantified using GIS techniques, as summarized in Fig. 19.3.

The plots of basin surface estimates (Fig. 19.3a) and of its aquatic vegetation cover (Fig. 19.3b) over time point to a substantial decrease in the size of the water body, in particular due to losses in the Southwest Basin, and to a steady increase, apparently reversed in the latest assessment, of the aquatic macrophytes. The growth of urban areas, around and inside the lagoon, is likely to have reduced the water surface and augmented the amount of sewage disposal (therefore increasing the total nutrient load off the inflow), in parts of the sub-basins considered.

19.4.2 Lake Edku

Data collected by SPOT HRG (2003 and 2009) show that Lake Edku is undergoing substantial alterations of its boundaries (Fig. 19.4, left panel). The same imagery shows that large quantities of aquatic macrophytes grow in the lagoon, covering about

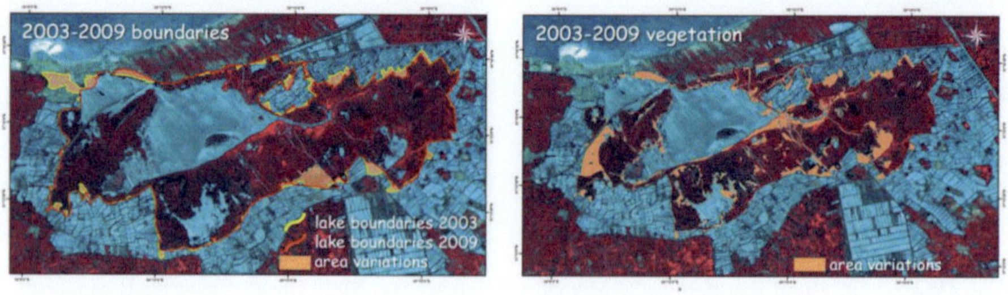

Fig. 19.4 Evolution of Lake Edku, as see by SPOT HRG (2003–2009): lagoon boundaries and surface variations (*left panel*); vegetated area variations (*right panel*). Background classification: water areas, including fish farms (*blue*); vegetated areas, including aquatic macrophytes (*red*)

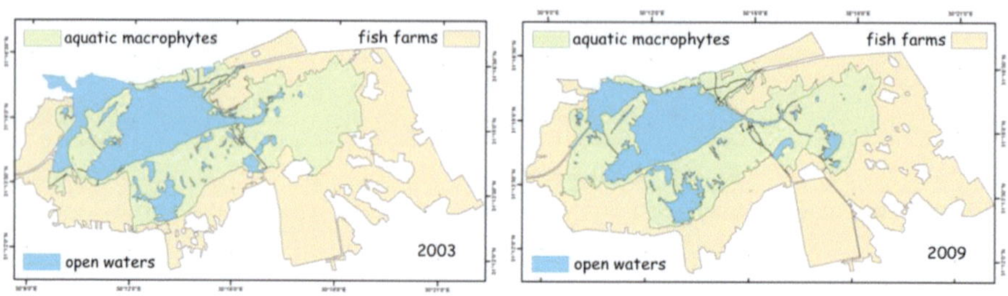

Fig. 19.5 Lake Edku, 2003 (*left panel*) and 2009 (*right panel*) quantitative area assessments of open waters, vegetation cover and aquaculture sites

50 % of its area (Fig. 19.4, right panel). The pattern of vegetation growth reflects that of the interaction between lagoon freshwater and saltwater intrusions through Boughaz El-Maadia. Quantitative estimates of the Lake Edku's total area (Fig. 19.5), from the same data set, indicate that the water surface went from about 62.12 km^2 in 2003–61.06 km^2 in 2009. As common for the coastal lagoons of the Nile river delta, this difference points once again to the expansion of various anthropic activities at the expense of the lagoon environment.Floating and submerged vegetation covers an area that went from 38.82 km^2 in 2003–40.04 km^2 in 2009 (Fig. 19.5).

Aquatic macrophytes are an important component of the food cycle in the lagoon habitat, contributing to the substantial fish production of Lake Edku (up to 19,000 tons/year). However, their presence limits open water areas and hinders fishing boat movements in the lagoon. Aquaculture sites in this region have increased from 59.17 km^2 in 2003–65.98 km^2 in 2009 (Fig. 19.5), suggesting that at least part of the area lost by the lake proper was replaced by fish farms.

19.4.3 Lake Burullus

A series of Landsat TM images was used to assess the main features of both water and land surroundings of Lake Burullus, and to determine the changes that have taken place in the lagoon system, from 1984–1997 and 2000. Figure 19.6 shows

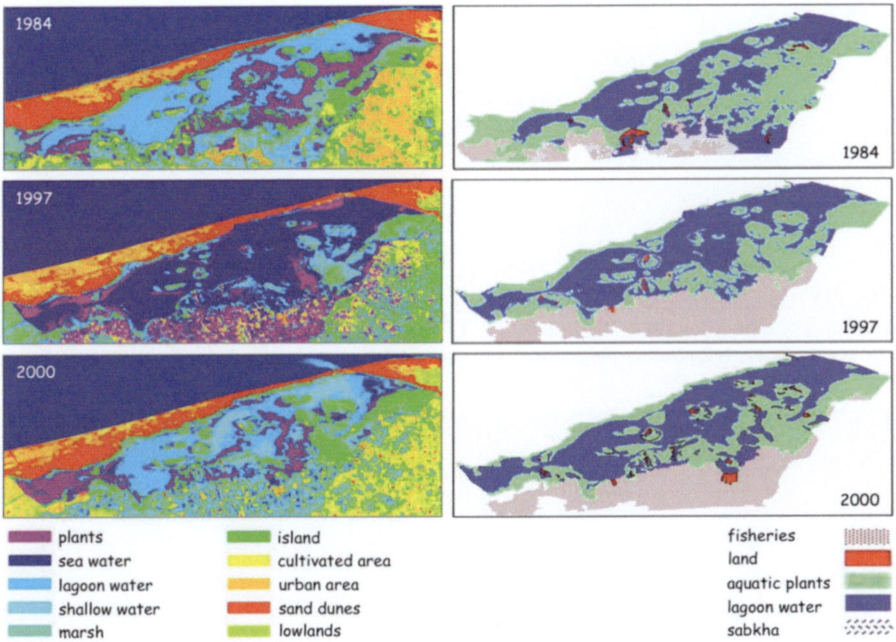

Fig. 19.6 Image classification (*left panels*) and GIS mapping (*right panels*) of Lake Burullus and surrounding area, based on 1984, 1997 and 2000 Landsat TM data

the classified images used (colour coded to represent seawater and lagoon waters, shallow water areas, marshes, islands, sand dunes, vegetated areas, cultivated areas, urban areas; left panels), as well as the GIS rendition of different surface cover/use (i.e. open waters, fisheries, aquatic plants, sabkha and land; right panels).

The data illustrate the spatial complexity of lagoon margins around Lake Burullus, and in particular the extent of vegetated marshes and wetlands important for birdlife conservation (Fig. 19.6, right panels). The lagoon is bordered to the south and east by aquaculture/agriculture areas and by recent land reclamation sites (Fig. 19.6, left panels), which have substantially decreased the size of the water surface. The inclusion in the time series of more recent Landsat ETM+ data, collected in 2003, 2006 and 2008 (not shown here), allowed to follow the evolution of Lake Burullus over the last 4 decades. The changes in the lagoon area are shown in Fig. 19.7.

The plots of basin surface estimates (Fig. 19.7a) and, in particular, of percent decrease with respect to the original estimate (note the polynomial 2nd order fit in Fig. 19.7b) suggest an escalation of area losses in the last decade. These results indicate that the lagoon has undergone dramatic deterioration, in recent times, possibly owing to diverse and intensive anthropogenic factors linked to the exponential growth of local population. In addition, ongoing natural factors, which include barrier erosion, inlet siltation and rising sea level, have also induced substantial changes in the lagoon environment (Dumont and El-Shabrawy 2007).

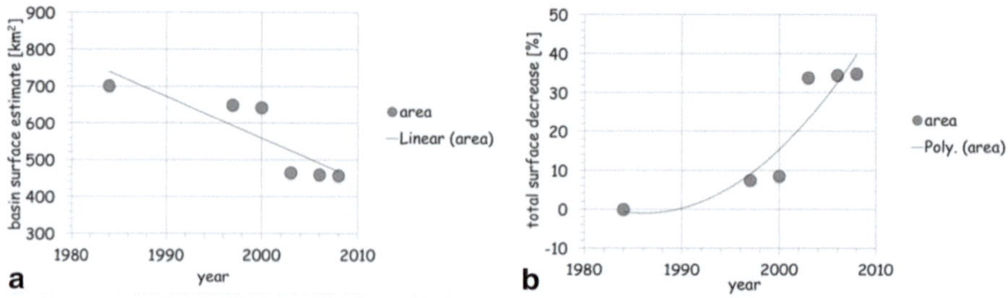

Fig. 19.7 Lake Burullus, **a** *basin surface* estimates and **b** percent decrease in total surface, from Landsat TM data from 1984, 1997, 2000 and ETM+ data from 2003, 2006, 2008. *Solid lines*: **a** linear regression, **b** 2nd order polynomial regression

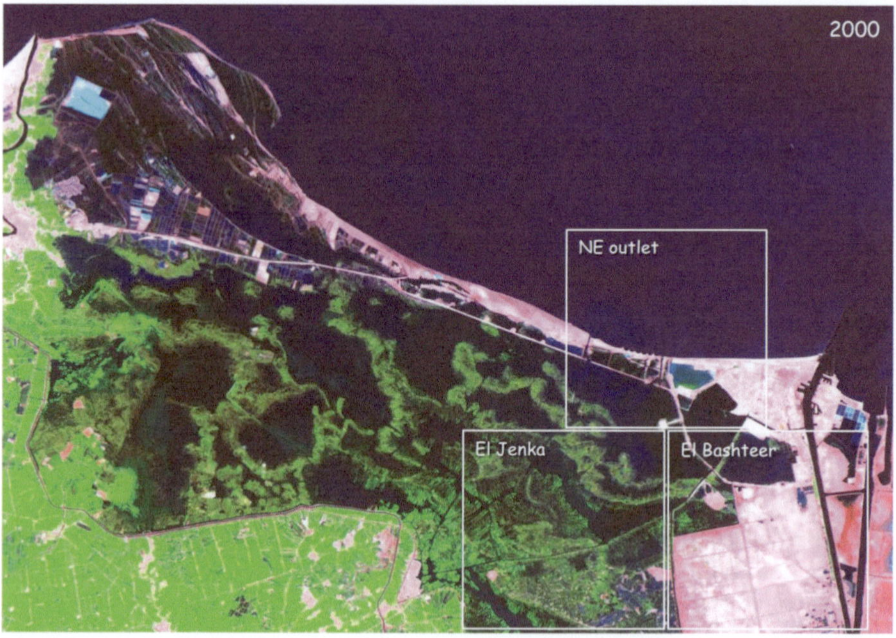

Fig. 19.8 Lake Manzala, classified Landsat ETM+ image, dated 05/02/2003, showing water surfaces (*blue hues*), vegetated zones, both inside and outside the lagoon (*green hues*), as well as agricultural sites and urban settlements (*red hues*). The most recent variations in the areas labeled El Bashteer, El Jenka and North East (NE) outlet are highlighted in Fig. 19.10

19.4.4 Lake Manzala

A multi-date set of Landsat MSS, TM and ETM + images, collected in 1973, 1978, 1983, 1987, 1990, 2000 and 2003, was used to detect the main trends of Lake Manzala (see sample ETM+ image, dated 05/02/2000, in Fig. 19.8). The aim was to evaluate the extent of anthropic activities, land reclamation in particular, as well as

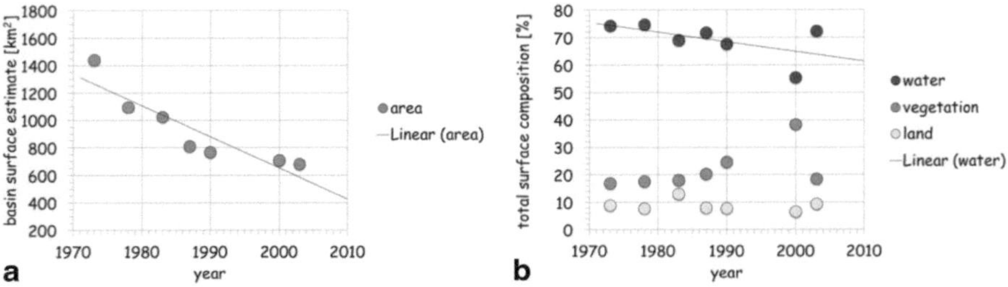

Fig. 19.9 Lake Manzala, **a** cumulative percent decrease of total lagoon area and **b** percent changes of the main environmental components within the lagoon area, i.e. water surface, floating/submerged vegetation and exposed land, in the period 1973–2003, from Landsat MSS, TM and ETM+ data. The solid lines represent a linear regression of **a** total area estimates and **b** water area estimates

their impact on the lagoon ecosystem. Unsupervised classification techniques were used to separate the lagoon area and estimate the temporal variations of its size. The resulting GIS analysis indicates that the total area of Lake Manzala has decreased by more than 50 % from 1973 until 2003 (Fig. 19.9a). The progression of this area loss suggests that about 40 % of land reclamation took place in the late 1970s and in the early 1980s, while only an additional 10 % has taken place since the 1990s.

Further analysis of the image time series allowed to identify the proportion of the lagoon's major environmental components, i.e. water, floating/ submerged vegetation and exposed land, within its boundaries (Fig. 19.9b). The relative percentages of the total area classified as water, vegetation or land appear to be remarkably constant (i.e. approximately 70 % water, 20 % land, 10 % vegetation), although the year 2000 estimate indicates a (temporary) decrease of the water surface, corresponding to a increase of the vegetated surface.

Several hotspots in the Lake Manzala region appear to have been affected by major anthropogenic changes. The region of El-Bashteer is one of the fast developing hotspots in the southeastern part of the basin. TM images of this region, collected in 1987 and 2000, were analyzed to assess the area lost from the lagoon because of land reclamation for agricultural purposes (Fig. 19.10, left panels). The comparison indicates for this area a loss of 100.73 km^2 of lagoon surface in 13 years, i.e. a land reclamation rate of about 7.75 km^2 per year. The region of El-Jenka is another hotspot that underwent different sorts of land filling for the (illegal) set up of small fish farms (named Hosha) destined to fish fry harvesting (Fig. 19.10, center panels). The comparison of the 1987 and 2000 TM images indicates for this area a loss of about about 8.45 km^2 of lagoon surface in 13 years.

The outflow from Lake Manzala into the Mediterranean Sea may also cause problems of water quality in the coastal area. The analysis of TM images, taken in both closed and open conditions of the northeastern lagoon outlet (Fig. 19.10, right panels), suggests that a considerable amount of dissolved and/or suspended matter can be discharged into the sea. These high concentrations may extend offshore and then be advected eastward by the prevailing current. Such a combination of bio-geo-chemical and dynamical features may result in severe changes of coastal water quality as far as the northern Sinai shoreline (Rasmussen et al. 2009).

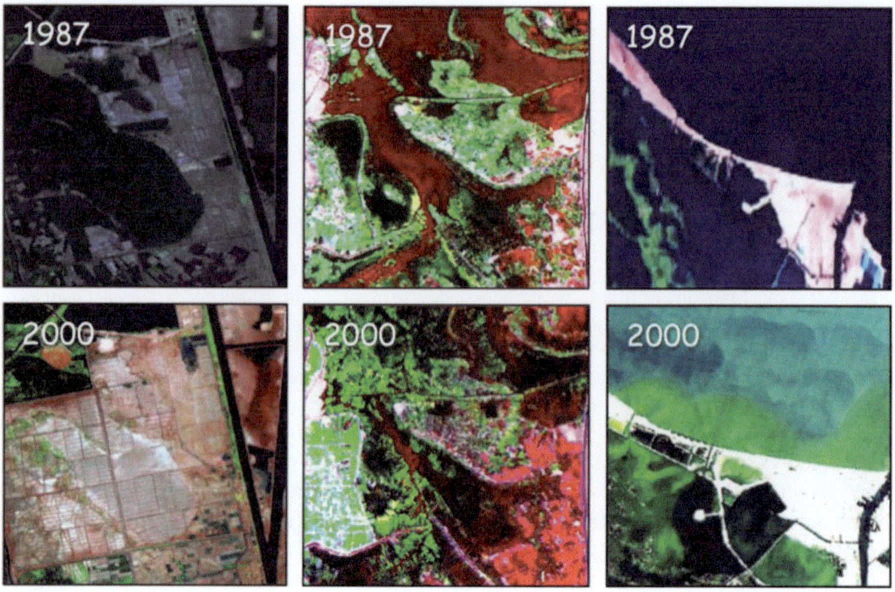

Fig. 19.10 Lake Manzala, evidence of anthropogenic impact at El-Bashteer (*left panels*), agricultural development (see *brown* and *green* tones, with regular *square* pattern, *lower left panel*); at El-Jenka (*center panels*), fish farming (*green* tones, *lower central panel*); and at the lagoon north-eastern outlet (*right panels*), suspended matter (*green* tones, lower right panel); from 1987 to 2000 Landsat TM data

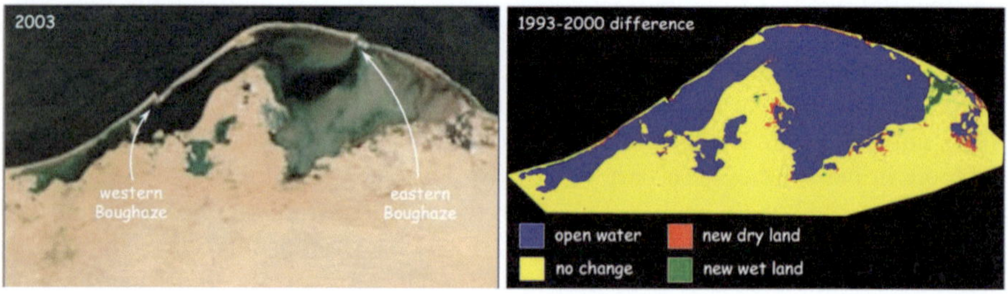

Fig. 19.11 Lake Bardawil: sample MODIS image dated 6/02/2003 (*left panel*); evolution observed in Landsat TM (1993, 2000) data (*right panel*)

19.4.5 Lake Bardawil

A series of Landsat TM images, collected in 1993, 1996 and 2000, was used to monitor the characteristics of both water and land surfaces around Lake Bardawil (see sample image from the Moderate Resolution Imaging Spectrometer, MODIS, in Fig. 19.11, left panel). The 1996 image was used as geo-reference for the accurate registration of other satellite data, while the 1993 and 2000 images were used to detect changes in lagoon size, as well as the evolution of the lagoon inlets and sea

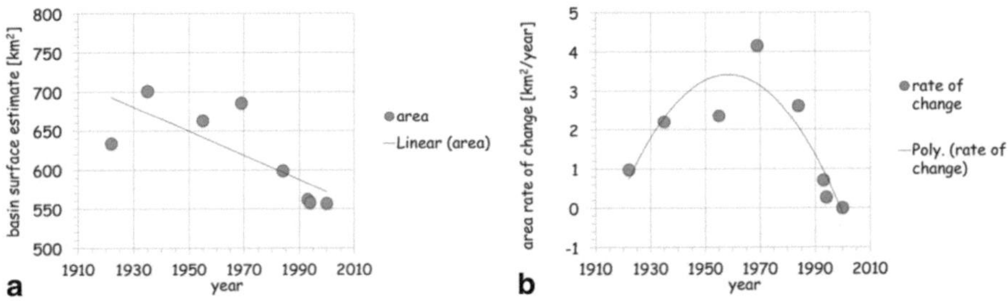

Fig. 19.12 Lake Bardawil evolution in the last century: **a** total surface estimates derived by means of historical data from the scientific literature (1922, 1935, 1955, 1969, 1984 and 1994) and from the 1993–2000 Landsat TM data; **b** rate of change of the surface area over various periods, from the available historical data, with respect to the 2000 conditions. The solid line represents **a** linear regression of area estimates, **b** 2nd order polynomial regression of rate of change

barrier. The 1993–2000 GIS inter-comparison, shown in Fig. 19.11 (right panel) highlights the lagoon border areas prone to accretion, erosion or inundation. The sea barrier appears to be eroded in the 2000 image, with respect to the 1993 image, providing a first indication that the lagoon might be decreasing in size.

The comparison suggests that from 1993–2000 the size of Lake Bardawil was decreasing at an overall rate of 0.71 km^2/year. This figure compares quite favorably with estimates obtained for the past century (i.e. 1922, 1935, 1955, 1969, 1984 and 1994) by means of historical data from the scientific literature (El Malky et al. 2003; Ahmed et al. 2000) and summarized in the plot of Fig. 19.12, together with the 1993 and 2000 TM estimates.

The long-term trend of the surface rate of change, shown in Fig. 19.12, indicates that the lagoon area variability has been substantial, somewhat reversing from the first to the second half of the last century, with a steep increase in the second half (i.e. a 0.98 km^2/year rate of change, from 1922 to 2000, going up to 4.15 km^2/year, in the period from 1969 to 2000), when compared with the more recent figure derived from 1993 to 2000 RS data alone.

Finally, the images in Fig. 19.13, also derived from the 1993–2000 Landsat TM data, show the shoreline changes for both the western and eastern Boughaze of Lake Bardawil. Dredging of these artificial inlets has been carried out repeatedly since 1927, to maintain water level, salinity and fish recruitment at sustainable levels, with the last campaign taking place in 1987. The present data indicate conditions similar to those appearing in recent reports (Co.R.I. 1988, 1990, 1992), concerning a series of field programs carried out in 1985–1986, and then again in 1990, 1991 and 1992, in order to investigate the recurrent siltation problem.

19.4.6 Near-coastal area of the Southeastern Mediterranean Sea

The near-coastal area off the Nile River delta, adjacent to the Egyptian coastal lagoons, is rather different from the remainder of the southeastern Mediterranean Sea. The climatological map in Fig. 19.14 shows the average CHL values (here intended

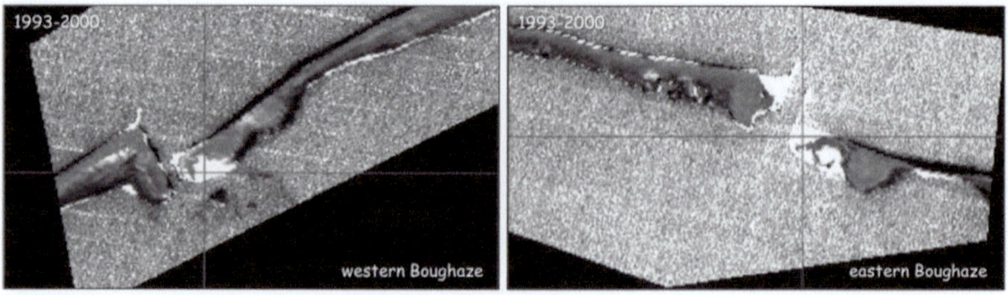

Fig. 19.13 Changes in western and eastern Boughaze of Lake Bardawil, as observed in 1993–2000 Landsat TM data. The "change detection" algorithm shows erosion in black and major accretion areas (centered over the crossing red lines) in white

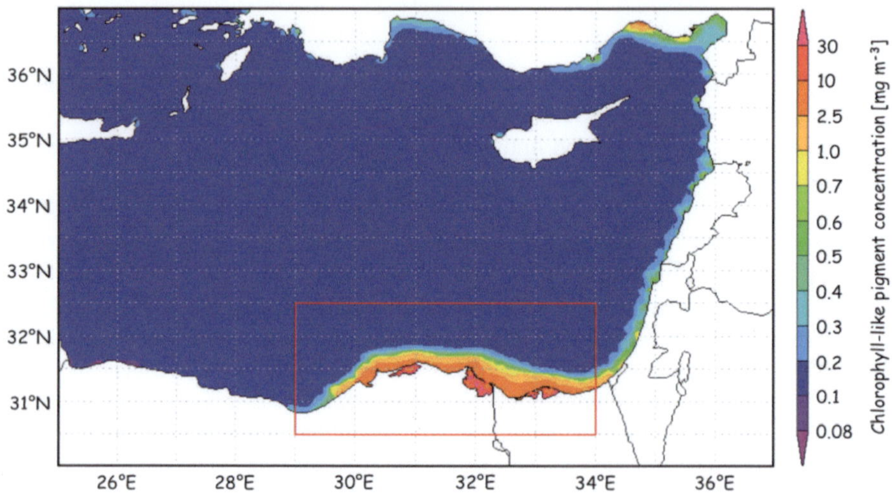

Fig. 19.14 SeaWiFS-derived climatological map of CHL [mg m^{-3}], computed over the September 1997 to August 2007 period, in the eastern Mediterranean Sea. The red box represents the impact area of the Nile river plume

as an index of ecosystem dynamics) derived for a 10-year period (September 1997 to August 2007) taken from the SeaWiFS dataset (Barale 2009). The oligotrophic basin interior exhibits very low values, with concentrations ranging from 0.1 to 0.2 mg m^{-3}. Higher (up to 1 order of magnitude) values appear in coastal zones, possibly due to the interaction with continental margins. The highest values pertain to the coast bordering the lagoons of the delta (although only Lake Bardawil, Lake Manzala, Lake Borullus, and possibly Lake Edku, can be seen in the image, while Lake Mariut falls below the spatial resolution of the image). Of course, the high CHL absolute values seen in these areas should be taken with care, due to the presence of other water constituents (suspended sediments and dissolved organics, in particular) in concentrations high enough to alter significantly the output of SeaWiFS processing algorithms. Yet, the plume described by such values provides clues about the area impacted by fluvial runoff and exchanges with the lagoons.

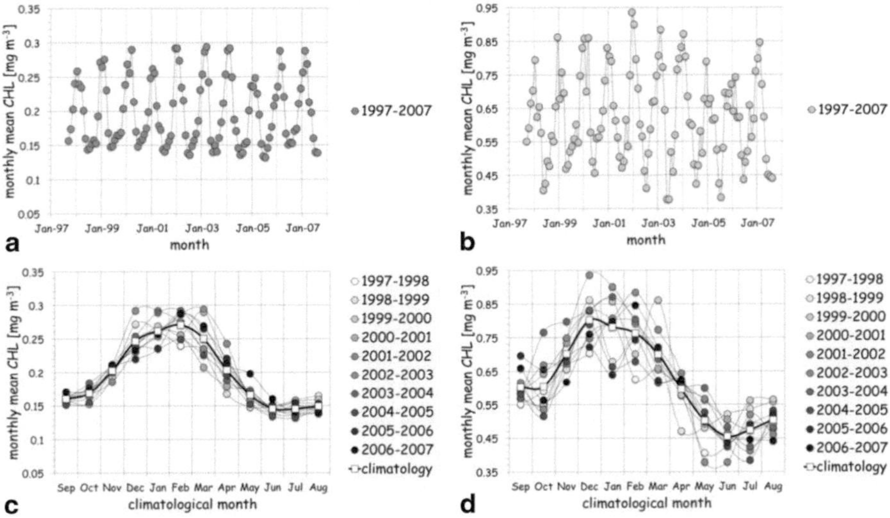

Fig. 19.15 SeaWiFS-derived monthly mean CHL [mg m^{-3}], computed as a single average value for the entire eastern Mediterranean basin (*left panels*) and the Nile river plume impact area (*right panels*): **a**, **b** multi-annual trend; **c**, **d** seasonal and climatological trend. Time coverage is September 1997–August 2007. Note the different CHL scale used for the Nile-only cases (**b**, **d**)

Variations in coastal interactions—due e.g. to changing amounts of coastal runoff, or to the changing load of water constituents in the runoff—can give rise to episodes of enhanced ecosystem response. Even small anomalies can be easily detected against the permanently oligotrophic background of this region, and can be monitored in a *quasi*-continuous manner by RS (Barale et al. 2005). Along the southern shoreline, the Nile River plume has by far the largest impact on the coastal ecosystem, providing a constant source of suspended or dissolved materials (including nutrients) of continental origin. Such impact was much reduced after the completion of the Aswan High Dam, in 1964, only to increase again in recent years, due to the high nutrient load carried by continental runoff in the agricultural area of the delta (Nixon 2004).

The multi-annual trends of the SeaWiFS-derived (monthly) mean chlorophyll-like pigments, computed as an average value for the entire eastern Mediterranean basin (30–37°N, 25–37°E, i.e. the whole marine area shown in Fig. 19.2) and for the Nile River plume impact area (30.5–32.5°N, 29–34°E, i.e. the marine area in the red box of Fig. 19.2) are shown in Fig. 19.15. At the basin level, maxima are systematically reached in winter (Fig. 19.15a). However, subtle, but significant, differences appear for the Nile river plume impact area (Fig. 19.15b). When the 10 seasonal cycles, from September to August, are plotted together with the climatological mean, the maxima reached at the basin scale spread evenly from December to March (Fig. 19.15c). But the Nile river outflow seems to foster an earlier peak, appearing first in December and then to a lesser extent in January, February and March (Fig. 19.15d). Considering CHL as an index of plankton biomass, even in the coastal area where the index absolute value cannot be accurate, such a distinction in blooming time suggests the existence of site-specific ecosystem dynamics in each area, as if different forcing functions induced different reactions at the sub-regional scale.

19.5 Conclusions

The integrated management of enclosed and coastal waters is a dynamical process, in which boundary conditions, problems to be solved and required approaches are changing continuously. Systematic monitoring, or at least a series of periodic assessments, is indispensable in order to make executive decisions at the proper time. RS (and GIS) techniques can help collect (and merge) data required by decision-makers. In recent years, such techniques have become an increasingly important element of many programs about the coastal area of northeastern Egypt, aiming both at addressing environmental change and supporting sustainable development. As summarized in the present review, information derived from (optical) RS indicate that the major near-coastal water bodies considered here have suffered intensive anthropogenic damages, possibly related to draining and land reclamation, uncontrolled urbanization and related waste dumping, aquaculture and even recreational uses. As the local population was expanding exponentially, most of these activities led to severe environmental problems, aggravating natural degradation processes and threatening the very existence of these fragile near-coastal ecosystems.

Knowledge of the basic ecosystem indicators provided by satellite observations—albeit with the limitations inherent in this technique, often due to the less-than-ideal spatial, temporal and spectral resolutions of the available orbital systems—is considered an important contribution to a sound decision-making process. The test cases above originate from a wide body of existing scientific and technical literature, based on RS (and GIS) applications and related *in situ* validation campaigns, which illustrate the ongoing efforts to gain an ecological understanding of the coastal and lagoon environments. Combining the potential of RS techniques with other indispensable information sources, the initial steps shown here will lead to the establishment of an integrated management program, ensuring both the protection of natural systems and the sustainability of human activities.

Acknowledgements The authors wish to express their gratitude to the many Colleagues from the Marine Sciences Department, National Authority Remote Sensing and Space Sciences (NARSSS), Cairo, Egypt, and from the Institute for Environment and Sustainability (IES), Joint Research Centre (JRC) of the European Commission (EC), Ispra, Italy, who contributed to the research activities summarized in the present paper. Work at IES, JRC EC, was carried out as part of the Framework Programme 7 (FP7) JRC Action no. 21203 "Protection and Conservation of the European Seas" (PROCAS). Some of the data analyses and visualizations used in this paper were produced with the online data services featured by the Goddard Earth Sciences (GES) and Data Information Services Center (DISC) Interactive Online Visualization ANd aNalysis Infrastructure (GIOVANNI), of the National Aeronautic and Space Administration (NASA). We acknowledge the Mission Scientists and Principal Investigators who provided the data used in this research effort. Thanks are also due to two anonymous reviewers of this paper, who contibuted substantially to the improvement of its content.

References

Ahmed MH (2000) Long-term changes along the Nile delta coast: Rosetta Promontory. A case study. Egyptian J Remote Sens Space Sci 3:125–134

Ahmed MH (2003) Erosion and accretion patterns along the coastal zone of Northern Sinai, Egypt. J Sedimentol Soc Egypt 11:281–290

Ahmed HM, Frihy OE, Yehia MA (2000) Environmental management of the Mediterranean coastal lagoons of Egypt. Ann Geol Surv Egypt XXIII:491–508

Ahmed MH, Abdel-Moati MAR, El-Bayomi G, Tawfik M, El-Kafrawy S (2001) Using geo-information and remote sensing data for environmental assessment of Burullus lagoon, Egypt. B Natil Institute of Oceanogra Fisheries ARE 27:241–263

Ahmed MH, Elaa AAA (2003) Study of Molluscan shells and their enclosed bottom sediments in Manzala lagoon, Nile delta, Egypt. B Natil Institute of Oceanogra Fisheries ARE 29:423–446

Ahmed MH, Noha D, Fahmy MA (2006) Eutrophication assessment of lake Manzala Egypt using GIS techniques. J Hydroinf 8(2):101–109

Ahmed MH, Donia NS (2007) Spatial investigation of water quality of lake Manzala using GIS techniques. Egyptian J Remote Sens Space Sci 10:63–86

Ahmed MH, El Leithy BM, Thompson JR, Flower RJ, Ramdani M, Ayache F, Hassan SM (2009) Application of remote sensing to site characterisation and environmental change analysis of North African coastal lagoons. Hydrobiologia 622:147–171

Baith K, Lindsay R, Fu G, McClain CR (2001) SeaDAS, a data analysis system for ocean-colour satellite sensors. EOS Transactions AGU 82:202

Barale V (2005) Satellite observations as indicators of the health of the Mediterranean Sea In: Saliot A (ed) The handbook of environmental chemistry, Vol 5. Water Pollution, Part K, "The Mediterranean Sea", Springer-Verlag, Berlin, pp 387–408

Barale V (2009) Assessing novel near-coastal ecosystem dynamics with satellite observations: case studies in the Eastern Mediterranean Sea. In: Ozhan E (ed) Proceedings of the Second International Conference/Workshop on State-of-the-Art of ICM in the Mediterranean and Black Sea, MED & BLACK SEA ICM 08, MEDCOAST, Middle East Technical University, 225–237

Barale V, Folving S (1996) Remote sensing of coastal interactions in the Mediterranean region. Ocean Coast Manag 30(2–3):217–233

Barale V, Gade M (eds) (2008) Remote sensing of the European seas. Springer, Hidelberg, p 513

Barale V, Larkin DJ (1998) Optical remote sensing of coastal plumes and run-off in the Mediterranean region. Journal of Coastal Conservation 4:51–68

Barale V, Jaquet JM, Allenbach K, Ndiaye M, Hoareau A (2005) Bio-optical environmental assessments of marginal seas. Progress Report 2. EUR 21899 EN, European Commission, Ispra, Italy, 63 pp

Barale V, Jaquet JM, Ndiaye M (2008) Algal blooming patterns and trends in the Mediterranean sea as derived from the SeaWiFS data set (1998–2003). Remote sensing of environment, special issue on: earth observations for marine and coastal biodiversity and. Ecosystems 112(8):3300–3313

Bernasconi MP, Stanley DJ (1994) Molluscan biofacies and their environmental implications, Nile delta lagoons, Egypt. J Coastal Res 10(2):440–465

Bosc E, Bricaud A, Antoine D (2004) Seasonal and interannual variability in algal biomass and primary production in the Mediterranean Sea, as derived from 4 years of SeaWiFS observations. Global Biogeochemical Cycles, 18, GB1005, doi:10.1029/2003GB002034

Brezonik PL, Menken K, Bauer ME (2005) Landsat-based remote sensing of lake water quality characteristics, including chlorophyll and colored dissolved organic matter (CDOM). Lake and Reserv Manage 21(4):373–382

Campbell JW, Blaisdell JM, Darzi M (1995) Level-3 SeaWiFS Data Products: Spatial and Temporal Binning Algorithms. In: Hooker SB, Firestone ER, Acker JG (eds) NASA technical memorandum 104566, SeaWiFS Technical Report Series 32, NASA Goddard Space Flight Center, Greenbelt, p 23

Carlson RE (1977) A trophic state index for lakes. Limnol Oceangra 22(2):361–369

Clark B, Suomalainen J, Pellikka P (2010) A comparison of methods for the retrieval of surface reflectance factor from multitemporal SPOT HRV, HRVIR, and HRG multispectral satellite imagery. Can J Remote Sens 36(4):397–411

Co.R.I. (1988) Project of Development of El Bardawil Fish Resources. Final Report, Vol 1, 2 and 3. Coastal Research Institute, Alexandria, Egypt.

Co.R.I. (1990) Updating the Coastal Processes Data of El-Bardawil Outlets, First Report, *and* Protective Works for El-Bardawil Outlets Alignment & Conceptual Design, Report no 2. Coastal Research Institute, Alexandria, Egypt.

Co.R.I. (1992) Alignment of El-Bardawil Outlets and Dredging Works, Report no 4. Coastal Research Institute, Alexandria, Egypt.

Cozar A, GarcIa CM, Galvez JA, Loiselle SA, Bracchini L, Cognetta A (2005) Remote sensing imagery analysis of the lacustrine system of Ibera wetland (Argentina). Ecol Modell 186:29–41

Dewider Kh, Khedr AA (2001) Water quality assessment with simultaneous Landsat-5 TM at Manzala lagoon, Egypt. Hydrobiologia 4577:49–58

Dewidar Kh, Khedr AA (2005) Remote sensing of water quality for Burullus Lake, Egypt. Geocarto International 20(3):43–49

Dumont HJ, El-Shabrawy GM (2007) Lake Borullus of the Nile delta: a short history and an uncertain future. Ambio 36(8):677–682

El Fishawi NM, Fanos AM (1989) Prediction of sea level rise by 2100, Nile delta coast. International Union for Quaternary Research (INQUA). Commission on Quat Shoreline Newsl 11:43–47

El Malky MG, Ahmed MH, El Marghany MA (2003) Heavy metals in water and sediments of Bardawil lagoon, Northern Coast of Sinai, Egypt. J Environ Sci 7(3):753–785

Elshehaby AR, Taha LGE-d (2010) Assessment of cartographic potential of EgyptSat-1 satellite image (case study in flat areas). Applied Geomatis 3:159–169

Fanos MA, Naffaa GM, Ali MM, Gewilli MY (1995) Coastal processes along Burullus headland Nile Delta, Egypt. In: Ozhan E (ed) MEDCOAST 95, Proceedings of the Second International Conference on the Mediterranean Coastal Environment, METU, Ankara, 2:1047–1061

Flower RJ, Appleby PG, Thompson JR, Ahmed MH, Ramdani M, Chouba L, Rose N, Rochester R, Ayache F, Kraiem MM, Elkhiati N, El Kafrawy S, Yang H, Rasmussen EK (2009) Sediment distribution and accumulation in lagoons of the Southern Mediterranean region (the MELMARINA Project) with special reference to environmental change and aquatic ecosystems. Hydrobiologia 622:85–112

Frihy OE (1992) Beach response to sea level rise along the Nile delta coast of Egypt. In: Woodworth PL (ed) Sea level changes: determination and effects, Geophysical Monograph 69, American Geophysical Union, IUGG 2:81–85

GAFRD (2005) General Authority for Fishery Resources Development (GAFRD) Year-Book of fishery statistics in Egypt (1989–2004). Cairo, Egypt.

GAFRD (2008) General Authority for Fisheries Resources Development (GAFRD) Annual Statistics Year Book. Cairo, Egypt.

GEF (1992) Global Environment Facility Chairman's Report, Part Two. UNDP/ UNEP/World Bank, Washington D.C., 182–199

Graetz R (1990) Remote sensing of terrestrial ecosystem structure: an ecologist's pragmatic view. In: Hobbs R, Mooney H (eds) Remote sensing of biosphere functioning, Springer-Verlag, New York, 5–30

Hess LL, Melack JM, Novo EMLM, Barbosa CCF, Mary GM (2003) Dual-season mapping of wetland inundation and vegetation for the central Amazon basin. Remote Sensing of. Environment 87:404–428

Kloiber SM, Brezonik PL, Bauer ME (2002) Application of Landsat imagery to regional-scale assessments of lake clarity. Water Res 36:4330–4340

Lathrop RG (1992) Landsat Thematic Mapper monitoring of turbid inland water quality. Photogramm Eng Remote Sens 58:465–470

Levy Y (1974) Sedimentary reflection of depositional environment in the Bardawil lagoon, northern Sinai. J Sediment Res 44(1):219–227

Markham BL, Storey JC, Williams DL, Irons JR (2004) Landsat sensor performance: history and current status. IEEE Trans Geosci Remote Sens 42(12):2691–2694
McClain CR, Feldman GC, Hooker SB (2004) An overview of the SeaWiFS project and strategies for producing a climate research quality global ocean bio-optical time series. Deep-Sea Res II 51:5–42
Mohsen MA (1992). Descriptive analysis of the north Delta Highway project. UNDP Workshop/Round-Table discussion on the impact of the Nile Delta coastal road and its effectiveness as defense measure against the expected sea-level rise, 10 pp
Neev D, Friedman GM (1978) Late Holocene tectonic activity along the margins of Sinai subplate. Science 202:427–429
Nixon SW (2003) Replacing the Nile: are anthropogenic nutrients providing the fertility once brought to the Mediterranean by a great river? Ambio 32:30–39
Nixon SW (2004) The Artificial Nile. American Scientist 92:158–165
Oczkowski AJ, Flower RJ, Thompson JR, Ayache F, Ahmed MH, Ramdani M, Turner S (2011) Evidence of North Africa's green revolution preserved in sedimentary organic matter deposited in three coastal lagoons. Ecol Appl 21:1708–1717
Okbah MA, El-Gohary SE (2002) Physical and chemical characteristics of Lake Edku Water, Egypt. Mediterranean Marine Science 3/2:27–39
Olmanson LG, Kloiber SM, Bauer ME Brezonik PL (2001) Image processing protocol for regional assessment of lake water quality. Water Resources Center Technical Report 14, University of Minnesota, St. Paul, MN 19 pp
Olmanson LG, Bauer ME, Brezonik PL, Kloiber SM (2002) Lake water quality image processing protocol. Water Resources Center Technical Report 146, University of Minnesota, St. Paul, MN 24 pp
Olmanson LG, Bauer ME, Brezonik PL (2008) A 20-year Landsat Water Clarity Census of Minnesot's 10,000 Lakes. Remote Sens Environ, Special Issue on Monitoring Freshwater and Estuarine Systems 112(11):4086–4097
Orlava G, Zenkovich V (1974) Erosion of the shores of the Nile Delta. Geoforum 18:68–72
Ozemi SL, Bauer ME (2004) Satellite remote sensing of wetlands. Wetl Ecol Manag 10:381–402
Rasmussen EK, Petersen OS, Thompson JR, Flower RJ, Ahmed MH (2009) Hydrodynamic-ecological model analyses of the water quality of Lake Manzala (Nile Delta, Northern Egypt). Hydrobiologia 622:195–220
Saad MA, Ezaat AA, El Rayis OA, Hafez H (1981) Occurrence and distribution of chemical pollutants in lake Mariut, Egypt, II Heavy metals. Water Air Soil Poll 16:401–407
Sathyendranath S et al (ed) (2000) Remote Sensing of Ocean Colour in Coastal, and other Optically-Complex, Waters. IOCCG Report, vol 3. IOCCG Project Office, Dartmouth, Nova Scotia, p 140
Sawaya K, Olmanson L, Heinert N, Brezonik P, Bauer M (2003) Extending satellite remote sensing to local scales: land and water resource monitoring using high-resolution imagery. Remote Sens Environ 88:144–156
Shaheen SE (1998) Geo-environmental studies on El-Bardawil lagoon and its surroundings, North Sinai, Egypt. PhD Thesis, Mansoura University, Mansoura, Egypt, pp 165
Siam E, Ghobrial M (1999) Pollution influence on bacterial abundance and chlorophyll—a concentration case study at Idku lagoon, Egypt. J Arab Acad Sci Technol Maritime Transport 24:18–26
Siegel FR, Slaboda ML, Stanley DJ (1994) Metal pollution loading, Manzala lake, Nile delta, Egypt: implications for aquaculture. Environ Geol 23:89–98
Stanley DJ, Warne AG (1993) Nile delta: recent geological evolution and human impact. Science 260:628–634
Volpe G, Santoleri R, Vellucci V, Ribera d'AlcalàM, Marullo S, D'Ortenzio F (2007) The colour of the Mediterranean Sea: global versus regional bio-optical algorithms evaluation and implication for satellite chlorophyll estimates. Remote Sens Environ 107:625–638
Zilioli E (2001) Lake water monitoring in Europe by means of remote sensing. Sci Total Environ 268:1–2

Chapter 20
Remote Sensing of African Lakes: A Review

Thomas J. Ballatore, Shane R. Bradt, Lydia Olaka, Andrés Cózar and Steven A. Loiselle

Abstract The optical complexity and small size of inland waters make the application of remote sensing more challenging than for the open ocean. However, in Africa, where *in situ* monitoring of important water bodies is financially, institutionally and spatially constrained, there is strong demand for remote sensing to fill the critical information gap. Here we review a wide range of applications of both passive and active remote sensing to African lakes. The applications fall into five main categories: (1) visible, NIR, thermal and microwave sensing of lake area; (2) altimetric and gravimetric sensing of lake level; (3) thermal sensing of lake surface temperature; (4) visible, NIR and microwave sensing of macrophytes; and (5) optical sensing of trophic conditions including chlorophyll-a and euphotic depth. Sensors used include Landsat MSS, TM and ETM+, MERIS, MODIS, SeaWiFS, AVHRR, Meteosat, TOPEX/Poseidon, Jason-1, OSTM/Jason-2, ERS-1, ERS-2, Envisat, GFO, ICESat, ALOS-PALSAR and GRACE. The majority of studies have been applied to the "great" lakes such as Chad, Malawi, Tanganyika and Victoria; however, there is a growing body of literature on smaller lakes. We examine the possibilities that remote sensing offers to monitoring and management of African lakes as well as the potential limitations of the technology using Lake Victoria as an illustrative case.

T. J. Ballatore (✉) · S. R. Bradt · L. Olaka
Lake Basin Action Network, Moriyama, Japan
e-mail: ballatore@lakebasin.org

S. R. Bradt
University of New Hampshire, Durham, NH, USA

L. Olaka
University of Nairobi, Nairobi, Kenya

A. Cózar
Universidad de Cádiz, Puerto Real, Spain

S. A. Loiselle
Università degli Studi di Siena, Siena, Italy

V. Barale, M. Gade (eds.), *Remote Sensing of the African Seas*,
DOI 10.1007/978-94-017-8008-7_20,

20.1 Introduction

The idea of remote sensing—the art and science of obtaining information about an object without being in physical contact with the object (Jensen 2007)—is in many ways the holy grail of lake management. The potential to look wide and often at conditions in a given lake has spurred much interest, resulting in a burgeoning body of research and applications including recent work by Gower et al. (2005), Ruiz-Verdú et al. (2008), Chipman et al. (2009), Gitelson et al. (2011), Bradt (2012) and Matthews et al. (2012).

The potential offered by remote sensing is particularly attractive in Africa where administrative capacity to carry out field-based monitoring is limited. Lake managers in Africa are confronted with a task of monitoring lakes that would challenge even the most well-funded monitoring programs.

Here, we review the applications of remote sensing to African lakes as described in the scientific literature. The applications are surprisingly broad: spatially, they include 29 lakes that cover all corners of the continent (see Fig. 20.1); thematically, they range from passive to active sensors looking at both water quantity and quality.

Overall, we find it convenient to divide the applications into two broad categories: those that look at the *amount* of water in a given lake and those that look at the *properties* of that water. The former category include studies that measure *lake area* through (1) visible and near-infrared (NIR) sensing of lake extent/shoreline, (2) thermal sensing of inundated areas, and (3) microwave sensing of water under cloud cover. They also include studies that examine *lake level* through (4) altimetric and (5) gravimetric observations. The latter category includes work on strictly physical aspects such as (6) thermal sensing for lake surface temperature as well as biological aspects such as (7) visible and NIR sensing of macrophytes and (8) microwave sensing of macrophytes and perhaps the most widely known application—(9) optical sensing of trophic conditions through measurement of chlorophyll-a, euphotic depth and a range of related indicators.

Few of the studies reviewed here fall solely into one of the above nine categories; most papers make use of a combination of techniques. For example, a paper using altimetric observations to estimate changes in lake level might also rely on visible and NIR imagery to confirm lake area (and hence, level) changes.

More importantly, it must be noted that remote sensing does not completely obviate the need for field observations and traditional sampling programs. In fact, most of the methods described here still require, or greatly benefit from, calibration with actual measurements taken in the field concurrent with the remotely-sensed images. Nevertheless, as the literature demonstrates, and as we explore in the discussion, remote sensing can indeed provide a wealth of information to greatly aid the work of those concerned with lake management in local communities, national governments and international organizations throughout Africa.

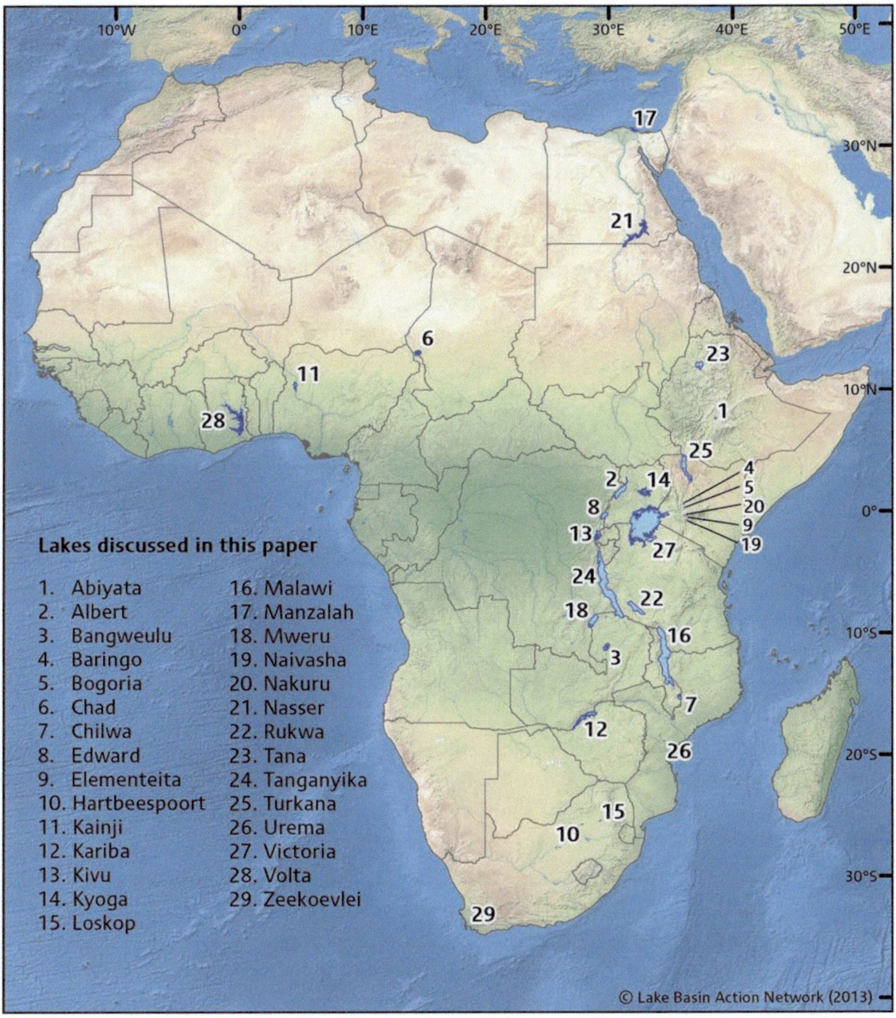

Fig. 20.1 Locator map of the lakes covered in the remote sensing literature reviewed here

20.2 Lake Area

20.2.1 Visible and NIR Applications

The one of the earliest and simplest applications of remote sensing in lake management is the use of visible and near-infrared (NIR) images to identify changes in a lake's extent. The wide field captured in a single image by remote sensors is especially useful for lakes in remote regions where field surveys of the shoreline are impractical because of distance or inaccessibility. The technique has been used mainly for relatively shallow lakes in endorheic drainage basins where changes in water balance are strongly reflected in changes in lake extent. The strong contrast between light reflected from land vs. water, along with the complete and fine spatial coverage offered by sensors such as the Multi Spectral Scanner (MSS), Thematic Mapper (TM) and Enhanced Thematic Mapper Plus (ETM+), onboard the Landsat satellite series, mean that this technique is not limited to large lakes and can be applied, in theory at least, to lakes of 1 km^2 or less.

Lake Chad has drawn much attention in the literature given its dramatic decrease in extent. Mohler et al. (1989) digitized photographs taken with onboard, handheld cameras during manned space flights (Gemini, Skylab, and the Space Shuttle) dating from 1966 to 1985 to document an approximate 90 % decrease in lake extent. Two main cameras were used: a 70 mm Hasselblad and a 230–460 mm Large Format Camera (LFC) and, except for one black-and-white photo and one color-infrared photo (both taken with the LFC), the images were captured on natural color film. After digitization, spatial resolution was reported as 11 m for LFC and 29 m for the Hasselblad. Water and land were classified with a maximum likelihood algorithm. It is important to note that the 1966 Gemini image, which showed an extent of 22,772 km^2, pre-dated the era of non-manned satellite observing systems which began with Landsat MSS in 1972, a time when the lake had already undergone most of the areal change. Wald (1990), building on the work of Mohler et al. (1989), used images from Meteosat in the visible and NIR range (0.4–1.1 μm) to document the lake's further shrinkage to 500 km^2 in April 1989.

Schneider et al. (1985) used the NIR Band 7 (0.8–1.1 μm) from Landsat MSS (spatial resolution $\sim$ 80 m) to document the lake area changes from 1972 to 1982. They also used the visible and NIR Channels 1 and 2 (0.55–0.68 and 0.725–1.1 μm) of the Advanced Very High Resolution Radiometer (AVHRR), onboard the NOAA-7 satellite, (spatial resolution 1.1 km^2) to develop a simple Normalized Difference Vegetation Index (NDVI) to detail open water and vegetation along transects of key portions of the lake from November 1981 to November 1982. Despite its lower spatial resolution, the higher temporal resolution of AVHRR (daily vs. 18 days for MSS) was an advantage in mapping the rapidly changing water area and vegetation.

Birkett (2000) used the AVHRR's NIR Channel 2 (0.725–1.1 μm) to map lake extent (inundated area) over a 4-yr period starting in 1995. A simple histogram technique of the Digital Number (DN) values allow classification of inundated areas with an overall relative error of $\sim$ 5 % (calculated from results of different algorithms given the lack of field data) with lower errors during peak flows and higher errors during other times, due in part to the presence of pools smaller than the AVHRR spatial resolution. The paper also made extensive use of satellite altimetry data.

Turada (2008) used a selection of visible and NIR bands from the Medium Resolution Imaging Spectrometer (MERIS), onboard the European Space Agency (ESA) satellite Envisat, (300 m spatial resolution) to look at inundation and land cover change in the Lake Chad basin through Spectral Mixture Analysis (endmembers: soil, vegetation and water). The MERIS-based classification was validated with data from the Advanced Spaceborne Thermal Emission and Reflection Radiometer (ASTER), a high resolution imaging instrument that is flying on the US National Aeronautics and Space Administration (NASA) orbital platform named Terra, (15 m spatial resolution; visible and NIR) as well as from IKONOS (in the 4 m spatial resolution version; visible and NIR) and showed that despite the relatively low spatial resolution, MERIS was able to detect changes in lake extent (kappa coefficients, as *per* Cohen 1960, ranging from 0.92 to 0.95).

Other lakes have also drawn attention. Booth and Oldfield (1988) used Landsat MSS NIR Band 7 to estimate declining Lake Kariba area in five images from 1981 to 1985 coincident with a drought in the region. Classification was done by manually judging an appropriate water cut off DN value in each image.

Ouma and Tateishi (2006) used visible and NIR bands from Landsat TM and ETM+ images to develop a new Water Index (WI) based on Tasseled Cap Wetness (TCW) and Normalized Difference Water Index (NDWI) to rapidly map shorelines in five Rift Valley lakes in Kenya (Baringo, Bogoria, Elementaita, Naivasha and Nakuru) for images in 1986 and 2001. Ground-based observations showed that the WI had an accuracy of 98.4 % outperforming TCW by $\sim$22 % and NWDI by $\sim$43 %.

Harris (1994) used the NIR band of AVHRR (Channel 2) to detect shoreline changes at Lake Abiyata (Ethiopia) with a local isoluminance contour (LIC) method. The technique also made use of a re-sampled NIR band from Landsat MSS (band 7). Also for Lake Abiyata, Legesse and Ayenew (2006) used Landsat MSS, TM, ETM+ and SPOT data, as well as data from the Moderate Resolution Imaging Spectrometer (MODIS)—which exists in two versions, carried onboard the NASA orbital platforms Terra (MODIS Terra) and Aqua (MODIS Aqua)—to visually display changes in the lake area from 1978 to 2005. The band used is not specified but the very dark appearance of the lake relative to bright land in the figures indicates it was an NIR band.

Finally, Böhme et al. (2006) examined intra- and inter-annual areal changes in Lake Urema (Mozambique) from 1979 to 2000 by a Maximum Likelihood Classification augmented with NDVI, Principal Component Analysis (PCA) and Tasseled-Cap-Transformation (TCT) based on visible and NIR bands from MSS, TM, ETM+, and ASTER.

20.2.2 Thermal Sensing Applications

The spectral differences between lake water and surrounding land in the visible and especially NIR bands are the basis for shoreline delineation discussed in the previous section. It is also possible to use the thermal part of the spectrum to do similar work. The main benefit of using thermal imagery is the ability of the technique to work at night, due to the fact that the electromagnetic radiation being received by the sensor is emitted from the target and not reflected from a source (sun).

Schneider et al. (1985), in addition to the work on visible and NIR bands discussed above, used two of the AVHRR thermal channels (4 and 5; wavelengths 10.3–11.3 and 11.5–12.5 μm, respectively) to differentiate Lake Chad's open water from flood plains, deltas and surrounding land.

Also for Lake Chad, Leblanc et al. (2011) used Meteosat 30-day Tmax data (10.5–12.5 μm) to thermally sense water under vegetative cover, validated by field surveys and AVHRR data over the period of 1986–2001. While they note that the thermal technique was difficult to apply during the rainy season because the high water content of non-lake areas produced a signature similar to that of open water, the high-temporal resolution (30 min) of the geosynchronous Meteosat meant that cloud cover was not a significant handicap.

20.2.3 Microwave Applications

In areas were cloud cover is more persistent, the visible, NIR and thermal techniques are less useful. Rebelo (2010) describes the application of the Phased Array type L-band Synthetic Aperture Radar (PALSAR), orbiting on the Advanced Land Observing Satellite (ALOS) of the Japan Aerospace Exploration Agency's (JAXA), an active microwave sensor, to identify surface changes in Lake Urema and surrounding wetlands from December 2006 to February 2008. The ALOS-PALSAR images ($\sim 12.5\,\text{m}$ spatial resolution) were smoothed with a 5×5 pixel median filter and classified using a PCA method to generate input for a supervised classification.

20.3 Lake Level

20.3.1 Altimetric Sensing (Microwave)

If bathymetry is known, lake level can be inferred from the shoreline changes discussed above; however, this is not always possible. Furthermore, cloud cover can greatly limit the time series data derived from visible, NIR and thermal products. Active altimetric sensing with microwaves was originally developed for ocean topographic mapping but has been successfully applied to lakes. Under certain conditions, lake level can be estimated to within a few cm of actual values. Additionally, cloud cover is not an issue. Given the important role that lake level plays in human use of lakes, this method has seen wide use. Unfortunately, the nature of the narrow active signal sent from a given sensor means that only lakes directly in the sensor path can be covered.

Use of altimetric data has been fostered by the development of online databases that process the raw data from various sensors and provide "finished" products. The main ones referred to in the literature below are (1) ESA's River and Lake website (ESA 2013), which uses data from Envisat, Jason-1 and Jason-2; (2) the Laboratoire d'Etude en Géophysique et Océanographie Spatiale (LEGOS) Hydroweb, a surface monitoring by satellite altimetry website (LEGOS 2013), which uses data from the satellites Envisat, European Remote Sensing 1 and 2 (ERS-1 and ERS-2), Geosat Follow-On (GFO), Jason-1, Jason-2 and TOPEX/Poseidon (Cretaux et al. (2011); and (3) the US Department of Agriculture (USDA) Global Reservoir and Lake Monitor (USDA 2013) which uses data from Envisat, GFO, Jason-1, Jason-2 and TOPEX/Poseidon.

Using data from the first 21 months of the TOPEX/Poseidon mission (September 1992–May 1994), Birkett (1995) estimated lake levels for climate sensitive lakes including the following African lakes: Chad, Kariba, Malawi, Nasser, Tana, Tanganyika, Turkana, Victoria and Volta. Field data were used to estimate the accuracy as $\sim$4 cm rms, thereby demonstrating the usefulness of satellite altimetry for lake level measurement. Merceir et al. (2002) used TOPEX/Poseidon data from 1993 to 1999 to look at the same lakes as Birkett (1995) as well as Kyoga, Mweru and Rukwa. They showed links between lake levels and the atmosphere over the Indian Ocean.

Ricko (2011) used lake level data from the LEGOS (2013) dataset for lakes Bangweulu, Chad, Kainji, Malawi, Mweru, Tana, Tanganyika and Turkana as part of a broader study of the effects of climate on tropical lake levels. Munyaneza et al. (2009) used Envisat and ERS-2 from ESA (2013) for water level monitoring of Lake Kivu. Velpuri et al. (2012) used TOPEX/Poseidon, Jason-1 and Envisat from USDA (2013) for validation of hydrological models for Lake Turkana.

Lake Chad has also drawn much interest. Birkett (2000) used TOPEX/Poseidon along with AVHRR to examine variability of basin inundation. Coe and Birkett (2004) used TOPEX/Poseidon along with ground-based information to estimate river discharge and lake level. In a review paper, Crétaux and Birkett (2006) noted that TOPEX/Poseidon data for Lake Chad were reported to have lower accuracy (tens of cm) compared with other lakes such as the Laurentian Great Lakes (a few cm). Turada (2008) used TOPEX/Poseidon derived levels from USDA (2013) to confirm MERIS-based areal extent changes. Lemoalle et al. (2012) used previous altimetric work as input for a hydrological model to examine management options for Lake Chad.

Finally, altimetric sensing has often played a key role in gravimetric approaches. Swenson and Wahr (2009) used lake level data from USDA (2013) for Lake Victoria, LEGOS (2013) data for lakes Albert, Edward and Kyoga, and data from Ice, Cloud, and land Elevation Satellite (ICESat), part of NASA's Earth Observing System (EOS), of later periods for Kyoga. They used altimetric data to show that excess water input to Lake Victoria since mid-2007 has been used for hydropower generation and not stored for future drought relief. Additionally, Lake Kyoga levels were shown to reflect an increase in discharge from Lake Victoria despite a regional drought. Additionally, Becker et al. (2010) used ESA (2013), LEGOS (2013) and USDA (2013) altimetric data along with data from the Gravity Recovery And Climate Experiment (GRACE), a joint mission of NASA and the German Aerospace Center, and precipitation data to study the recent hydrological behavior of Lakes Malawi, Tanganyika, Turkana and Victoria.

20.3.2 Gravimetric Sensing

The launch of the twin GRACE sensors in 2002 added gravimetry, a non-electromagnetic radiation based form of remote sensing, to the options available. Differences in mass on earth can be detected by examining small changes in the distance between the orbiting GRACE sensors as they pass over a given point. With a repeat time of 15 days, this technique has been applied successfully to determine water level (volume) changes in some of the larger African lakes.

A number of papers have looked at the water balance of Lake Victoria whose dramatic decline in lake water coinciding with the completion of a second dam on the outflowing river in 1999 in Uganda. Awange et al. (2008) used changes in geoidal variation as measured by GRACE to show an annual loss of 1.6 mm/yr in the lake basin from 2002 to 2006. This information, coupled with Tropical Rainfall Measuring Mission (TRMM) and the Challenging Minisatellite Payload (CHAMP) indicated that dam operations were the likely cause in the decline in lake level with respect to precipitation (TRMM) or evaporation (CHAMP). This conclusion was confirmed by Swenson and Wahr (2009) using GRACE, altimetric lake level data remotely-sensed data on precipitation (TRMM) and evaporation, including wind speed from the scatterometer SeaWinds, onboard the QuikSCAT satellite; temperature from the MODIS Aqua; and near-surface temperature and humidity from the Advanced Microwave Sounding Unit (AMSU-A), the Special Sensor Microwave Imager (SSM/I) and the Special Sensor Microwave/ Temperature & Humidity Profile (SSM/T2).

Becker et al. (2010) used GRACE measurements of changes in terrestrial water storage, altimetrically-sensed changes in lake level, and Global Precipitation Climatology Centre (GPCC) data on precipitation to show common modes of variability among the levels of Lakes Malawi, Tanganyika, Turkana and Victoria due to the effect of the Indian Ocean Dipole (IOD) and El Niño-Southern Oscillation from 2003 to 2008.

As part of a large-scale study of hydrological variations in Africa, Boy et al. (2012) compared GRACE-based estimates of water level change with altimetric data for Lakes Malawi, Tanganyika and Victoria. They showed that spherical harmonic solutions, while suitable for Lake Victoria, do not perform as well for lakes with a narrow east-west extent such as Malawi and Tanganyika. For the latter, they recommend localized approaches such as mascons. Work was also done by Boy et al. (2012) and Hinderer et al. (2012) on the Lake Chad basin but the GRACE data were not linked to lake level, undoubtedly due to the very small volume of water contained in Lake Chad.

20.4 Lake Properties: Physical

20.4.1 Thermal Sensing of Surface Temperature

Schneider et al. (1985), in addition to visible, NIR and thermal work to delineate water extent discussed above, also produced a surface water temperature map for Lake Chad and the surrounding area using daytime observations with Channels 4 and 5 of the AVHRR sensor. They applied a two atmospheric window approach from McClain (1981) and were able to demonstrate distinct and large temperature differences among the lake regions and surrounding terrain including inflowing river deltas.

Wooster et al. (2001) used Channels 3, 4 and 5 from nighttime AVHRR for Lake Malawi to estimate lake surface temperatures from May 1993 to September 2004. Using nighttime images and including information from Channel 3, they were able to show that, based on field vertical temperature profiles, the near-surface values were indeed representative of the temperature of the mixed epilimnion.

Chavula et al. (2009a) used MODIS images to infer lake circulation patterns from surface temperature. MODIS Terra daytime images for the period May to November 2006 were downloaded as both Land Surface Temperature (i.e. a data product coded as MOD11a1) and SST. Compared with field observations, SST was a better estimator ($r^2 = 0.75$) of actual lake surface temperature than MOD11a1 ($r^2 = 0.68$). Based on this agreement, they were able to infer the location of an upwelling zone with potential for fisheries.

20.5 Lake Properties: Biological

20.5.1 Visible and NIR Sensing of Macrophytes

Proliferation of invasive aquatic macrophytes has significantly impacted several African lakes. The rapid expansion of *Eichhornia spp.* (water hyacinth) and other plants in Lake Victoria has received much attention. Given the vast spatial dimensions of floating plant biomass and their elevated reflectance in visible and infrared bands, there have been numerous uses of remote sensing to monitor macrophyte extent.

Cavalli et al. (2009a) used MERIS and ASTER to map aquatic weed infestations in Winam Gulf, Lake Victoria, after building a spectral library (350–2500 nm) of nine common weed species. Two ETM+ scenes (1999 and 2001) were used to perform spatial and temporal mapping of the major invasive aquatic plants (water hyacinth, Nile lily, Nile cabbage). The authors then used 3-band (visible and NIR) ASTER images to follow the evolution of weed cover from 2002 to 2006 in selected parts of Winam Gulf. Because of the limited spectral resolution of ASTER compared with ETM+, individual species could not be identified but the clear patterns and changes in weed cover were identified.

Cavalli et al. (2009b) used an 18-band (410–1040 nm) image generated by the Compact High Resolution Imaging Spectrometer (CHRIS), onboard the Project for On-Board Autonomy (PROBA) satellite, (spatial resolution = 17 m) along with a contemporary ETM+ image to identify specific plant species based on the spectral library described in Cavalli et al. (2009a). The much wider swath width of ETM+ allowed mapping of macrophyte distribution over the whole gulf.

Fusilli et al. (2011) used MODIS images to map floating vegetation in Winam Gulf at a minimum half-weekly frequency over a 10-year period of 2000–2009. NDVI was calculated and used to distinguish floating vegetation from water and land. A Landsat TM image with higher spatial resolution was used to help determine the appropriate NDVI cut off value for floating vegetation (> 0.4). Additional information on rainfall, lake height, and suspended solids among others allowed the authors to tie the cause of the abnormal proliferation from October 2006 to January 2008 to an unusually heavy rainy season and the consequent large inflows of nutrients and sediment from surrounding land.

Rebelo (2010) used blue, red and NIR bands from Landsat TM to calculate an Enhanced Vegetation Index (EVI) for Lake Chilwa (Malawi, Mozambique) for May 2006. This was combined with SAR data to achieve a high classification accuracy.

20.5.2 Microwave Sensing of Macrophytes

Rebelo (2010) used Fine Beam Single, Fine Beam Dual, and polarimetric images (all 12.5 m spatial resolution) from the L-band ALOS-PALSAR instrument to classify vegetative cover in and around lakes Chilwa and Urema to an accuracy of 89 and 84 %, respectively, compared with field observations. PALSAR was notably useful in these two cases because of the ability of L-band SAR to penetrate frequent cloud cover and smoke, especially in Mozambique where visible and NIR images were limited during the study period (2006–2008).

20.5.3 Optical Sensing of Trophic Conditions

One of the most challenging areas of remote sensing for lake management is the application of optical (visible wavelength) images for the determination of trophic conditions. The presence of elevated concentrations of Colored Dissolved Organic Matter (CDOM) and Total Suspended Matter (TSM) as well as the difficulty of atmospheric correction over African lakes with a range of elevations makes the process inherently more difficult than typical ocean color approaches. Nevertheless new algorithms and techniques promise to make optical sensing of trophic conditions a growing application.

Bergamino et al. (2007) used SeaWiFS imagery to assess the changes in chlorophyll concentration in Lake Tanganyika during a 7-year period using the OC4.v4 algorithm. Empirical Orthogonal Function (EOF) analysis was used to define regions of the lake with the same temporal dynamics of phytoplankton biomass. Three co-varying regions (north, south and central) were identified which displayed different seasonal and inter-annual dynamics in estimated chlorophyll concentrations.

Sørensen et al. (2008) evaluated the performance of MERIS products processed using the Case2R and the EUL processors, including signal depth (Z90_max), to estimate chlorophyll-a concentrations and TSM in Lake Victoria. Reference data consisted of actual Secchi Disk Depth (SDD) measurements. The Z90_max was found to correlate well with measured SDD when using both the Case2R and EUL processors, but the relationship was different in each case. For the few chlorophyll-a data available, the OC4.v4 algorithm analyzed with the Case2R processor produced chlorophyll estimates in the same order of magnitude as those measured. In the case of the few available TSM values, the MERIS-estimated data were found to be in the same range as those measured in the field.

Koponen et al. (2008) evaluated the performance of the chlorophyll-a MERIS products in Lake Manzalah, Egypt, using the EUL processor both with and without the use of the Improve Contrast between Ocean & Land (ICOL) technique.

While the correlation between estimated chlorophyll and measured chlorophyll was linear, large errors were evident in slope and bias. Based on the observation of the imagery products, it was clear that local water quality variation present in Lake Manzalah exceeded the processor capability. Nearly all pixels were flagged "out of scope" for estimated chlorophyll concentration product, largely due to failed atmospheric correction techniques.

Kneubühler et al. (2007) used MERIS imagery to map chlorophyll-a in Lake Kivu during the dry season over the course of three years. The images were atmospherically corrected using both the Atmospheric & Topographic Correction (ATCOR) model, version 2 (Atmospheric Correction for Flat Terrain, ATCOR2), and Simplified Method for Atmospheric Correction (SMAC) processor. The study tested five band radio algorithms using MERIS bands at 443, 490, 560, 620, 665, 709 and 754 nm, as well as, an algorithm using 443, 560 and 665 nm modified for Landsat bands. Linear regression models were applied to seven MERIS images to investigate spatial and inter-annual variability in the lake, particularly the increased chlorophyll concentrations in bays and coastal areas. The authors note that CDOM variability might have influenced these estimates.

Chavula et al. (2009b) used MODIS imagery products to estimate chlorophyll in Lake Malawi. Lake water was collected at three lake sampling locations during numerous sampling cruises and processed to determine chlorophyll-a concentration ($n = 136$). Level 1 MODIS Terra imagery for the sampling dates was obtained from the NASA Level 1 and Atmospheres Archive and Distribution System (LAADS) system, while Level 2 and Level 3 MODIS Aqua imagery was obtained from the NASA Ocean Color Group. Terra algorithms were found to be of limited use, a situation attributed to the adverse effects of atmospheric effects on reflectance. The most effective Aqua chlorophyll algorithm was $R_{rs}443/R_{rs}551$.

Majozi (2011) used a combination of field-collected spectral attenuation measurements with MERIS products to examine eutrophic depth distribution in Lake Naivasha. Field measurements of downwelling irradiance at 0.1 and 0.6 m were used to calculate attenuation (K_d) for three wavelength ranges (430–600, 600–800 and 800–930 nm). Lake-specific attenuation models were tested against MERIS products [$K_d(490)$, $K_d(620)$] and found to be fairly accurate [rmse = 0.044 m for $K_d(620)$]. The attenuation model for euphotic depth was applied to MERIS images of the lake, revealing the spatial variety and range of light penetration in Lake Naivasha.

Oberholster and Botha (2010) used Landsat 5 imagery in combination with molecular markers to successfully detect cyanobacteria in Lake Hartbeespoort. A red to NIR algorithm was used to determine when cyanobacteria populations were present. Lake sampling to determine the presence and density of cyanobacteria relied on a combination of chlorophyll concentration, PCR analysis and microcystin toxin detection using an Enzyme-Linked Immunosorbent Assay (ELISA).

Horion et al. (2010) used standard and adapted MODIS Aqua Ocean Color products to create regionally calibrated products for chlorophyll-a concentration and attenuation coefficients at 490 nm [$K_d(490)$] for Lake Tanganyika. The examination of known targets using clear sky images revealed the standard atmospheric correction method to be inadequate for Lake Tanganyika. An aerosol correction model designed for coastal regions with 90 % humidity proved the most effective approach. When using this atmospheric correction, the regression between the standard Ocean Color Group algorithm for chlorophyll-a concentration and *in situ* chlorophyll concentration had an offset near 0, a slope near 1, and an rmse around 20 %.

MODIS-derived chlorophyll-a estimations revealed a pattern of higher concentration events compared to field sampling. The high chlorophyll values seemed linked to increases in cyanobacteria abundance in surface waters, and often occurred near the onset of the rainy season. The authors concluded that their MODIS-based estimates of chlorophyll-a provided a good approximation of actual concentrations in the upper water column. Bergamino et al. (2010) relied on the regionally calibrated chlorophyll-a product developed by Horion et al. (2010) to examine the spatio-temporal dynamics of phytoplankton in Lake Tanganyika. The daily phytoplankton photosynthesis per unit area of Lake Tanganyika was modeled and the daily primary production of the whole lake was estimated by weighting the averages according to the relative area of each region.

Based on an EOF analysis, 13 co-varying regions in Lake Tanganyika were identified. These data show that the traditional division of the lake into three lake basins is largely correct, and the observed temporal patterns revealed significant differences in estimated concentrations between the regions. The dry season was found to be more productive than the wet season, with the southern section exhibiting two peaks in primary production and the central and northern sections exhibiting a single, smaller peak. The observations of these trends indicated the usefulness of satellite-based analysis for Lake Tanganyika, providing the possibility to examine long-term and short-term changes in lake dynamics.

Matthews et al. (2010) used MERIS imagery to assess the suitability of a variety of algorithms for use in cyanobacteria-rich waters. Empirical algorithms resulted in high correlation values, but were found to be of limited use to separate signals from co-varying in-water constituents. Semi-analytical algorithms were not successful, due to poor atmospheric correction techniques and training ranges of key optical variables not being appropriate for a water body of this type. A major conclusion of this study was the observation that top of atmosphere satellite data could be used to detect hypereutrophic conditions in inland waters.

Cózar et al. (2012) used water-leaving radiances from the SeaWiFS sensor to estimate chlorophyll-a concentration in Lake Victoria by way of the Maximum Band Ratio (MBR) using the OC4.v4 algorithm. Statistical analysis of the relative optical variability within the lake provided the means for assessing bio-optical, and thus, limnological variability and lake dynamics. In addition, AVHRR imagery was used in this study to calculate lake surface temperature to examine the changes in water column stability. Seasonal increases in lake surface temperature in the North Region coincided with rises in the estimated chlorophyll concentration, while the South

Region experienced increases in estimated chlorophyll two months after the onset of lake surface warming. Differential cooling in the north and south were shown to drive the large-scale convective circulation observed in the lake, a pattern that offers a possible explanation for the observed bio-optical patterns. This study revealed how year-to-year variability in physical conditions may play a major role in phytoplankton dynamics, irrespective of the influence of nutrient dynamics.

Matthews et al. (2012) developed a novel algorithm for detecting chlorophyll-a, cyanobacterial blooms, surface scum and floating vegetation using MERIS imagery of three diverse African lakes (Loskop dam, Hartbeespoort dam, Zeekoevlei Lake) and coastal ocean (southern Benguela). The Maximum Peak Height (MPH) algorithm uses the top-of-atmosphere MERIS spectral data, not the atmospherically corrected data commonly used by bio-optical algorithms targeting inland water bodies.

The MPH is designed to provide accurate chlorophyll estimation in three cases: (1) mixed oligotrophic/mesotrophic low-medium biomass waters, (2) high biomass eutrophic/hypereutrophic waters and (3) extremely high biomass waters with surface scums or floating vegetation. The parameterization of the algorithm for each potential peak provided reasonable chlorophyll estimates across a vast range of trophic states (mean absolute percentage error = 59.9 %, $r^2 = 0.71$), providing a promising new tool for remote sensing of lakes throughout Africa and beyond.

20.6 Discussion

Remote sensing offers the potential to provide a rich body of data for lake management. Whether the concern about a lake relates to area, volume, temperature or trophic state, remote sensing has shown to be appropriate for use in African lakes. In the studies covered here, a large array of sensors (17) have been used to measure features of 29 African lakes, providing technical developments, real application, or both.

While these 29 lakes represent only a fraction of the more than 17,000 lakes in Africa identified in the Shuttle Radar Topography Mission (SRTM) Waterbody Dataset (SWBD) of NASA (NASA 2013), they cover 76 % (186,192 km^2) of the total lake area (245,137 km^2) in Africa. In addition, 13 of the largest 15 lakes in terms of area are included. These larger lakes are those for which remote sensing can be most useful, providing information on a spatial and temporal scale simply not possible with traditional sampling methods. As physical and biological information becomes more available through remote monitoring and analysis, a better understanding of lake conditions and lake dynamics will emerge, paving the way for better management of these important resources. It should also be noted that many applications of remote sensing are used for internal policy documents and unpublished land cover analyses and do not make it into the literature. These practical, informal uses of remote sensing are often not fully appreciated.

20.6.1 An Illustrative Case: Lake Victoria

The case of Lake Victoria (Fig. 20.2) provides a concrete example of the ability of remote sensing techniques to contribute to improved lake management and should serve as an inspiration to African scientists and policy makers.

The lake's importance as well as its large areal extent (which makes direct sampling particularly difficult) mean that Lake Victoria has seen more applications of remote sensing than most African lakes. As discussed above, these applications cover a wide spatial range from enclosed areas of the lake to the drainage basin scale; they also illustrate the wide variety of sensors and techniques that can be brought to bear on policy-relevant questions.

One of issues attracting much concern over the years has been the infestation by water hyacinth in shoreline areas along the Ugandan and Kenyan coasts (Williams et al. 2005). Cavalli et al. (2009a, b) and Fusilli et al. (2011) were able to use a range of sensors (MODIS, MERIS, ASTER and CHRIS) to map aquatic weed infestations in Winam Gulf. The ability of remote sensing to elucidate the evolution of weed cover, including the abnormal proliferation from October 2006 to January 2008 which was attributed to an unusually heavy rainy season and the consequent large inflows of nutrients and sediment from surrounding land, has proven important for policy makers.

Related to the localized water hyacinth infestation is the lake-wide eutrophication which has been driven by nutrient loadings which began increasing as far back as the 1930s (Verschuren et al. 2002, Hecky et al. 2010). Loiselle et al. (2006) used SeaWiFS-derived chlorophyll-a measurements to show the influence of littoral population on the inshore areas of the lake (see Fig. 20.2). Sørensen et al. (2008) showed the utility of MERIS for lake-wide chlorophyll-a measurements. Cózar et al. (2012) used SeaWIFS for chlorophyll-a and AVHRR for surface temperature to show how interannual changes in physical conditions can greatly affect phytoplankton dynamics across the lake. The important implication is that policy makers need to look beyond just phytoplankton biomass as an indicator of eutrophication and expand their consideration of climatic effects.

Finally, the water balance of Lake Victoria has been a contentious policy issue which altimetric and gravimetric data have been able to impartially elucidate. In particular, Swenson and Wahr (2009) used changes in catchment water storage, as measured by GRACE, along with altimetric lake level data to show that increased lake discharge (and not drought alone) contributed to the decline in lake level observed in the early 2000s.

Overall, these examples demonstrate how remote sensing can be applied to policy-relevant questions in an important African lake. They should serve as motivation, along with all the other cases discussed in this review, for those hoping to make similar contributions at other lakes.

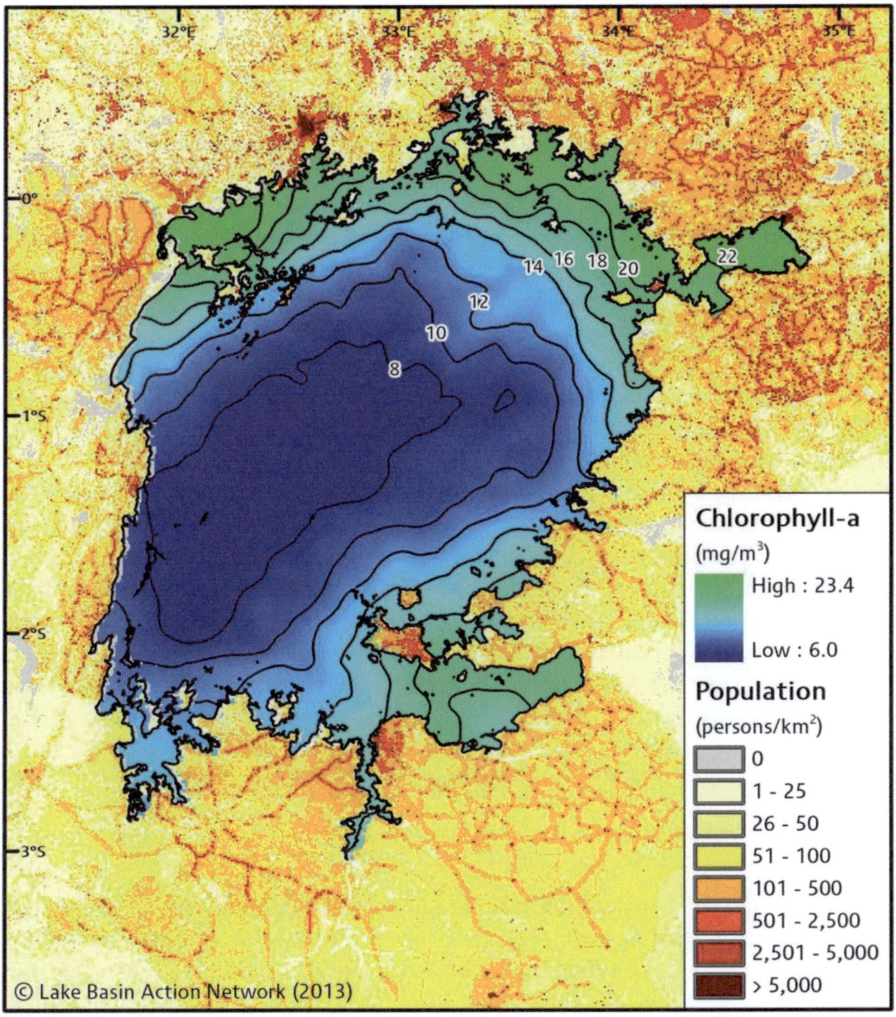

Fig. 20.2 Average SeaWiFS-derived chlorophyll-a concentrations (mg/m^3) in Lake Victoria during the period 1997–2004, based on Loiselle et al. (2006) and Cózar et al. (2012), with population density from LandScan 2008 (Bright et al. 2009)

20.6.2 Institutional Challenges

The studies reviewed indicate the high potential of remote sensing for scientific study and integrated management of African lakes. To truly fulfill the potential of remote sensing, a fair amount of scientific investigation and capacity building is still necessary. Many existing remote sensing algorithms suffer from the lack of suitable calibration datasets for inland waters, making regional use of these algorithms problematic. The shortage of calibration data is especially limiting for the African lakes, where financial and logistic constraints limit regular monitoring.

In addition, the organizations responsible for managing lakes in Africa commonly lack the sufficient expertise and technology (hardware, software and training) to develop appropriate remote sensing methods. Slow and intermittent Internet connections are a particular problem as the size and availability of remotely-sensed images grow. As a result, the full potential of remote sensing to aid in lake monitoring efforts has not been realized.

To bring the full promise of this technology to bear on African lakes, further collaboration between international and local research teams is essential. Relatively few African researchers are currently involved in remote sensing research, in spite of their critical role in advancing this discipline for the African lakes. It is interesting to note that out of the 40 Africa-specific studies in our reference list, 9 (22.5 %) have lead authors with either African affiliations or backgrounds. If we consider all authors (some counted more than once if more than one publication), then Africans make up 15 % of the total (26 out of 169). Clearly, the portion of African-led and African-authored papers has room to grow.

For remote sensing to become standard in African lake monitoring and management, a greater exchange between African and non-African researchers and managers must occur. This exchange could take place through remote scientific collaboration (research, grants), gatherings for team building and knowledge dissemination (conferences, meetings) and/or on-the-ground in-country working relationships (trainings, projects). These collaborative activities would act to increase awareness of the potential of the technology and develop the technical and analytical skills needed for deployment in country.

20.6.3 Technological Challenges

The high altitude of many of these lakes requires further development of atmospheric correction methodologies developed for oceans and low-altitude lakes, sometimes necessitating region-specific calibrations (i.e. Horion et al. 2010). The characteristics of atmospheric aerosols over sub-Sahara Africa are significantly different than those considered in most ocean-based atmospheric calibration algorithms (Pinker et al. 2010).

The application of remote sensing technology is more challenging in optically complex inland waters, where the effects of CDOM and suspended sediments can reduce the utility of standard algorithms. Fortunately, the offshore sections of many African lakes are fairly well suited for the remote sensing applications as phytoplankton abundance generally controls the variability of the optical properties in these areas (Silsbe et al. 2006; Loiselle et al. 2008; Bergamino et al. 2010; Cózar et al. 2012). Further investigations of the optical properties of African lakes are needed to favor the development of regionally valid algorithms.

20.6.4 A Field Progressing

The field of remote sensing, in general, and the applications of remote sensing to Africa specifically, are rapidly progressing. On a practical "access" level, much of the imagery discussed here can be downloaded for free. Where Internet connections are an issue, the ESA has launched the "TIGER Initiative" to ensure distribution of images in addition to other capacity building. Additionally, more undersea cables will increase bandwidth.

Technically, there are many interesting developments. The statistical analysis of the relative variability of remotely-sensed variables (climate, hydrological and biological), has allowed for new insights in lake functioning based on limited calibration data (Bergamino et al. 2007; Cózar et al. 2012). Using multiannual optical time series with monthly or shorter periodicity, this approach has proven successful for identifying important basin-scale features of several of the larger lakes. The analysis of common patterns in lacustrine systems remains still unexplored, but it is apparently suitable for the Eastern Africa region, where the spatial density of lakes is relatively high. The analysis of the relative optical variability will likely provide further valuable information on the processes controlling the dynamics of African lakes. However, the advancement and successful deployment of remote sensing techniques for the study and management of African lakes require more accurate estimates of specific biogeochemical water properties (e.g. Matthews et al. 2012).

As progress is made, it is interesting to imagine what an ideal future might look like for remote sensing of African lakes. More sensors would be available, not only to replace past and current ones (such as the Landsat Data Continuity Mission) but to expand spatial, spectral, temporal and radiometric capacity for satellite-based monitoring. The capacity of African scientists and lake managers would be increased through local and international cooperation. The ability to download large files would approach current high-income nation standards through more bandwidth and more stable connections (including electricity). The availability of regular standardized field data for algorithm development and validation would be increased. The understanding of the advantages of remote sensing technology for integrated lake-catchment management by national and international decision makers and stakeholders would improve. Finally, links between the remote sensing of lakes and the remote sensing of the land in their drainage basins would be strengthened.

References

Awange JL, Sharifi MA, Ogonda G, Wickert J, Grafarend EW, Omulo MA (2008) The falling lake Victoria water levels: GRACE, TRIMM and CHAMP satellite analysis of the lake Basin. Water Resour Manage 22:775–796. doi:10.1007/s11269-007-9191-y

Becker M, Llovel W, Cazenave A, Güntner A, Crétaux J-F (2010) Recent hydrological behavior of the East African great lakes region inferred from GRACE, satellite altimetry and rainfall observations. C.R. Geosci 342:223–233. doi:10.1016/j.crte.2009.12.010

Bergamino N, Horion S, Stenuite S, Cornet Y, Loiselle S, Plisner P-D, Descy J-P (2010) Spatio-temporal dynamics of phytoplankton and primary production in lake Tanganyika using a MODIS based bio-optical time series. Remote Sens Environ 114:772–780. doi:10.1016/j.rse.2009.11.013

Bergamino N, Loiselle SA, Cózar A, Dattilo AM, Bracchini L, Rossi C (2007) Examining the dynamics of phytoplankton biomass in lake Tanganyika using Empirical Orthogonal Functions. Ecol Modell 204:156–162. doi:10.1016/j.ecolmodel.2006.12.031

Birkett CM (1995) The contribution of TOPEX/POSEIDON to the global monitoring of climatically sensitive lakes. J Geophys Res 100(C12):25179–25204

Birkett CM (2000) Synergistic remote sensing of lake Chad: variability of Basin inundation. Remote Sens Environ 72:218–236

Böhme B, Steinbruch F, Gloaguen R, Heilmeier H, Merkel B (2006) Geomorphology, hydrology, and ecology of Lake Urema, central Mozambique, with focus on lake extent changes. Phys Chem Earth 31:745–752. doi:10.1016/j.pce.2006.08.010

Booth DJ, Oldfield RB (1988) Estimation of the area of lake Kariba, Zimbabwe, using Landsat MSS imagery. Proceedings of IGRASS '88 Symposium, Edinburgh, Scotland, 13–16 September 1988:1565–1567

Boy J-P, Hinderer J, de Linage C (2012) Retrieval of large-scale hydrological signals in Africa from GRACE time-variable gravity fields. Pure Appl Geophys 169:1373–1390. doi:10.1007/s00024-011-0416-x

Bradt SR (2012) Development of bio-optical algorithms to estimate chlorophyll in the Great Salt Lake and New England lakes using *in situ* hyperspectral measurements. Dissertation, University of New Hampshire

Bright EA, Coleman PR, King AL, Rose AN, Urban ML (2009) LandScan 2008. Oak Ridge National Laboratory, Oak Ridge, TN. http://www.ornl.gov/sci/landscan/. Accessed 19 May 2013

Cavalli RM, Laneve G, Fusilli L, Pignatti S, Santini F (2009a) Remote sensing water observation for supporting lake Victoria weed management. J Environ Manage 90:2199–2211. doi:10.1016/j.jenvman.2007.07.036

Cavalli RM, Fusilli L, Laneve G, Pascucci S, Palombo A, Pignatti S, Santini F (2009b) Lake Victoria aquatic weeds monitoring by high spatial and spectral resolution satellite imagery. Proceedings of IGARSS 2009 Symposium, Cape Town, South Africa, 12–17 July 2009:II-1048-II–1051. doi:10.1109/IGARSS.2009.5418284

Chavula G, Brezonik P, Thenkabail P, Johnson T, Bauer M (2009a) Estimating the surface temperature of Lake Malawi using AVHRR and MODIS satellite imagery. Phys Chem Earth 34:749–754. doi:10.1016/j.pce.2009.08.001

Chavula G, Brezonik P, Thenkabail P, Johnson T, Bauer M (2009b) Estimating chlorophyll concentration in Lake Malawi from MODIS satelitte imagery. Phys Chem Earth 34:755–760. doi:10.1016/j.pce.2009.07.015f

Chipman JW, Olmanson LG, Gitelson AA (2009) Remote sensing methods for Lake management: a guide of resource managers and decision-makers. Developed by the North American lake management society in collaboration with Dartmouth College, University of Minnesota and University of Nebraska for the United States Environmental Protection Agency

Coe M, Birkett C (2004) Calculation of river discharge and prediction of lake height from satellite radar altimetry: Example for the Lake Chad basin. Water Resour Res 40(10). doi: 10.1029/2003WR002543

Cohen J (1960) A coefficient of agreement for nominal scales. Educ Psychol Measur 20(1):37–46

Cózar A, Bruno M, Bergamino N, Übeda B, Bracchini L, Dattilo AM, Loiselle SA (2012) Basin-scale control on the Phytoplankton Biomass in lake Victoria, Africa. PLoS ONE 7(1):e29962. doi:10.1371/journal.pone.0029962

Crétaux J-F, Bickett C (2006) Lake studies from satellite radar altimetry. C.R. Geosci 338:1098–1112. doi:10.1016/j.crte.2006.08.002

Crétaux J-F, Jelinski W, Calmant S, Kouraev A, Vuglinski V, Bergé-Nguyen M, Gennero M-C, Nino F, Abarca Del Rio R, Cazenave A, Maisongrande P (2011) SOLS: a lake database to monitor in the near real time water level and storage variations from remote sensing data. Adv Space Res 47:1497–1507. doi:10.1016/j.asr.2011.01.004

ESA (European Space Agency) (2013) River and lake website http://tethys.eaprs.cse.dmu.ac.uk/RiverLake/shared/main. Accessed 19 May 2013

Fusilli L, Collins MO, Laneve G, Palombo A, Pignatti S, Santini F (2011) Assessment of the abnormal growth of floating macrophytes in Winam Gulf (Kenya) by using MODIS imagery time series. Int J Appl Earth Obs Geoinf doi:10.1016/j.jag.2011.09.002

Gitelson A, Gurlin D, Moses W, Yacobi Y (2011) Remote estimation of Chlorophyll-a concentration in Inland, Estuarine, and Coastal waters. In: Weng Q (ed) Advances in environmental remote sensing: sensors, algorithms, and applications. CRC Press, Boca Raton

Gower J, King S, Borstad G, Brown L (2005) Detection of intense plankton blooms using the 709 nm band of the MERIS imaging spectrometer. Int J Remote Sens 26:2005–2012

Harris AR (1994) Time series remote sensing of a climatically sensitive lake. Remote Sens Environ 50:83–94

Hecky RE, Mugidde R, Ramlal PS, Talbot MR, Kling GW (2010) Multiple stressors cause rapid ecosystem change in lake Victoria. Freshwater Biol. 55(Suppl. 1):19–42. doi:10.1111/j.1365-2427.2009.02374.x

Hinderer J, Pfeffer J, Boucher M, Nahmani S, de Linage C, Boy J-P, Genthon P, Seguis L, Favreau G, Bock O, Descloitres M, GHYRAF Team (2012) Land Water storage changes from ground and space Geodesy: first results from the GHYRAF (Gravity and Hydrology in Africa) Experiment. Pure Appl Geophys 169:1391–1410. doi:10.1007/s00024-011-0417-9

Horion S, Bergamino N, Stenuite S, Descy J-P, Plisnier P-D, Loiselle SA, Cornet Y (2010) Optimized extraction of daily bio-optical time series derived from MODIS/Aqua imagery for Lake Tanganyika, Africa. Remote Sens. Environ 114:781–791. doi:10.1016/j.rse.2009.11.012

Jensen JR (2007) Remote sensing of the environment: an earth resource perspective, 2nd edn. Pearson Prentice Hall, Upper Saddle River

Kneubühler M, Frank T, Kellenberger TW, Pasche N, Schmid M (2007) Mapping chlorophyll-a in Lake Kivu with remote sensing methods. Proceedings of Envisat Symposium 2007, Montreux, Switzerland, 23–27 April 2007. ESA Publications Division, Noordwijk

Koponen S, Ruiz-Verdú A, Heege T, Heblinski J, Sorensen K, Kallio K, Pyhälahti T, Doerffer R, Brockmann C, Peters M (2008) Development of MERIS lake water algorithms: validation report. ESRIN contract No. 20436/06/I–LG

Leblanc M, Lemoalle J, Bader J-C, Tweed S, Mofor L (2011) Thermal remote sensing of water under flooded vegetation: new observations of inundation patterns for the 'Small' lake Chad. J Hydrol 404:87–98. doi:10.1016/j.jhydrol.2011.04.023

Legesse D, Ayenew T (2006) Effect of improper water and land resource utilization on the central main Ethiopian Rift lakes. Quat. Int 148:8–18. doi:10.1016/j.quaint.2005.11.003

LEGOS (Laboratoire d'Etudes en Géophysique et Océanographie Spatiales) (2013) Hydroweb: surface monitoring by satellite altimetry. http://www.legos.obs-mip.fr/en/soa/hydrologie/hydroweb/. Accessed 19 May 2013

Lemoalle J, Bader J-C, Leblanc M, Sedick A (2012) Recent changes in lake Chad: observations, simulations and management options (1973–2011). Glob Planet Change 80–81:247–254. doi:10.1016/j.gloplacha.2011.07.004

Loiselle SA, Azza N, Cózar A, Bracchini L, Tognazzi A, Dattilo AM, Rossi C (2008) Variability in factors causing light attenuation in lake Victoria. Freshwater Biol 53:535–545. doi:10.1111/j.1365-2427.2007.01918.x

Loiselle S, Cózar A, van Dam A, Kansiime F, Kelderman P, Saunders M, Simonit S (2006) Tools for wetland ecosystem resource management in East Africa: focus on the lake Victoria Papyrus wetlands. In Verhoeven JTA, Beltman B, Bobbink R, Whigham DF (eds) Wetlands and Natural Resource Management. Ecol Stud 190:97–121. doi:10.1007/978-3-540-33187-2_6

Majozi NP (2011) Remote sensing of euphotic depth in Lake Naivasha. Master's Thesis, University of Twente

Matthews MW, Bernard S, Winter K (2010) Remote sensing of cyanobacteria-dominant algal blooms and water quality parameters in Zeekoevlei, a small hypertrophic lake, using MERIS. Remote Sens Environ 114(9):2070–2087. doi:10.1016/j.rse.2010.04.013

Matthews MW, Bernard S, Robertson L (2012) An algorithm for detecting trophic status (chlorophyll-*a*) cyanobacterial-dominance, surface scums and floating vegetation in inland and coastal waters. Remote Sens Environ 124:637–652. doi:10.1016/j.rse.2012.05.032
McClain EP (1981) Multiple atmospheric-window techniques for satellite-derived sea surface temperatures. In proceedings of symposium on ocean oceanography from Space, Venice, Italy, 26–30 May 1980:73–85
Mercier F, Cazenave A, Maheu C (2002) Interannual lake level fluctuations (1993–1999) in Africa from Topex/Poseidon: connections with ocean-atmosphere interactions over the Indian Ocean. Glob Planet Change 32:141–163
Mohler RRJ, Helfert MR, Giardino JR (1989) The decrease of lake Chad as documented during twenty years of manned space flight. Geocarto Int 1:75–79
Munyaneza O, Wali UG, Uhlenbrook S, Maskey S, Mlotha MJ (2009) Water level monitoring using radar remote sensing data: application to Lake Kivu, central Africa. Phys Chem Earth 34:722–728. doi:10.1016/j.pce.2009.06.008
NASA (National Aeronautics and Space Agency) (2013) SRTM water body dataset (SWBD). http://dds.cr.usgs.gov/srtm/version2_1/SWBD. Accessed 19 May 2013.
Oberholster PJ, Botha AM (2010) Use of remote sensing and molecular markers to detect toxic cyanobacterial hyperscum crust: a case study on lake Hartbeespoort, South Africa. Afr J Biotechnol 9(51):8791–8799. doi:10.5897/ajb10.530
Ouma Y, Tateishi R (2006) A water index for rapid mapping of shoreline changes of five East African Rift Valley lakes: an empirical analysis using Landsat TM and ETM+ data. Int J Remote Sens 27(15):3153–3181. doi:10.1080/01431160500309934
Pinker RT, Liu H, Osborne SR, Akoshile C (2010) Radiative effects of aerosols in sub-Sahel Africa: dust and biomass burning. J Geophys Res 115:D15205. doi:10.1029/2009jd013335
Rebelo L-M (2010) Eco-hydrological characterization of Inland wetlands in Africa using L-Band SAR. IEEE J Sel Topics Appl Earth Obser 3(4):554–559. doi:10.1109/jstars.2010.2070060
Ricko M, Carton JA (2011) Climate effects on lake Basins. Part I: modeling tropical lake levels. J Climate 24:2983–2999. doi:10.1175/2010jcli3602.1
Ruiz-Verdú A, Simis SGH, de Hoyos C, Gons HJ, Pea-Martínez R (2008) An evaluation of algorithms for the remote sensing of cyanobacterial biomass. Remote Sens Environ 112:3996–4008
Schneider SR, McGinnis DF Jr, Stephens G (1985) Monitoring Africa's lake Chad basin with LANDSAT and NOAA satellite data. Int J Remote Sens 6(1):59–73
Silsbe GM, Hecky RE, Guildford SJ, Mugidde R (2006) Variability of chlorophyll a and photosynthetic parameters in a nutrient saturated tropical great lake. Limnol Oceanogr 51:2052–2063
Sørensen K, Folkestad A, Stelzer K, Brockmann C, Doerffer R, Okullo W, Schouten L (2008) Performance of MERIS products in Lake Victoria. Proceedings of the 2nd MERIS I (A) ATSR User Workshop, Frascati, Italy, 22–26 September 2008
Swenson S, Whar J (2009) Monitoring the water balance of Lake Victoria, East Africa, from space. J Hydrol 370:163–176
Turada IA (2008) The use of MERIS data to detect the impact of flood inundation on land cover changes in the lake Chad basin. Dissertation, The Hong Kong Polytechnic University
USDA (United States Department of Agriculture) (2013) Global reservoir and lake Monitor. http://www.pecad.fas.usda.gov/cropexplorer/global_reservoir. Accessed 19 May 2013
Velpuri NM, Senay GB, Asante KO (2012) A multi-source satellite data approach for modelling lake Turkana water level: calibration and validation using satellite altimetry data. Hydrol Earth Syst Sci 16:1–18. doi:10.5194/hess-16-1-2012
Verschuren D, Johnson TC, Kling HJ, Edgington DN, Leavitt PR, Brown ET, Talbot MR, Hecky RE (2002) History and timing of human impact on Lake Victoria, East Africa. Proc R Soc Lond B 269:289–294. doi: 10.1098/rspb.2001.1850
Wald L (1990) Monitoring the decrease of lake Chad from space. Geocarto Int 3:31–36
Williams AE, Duthie HC, Hecky RE (2005) Water hyacinth in lake Victoria: why did it vanish so quickly and will it return? Aquat Bot 81:300–314. doi: 10.1016/j.aquabot.2005.01.003
Wooster M, Patterson G, Loftie R, Sear C (2001) Derivation and validation of the seasonal thermal structure of lake Malawi using multi-satellite AVHRR observations. Int J Remote Sens 22(15):2953–2972

Index

V. Barale, M. Gade (eds.), *Remote Sensing of the African Seas*,
DOI 10.1007/978-94-017-8008-7,

MIX
Papier aus verantwortungsvollen Quellen
Paper from responsible sources
FSC® C105338

If you have any concerns about our products,
you can contact us on
ProductSafety@springernature.com

In case Publisher is established outside the EU,
the EU authorized representative is:
Springer Nature Customer Service Center GmbH
Europaplatz 3, 69115 Heidelberg, Germany

Printed by Libri Plureos GmbH
in Hamburg, Germany